Fractality of Wave Fields and Processes in Radar and Control

雷达与控制系统中
波动场的分形学及其应用

[俄罗斯] 亚历山大 · A. 波塔波夫（Alexander A. Potapov）
吴 浩
熊 珊
著

华南理工大学出版社
SOUTH CHINA UNIVERSITY OF TECHNOLOGY PRESS
·广州·

图书在版编目（CIP）数据

雷达与控制系统中波动场的分形学及其应用 = Fractality of Wave Fields and Processes in Radar and Control：英文/（俄罗斯）亚历山大·A. 波塔波夫（Alexander A. Potapov），吴浩，熊珊著. —广州：华南理工大学出版社，2020. 11

ISBN 978-7-5623-6276-0

Ⅰ. ①雷… Ⅱ. ①亚… ②吴… ③熊… Ⅲ. ①雷达探测-分形学-研究-英文 Ⅳ. ①TN953

中国版本图书馆 CIP 数据核字（2020）第 036411 号

雷达与控制系统中波动场的分形学及其应用

亚历山大·A. 波塔波夫（Alexander A. Potapov），吴浩，熊珊　著

出 版 人：卢家明

出版发行：华南理工大学出版社

（广州五山华南理工大学 17 号楼，邮编 510640）

http://www.scutpress.com.cn　E-mail：scutc13@scut.edu.cn

营销部电话：020-87113487　87111048（传真）

策划编辑：吴翠微

责任编辑：陈　蓉　张　楚

责任校对：袁桂香

印 刷 者：佛山家联印刷有限公司

开　　本：787mm×1092mm　1/16　印张：18.25　字数：577 千

版　　次：2020 年 11 月第 1 版　2020 年 11 月第 1 次印刷

定　　价：168.00 元

About the Book

The book is devoted to contemporary issues of radar, control, fractal theory and the detection of weak radar signals in intense real non-Gaussian interference.

The core of the book is the lectures and reports of Prof. Alexander Potapov (V. A. Kotelnikov Institute of Radio Engineering and Electronics RAS, Moscow, Russia) for the period 2011 – 2019, read at universities in China and at many international conferences in China.

The theory of wave scattering by anisotropic statistically rough surfaces, which is an important part of statistical radiophysics and radar, is considered. A new analytic method is developed and generalized for solving problems of radar imaging. The method involves analytic determination of the functionals of stochastic backscattered fields and can be applied to solve a wide class of physical problems with allowance for the finite width of an antenna's pattern. The unified approach based on this method is used to analyze the generalized frequency response of a scattering radio channel, a generalized correlator of scattered fields, spatial correlation functions of stochastic backscattered fields, frequency coherence functions of stochastic backscattered fields, the coherence band of a spatial-temporal scattering radar channel, the kernel of the generalized uncertainty function, and the measure of noise immunity characterizing radar probing of the Earth's surface or extended targets. The introduced frequency coherence functions are applied for thorough and consistent study of techniques for measuring the characteristics of a rough surface, aircraft altitude, and distortions observed when radar signals are scattered by statistically rough surfaces, including fractal ones. To exemplify urgent applications, radiophysical synthesis of detailed digital reference radar terrain maps and microwave radar images that was proposed earlier is considered and improved with the use of the theory of fractals.

Problems and methods of development of the general theory of multiple scattering of electromagnetic waves in fractal randomly inhomogeneous media, based on modifications of the classical theory of Foldy-Tversky, are considered. The backscattering processes which are characteristic for radar sensing of a fractal medium or a set of fractal targets are studied. Examples of numerous fractal radio systems, sensors and elements are presented.

The book presents the strong large deviations principles (SLDP) of non-Freidlin-Wentzell type, corresponding to the solutions Colombeau-Ito's SDE. Exact quasi-classical asymptotic beyond WKB-theory and beyond Maslov canonical operator to the Colombeau solutions of the-dimensional Schrodinger equation is presented. Quantum jumps nature is considered successfully. An explanation of quantum jumps can be found to result from Colombeau solutions of the Schrödinger equation alone without additional postulates. (These sections were prepared jointly with Prof. J. Foukzon, Israel Institute of Technologies, Haifa, Israel based on our works.)

The book is addressed to engineers and researchers in the field of physics, applied mathematics, radar, control, fractal theory and related fields.

PREFACE

The book presents the results obtained by the authors, and united by a common theme of modern radar. The new kind and approach of up-to-date radiolocation: fractal-scaling or scale-invariant radiolocation has been proposed.

A fractal approach to solving theoretical and practical problems of detection, measurement of parameters and classification of spatially distributed targets is considered. These problems arise in radar imaging of air and outer space, land and water surface using various radio systems: early warning radars, synthetic aperture radars (SAR) on air and space carriers, and many other similar systems.

The efficiency of radar, radio-wave imaging, remote sensing, and many other systems has been enhanced by the intense application of millimeter waves (MMWs). Therefore, it is necessary to study the processes and characteristic features of MMW scattering by terrain. This study has adequately described the situation in this field, which, today, is again attracting the attention of numerous radiophysicists and radio engineers.

The mathematical approach which is continually developed by the authors involves functionals of stochastic backscattered fields and allows qualitative estimation of the spatial-temporal and spatial-frequency characteristics of a scattered signal in the case of combined or spaced radio systems and an adequate description of the formation of observed radiophysical fields. The expressions obtained take into account the effects of irregularities' slopes and the dimensions of the antenna's aperture more accurately.

Fluctuation effects occurring during random and chaotic nonstationary reflection of waves by a medium's boundary have been investigated in the case when the boundary has an arbitrary (integer or fractal) topological dimension.

The approach proposed makes it possible to study spatial and frequency functionals of stochastic fields scattered by rough anisotropic surfaces with allowance for the mutual statistical relationship between irregularities' slopes and to specify the limits of applicability of the approximations employed. The developed generalizations and new solutions substantially extend the scope of problems of the statistical theory of wave diffraction.

The application of the frequency coherence function(FCF) absolute value and phase characteristic for precision estimation of an aircraft's flight altitude and the height of large irregularities have been investigated more accurately and justified quantitatively for various cross-correlation factors of irregularities' slopes.

The measurement errors of the aforementioned quantities increase for large probing angles and wide antenna patterns. It has been shown that measurement techniques using the absolute values and phase characteristics of FCFs are competitive with standard shortpulse methods for measuring flight altitudes and the characteristics of large-height irregularities of terrain.

Analytic expressions for the kernel of the generalized ambiguity function(AF) have been derived with allowance for the antenna's characteristics and angular orientation and the characteristics of an isotropic or anisotropic scattering surface. These expressions are valid over the entire range of the correlation factor of large-height irregularities' slopes. It has been shown that the kernel of the generalized AF can be expressed through elementary functions.

The results obtained can be used to choose the antenna's parameters, the type of modulation of the probing signal, the detection characteristics, and the measure of noise immunity for specified ranges of the statistical characteristics of a scattering surface. More general solutions that fit known results can be obtained with the use of FCFs.

The investigations have shown that the generalized complex radiophysical model and the method developed by Professor Alexander A. Potapov for formation of MMW reference detailed digital radar maps(DDRMs) of an inhomogeneous terrain and for DDRM synthesis, respectively, are promising and highly effective.

Combination of the developed methods and fractal description of wave scattering will undoubtedly result in the discovery of new physical laws in the wave theory. We are sure that, combined with the formalism of fractal operators, the theory of fractals and deterministic chaos applied to the problems considered above will make it possible to synthesize more adequate radiophysical and radar models that will substantially reduce discrepancies between theoretical predictions and experimental results. This study rather comprehensively covers the variety of modern problems of wave scattering that arise in theoretical fields and applications of radiophysics and radar and, generally, involve the theory of integer and fractal measures. Thus, the use of the formalism of dynamics of dissipative systems (the fractal character, fractal operators, a non-Gaussian statistic, distributions with heavy tails, the mode of deterministic chaos, the existence of strange attractors in the phase space of reflected signals, their topology, etc.) makes it possible to expect that the classical problem of wave scattering by random media will remain an area of fruitful future investigations.

The results obtained can be widely applied for designing various modern radio systems in the microwave, optical, and acoustic bands.

The book presents the strong large deviations principles (SLDP) of non-Freidlin-Wentzell type, corresponding to the solutions Colombeau-Ito's SDE. Using SLDP we present a new approach to construct the Bellman function $v(t,x)$ and optimal control $u(t,x)$ directly by way of using strong large deviations principle for the solutions Colombeau-Ito's SDE. As important application of SLDP, the generic imperfect dynamic models of air-to-surface missiles are given in addition to the related simple guidance law. Four, examples have been illustrated, and corresponding numerical simulations have been illustrated and analyzed. Using SLDP approach, jump phenomena in financial markets is considered and explained from the first principles, without any reference to Poisson jump process. (This section was prepared jointly with Prof. J. Foukzon, Israel Institute of Technologies, Haifa, Israel based on our works.)

In this book, exact quasi-classical asymptotic beyond WKB-theory and beyond Maslov canonical operator to the Colombeau solutions of the n-dimensional Schrödinger equation is presented. Quantum jump nature is considered successfully. We pointed out that an explanation of quantum jumps can be found to result from Colombeau solutions of the Schrödinger equation alone without additional postulates. (This section was prepared jointly with Prof. J. Foukzon, Israel Institute of Technologies, Haifa, Israel based on our works.)

The main purpose of the work described in Chapter 4 – 6 is to interpret the main directions of radio physics, radio engineering and radio location in "fractal" language that makes new ways and generalizations for future promising radio systems. The new kind and approach of up-to-date radiolocation: fractal-scaling or scale-invariant radiolocation has been proposed by Professor Alexander A. Potapov. It leads to basic changes in the theoretical radiolocation structure itself and also in its mathematical apparatus. The fractal radio systems conception, sampling topology, global fractal-scaling approach and the fractal paradigm underlies in the scientific direction established by Professor Alexander A. Potapov in Russia and all over the world for the first time. Professor Alexander A. Potapov has been investigating these issues for exactly 40 years. The results of big practical and scientific importance obtained by him were published in four summary reports of the Presidium of Russian Academy of Sciences (2008, 2010, 2012, 2013) and in the Report for the Government of Russian Federation (2012).

All these results of priority in the world allow one to move to a new level of information structure of real non-Markov ("with memory") signals and fields. Accordingly, the proposed fractal radio systems, sensors and elements open up new opportunities in modern radio electronics and can have the broadest prospects for practical application.

The proposed work is written to fill in the gap between abundance of journal articles and deficiency of books in the investigation of fractal radars. *It should be noted that fractal radar is a necessary step in the future transition to cognitive radar and quantum radar.* In considerable part, the book is the generalization of published works by authors in Russia and in abroad.

CONTENTS

Chapter 1 The Theory of Functionals of Stochastic Backscattered Fields

The theory of wave scattering by anisotropic statistically rough surfaces, which is an important part of statistical radiophysics, is considered. A new analytic method is developed and generalized for solving problems of radar imaging. The method involves analytic determination of the functionals of stochastic backscattered fields and can be applied to solve a wide class of physical problems with allowance for the finite width of an antenna's pattern. The unified approach based on this method is used to analyze the generalized frequency response of a scattering radio channel, a generalized correlator of scattered fields, spatial correlation functions of stochastic backscattered fields, frequency coherence functions of stochastic backscattered fields, the coherence band of a spatial-temporal scattering radar channel, the kernel of the generalized uncertainty function, and the measure of noise immunity characterizing radar probing of the Earth's surface or extended targets. The introduced frequency coherence functions are applied for thorough and consistent study of techniques for measuring the characteristics of a rough surface, aircraft altitude, and distortions observed when radar signals are scattered by statistically rough, including fractal, surfaces. To exemplify urgent applications, radiophysical synthesis of detailed digital reference radar terrain maps and microwave radar images that were proposed earlier is considered and improved with the use of the theory of fractals.

1.1 Introduction

At present, interest in processing of fields (spatial-temporal signals) scattered by statistically rough surfaces has grown substantially because of the following factors. In contrast to the past situation, today, the possibilities of solving various applied problems with the use of millimeter waves (MMWs) have again attracted the attention of researchers[1-7]. Lying between two classical (centimeter and optical) bands, MMWs often are more suitable for specific applications. This circumstance results from the progress of the MMW circuit technology and from the characteristic features of this band, in particular, the higher immunity of MMW radars to electromagnetic countermeasures[6,7].

Digital processing of spatial-temporal signals and digital control of an antenna's aperture (when an antenna is considered as a dynamic spatial-temporal filter) allows formation and reception of radar images (RIs) in almost real time. Thus, it attempts to effectively use the spatialtemporal structure of the electromagnetic field for maximizing the amount of information obtained from received signals have stimulated the development of systems forming millimeter- and centimeter-wave RIs. The main problem of radar—detection and discrimination of targets in the presence of reflections from terrain and of intrinsic radar noise—has been solved simultaneously.

An RI can generally be interpreted as the map (matrix) of specific radar cross sections (RCSs) σ_* of a probed object or the signature (portrait) of a probed object in the case of a high angular resolution. When a probing beam is wide, a real RI is associated with an RCS map with smeared contours. Enhancement of RI resolution necessitates complex probing signals, such as chirp, nonlinearly frequency-modulated, or phasecode-shift keyed signals[1-9]. Processing of current images often yields specific detailed digital radar maps (DDRMs) of terrains or reference maps[1,2,7]. At present, rapid progress in the development of methods of DDRM synthesis is taking place[7,10].

Spatial-temporal signal processing requires the use of array antennas. However, theoretical investigations of problems of radar imaging and detailed digital radar mapping often deal with a continuous antenna aperture. This assumption simplifies solution of the aforementioned problems and allows determination of potential radar characteristics exhibited in the case when the entire space assigned for observations is employed[4,7,11,12].

When an RI is formed, the structure and parameters of the wave field induced by a distant statistically rough surface in the region where this field is analyzed depend on the reception point and surface characteristics. If the above factors are taken into account, it is fundamentally important to obtain a complete mathematical description of the scattered field in the space-time continuum[7]. Therefore, in the late 1970s, I formulated the problem of theoretical simulation of a spatial-temporal MMW signal with allowance for a linear radio channel consisting of an antenna aperture, the atmosphere, and a chaotic vegetationless surface and the problem of forming new classes of radar signatures[7,13].

This study is the first to systemically consider a unified approach to mathematical description of spatial-temporal scattering radio channels. The approach proposed allows investigations of wide classes of various problems in the statistical theory of diffraction. In the study, methods of investigation and analysis are developed that provide for explicit formulas and noticeable results in specific physical problems. Some of these techniques are new[7] and have not been reported in the literature. The method developed in the study is used for solution of certain physical problems. Thus, the purpose of the study is to present numerous results consistently and as comprehensively as possible and to demonstrate application of these results to solution of practical problems that often arise in radiophysics, radar, acoustics, and optics. The presented results were obtained at the Institute of Radio Engineering and Electronics of Russian Academy of Sciences (IRE RAS) before the 1990s, and some of the results have been employed at a number of organizations. Most of the results are reported in my doctoral thesis[7] and summarized in Monographs[1,2,5,6]. Later, my scientific interests concentrated exclusively on investigations, development, and application of fractal methods of radiophysical data processing based on the theory of fractal measures, fractal operators, and scaling relationships.

The effectiveness of radiophysical investigations can be enhanced substantially if the fractal character of wave phenomena that occur during all the stages of wave radiation, scattering, and propagation in various media is taken into account. In addition to pure theoretical significance, fractal methods are important for solution of practical radar and telecommunication problems and problems of medium monitoring on various space-time scales[1-3].

1.2 The angular spectrum of wave fields

Since the microwave band is located closely to the optical band, it is suitable to solve problems of diffraction of microwave fields with the use of integral Fourier transforms, which is an optical technique, and spatial (angular) and temporal spectra[7,14-19]. Assume that a monochromatic wave propagates in free space along the z axis in the region where $z > 0$. In a homogeneous medium without currents or charges, complex amplitude $E \equiv E\ (x,\ y,\ z)$ of a monochromatic wave satisfies the Helmholtz equation

$$\nabla^2 E + k^2 E = 0 \tag{1.1}$$

A partial solution to Eq. (1.1) has the form

$$E(x,y,z) = E_0(x,y,0)\exp(\mathrm{i}\vec{k}\cdot\vec{r}) = E_0(x,y,0)\exp[\mathrm{i}(k_x x + k_y y + k_z z)] \tag{1.2}$$

and describes a plane wave.

In Eq. (1.2), $E_0(x,y,0)$ is the complex wave amplitude in the plane $z=0$, is the wave vector with the rojections $\{k_x, k_y, k_z = \sqrt{k^2 - k_x^2 - k_y^2}\}$, and $\vec{r} = \vec{x}_0 x + \vec{y}_0 y + \vec{z}_0 z$ is the radius vector of an observation point. If, in Cartesian coordinates (x,y,z), the direction of wave propagation is specified by angles α, β, and γ, we have

$$\cos^2\alpha = k_x/k,\ \cos^2\beta = k_y/k,\ \cos^2\gamma = k_z/k,\ \cos^2\alpha + \cos^2\beta + \cos^2\gamma = 1. \tag{1.3}$$

The 2D Fourier transform of function $E(x,y,z)$ has the form

$$E_0(x,y,0) = \frac{1}{4\pi^2}\int_{-\infty}^{\infty}\int_{-\infty}^{\infty} F_0(\omega_x,\omega_y)\exp[\mathrm{i}(\omega_x x + \omega_y y)]\mathrm{d}\omega_x \mathrm{d}\omega_y, \tag{1.4}$$

where

$$F_0(\omega_x,\omega_y) = \int_{-\infty}^{\infty}\int_{-\infty}^{\infty} E_0(x,y)\exp[-\mathrm{i}(\omega_x x + \omega_y y)]\mathrm{d}x\mathrm{d}y. \tag{1.5}$$

Comparing Eqs. (1.2) and (1.4) and taking into account Eq. (1.3), we see that the integrand in Eq. (1.4) can be regarded as a plane wave with the direction cosines

$$\cos\alpha = \lambda f_x,\ \cos\beta = \lambda f_y,\ \cos\gamma = \sqrt{1 - (\lambda f_x)^2 - (\lambda f_y)^2}. \tag{1.6}$$

The independent variables

$$f_x = \frac{\omega_x}{2\pi} = \frac{k_x}{2\pi} = \frac{\cos\alpha}{\lambda},\ f_y = \frac{\omega_y}{2\pi} = \frac{k_y}{2\pi} = \frac{\cos\beta}{\lambda} \tag{1.7}$$

are linear spatial frequencies of the dimension that is the reciprocal of the unit length. For any fixed two frequencies $\{f_x, f_y\}$, the phase of the elementary function $\exp[i 2\pi (f_x x + f_y y)]$ is zero along the line described by the equation

$$y = \frac{f_x}{f_y} x + \frac{n}{f_y}, \tag{1.8}$$

where n is an integer.

These parallel straight lines form a set with the spatial period

$$L = (f_x^2 + f_y^2)^{-1/2}, \tag{1.9}$$

and their slope with respect to the x axis is specified by the angle

$$\theta = \arctan f_y/f_x . \tag{1.10}$$

The complex amplitude of a plane wave from Eq. (1.4) equals $F_0(\omega_x, \omega_y)\mathrm{d}\omega_x \mathrm{d}\omega_y$. Therefore, Function (1.5), which can be represented as

$$F_0\left(\frac{\cos\alpha}{\lambda}, \frac{\cos\beta}{\lambda}\right) = \int_{-\infty}^{\infty}\int_{-\infty}^{\infty} E_0(x,y,0)\exp\left[-\mathrm{i}\left(\frac{\cos\alpha}{\lambda}x + \frac{\cos\beta}{\lambda}y\right)\mathrm{d}x\mathrm{d}y\right]. \tag{1.11}$$

is the angular (spatial) spectrum of perturbation $E_0(x, y, 0)$.

In order to find a general solution, let us define the angular spectrum of the perturbation in a plane that is parallel to the XOY plane and has arbitrary coordinate z in the following form:

$$F_0(\omega_x, \omega_y, z) = \int_{-\infty}^{\infty}\int_{-\infty}^{\infty} E_0(x,y,z)\exp[-\mathrm{i}(\omega_x x + \omega_y y)\mathrm{d}x\mathrm{d}y]. \tag{1.12}$$

Then, perturbation $E(x, y, z)$ can be represented as

$$E_0(x,y,z) = \frac{1}{4\pi^2}\int_{-\infty}^{\infty}\int_{-\infty}^{\infty} F(\omega_x, \omega_y, z)\exp[\mathrm{i}(\omega_x x + \omega_y y)\mathrm{d}\omega_x \mathrm{d}\omega_y]. \tag{1.13}$$

The substitution of Eq. (1.13) into Helmholtz equation (1.1) yields the solution

$$F(\omega_x, \omega_y, z) = F_0(\omega_x, \omega_y)\exp(\mathrm{i}z\sqrt{k^2 - \omega_x^2 + \omega_y^2}) \tag{1.14}$$

for the traveling-wave mode.

Eq. (1.14) implies that, as the distance of point z from the origin grows, the angular spectrum changes. When

$$k^2 - \omega_x^2 - \omega_y^2 > 0, \tag{1.15}$$

this change manifests itself in changes of relative phases of different components of the angular spectrum. These phase shifts result from the fact that plane waves propagating at different angles to the z axis cover different distances by the moment when they reach a point considered. When

$$k^2 - \omega_x^2 - \omega_y^2 < 0, \tag{1.16}$$

Eq. (1.14) describes plane nonuniform waves whose amplitudes decrease exponentially as the distance from the point $z = 0$ grows. The limiting case

$$k^2 - \omega_x^2 - \omega_y^2 = 0 \tag{1.17}$$

corresponds to plane waves that propagate perpendicularly to the z axis and do not transfer energy in the z direction. A perturbation occurring at arbitrary point (x, y, z) can be expressed through the angular spectrum as

$$E(x,y,z) = \frac{1}{4\pi^2}\int_{-\infty}^{\infty}\int_{-\infty}^{\infty} F_0(\omega_x,\omega_y)\exp(\mathrm{i}z\sqrt{k^2-\omega_x^2-\omega_y^2})\exp[\mathrm{i}(\omega_x x+\omega_y y)]\mathrm{d}\omega_x\mathrm{d}\omega_y . \tag{1.18}$$

Despite the formal resemblance, Eq. (1.18) and the Fourier representation should not be confused. In contrast to Eq. (1.18) which involves the angular spectrum, the Fourier representation of a function of three real arguments contains triple integrals rather than double integrals. The assumption that field $E(x, y, z)$ is known only in a finite region rather than in the entire space implies that the Fourier decomposition (in contrast to Eq. (1.18)) is not unique and does not yield a representation in terms of wave-field modes.

In radiophysics, integral of Eq. (1.18) is known as the Rayleigh representation. Spatial frequencies must satisfy the single condition

$$\omega_x^2 + \omega_y^2 + \omega_z^2 = \frac{\omega^2}{c^2}. \tag{1.19}$$

The function under consideration satisfies the wave equation only when condition (1.19) is fulfilled.

It is seen from Eq. (1.14) that a spatial layer of thickness z performs signal transformation as a linear dispersion filter with the transfer function

$$K(\omega_x,\omega_y) = \exp(\mathrm{i}z\sqrt{k^2-\omega_x^2-\omega_y^2}). \tag{1.20}$$

Amplitude-frequency and phase-frequency characteristics $|K(\omega_x, \omega_y)|$ and $\varphi(\omega_x, \omega_y)$ have the form

$$|K(\omega_x,\omega_y)| = 1, \varphi(\omega_x,\omega_y) = z\sqrt{k^2-\omega_x^2-\omega_y^2}, \tag{1.21}$$

in frequency range in Eq. (1.15) and

$$|K(\omega_x,\omega_y)| = \exp(-z\sqrt{\omega_x^2+\omega_y^2-k^2}), \quad \varphi(\omega_x,\omega_y) = 0 \tag{1.22}$$

in frequency range in Eq. (1.16). The passband of such a filter is limited by the cutoff frequency

$$\sqrt{f_x^2+f_y^2} = 1/\lambda . \tag{1.23}$$

Hence, the passband of a free-space region of length z decreases as wavelength λ grows (at fixed z) or as length z grows (at fixed λ).

The well-known Weyl decomposition[20] can be regarded as a representation in the form of the angular spectrum of the wave field produced in free space by a point source located at the origin. The Hankel transformation and formula (1.18) yield the plane-wave decomposition of the field of an outgoing spherical wave,

$$\frac{\exp(\mathrm{i}kr)}{r} = \frac{\mathrm{i}}{2\pi}\int_{-\infty}^{\infty}\int_{-\infty}^{\infty}\frac{\exp[\mathrm{i}(\omega_x x+\omega_y y)+\mathrm{i}z\sqrt{k^2-\omega_x^2-\omega_y^2}]}{\sqrt{k^2-\omega_x^2-\omega_y^2}}\mathrm{d}\omega_x\mathrm{d}\omega_y . \tag{1.24}$$

Interestingly[19], the well-known formula

$$\frac{\sin(kr)}{kr} = \frac{1}{4\pi}\int_{4\pi}\exp(\mathrm{i}\vec{k}\cdot\vec{r})\mathrm{d}\Omega \tag{1.25}$$

can be obtained with the use of the Weyl decomposition for outgoing and incoming spherical waves. Therefore, from the physical viewpoint, the function sin[$(kr)/(kr)$] is a superposition of uniform plane waves that propagate omnidirectionally and have the amplitudes $1/4\pi$.

Dirichlet and Neumann boundary value problems for the Helmholtz equation in a half-space are solved in Monograph [19] with the help of the Weyl transformation and the concept of an angular spectrum.

A solution to the Dirichlet problem or the Rayleigh diffraction formula of the first kind for the half-space $z>0$ has the form

$$E(x,y,z) = -\frac{1}{2\pi}\int_{-\infty}^{\infty}\int_{-\infty}^{\infty} E_0(x',y',0)\frac{\partial}{\partial z}\left(\frac{e^{ikR}}{R}\right)dx'dy', \tag{1.26}$$

where boundary value $E_0(x,y,0)$ is specified on the plane $z=0$.

A solution to the Neumann problem or the Rayleigh diffraction formula of the second kind for the halfspace $z>0$ has the form

$$E(x,y,z) = -\frac{1}{2\pi}\int_{-\infty}^{\infty}\int_{-\infty}^{\infty}\left[\frac{\partial E(x',y',z)}{\partial z}\right]_{z=0}\left(\frac{e^{ikR}}{R}\right)dx'dy' \tag{1.27}$$

where the boundary value of the derivative $\partial E(x,y,z)/\partial z$ is specified on the plane $z=0$.

Note once again the dualism of variables ω_x and ω_y[16]. In Formulae (1.4) and (1.5), variables ω_x and ω_y have the meaning of spatial frequencies. At the same time, in Formula (1.18), variables ω_x and ω_y determine the direction of propagation of plane waves. This direction is specified by angles in Eqs. (1.3) and (1.6). The spatial spectrum of complex amplitudes distributed over a plane and the angular spectrum of radiation may differ substantially.

1.3 The angular spectrum of modulated waves

The above solutions have been obtained under the assumption that the field under consideration is monochromatic. In a linear approximation, a solution can be constructed for the general case of nonmonochromatic radiation[7,15,16]. Let arbitrary field $E(x,y,t)$ be specified on the plane $z=0$. By representing this field as a sum of harmonic components, we can consider each of these as a monochromatic field satisfying Eq. (1.18). Then, a general solution to the wave problem is obtained via the Fourier transform where integration is performed over all frequency components:

$$E(x,y,z,t) = \frac{1}{8\pi^3}\int_{-\infty}^{\infty}\int_{-\infty}^{\infty}\int_{-\infty}^{\infty} F_\omega(\omega_x,\omega_y)\exp\left(iz\sqrt{\frac{\omega^2}{c^2}-\omega_x^2-\omega_y^2}\right)\cdot$$

$$\exp[i(\omega_x x+\omega_y y-\omega t)]d\omega_x d\omega_y d\omega, \tag{1.28}$$

$$F_\omega(\omega_x,\omega_y) = \int_{-\infty}^{\infty}\int_{-\infty}^{\infty} E_\omega(x,y,0)\exp[-i(\omega_x x+\omega_y y)]dxdy, \tag{1.29}$$

here, $E_\omega(x, y, 0)$ is the spectral density of the complex amplitude.

If the functions of time and coordinates involved in the above relationships are factorable,

$$E(x,y,t) = E(x,y)E(t) , \quad (1.30)$$

then solution (1.28) can be simplified to take the form

$$E(x,y,z,t) = \frac{1}{2\pi}\int_{-\infty}^{\infty} F(\omega)E_{\omega}(x,y,z)\exp(-\mathrm{i}\omega t)\mathrm{d}\omega , \quad (1.31)$$

where field $E_{\omega}(x,y,z)$, which is a function of ω as well, can be found from integral relationship of (1.18).

Furthermore, this description is valid when a function is not factorable but can be represented as a sum of factorable functions[16]:

$$E(x,y,t) = \sum_{n} E_n(x,y)E_n(t) . \quad (1.32)$$

In Eq. (1.32), value $E_n(x,y)$ is the distribution of an individual source producing oscillation $E_n(t)$. For each component, it is necessary to use Eq. (1.31) and then sum the results obtained.

In various radio systems, information on an object observed is extracted from the field reradiated or radiated by the object. It is always a relatively small region with this field that is incident on an antenna's aperture and examined. The pattern (i. e., the main characteristic) of an antenna is the Fourier transform of amplitude-phase distribution $A(s)$ over aperture d:

$$F(u) = \int A(s)\mathrm{e}^{-\mathrm{i}us}\mathrm{d}s . \quad (1.33)$$

In Eq. (1.33), integration is performed over the plane of the antenna's aperture with respect to coordinates such that the antenna's aperture coincides with a coordinate plane. The physical meaning of an antenna's pattern is the frequency response of a spatial filter connected in series with free space[16]. Scanning performed by means of antenna rotation is equivalent to tuning the corresponding spatial filter to another frequency without changing the filter's shape. Neither the frequency response of an antenna regarded as a spatial filter nor the antenna's pattern can be specified arbitrarily. This function has a bounded spatial spectrum and belongs to a certain class of functions[21].

In practice, the angular spatial frequencies

$$\alpha_x = \frac{x}{\lambda} = \frac{fx}{c}, \quad \alpha_y = \frac{y}{\lambda} = \frac{fy}{c}, \quad (1.34)$$

are often used[11]. These frequencies characterize the rate of variation of a plane harmonic wave with frequency f along the directions specified by direction cosines that are measured relative to the x and y axes of the aperture, respectively:

$$\varphi(\cos\alpha) = 2\pi\alpha_x\cos\alpha = 2\pi\frac{x\cos\alpha}{\lambda}, \quad \varphi(\cos\beta) = 2\pi\alpha_y\cos\beta = 2\pi\frac{y\cos\beta}{\lambda}. \quad (1.35)$$

For the linear spatial frequencies specified by Eqs. (1.6) and (1.7), we have

$$\varphi(x) = 2\pi f_x x = 2\pi\frac{x\cos\alpha}{\lambda}, \quad \varphi(y) = 2\pi f_y y = 2\pi\frac{y\cos\beta}{\lambda}. \quad (1.36)$$

Since the right-hand sides of Eqs. (1.35) and (1.36) coincide, it is possible to introduce frequencies $\alpha_{x,y}$ or $f_{x,y}$ the choice depends on specific circumstances[7,11]. Both of the approaches can be applied to analyze spectra of narrowband signals, because the form and width of the spatial-frequency spectrum are determined by an antenna. For example, the width of the spatial-frequency spectrum is determined by the antenna's pattern for linear spatial frequencies $f_{x,y}$ and by the aperture function for angular spatial frequencies $\alpha_{x,y}$. These characteristics of the antenna are related through the Fourier transforms; therefore, under the aforementioned conditions, both of the approaches are equivalent.

When angular spatial frequencies $\alpha_{x,y}$ are used, the analysis of wideband signals is impeded by the ambiguous correspondence between the generalized angular coordinate and spatial frequency. For different components of the frequency spectrum of a signal, one value of the angular coordinate is associated with different values of the spatial frequency. In the case of spherical waves, the analysis is impeded more substantially when frequencies $\alpha_{x,y}$ are used, because each $\alpha_{x,y}$ component depends on all coordinates of the antenna's aperture and on all coordinates of a target.

The spatial and temporal characteristics of a radio system cannot be considered independently when signals with wider spectra and ultrawideband (UWB) signals or ultrashort electromagnetic pulses are analyzed[5]. For example, one can speak about the pattern formed by a given instant. Thus, when UWB signals are used, the dimension of the space of radar characteristics and features, i. e., radar signatures, abruptly increases owing to the dynamics of the processes developing in a system[1-4,7,22].

1.4 Simulation of the spatial-temporal structure of the field scattered by a statistically rough anisotropic surface: a mathematical model taking into accout the effect of an antenna

Let us apply the spectral approach presented above and take into account an antenna's pattern, G, to obtain, for a wide range of incidence angles θ, a general analytic solution in the 3D case, $R \in R^3$, for the field of a monochromatic wave scattered by a statistically rough surface[7,13]. In a real situation when RIs of terrains with objects or DDRMs are formed, probing angles often range in the interval of $30° - 40°$.

Let us analytically solve the 2D scattering problem in the Kirchhoff approximation or with the method of a tangent plane (MTP)[5-7,23,24]. The applicability conditions for this method can be generalized as follows: the height of surface irregularities is much greater than unity, $l_\xi/\lambda \gg 1$; the smoothness of a surface is specified by the inequality $\alpha/\lambda \gg 1$; and the flatness of a statistically rough surface (X', Y', Z') that is flat on the average is determined by the inequality $\sqrt{\langle\gamma^2\rangle} \ll 1$. Here, $\sqrt{\langle\gamma^2\rangle}$ is the rms value of the slope ratio of the surface, l_ξ is the correlation radius of the surface, $a = [1 + (\xi')^2]^{3/2}/\xi''$ is the local curvature radius of surface ξ, and the angle brackets denote statistical averaging over the set of realizations. Curvature radii must satisfy the condition $ka\cos^3\theta \gg 1$ [23] or $ka\cos\theta \gg 1$ [24].

Fig. 1. 1 displays the geometry of the problem of radar imaging. Fixed coordinate frame (X, Y, Z) is associated with the mean level of scattering surface $z = \xi(x, y)$. Reference coordinate frame (X'', Y'', Z'') is associated with an aircraft. The position of the antenna relative to the reference coordinate frame is determined by coordinate frame (X', Y', Z') that is specified by nutation angle θ, precession angle φ, and pure rotation angle ψ. With respect to the surface, the antenna of an aircraft located at the point $\{\vec{r}_0\} = \{\vec{\rho}_0, z_0\} = \{x_0, y_0, z_0\}$ is in the Fraunhofer zone. It is assumed that the scattering surface $z = \xi(x, y)$ is homogeneous and isotropic and has the mean $\langle \xi(\vec{\rho}) \rangle = 0$ and the variance $\langle \xi^2(\vec{\rho}) \rangle = \sigma_\xi^2$. For small precession and pure rotation angles and arbitrary θ (which is a usual situation in practice), the transformation of the coordinates of the antenna's center has the following form (with the terms of the second order of smallness disregarded):

$$\begin{cases} x' = x + y(\varphi + \psi\cos\theta) + z\psi\sin\theta, \\ y' = -x(\varphi\cos\theta + \psi) + y\cos\theta + z\sin\theta, \\ z' = x\varphi\sin\theta - y\sin\theta + z\cos\theta. \end{cases} \tag{1.37}$$

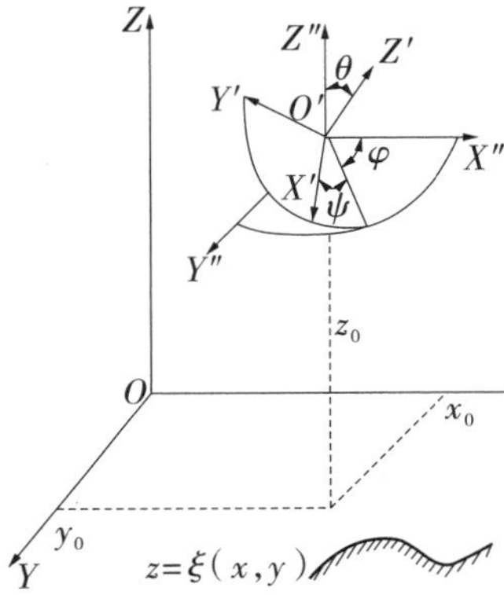

Fig. 1. 1 Geometry of the problem of radar imaging

The further analysis necessitates the scheme of formation of the scattered field. Fig. 1. 2 illustrates radar imaging as a result of the following process: the receiving-transmitting antenna radiates a field; the field propagates in a medium, undergoes reflection, and propagates in the backward direction toward the antenna[7]. Below, we consider the scalar case; i. e., we analyze the component of the electromagnetic field that corresponds to the polarization of the antenna.

According to Rayleigh representation Eq. (1. 18), the field at an arbitrary point of the half-space $z > \xi(\vec{\rho})$ is a superposition of plane waves:

$$E(\vec{\rho}', z') = \frac{1}{4\pi^2}\int_{-\infty}^{\infty} F_0(\vec{\omega}'_\perp)\exp(i\vec{\omega}'_\perp \cdot \vec{\rho}' - ik_{R}, z')\, d^2\vec{\omega}'_\perp . \tag{1.38}$$

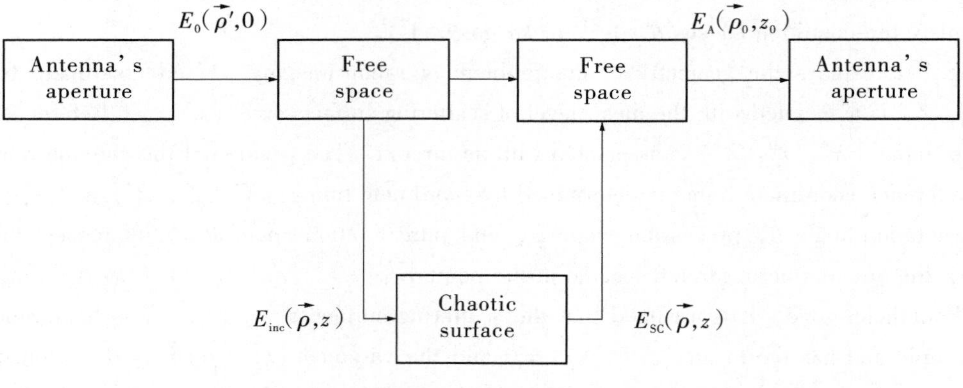

Fig. 1. 2 Formation of the continuum of reflected signals

Here,

$$F(\vec{\omega}'_\perp) = \int_{-\infty}^{\infty} E_0(\vec{\rho}',0)\exp(-\mathrm{i}\vec{\omega}'_\perp\cdot\vec{\rho}')\mathrm{d}^2\vec{\rho}' \quad (1.39)$$

is the angular spectrum of the field radiated by the antenna, E_0 $(\vec{\rho}',\ 0)$ is the field distribution in plane ρ' of the antenna's aperture,

$$\exp(\mathrm{i}k_R, z') = K(z',\vec{\omega}'_\perp) = \exp[\mathrm{i}z'(k^2-\vec{\omega}'^2_\perp)^{1/2}] \quad (1.40)$$

is the frequency response of the free-space layer with transfer function (1.20), $\vec{\omega}_\perp = \{\omega_x,\ \omega_y\}$ are spatial frequencies, and

$$k_z = \sqrt{k^2 - \vec{\omega}'^2_\perp} \quad (1.41)$$

With allowance for expressions Eq. (1.37), we have

$$\omega'_x = \omega_x + \omega_y(\varphi+\psi\cos\theta) + k_z\psi\sin\theta,$$
$$\omega'_y = -\omega_x(\varphi\cos\theta+\psi) + \omega_y\cos\theta + k_z\sin\theta. \quad (1.42)$$

In coordinates $(X,\ Y,\ Z)$, field E_{inc} $(\vec{\rho},\ z)$ incident on surface ξ $(\vec{\rho})$ has the form

$$E_{\text{inc}}(\vec{\rho}',z') = \frac{\cos\theta}{4\pi^2}\int_{-\infty}^{\infty}\int_{-\infty}^{\infty} F_0(\vec{\omega}_\perp)K(z_0-z,\vec{\omega}_\perp)\exp\{\mathrm{i}[\omega_x(x-x_0)+$$
$$\omega_y(y-y_0)]\}\mathrm{d}\omega_x\mathrm{d}\omega_y. \quad (1.43)$$

In integral Eq. (1.43), the arguments of angular spectrum F_0 $(\vec{\omega}_\perp)$ are expressed according to formulae of (1.42) and $\Delta \approx \cos\theta$ is the Jacobian of Eq. (1.42).

For $z \geqslant \xi(\vec{\rho})$, scattered field $E_{SC}(\vec{\rho},z)$ is expressed in a form similar to Eq. (1.38) with allowance for the fact that the scattered field propagates in the opposite direction:

$$E_{\text{SC}}(\vec{\rho},z) = \frac{1}{4\pi^2}\int_{-\infty}^{\infty} F_p(\vec{\omega}_\perp)\exp(\mathrm{i}\vec{\omega}_\perp\cdot\vec{\rho})K(z,\omega'_\perp)\mathrm{d}^2\vec{\omega}_\perp, \quad (1.44)$$

where

$$F_{\text{SC}}(\vec{\omega}_\perp) = \int_{-\infty}^{\infty} E_p(\vec{\rho},0)\exp(-\mathrm{i}\vec{\omega}_\perp\cdot\vec{\rho})\mathrm{d}^2\vec{\rho}, \quad (1.45)$$

is the angular spectrum of the scattered field.

When the above conditions for the MTP applicability are fulfilled and $V(\vec{\omega}_\perp) \approx V$, the field on the rough surface is $E_{SC} = VE_{inc}$. It can be shown readily that

$$E_{SC}(\vec{\rho},0) = VE_{inc}(\vec{\rho},0)\exp[-i2k\xi(\vec{\rho})\cos\theta]. \tag{1.46}$$

Then, taking into account Eqs. (1.43) and (1.46), we obtain

$$F_{SC}(\vec{\omega}_\perp) = \frac{V\cos\theta}{4\pi^2}\int_{-\infty}^{\infty}\exp[-i2k\xi(\vec{\rho})\cos\theta]\int_{-\infty}^{\infty}F_0(\vec{\omega}'_\perp)\exp(i\vec{\omega}'_\perp\cdot\vec{\rho})\cdot K(z_0,\vec{\omega}'_\perp)\exp[i\vec{\omega}'_\perp(\vec{\rho}-\vec{\rho}_0)]d^2\vec{\rho}d^2\vec{\omega}'_\perp. \tag{1.47}$$

The field received by the antenna is expressed by the formula

$$E_A(\vec{\rho}_0,z_0) = \frac{1}{4\pi^2}\int_{-\infty}^{\infty}F_0(-\vec{\omega}_\perp)F_p(\vec{\omega}_\perp)K(z_0,\vec{\omega}'_\perp)\exp(i\vec{\omega}_\perp\cdot\vec{\rho}_0)d^2\vec{\omega}_\perp \tag{1.48}$$

Following the technique from Ref. [16], we determine the pattern of the receiving-transmitting antenna from the relationship

$$F(\vec{\omega}_\perp) = \int_{-\infty}^{\infty}E_0(\vec{\rho},z)\exp(\pm i\vec{\omega}'_\perp\cdot\vec{\rho}')d^2\vec{\rho}' \equiv F_0(-\vec{\omega}_\perp). \tag{1.49}$$

Upon the substitutions of Eqs. (1.47) and (1.49), integral (1.48) can be calculated with the use of the stationary-phase method[7,13]. Then, the expression for field $E_A(k,z_0)$ can be specified as

$$E_A(k,z_0) = -\frac{Vk^2\cos\theta}{4\pi^2 z_0^2}\exp(i2kz_0)\int_{-\infty}^{\infty}\exp[-i2k\xi(\vec{\rho})\cos\theta]\cdot \exp\left[i\frac{k}{z_0}(\vec{\rho}-\vec{\rho}_0)^2\right]F[a_1(\omega\vec{\rho}),b_1(\omega\vec{\rho})]F[a_2(\omega\vec{\rho}),b_2(\omega\vec{\rho})]d^2\vec{\rho}, \tag{1.50}$$

the stationary-phase points are determined from the equalities

$$\omega_{x0} = \omega'_{x0} = -\frac{k(x-x_0)}{z_0},\quad \omega_{y0} = \omega'_{y0} = -\frac{k(y-y_0)}{z_0}, \tag{1.51}$$

and the coefficients of the pattern have the form

$$\begin{cases} a_{1,2} = \pm\dfrac{k(x-x_0)}{z_0} \pm \dfrac{k(y-y_0)}{z_0}(\varphi+\psi\cos\theta) + k\psi\sin\theta, \\ b_{1,2} = \mp\dfrac{k(x-x_0)}{z_0}(\varphi\cos\theta+\psi) \pm \dfrac{k(y-y_0)}{z_0}\cos\theta + k\sin\theta. \end{cases} \tag{1.52}$$

In the case of spaced antennas, quantities $F(a_1,b_1)$ and $F(a_2,b_2)$ are the patterns of the transmitting and receiving antennas, respectively. It follows from Eq. (1.52) that, with allowance for the antenna's pattern, the scattered field depends on the location and orientation of the antenna. This dependence is equivalent to tuning of a spatial filter whose frequency response is the antenna's pattern[16]. The spatial frequency changes when angle θ changes, i. e., when antenna scanning is performed. Antennas with a continuous aperture and a pattern with a single main lobe of width θ_A are characterized by a continuous spectrum of spatial frequencies bounded by the value $\Delta\alpha = \theta_A^{-1} = d/\lambda$.

Consider the particular case of small angular fluctuations of an aircraft and almost vertical probing. Then, taking into account that $\varphi \approx \psi \approx 0$ and $\theta \ll 1$, we see that Eqs. (1.50) –

(1.52) for the backscattered field coincide with the results from Ref. [25] and field $E_A(k,z_0)$ is expressed in the form

$$E_A(k,z_0) = -\frac{Vk^2}{4\pi^2 z_0^2}\exp(\mathrm{i}2kz_0)\int_{-\infty}^{\infty}\int_{-\infty}^{\infty}\exp[-\mathrm{i}2k\xi(\vec{\rho})]\cdot$$

$$\exp\left\{\mathrm{i}\frac{k}{z_0}[(x-x_0)^2+(y-y_0)^2]\right\}F^2\left[\frac{k(x-x_0)}{z_0},\frac{k(y-y_0)}{z_0}+k\theta\right]\mathrm{d}x\mathrm{d}y\,. \quad (1.53)$$

Note that the MMW (λ =2, 2 and 8.6 mm) field scattered by a rough terrain has been simulated comprehensively in Ref. [26] according to my Eqs. (1.50) – (1.52) with allowance for the Gaussian pattern of an antenna and analyzed as a function of the altitude of an aircraft.

1.5 Generalized frequency response of a scattering radio channel

Analysis of a radio channel by means of the theory of linear systems and transformations is based on the linearity of the Maxwell equations. Of the variety of functions describing linear radio channels with variable parameters[5,7,27-30], generalized frequency response $K(\omega,t,\vec{r})$ is used most often. This characteristic is related to generalized impulse response (the Green's function) $G(t,\tau,\vec{r})$ by the Fourier transform:

$$K(\omega,t,\vec{r}) = \int_{-\infty}^{\infty}G(t,\tau,\vec{r})\mathrm{e}^{-\mathrm{i}\omega t}\mathrm{d}\tau\,. \quad (1.54)$$

As a function of sampling instant t of a radio channel's response, function $G(t,\tau,\vec{r})$ is the response to an input pulse that is the Dirac delta function and arrives at instant τ. The function $\mathrm{e}^{\mathrm{i}\omega t}$ is an eigenfunction of a linear system. Eigenvalue $K(\omega,t,\vec{r})$ corresponding to this function is the frequency response of a radio channel. Hence, expression (1.50) for a monochromatic field scattered by a statistically rough surface should be regarded as generalized characteristic $K(\omega,t,\vec{r})$.

In real situations, $K(\omega,t,\vec{r})$ is a random function. Averaging expression (1.50) over all possible realizations of random quantity $\xi(\vec{\rho})$, we obtain mean $\langle K(\omega,t,\vec{r})\rangle$ for $\varphi=\psi=0$ and the Gaussian pattern:

$$\langle K(\omega,t,\vec{r})\rangle = \frac{V\theta_A^2\omega^2\cos\theta}{4\pi c^2[(1+m^2)(\cos^4\theta+m^2)]^{1/4}}\exp\left[-2\sigma_\xi^2\frac{\omega^2}{c^2}\cos^2\theta-\frac{m^2\sin^2\theta}{\theta_A^2(\cos^4\theta+m^2)}\right]\cdot$$

$$\exp\left\{\mathrm{i}\left[2\frac{\omega}{c}z_0+\frac{1}{2}\arctan m+\frac{1}{2}\arctan\frac{m}{\cos^2\theta}+\frac{m\sin^2 2\theta}{4\theta_A^2(\cos^4\theta+m^2)}\right]\right\}. \quad (1.55)$$

Here,

$$m = \frac{z_0\theta_A^2\omega}{c} = 2\pi p^2,\ p = \frac{\sqrt{z_0\lambda}}{d}, \quad (1.56)$$

p is the wave parameter[31], $\theta_A=\lambda/d$ is the width of the antenna's pattern, λ is the wavelength, and d is the characteristic dimension of the antenna's aperture.

In addition, Eq. (1.55) characterizes the coherent (regular) component of the random field. As the height of irregularities and the incidence angle grow and the width of the pattern decreases, the coherent component rapidly decreases.

As follows from Eq. (1.55), the amplitude of the received signal is proportional to absolute value $|\omega|$. This dependence is a frequency-selective equivalent of differentiation. Ultrawideband signals are differentiated, and a zero-frequency signal cannot be transmitted.

Thus, the shape of a UWB signal that is neither a high-frequency signal nor a video signal should meet special requirements[5,7]. This frequency selectivity affects narrowband signals only slightly.

1.6 Generalized correlator of fields scattered by a statistically rough anisotropic surface

The fluctuation properties of a scattered field forming an RI are determined by cross moments. In the limiting case of absolutely coherent waves, signal intensity $I(\vec{r})$ is related with the amplitude via the formula $I(\vec{r}) = E(\vec{r})E^*(\vec{r})$, where the asterisk denotes complex conjugation. This relationship is invalid for partially coherent waves because the intensity of a partially coherent field is expressed through the field coherence function $\Gamma(\tau) = \langle E_l(t+\tau)E_2^*(t)\rangle$ rather than field amplitude $E(\vec{r})$. The coherence function determines correlation between complex fields observed at two spatial points.

Let us consider the general case of scattering of modulated waves by a chaotic surface and analyze the following functional of backscattered stochastic fields observed at different spatial points $(\vec{\rho}_1, z_1)$ and $(\vec{\rho}_2, z_2)$ at different instants t and $t+\Delta t$[7,13,32,33]:

$$\langle E_A(t_1,\vec{r}_1)E_A^*(t_2,\vec{r}_2)\rangle = \frac{1}{4\pi^2}\int_{-\infty}^{\infty}\int_{-\infty}^{\infty} F(\omega_1)F^*(\omega_2)\Psi_\omega(t_1,t_2,\vec{r}_1,\vec{r}_2,\omega_1,\omega_2)\cdot$$
$$\mathrm{e}^{-\mathrm{i}(\omega_1 t_1-\omega_2 t_2)}\mathrm{d}\omega_1\mathrm{d}\omega_2\,. \tag{1.57}$$

It is seen from Eq. (1.57) that this functional is completely determined by the frequency coherence function (FCF)

$$\Psi_\omega = \Psi_\omega(t_1,t_2,\vec{r}_1,\vec{r}_2,\omega_1,\omega_2) = \langle E_A(t_1,\vec{r}_1,\omega_1)E_A^*(t_2,\vec{r}_2,\omega_2)\rangle\,, \tag{1.58}$$

which describes correlation of two monochromatic fields excited at frequencies ω_1 and ω_2[7,13,34-37].

Functional (1.57) determines a spatial-temporal correlator of backscattered stochastic fields. The cross statistical moments of the scattered field determine its spatial and temporal correlation and the frequency coherence. The following important circumstance should be emphasized. In the theory of random waves, for simplicity and convenience, the aforementioned correlation functions often are analyzed independently. The mathematical apparatus of multidimensional semi-invariants (cumulants) or correlators taking into account the characteristic functional[1,2,5,7,38-45] ensures a comprehensive description of the scattering process for an arbitrary distribution of the components of the resulting field rather than for only the Gaussian distribution (!).

Note that Thiele (1903) was the first to consider semi-invariants. The mathematical apparatus of the cumulant analysis of non-Gaussian random quantities and processes (language of cumulants) is described comprehensively in Monographs [38-41, 44]. Semi-invariants are important mainly because they are involved in determination of correlation functions of various variables.

This circumstance is especially important for investigation of fractal processes in radiophysics and radio electronics[1-3].

The asymmetry coefficients $K_a = \mu_3/\sigma^3 = \kappa_3/\kappa_2^{3/2}$ and the excess coefficients $K_e = (\mu_4/\sigma^4) - 3 = \kappa_4/\kappa_2^2$ have been used to find, on the Pearson plane, experimental probability distributions of MMWs scattered by surfaces with vegetation[1,2,7,46-49]. In the aforementioned studies, this approach is applied to develop new models of scattering, including fractal models. In the expressions for K_a and K_e, the following notation is used: μ_3 and μ_4 are the third and fourth central moments, respectively; σ^2 is the variance; and κ_2, κ_3, and κ_4 are the second-, third-, and fourth-order semi-invariants, respectively. For normalized distributions (characterized by the zero mean and the unit variance), coefficients K_a and K_e are equal to the third and fourth semi-invariants, respectively.

Since the late 1970s, in cooperation with the researchers from the Almaz Central Design Office and the Central Research Institute of Automation and Hydraulics, Prof. A. Potapov has theoretically obtained and experimentally analyzed various approximations of the FCF of a radar channel employed for terrain probing with simple and complex microwave (for example, MMWband) signals. Some of these results (approximate FCFs and experimental data) are presented in [5, 7, 13, 32, 33, 37, 46].

Consider the proposed method, which is based on generalized correlator $\Psi(k_{1,2}, \vec{r}_{1,2})$ of backscattered fields[7,13,32,33,50,51]. We solve the 3D scattering problem ($R \in \mathbf{R}^3$) and derive explicit analytic relationships that allow for the finite width of the pattern of the receiving-transmitting antenna and the correlation between slope angles of irregularities of an anisotropic Gaussian surface. Let an aircraft move in space and illuminate the surface $z = \xi(x,y)$ by two monochromatic waves (Fig. 1.3) at the frequencies $\omega_1(k_1 = 2\pi/\lambda_1)$ and $\omega_2(k_2 = 2\pi/\lambda_2)$. With respect to the mean plane, the receiving-transmitting antenna is located in the Fraunhofer zone.

In the MTP approximation, the backscattered field at spaced points $A_1(x_{10}, y_{10}, z_{10})$ and $A_2(x_{20}, y_{20}, z_{20})$ is expressed as

$$E_{1,2}(k_{1,2}, \vec{R}_{1,2}) = \alpha_{1,2}\exp(\mathrm{i}2k_{1,2}z_{1,2}\sec\theta_{1,2})\int_{S_{1,2}} \mathrm{d}x_{1,2}\mathrm{d}y_{1,2}G(\theta_{1,2}, \varphi_{1,2}) \cdot$$

$$\exp[-\mathrm{i}2k_{1,2}\xi(x_{1,2}, y_{1,2})\cos\theta_{1,2}]\exp\left\{\mathrm{i}\frac{k_{1,2}}{z_{1,2}}[(x_{1,2} - x_{10,20})^2 + (y_{1,2} - y_{10,20})^2]\right\} \quad (1.59)$$

Here, $\alpha_{1,2} = -\mathrm{i}V(k_{1,2}, \vec{\rho}_{1,2})\frac{k_{1,2}}{z}Q\cos\theta_{1,2}$, V is the Fresnel coefficient, Q is the energy factor, $z_{1,2}$ is the altitude of the antenna of a radiation source over the mean plane $z = 0$, $G(\theta, \varphi) =$

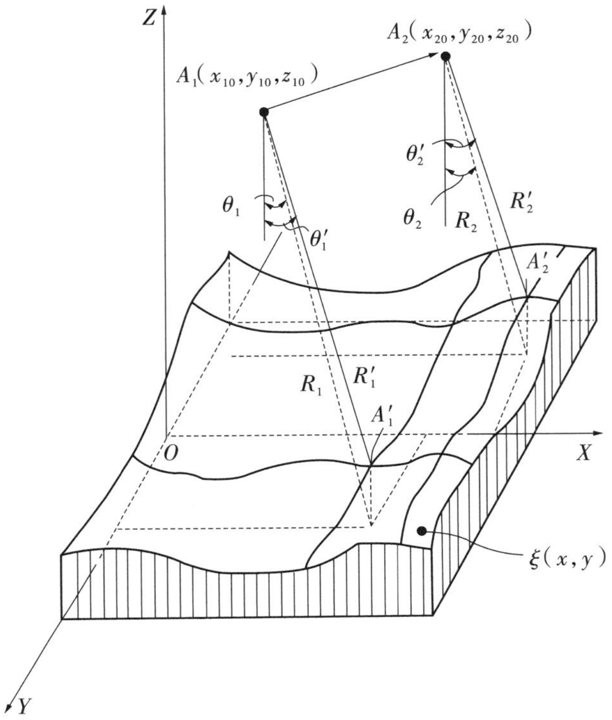

Fig. 1. 3 Geometry of the problem of determining a generalized functional of stochastic backscattered fields

$\langle g(\theta,\varphi)g^*(\theta,\varphi)\rangle$ is the normalized field ($g(\theta,\varphi)$) or power ($G(\theta,\varphi)$) pattern of the antenna, and $S_{1,2}$ is the projection of the illuminated surface on the XOY plane.

Then, the complete formulation of the considered problem of radar imaging is based on a correlator of backscattered fields. The correlator can be represented as

$$\Psi(k_{1,2},\vec{r}_{1,2}) = \langle E_1(k_1,\vec{r}_1)E_2^*(k_2,\vec{r}_2)\rangle, \tag{1.60}$$

or, with allowance for Eq. (1. 59), in the form of an integral functional of stochastic backscattered fields:

$$\Psi(k_{1,2},\vec{r}_{1,2}) = \alpha_1\alpha_2\exp[\mathrm{i}2(k_1z_1\sec\theta_1 - k_2z_2\sec\theta_2)]\iint\limits_{S_1S_2}\mathrm{d}x_1\mathrm{d}x_2\mathrm{d}y_1\mathrm{d}y_2G(\theta_1,\varphi_1)G(\theta_2,\varphi_2)\cdot$$
$$\exp\left\{\mathrm{i}\frac{k_1}{z_1}[(x_1 - x_{10})^2 + (y_1 - y_{10})^2] - \mathrm{i}\frac{k_2}{z_2}[(x_2 - x_{20})^2 + (y_2 - y_{20})^2]\right\}\cdot$$
$$\langle\exp\{-\mathrm{i}2[k_1\xi(x_1,y_1)\cos\theta_1 - k_2\xi(x_2,y_2)\cos\theta_2]\}\rangle. \tag{1.61}$$

In integral functional (1. 61), the fourth multiplier of the integrand describes the phase modulation of probing waves caused by the presence of a rough surface. The integrals in Eq. (1. 61) should be calculated accurately.

The terrain area that is covered by the antenna's axially symmetric pattern $\theta_v = \theta_\varphi = \theta_A$ and participates in scattering generally is an ellipse with the major axis

$$a = z\sin\theta_A \Big/ \left(\cos^2\frac{\theta_A}{2} - \sin^2\theta \right) = 2z\sin\theta_A / (\cos\theta_A + \cos 2\theta) \tag{1.62}$$

and the minor axis

$$b = 2z\sin\frac{\theta}{2} \Big/ \left(\cos^2\frac{\theta_A}{2} - \sin^2\theta \right)^{1/2} = 2z(1 - \cos\theta_A)^{1/2} / (\cos\theta_A + \cos 2\theta)^{1/2}. \tag{1.63}$$

If the antenna's beam is rather narrow, we have in the first approximation

$$a \approx z\theta_A / \cos^2\theta, \quad b \approx z\theta_A / \cos\theta. \tag{1.64}$$

Factor $< \cdots >$ in Eq. (1.61) is 2D characteristic field function ξ that depends on $\xi_1 = \xi(x_1, y_1)$ and $\xi_2 = \xi(x_2, y_2)$ as it depends on the parameters. For a surface characterized by the Gaussian distribution of the irregularities' heights,

$$p_2(\xi_1, \xi_2) = \frac{1}{2\pi\sigma_\xi^2 \sqrt{1 - R^2}} \exp\left[\frac{\xi_1^2 - 2R\xi_1\xi_2 + \xi_2^2}{2\sigma_\xi^2 (1 - R^2)} \right] \tag{1.65}$$

having the zero mean $< \xi(x, y) > = 0$ and variance σ_ξ^2 and by spatial correlation factor $R(\Delta x = x_2 - x_1, \Delta y = y_2 - y_1)$, we find

$$< \cdots > = \exp\{ -2\sigma_\xi^2 [k_1^2 \cos^2\theta_1 - 2k_1 k_2 R(\Delta x, \Delta y) \cos\theta_1 \cos\theta_2 + k_2^2 \cos^2\theta_2] \}. \tag{1.66}$$

Let us expand the correlation factor in the Taylor series about the point ($\Delta x = 0$, $\Delta y = 0$), retain the terms on the order of smallness no higher than two, and take into account the known relationship between the correlation functions of the random process and its derivative. Then, we obtain

$$R(\Delta x, \Delta y) = 1 - \frac{\langle \gamma_x^2 \rangle}{2\sigma_\xi^2} \Delta x^2 - K_{xy} \frac{\Delta x \Delta y}{\sigma_\xi^2} - \frac{\langle \gamma_y^2 \rangle}{2\sigma_\xi^2} \Delta y, \tag{1.67}$$

where

$$K_{xy} = R_{xy} \sqrt{\langle \gamma_x^2 \rangle \langle \gamma_y^2 \rangle} \tag{1.68}$$

is the cross-correlation moment of the slope ratios $\gamma_x = \partial\xi/\partial x$ and $\gamma_y = \partial\xi/\partial y$ with the Gaussian distribution

$$p_2(\gamma_x, \gamma_y) = \frac{1}{2\pi \sqrt{\langle \gamma_x^2 \rangle \langle \gamma_y^2 \rangle} \sqrt{1 - R_{xy}^2}} \exp\left\{ -\frac{1}{2(1 - R_{xy}^2)} \right\} \cdot$$
$$\left[\frac{\gamma_x^2}{\langle \gamma_x^2 \rangle} - 2R_{xy} \frac{\gamma_x \gamma_y}{\sqrt{\langle \gamma_x^2 \rangle \langle \gamma_y^2 \rangle}} + \frac{\gamma_y^2}{\langle \gamma_y^2 \rangle} \right] \tag{1.69}$$

with variances $\langle \gamma_x^2 \rangle$ and $\langle \gamma_y^2 \rangle$ determined along the OX and OY axes, respectively.

Then, we have

$$< \cdots > = \exp[-2\sigma_\xi^2 (k_1 \cos\theta_1 - k_2 \cos\theta_2)^2] \cdot$$
$$\exp[-2k_1 k_2 \cos\theta_1 \cos\theta_2 (\langle \gamma_x^2 \rangle \Delta x^2 + 2R_{xy} \sqrt{\langle \gamma_x^2 \rangle \langle \gamma_y^2 \rangle} \Delta x \Delta y + \langle \gamma_y^2 \rangle \Delta y^2)] \tag{1.70}$$

Eq. (1.70) can be simplified in the case of isotropic random irregularities when $\langle \gamma_x^2 \rangle = \langle \gamma_y^2 \rangle = \langle \gamma^2 \rangle$ at $R_{xy} = 1$:

$$< \cdots > = \exp[-2\sigma_\xi^2 (k_1 \cos\theta_1 - k_2 \cos\theta_2)^2] \exp[-2k_1 k_2 \langle \gamma^2 \rangle (\Delta x + \Delta y)^2 \cdot$$
$$\cos\theta_1 \cos\theta_2] \tag{1.71}$$

1.7 Generalized correlator of anisotropically scattered fields for the Gaussian pattern of the antenna

For the further analysis, we choose the Gaussian approximation of the antenna's pattern. Although this approximation is not realized in practice, it ensures sufficiently accurate analysis of radar imaging. When the antenna is deflected by angle θ, the pattern is expressed in the coordinates of the mean plane as follows[52]:

$$G(x,y) = \exp\left\{ -(4\ln 2)\frac{\cos^2\theta}{z_0^2\theta_A^2}[(x-x_0)^2\cos^2\theta + y^2]\right\} \tag{1.72}$$

where $\theta_A = \theta_\theta = \theta_\varphi$ is the half-power width of the pattern.

Assume that the condition for quasi-monochromatism is fulfilled, angles θ and φ change only slightly within areas $S_{1,2}$, correlation radii of irregularities are small compared to axes Eqs. (1.62) and (1.63), and reflection coefficients $V(k_{1,2})$ are frequency-independent. Then, in Eqs. (1.61) and (1.70), we can set

$$V(k_1,\vec{\rho}_1) \approx V(k_2,\vec{\rho}_2) \equiv V,\ G(\theta_1,\varphi_1) \approx G(\theta_2,\varphi_2) \equiv G,\ \cos\theta_1 \approx \cos\theta_2 \equiv \cos\theta, \tag{1.73}$$

and extend the limits of integration in Eq. (1.61) to infinity.

Let us apply approximations Eqs. (1.72) and (1.73), take into account Eq. (1.70), and pass to the new variables $u = x_2 - x_1$ and $v = y_2 - y_1$. Then, we can represent solution (1.61) in the final explicit form

$$\Psi(k_{1,2},\vec{r}_{1,2}) = A(k_{1,2},\vec{r}_{1,2})\exp[B(k_{1,2},\vec{r}_{1,2})], \tag{1.74}$$

where

$$A(k_{1,2},\vec{r}_{1,2}) = \alpha_1\alpha_2\pi^2[A_1A_2 - 4k_1^2k_2^2K_{xy}^2(a_1+a_2)(a_3+a_4)\cos^4\theta]^{-1/2}, \tag{1.75}$$

$$\begin{aligned}B(k_{1,2},\vec{r}_{1,2}) = &-\mathrm{i}2(k_2z_2 - k_1z_1)\sec\theta - 2\sigma_\xi^2(k_2-k_1)^2\cos^2\theta + \\ &\mathrm{i}\left[k_1\frac{(x_{10}^2+y_{10}^2)}{z_1} - k_2\frac{(x_{20}^2+y_{20}^2)}{z_2}\right] - b\left(\frac{x_{10}^2\cos^2\theta + y_{10}^2}{z_1^2} + \frac{x_{20}^2\cos^2\theta + y_{20}^2}{z_2^2}\right) + \\ &\frac{(a_1x_{10}+a_2x_{20})^2}{a_1+a_2} + \frac{(a_3y_{10}+a_4y_{20})^2}{a_3+a_4} + \frac{a_1^2a_2^2(x_{20}-x_{10})^2}{A_1(a_1+a_2)} + \\ &\frac{[A_1a_3a_4(y_{20}-y_{10}) - 2k_1k_2K_{xy}a_1a_2(a_3+a_4)(x_{20}-x_{10})\cos^2\theta]^2}{A_1(a_3+a_4)[A_1A_2 - 4k_1^2k_2^2K_{xy}^2(a_1+a_2)(a_3+a_4)\cos^4\theta]}.\end{aligned} \tag{1.76}$$

In Eqs. (1.75) and (1.76), the following notation is used:

$$\begin{aligned}&A_1 = 2k_1k_2\langle\gamma_x^2\rangle(a_1+a_2)\cos^2\theta + a_1a_2,\ A_2 = 2k_1k_2\langle\gamma_y^2\rangle(a_3+a_4)\cos^2\theta + a_3a_4, \\ &a_1 = b\frac{\cos^2\theta}{z_1^2} - \mathrm{i}\frac{k_1}{z_1},\ a_2 = b\frac{\cos^2\theta}{z_2^2} + \mathrm{i}\frac{k_2}{z_2},\ a_3 = \frac{b}{z_1^2} - \mathrm{i}\frac{k_1}{z_1},\ a_4 = \frac{b}{z_2^2} + \mathrm{i}\frac{k_2}{z_2}, \\ &b = \frac{4(\ln 2)\cos^2\theta}{\theta_A^2}.\end{aligned} \tag{1.77}$$

Eq. (1.77) imply that, for any θ, the inequalities $|a_1| \leqslant |a_3|$ and $|a_2| \leqslant |a_4|$ are valid. It is convenient to perform the further analysis of correlators of backscattered fields with the use of the following identities:

$$(a_1 + a_2)(a_3 + a_4) = \left(\frac{1}{z_1^2} + \frac{1}{z_2^2}\right)^2 b^2\cos^2\theta - \left(\frac{k_1}{z_1} - \frac{k_2}{z_2}\right)^2 + \mathrm{i}b\left(\frac{1}{z_1^2} + \frac{1}{z_2^2}\right)\left(\frac{k_2}{z_2} - \frac{k_1}{z_1}\right)(1 + \cos^2\theta), \tag{1.78}$$

$$a_1a_2a_3a_4 = \frac{1}{z_1^2z_2^2}\left[\left(k_1k_2 + \frac{b^2}{z_1z_2}\right)\left(k_1k_2 + \frac{b^2\cos^4\theta}{z_1z_2}\right) - b^2\left(\frac{k_2}{z_1} - \frac{k_1}{z_2}\right)^2\cos^2\theta\right] + \mathrm{i}\frac{b}{z_1^2z_2^2}\left(k_1k_2 + \frac{b^2\cos^2\theta}{z_1z_2}\right)\left(\frac{k_2}{z_1} - \frac{k_1}{z_2}\right)(1 + \cos^2\theta). \tag{1.79}$$

For arbitrary polynomials L and M, the equality

$$L(a_1 + a_2)a_3a_4 + Ma_1a_2(a_3 + a_4) = (L + M)\cdot \left\{\frac{bk_1k_2}{z_1z_2}\left(\frac{1}{z_1^2} + \frac{1}{z_2^2}\right)(1 + \cos^2\theta) - \mathrm{i}\frac{1}{z_1z_2}\left[k_1k_2\left(\frac{k_1}{z_1} - \frac{k_2}{z_2}\right) + \left(\frac{k_1}{z_2} - \frac{k_2}{z_1}\right)\left(\frac{1}{z_1^2} + \frac{1}{z_2^2}\right)b^2\cos^2\theta\right]\right\} - (L + M\cos^2\theta)\frac{b}{z_1^2z_2^2}\left[k_1^2 + k_2^2 - b^2\left(\frac{1}{z_1^2} + \frac{1}{z_2^2}\right)\cos^2\theta\right] - \mathrm{i}(L + M\cos^4\theta)\frac{b^2}{z_1^2z_2^2}\left(\frac{k_1}{z_1} - \frac{k_2}{z_2}\right) \tag{1.80}$$

holds as well.

After bulky calculations, we obtain one more explicit equivalent form of the unnormalized generalized correlator of stochastic fields scattered by an anisotropic surface:

$$A(k_{1,2}, \vec{r}_{1,2}) = \alpha_1\alpha_2\pi^2 z_1z_2\{C_0^2C_{11}\langle\gamma_x^2\rangle\langle\gamma_y^2\rangle(1 + R_{xy}^2) + C_0(C_{12}\langle\gamma_x^2\rangle + C_{13}\langle\gamma_y^2\rangle) + C_{14} + \mathrm{i}[C_0^2C_{15}\langle\gamma_x^2\rangle\langle\gamma_y^2\rangle(1 - R_{xy}^2) + C_0(C_{16}\langle\gamma_x^2\rangle + C_{17}\langle\gamma_y^2\rangle) + C_{18}]\}^{-1/2}, \tag{1.81}$$

$$B(k_{1,2}, \vec{r}_{1,2}) = -\mathrm{i}2(k_2z_2 - k_1z_1)\sec\theta - 2\sigma_\xi^2(k_2 - k_1)^2\cos^2\theta + \mathrm{i}\left[k_1\frac{(x_{10}^2 + y_{10}^2)}{z_1} - k_2\frac{(x_{20}^2 + y_{20}^2)}{z_2}\right] - b\left(\frac{x_{10}^2\cos^2\theta + y_{10}^2}{z_1^2} + \frac{x_{20}^2\cos^2\theta + y_{20}^2}{z_2^2}\right) + \frac{(a_1x_{10} + a_2x_{20})^2}{a_1 + a_2} + \frac{(a_3y_{10} + a_4y_{20})^2}{a_3 + a_4} - \frac{W_1}{W_2}, \tag{1.82}$$

where

$$W_1 = C_1\Delta x^2 + C_2\Delta y^2 + C_0(C_4\langle\gamma_x^2\rangle\Delta y^2 + C_3\langle\gamma_y^2\rangle\Delta x^2) + 2C_0C_5\sqrt{\langle\gamma_x^2\rangle\langle\gamma_y^2\rangle}R_{xy}\Delta x\Delta y + \mathrm{i}[C_6\Delta x^2 + C_7\Delta y^2 + 2C_0(C_8\langle\gamma_x^2\rangle\Delta y^2 + C_9\langle\gamma_y^2\rangle\Delta x^2) - 2C_{10}\sqrt{\langle\gamma_x^2\rangle\langle\gamma_y^2\rangle}R_{xy}\Delta x\Delta y], \tag{1.83}$$

$$W_2 = (C_{11} + \mathrm{i}C_{15})z_1z_2\{C_0^2C_{11}\langle\gamma_x^2\rangle\langle\gamma_y^2\rangle(1 - R_{xy}^2) + C_0(C_{12}\langle\gamma_x^2\rangle + C_{13}\langle\gamma_y^2\rangle) + C_{14} + \mathrm{i}[C_0^2C_{15}\langle\gamma_x^2\rangle\langle\gamma_y^2\rangle(1 - R_{xy}^2) + C_0(C_{16}\langle\gamma_x^2\rangle + C_{17}\langle\gamma_y^2\rangle) + C_{18}]\}. \tag{1.84}$$

In Eqs. (1.81) – (1.84), the following notation is used:

$$C_0 = 2k_1k_2z_1z_2\cos^2\theta,$$

$$C_1 = b[(p_1p_4 - p_2p_3)(p_5^2 - b^2p_3^2\cos^4\theta) - 2p_3p_5(b^2p_1p_3 + p_2p_4)\cos^2\theta],$$

$$
\begin{aligned}
C_2 &= b[(p_1p_5 - p_2p_3)(p_4^2 - b^2p_3^2)\cos^2\theta - 2p_3p_4(b^2p_1p_3\cos^4\theta + p_2p_5)],\\
C_3 &= (p_5^2 - b^2p_3^2\cos^4\theta)(b^2p_1^2 + p_2^2) - 4b^2p_1p_2p_3p_5\cos^2\theta,\\
C_4 &= (p_4^2 - b^2p_3^2)(b^2p_1^2\cos^4\theta - p_2^2) - 4b^2p_1p_2p_3p_4\cos^2\theta,\\
C_5 &= C_{11}C_{14} - C_{15}C_{18}(1+\cos^2\theta)^{-1},\\
C_6 &= (b^2p_1p_3 + p_2p_4)(p_5^2 - b^2p_3^2\cos^4\theta) + 2b^2p_3p_5(p_1p_4 - p_2p_3)\cos^2\theta,\\
C_7 &= (b^2p_1p_3\cos^4\theta + p_2p_5)(p_4^2 - b^2p_3^2) + 2b^2p_3p_4(p_1p_5 - p_2p_3)\cos^2\theta,\\
C_8 &= b[(p_4^2 - b^2p_3^2)p_1p_2\cos^2\theta + p_3p_4(b^2p_1^2\cos^4\theta - p_2^2)],\\
C_9 &= b[(p_5^2 - b^2p_3^2\cos^4\theta)p_1p_2 + p_3p_5(b^2p_1^2 - p_2^2)\cos^2\theta],\\
C_{10} &= b(C_{11}p_3p_6 + C_{14}p_1p_2)(1+\cos^2\theta),\\
C_{11} &= p_1^2b^2\cos^2\theta - p_2^2,\quad C_{12} = b(p_1p_4\cos^2\theta - p_2p_3),\\
C_{13} &= b(p_1p_5 - p_2p_3\cos^2\theta),\quad C_{14} = p_4p_5 - b^2p_3^2\cos^2\theta,\\
C_{15} &= bp_1p_2(1+\cos^2\theta),\quad C_{16} = p_2p_4 + b^2p_1p_3\cos^2\theta,\\
C_{17} &= p_2p_5 + b^2p_1p_3\cos^2\theta,\quad C_{18} = bp_3p_6(1+\cos^2\theta),
\end{aligned}
\tag{1.85}
$$

where the coefficients are determined by the formulas

$$
\begin{aligned}
p_1 &= \frac{1}{z_1^2} + \frac{1}{z_2^2},\quad p_2 = \frac{k_2}{z_2} - \frac{k_1}{z_1},\quad p_3 = \frac{k_2}{z_1} - \frac{k_1}{z_2},\quad p_4 = \frac{b^2}{z_2z_1} + k_1k_2,\\
p_5 &= \frac{b^2\cos^4\theta}{z_1z_2} + k_1k_2,\quad p_6 = \frac{b^2\cos^2\theta}{z_1z_2} + k_1k_2.
\end{aligned}
\tag{1.86}
$$

1.8 Normalized generalized correlator of fields scattered by a statistically rough anisotropic surface

The normalized correlator of fields scattered by an anisotropic rough surface (Fig. 1.3) is defined as

$$
\begin{aligned}
\Psi(\Delta k,\Delta\vec{r}) &= \frac{\Psi(\Delta k_{1,2},\vec{r}_{1,2})}{[\langle|E_1(k_1)|^2\rangle\langle|E_2(k_2)|^2\rangle]^{1/2}}\\
&= \frac{\Psi(\Delta k_{1,2},\vec{r}_{1,2})}{[\langle E_1(k_1)E_1^*(k_1)\rangle\langle E_2(k_2)E_2^*(k_2)\rangle]^{1/2}}\\
&= \frac{\Psi(\Delta k_{1,2},\vec{r}_{1,2})}{[\Psi(k_1,k_1)\,|_{\vec{R}_1}\Psi(k_2,k_2)\,|_{\vec{R}_2}]^{1/2}}.
\end{aligned}
\tag{1.87}
$$

Determination of normalization coefficients $\Psi(k_i,k_i)\,|_{\vec{R}_i}$, where $i = 1,2$, necessitates allowance for the fact that coefficients (1.86) for the index $i=1$ are $p_1 = 2/z_1^2$, $p_2 = p_3 = 0$, $p_4 = p_{4,1} = (b^2/z_1^2) + k_1^2$, $p_5 = p_{5,1} = (b^2\cos^4\theta/z_1^2) + k_1^2$, and $p_6 = (b^2\cos^2\theta/z_1^2) + k_1^2$. Then, according to (1.85), we have

$$
\begin{aligned}
C'_0 &= 2k_1^2z_1^2\cos^2\theta,\ C'_1 = bp_1p_4p_5^2,\ C'_2 = bp_1p_4^2p_5,\ C'_3 = b^2p_1^2p_5^2,\\
C'_4 &= b^2p_1^2p_4^2\cos^4\theta,\ C'_5 = C'_{11}C'_{14} = b^2p_1^2p_4p_5\cos^2\theta,
\end{aligned}
$$

$$C'_6 = C'_7 = C'_8 = C'_9 = C'_{10} = C'_{15} = C'_{16} = C'_{17} = C'_{18} = 0,$$

$$C'_{11} = b^2 p_1^2 \cos^2\theta \,,\ C'_{12} = bp_1 p_4 \cos^2\theta \,,\ C'_{13} = bp_1 p_5 \,,\ C'_{14} = p_4 p_5 \,. \tag{1.88}$$

Relationships (1.81) – (1.84) imply that, since equalities (1.88) hold, the expression for $\Psi(k_1, k_1) \mid_{\vec{R}_1}$ is no longer a complex function:

$$\Psi(k_1,k_1) \mid_{\vec{R}_1} = A \mid_{\vec{R}_1} = \alpha_1^2 \pi^2 z_1^2 [C'_{14} + C'^2_0 C'_{11} \langle \gamma_x^2 \rangle \langle \gamma_y^2 \rangle (1 - R_{xy}^2) + C'_0 (C'_{12} \langle \gamma_x^2 \rangle + C'_{13} \langle \gamma_y^2 \rangle)]^{-1/2}. \tag{1.89}$$

The substitution of the values of coefficients $C'_0, \cdots, C'_{18}$ from (1.88) into (1.89) yields the expression for the normalization coefficient:

$$\Psi(k_1,k_1) \mid_{\vec{R}_1} = \alpha_1^2 \pi^2 z_1^2 \left\{ 16 k_1^4 \langle \gamma_x^2 \rangle \langle \gamma_y^2 \rangle (1 - R_{xy}^2) b^2 \cos^6\theta + 4k_1^2 b \left[\langle \gamma_x^2 \rangle \left(k_1^2 + \frac{b^2}{z_1^2} \right) \cos^2\theta + \langle \gamma_y^2 \rangle \left(k_1^2 + \frac{b^2 \cos^4\theta}{z_1^2} \right) \right] \cos^2\theta + \left(k_1^2 + \frac{b^2 \cos^4\theta}{z_1^2} \right) \left(k_1^2 + \frac{b^2}{z_1^2} \right) \right\}^{-\frac{1}{2}}. \tag{1.90}$$

In a similar manner, normalization coefficient $\Psi(k_2, k_2) \mid_{\vec{R}_2}$ is determined upon the corresponding replacement of indices and substitution of the coefficients $p_1 = 2/z_2^2$, $p_{4,2} = (b^2/z_2^2) + k_2^2$, $p_{5,2} = (b^2 \cos^4\theta / z_2^2) + k_2^2$, $p_6 = (b^2 \cos^2\theta / z_2^2) + k_2^2$, and $C''_0, \cdots, C''_{18}$.

Using the obtained normalization coefficients in Eq. (1.87), we arrive at the final explicit expression for the normalized generalized correlator of fields scattered by anisotropic rough surface[7]:

$$\Psi(\Delta k, \vec{r}) = A_0 \exp B_0 \,,\ A_0 = U_1 / U_2 \,,\ B_0 = (Q_1 / Q_2) + f(x_0, y_0) \,, \tag{1.91}$$

where

$$U_1 = [C'^2_0 C'_{11} \langle \gamma_x^2 \rangle \langle \gamma_y^2 \rangle (1 - R_{xy}^2) + C'_{14} + C'_0 (C'_{12} \langle \gamma_x^2 \rangle + C'_{13} \langle \gamma_y^2 \rangle)]^{\frac{1}{4}} \cdot [C''^2_0 C''_{11} \langle \gamma_x^2 \rangle \langle \gamma_y^2 \rangle (1 - R_{xy}^2) + C''_{14} + C''_0 (C''_{12} \langle \gamma_x^2 \rangle + C''_{13} \langle \gamma_y^2 \rangle)]^{\frac{1}{4}},$$

$$U_2 = \{ C_0^2 C_{11} \langle \gamma_x^2 \rangle \langle \gamma_y^2 \rangle (1 - R_{xy}^2) + C_{14} + C_0 (C_{12} \langle \gamma_x^2 \rangle + C_{13} \langle \gamma_y^2 \rangle) + \mathrm{i} [C_0 (C_{16} \langle \gamma_x^2 \rangle + C_{17} \langle \gamma_y^2 \rangle) + C_0^2 C_{15} \langle \gamma_x^2 \rangle \langle \gamma_y^2 \rangle (1 - R_{xy}^2) + C_{18}] \}^{\frac{1}{2}},$$

$$Q_1 = -2\sigma_\xi^2 (k_2 - k_1)^2 \cos^2\theta - \mathrm{i}2(k_2 z_2 - k_1 z_1) \sec\theta - \{ C_1 \Delta x^2 + C_2 \Delta y^2 + C_0 (C_3 \langle \gamma_y^2 \rangle \Delta x^2 + C_4 \langle \gamma_x^2 \rangle \Delta y^2) + 2 C_0 C_5 \sqrt{\langle \gamma_x^2 \rangle \langle \gamma_y^2 \rangle} R_{xy} \Delta x \Delta y + \mathrm{i} [C_6 \Delta x^2 + C_7 \Delta y^2 + 2 C_0 (C_8 \langle \gamma_x^2 \rangle \Delta y^2 + C_9 \langle \gamma_y^2 \rangle \Delta x^2) - 2 C_{10} \sqrt{\langle \gamma_x^2 \rangle \langle \gamma_y^2 \rangle} R_{xy} \Delta x \Delta y] \},$$

$$Q_2 = (C_{11} + \mathrm{i} C_{15}) z_1 z_2 \{ C_0^2 C_{11} \langle \gamma_x^2 \rangle \langle \gamma_y^2 \rangle (1 - R_{xy}^2) + C_0 (C_{12} \langle \gamma_x^2 \rangle + C_{13} \langle \gamma_y^2 \rangle) + C_{14} + \mathrm{i} [C_0^2 C_{15} \langle \gamma_x^2 \rangle \langle \gamma_y^2 \rangle (1 - R_{xy}^2) + C_0 (C_{16} \langle \gamma_x^2 \rangle + C_{17} \langle \gamma_y^2 \rangle) + C_{18}] \},$$

$$f(x_0, y_0) = \mathrm{i} \left[k_1 \frac{(x_{10}^2 + y_{10}^2)}{z_1} - k_2 \frac{(x_{20}^2 + y_{20}^2)}{z_2} \right] - b \left(\frac{(x_{10}^2 \cos^2\theta + y_{10}^2)}{z_1^2} + \frac{(x_{20}^2 \cos^2\theta + y_{20}^2)}{z_2^2} \right) + \frac{[b(x_{10} z_1^{-2} + x_{20} z_2^{-2}) \cos^2\theta + \mathrm{i}(k_2 x_{20} z_2^{-1} - k_1 x_{10} z_1^{-1})]^2}{b(z_1^{-2} + z_2^{-2}) \cos^2\theta + \mathrm{i}(k_2 z_2^{-1} - k_1 z_1^{-1})} +$$

$$\frac{[b(y_{10}z_1^{-2}+y_{20}z_2^{-2})+\mathrm{i}(k_2y_{20}z_2^{-1}-k_1y_{10}z_1^{-1})]^2}{b(z_1^{-2}+z_2^{-2})\cos^2\theta+\mathrm{i}(k_2z_2^{-1}-k_1z_1^{-1})}. \tag{1.92}$$

The above results allow investigation of various sets of spatial and frequency functionals of stochastic fields[1,2,5,7,12,19] scattered by anisotropic rough surfaces. A large number of problems formulated in similar forms can be studied on the basis of the proposed approach[29,53-56].

1.9 Spatial correlations of stochastic backscattered fields

During wave scattering by rough surfaces, reflected waves exhibit amplitude and phase fluctuations. Violation of the coherence of a signal results in the circumstance that the reception of the signal provided by a receiver matched with a probing signal becomes nonoptimal; i. e., signal-to-noise ratio q_0^2 does not take its maximum value. When the interval of spatial correlation of reflected-signal fluctuations is smaller than the spatial duration of the probing signal and the interval of frequency coherence is smaller than the width of the signal's spectrum, a linear correlation receiver effectively uses only a portion of the received energy[5,7].

Theoretical estimation of these effects is important for both designing radars for practical purposes and solution of inverse problems. Analysis of the solution for the generalized correlator obtained on the basis of the functional approach developed in [7] allows simultaneous study of all correlation dependences of fields scattered by isotropic or anisotropic surfaces. Below, we theoretically analyze the spatial correlation functions of stochastic backscattered fields with allowance for the finite width of the antenna's pattern and the statistical relationship between slopes of large-height irregularities of isotropic and anisotropic surfaces. The problem has been solved in the scalar Kirchhoff approximation for large values of the Rayleigh parameter. Below, we consider practically important special cases of general Eqs. (1.85), (1.86), (1.91), and (1.92). In order to find transverse coefficient $R_\perp$ of the spatial correlation of a stochastic backscattered field (Fig. 1.3) it suffices to assume that

$$z_1=z_2=z_0,\ k_1=k_2=k,\ p_1=\frac{2}{z^2},\ p_2=p_3=0,\ p_4=p_{4,1}=p_{4,2}=\frac{b^2}{z^2}+k^2,$$

$$p_5=p_{5,1}=p_{5,2}=\frac{b^2\cos^4\theta}{z^2}+k^2,\ p_{5,1}=\frac{b^2\cos^4\theta}{z^2}+k_1^2,\ p_{5,2}=\frac{b^2\cos^4\theta}{z^2}+k_2^2,$$

$$p_6=\frac{b^2\cos^2\theta}{z^2}+k^2,\ b=\frac{4(\ln 2)\cos^2\theta}{\theta_A^2} \tag{1.93}$$

in the formulae of (1.86). The substitution of expressions of (1.93) into formulae of (1.85) followed by the substitution of the result into formulae of (1.91) and (1.92) yields:

$$R_\perp(A_1,A_2)=\exp\left(-\frac{V_1}{V_2}\right),$$

$$V_1=[p_5^2(p_4+4k^2b\langle\gamma_y^2\rangle\cos^2\theta)\Delta x^2+p_4^2(p_5+4k^2b\langle\gamma_x^2\rangle\cos^4\theta)\Delta y^2\cos^2\theta+8k^2b\cdot\sqrt{\langle\gamma_x^2\rangle\langle\gamma_y^2\rangle}R_{xy}p_4p_5\Delta x\Delta y\cos^4\theta],$$

$$V_2 = 2b[16k^4\langle\gamma_x^2\rangle\langle\gamma_y^2\rangle(1-R_{xy}^2)b\cos^6\theta + 4k^2b\cos^2\theta(\langle\gamma_x^2\rangle p_4\cos^2\theta + \langle\gamma_y^2\rangle p_5) + p_4p_5]\cos^2\theta. \quad (1.94)$$

Under the condition $k^2 \gg (b^2/z_0^2)$ which corresponds to large wave parameter (1.56), $p = \sqrt{z_0\lambda/d} \gg 1$, we find from Eq. (1.94) for $R_\perp(A_1, A_2) = \exp\left(-\frac{V'_1}{V'_2}\right)$ that

$$V'_1 = [k^2(1+4b\langle\gamma_y^2\rangle\cos^2\theta)\Delta x^2 + (1+4b\langle\gamma_x^2\rangle\cos^4\theta)\Delta y^2\cos^2\theta + 8b\sqrt{\langle\gamma_x^2\rangle\langle\gamma_y^2\rangle}R_{xy}\Delta x\Delta y\cos^4\theta],$$

$$V'_2 = 2b[16\langle\gamma_x^2\rangle\langle\gamma_y^2\rangle(1-R_{xy}^2)b\cos^6\theta + 4b\cos^2\theta(\langle\gamma_x^2\rangle\cos^2\theta + \langle\gamma_y^2\rangle) + 1]\cos^2\theta. \quad (1.95)$$

Next, let us consider the two limiting cases of the narrow and wide patterns of the antenna θ_A^2. When the antenna has the narrow pattern ($b \gg 1$), transverse coefficient $R_\perp(A_1, A_2)$ of spatial correlation takes the form

$$R_\perp(A_1,A_2) = \exp\left\{-\frac{k^2\theta_A^2[\langle\gamma_x^2\rangle\Delta y^2\cos^4\theta + \langle\gamma_y^2\rangle\Delta x^2 + 2\sqrt{\langle\gamma_x^2\rangle\langle\gamma_y^2\rangle}R_{xy}\Delta x\Delta y]\cos^2\theta}{8\ln 2[4\langle\gamma_x^2\rangle\langle\gamma_y^2\rangle(1-R_{xy}^2)\cos^4\theta + \langle\gamma_x^2\rangle\cos^2\theta + \langle\gamma_y^2\rangle]\cos^4\theta}\right\} \quad (1.96)$$

When surface slopes are strongly correlated, i.e., when $R_{xy} = 1$, Eq. (1.96) yields the following expression for transverse coefficient $R_\perp(A_1, A_2)$ of spatial correlation:

$$R_\perp(A_1,A_2) = \exp\left\{-\frac{k^2\theta_A^2(\sqrt{\langle\gamma_y^2\rangle}\Delta x + \sqrt{\langle\gamma_x^2\rangle}\Delta y\cos^2\theta)^2}{8\ln 2(\langle\gamma_x^2\rangle\cos^2\theta + \langle\gamma_y^2\rangle)\cos^4\theta}\right\}. \quad (1.97)$$

When surface slopes $\langle\gamma_x^2\rangle$ and $\langle\gamma_y^2\rangle$ are statistically independent ($R_{xy} = 0$) and the antenna has a narrow pattern, Eq. (1.96) implies that

$$R_\perp(A_1,A_2) = \exp\left\{-\frac{k^2\theta_A^2(\langle\gamma_x^2\rangle\Delta y^2\cos^4\theta + \langle\gamma_y^2\rangle\Delta x^2)}{8\ln 2[4\langle\gamma_x^2\rangle\langle\gamma_y^2\rangle\cos^4\theta + \langle\gamma_y^2\rangle + \langle\gamma_x^2\rangle\cos^2\theta]\cos^4\theta}\right\}. \quad (1.98)$$

Note that, when the scattering surface is isotropic and the condition

$$\langle\gamma_x^2\rangle + \langle\gamma_y^2\rangle = \langle\gamma^2\rangle \quad (1.99)$$

is satisfied, Eqs. (1.97) and (1.98) are similar and e^{-1}-level transverse correlation radius $\rho_\perp$ of the stochastic backscattered field is specified by the expression

$$\rho_\perp \approx \pi^{-1}\sqrt{2\ln 2}\, d\cos^2\theta. \quad (1.100)$$

Thus, when the antenna has a narrow pattern, the transverse spatial correlation of the field scattered by an isotropic rough surface is determined by the antenna's dimension. This result coincides with the conclusions drawn in Ref. [23, 25, 53, 54, 56-60].

Now, consider the case $b \ll 1$, i.e., the case where the antenna has a wide pattern. In this case, transverse coefficient Eq. (1.95) of spatial correlation can be represented as

$$R_\perp(A_1,A_2) = \exp\left[-2k^2\left(\langle\gamma_x^2\rangle\Delta y^2\cos^4\theta + \langle\gamma_y^2\rangle\Delta x^2 + 2\sqrt{\langle\gamma_x^2\rangle\langle\gamma_y^2\rangle}R_{xy}\Delta x\Delta y\cos^2\theta\right)\right]. \quad (1.101)$$

When $R_{xy} = 1$, Eq. (1.101) can be modified to obtain

$$R_{\perp}(A_1,A_2) = \exp\left[-2k^2\left(\sqrt{\langle\gamma_x^2\rangle}\Delta y\cos^2\theta + \sqrt{\langle\gamma_y^2\rangle}\Delta x\right)^2\right]. \qquad (1.102)$$

When surface slopes $\langle\gamma_x^2\rangle$ and $\langle\gamma_y^2\rangle$ are statistically independent, i. e., when $R_{xy}=0$, we have

$$R_{\perp}(A_1,A_2) = \exp\left[-2k^2\left(\langle\gamma_x^2\rangle\Delta y^2\cos^4\theta + \langle\gamma_y^2\rangle\Delta x^2\right)\right]. \qquad (1.103)$$

Hence, when the antenna has a wide pattern, Eqs. (1.102) and (1.103) imply that the transverse spatial correlation of the scattered field depends on characteristic slope ratios of large-height irregularities. When the antenna has a wide pattern, correlation radius $\rho_{\perp}$ of the field scattered by an isotropic surface is estimated as the value

$$\rho_{\perp} \approx \frac{1}{k\sqrt{2\langle\gamma^2\rangle(1+\cos^2\theta)}} \cdots \frac{1}{k\sqrt{2\langle\gamma^2\rangle}\sqrt{1+\cos^4\theta}}, \qquad (1.104)$$

which does not contradict the data from [25, 57, 58].

When observation points are spaced along the vertical direction, it is necessary to set $\Delta x = \Delta y = 0$ in formulae of (1.91) and (1.92). Thus, we arrive at the following known expression for transverse coefficient $R_{\parallel}(A_1,A_2)$ of spatial correlation:

$$R_{\parallel}(A_1,A_2) = \exp(-\mathrm{i}2k\Delta z). \qquad (1.105)$$

Transverse coefficient $R_{\parallel}(A_1,A_2)$ of spatial correlation of stochastic scattered field Eq. (1.105) is a spatial harmonic function, equals unity for any values z_1 and z_2 corresponding to the far zone, and carries no information on the characteristics of a scattering surface[53]. Thus, the transverse radius of spatial correlation is infinite: $\rho_{\parallel} \to \infty$.

1.10 Frequency coherence functions of stochastic backscattered fields

In many problems of radar, radio navigation, remote sensing of the environment, and radio communications, scattering of modulated waves by the Earth's surface plays the main role. Depending on the problem to be solved, this effect of the terrain is a source of desired information or an impairing factor.

Effects of frequency correlation of modulated waves scattered by a rough surface are especially important for estimation of the spectral widths of complex probing signals[5,7,13,32,33,37,61], frequency spacing in multifrequency radio systems and the bandwidth of wideband-signal transmission[5,7,13,27,30,32,33,37,62,63], the frequency-averaged intensity of a wave field[5,7,13,37,64], distortions of the shapes of probing signals[5,7,13,32-34,37], an aircraft's altitude and oceanographic characteristics[5,7,13,32,33,65-68], and the wave phase (the so-called phase problem)[69,70].

The statistical description of the fluctuation characteristics of modulated waves scattered by a rough surface is based on an FCF (see, e. g. [7, 34]). In Ref. [66], FCFs were studied for the 2D scattering problem at small incidence angles in the case when a statistical surface is modeled by a set of bright points. Later, 3D scattering problems were analyzed as well (including the problem with a two-scale model of a surface), but these problems were investigated only for decimeter and centimeter wavelengths[61,62,68]. The expression derived in this study for the functional of stochastic backscattered fields can be applied to determine the normalized FCF $\Psi_\omega = \Psi_\omega(\Delta k)$ with allowance for finite width θ_A of the antenna's pattern, surface slopes $\langle\gamma_x^2\rangle$ and $\langle\gamma_y^2\rangle$, and their correlation K_{xy} in perpendicular directions[5,7,13,32,33,65,71].

The case (Fig. 1.3) where the altitudes of flight and shifts are specified by the equalities $z_1 = z_2 = z$ and $\Delta x = \Delta y = 0$, respectively, is of greatest interest for applications. In this situation, coefficients (1.86) in Eqs. (1.85), (1.91), and (1.92) are as follows:

$$p_1 = \frac{2}{z^2},\ p_2 = p_3 = \frac{\Delta k}{z},\ p_4 = \frac{b^2}{z^2} + k_1k_2,\ p_{4,1} = \frac{b^2}{z^2} + k_1^2,\ p_{4,2} = \frac{b^2}{z^2} + k_2^2,$$

$$p_5 = \frac{b^2\cos^4\theta}{z^2} + k_1k_2,\ p_{5,1} = \frac{b^2\cos^4\theta}{z^2} + k_1^2,\ p_{5,2} = \frac{b^2\cos^4\theta}{z^2} + k_2^2,$$

$$p_6 = \frac{b^2\cos^4\theta}{z^2} + k_1k_2,\ b = \frac{4(\ln 2)\cos^2\theta}{\theta_A^2},\ \Delta k = k_2 - k_1. \tag{1.106}$$

Let us represent the normalized FCF as

$$\Psi_\omega(\Delta k) = A_{\Delta\omega}\exp(B_{\Delta\omega}), \tag{1.107}$$

and determine decorrelating factor $A_{\Delta\omega}$ under the condition

$$4b(z^2\Delta k^2)^{-1} > 1 \tag{1.108}$$

in the form

$$A_{\Delta\omega}^{-1} = 1 + \mathrm{i}\frac{2\Delta kz\cos^2\theta}{1+4b\cos^2\theta}\cdot \frac{[4\langle\gamma_x^2\rangle\langle\gamma_y^2\rangle(1-R_{xy}^2)b(1+\cos^2\theta)\cos^2\theta + \langle\gamma_x^2\rangle + \langle\gamma_y^2\rangle]}{[4\langle\gamma_x^2\rangle\langle\gamma_y^2\rangle(1-R_{xy}^2)b\cos^4\theta + \langle\gamma_x^2\rangle\cos^2\theta + \langle\gamma_y^2\rangle]}. \tag{1.109}$$

The second multiplier in Eq. (1.107) can be expressed as

$$\exp B_{\Delta\omega} = \exp(-\mathrm{i}2z\Delta k\sec\theta)\exp(-2\sigma_\xi^2\Delta k^2\cos^2\theta). \tag{1.110}$$

As in the foregoing, we discuss the two limiting cases in which the antenna has a narrow pattern and a wide pattern (θ_A^2). When the antenna's pattern is narrow ($b \gg 1$) and surface slopes are strongly correlated ($R_{xy} = 1$), we find from Eq. (1.109) that

$$A_{\Delta\omega}^{-1} = 1 + \mathrm{i}\frac{\Delta kz}{2b}\cdot\frac{[\langle\gamma_x^2\rangle + \langle\gamma_y^2\rangle]}{[\langle\gamma_x^2\rangle\cos^2\theta + \langle\gamma_y^2\rangle]}; \tag{1.111}$$

then, absolute value $|A_{\Delta\omega}|$ of the decorrelating factor in Eq. (1.107) has the value

$$|A_{\Delta\omega}| = \left\{1 + \left[\frac{\Delta kz(\langle\gamma_x^2\rangle + \langle\gamma_y^2\rangle)\theta_A^2}{8(\ln 2)(\langle\gamma_x^2\rangle\cos^2\theta + \langle\gamma_y^2\rangle)\cos^2\theta}\right]^2\right\}^{-\frac{1}{2}}. \tag{1.112}$$

If surface slopes $\langle\gamma_x^2\rangle$ and $\langle\gamma_y^2\rangle$ are statistically independent ($R_{xy} = 0$), Eq. (1.109) implies that

$$A_{\Delta\omega}^{-1} = 1 + \mathrm{i}\frac{2\Delta kz\cos^2\theta}{1+4b\cos^2\theta}\frac{[4\langle\gamma_x^2\rangle\langle\gamma_y^2\rangle b(1+\cos^2\theta)\cos^2\theta + \langle\gamma_x^2\rangle + \langle\gamma_y^2\rangle]}{[4\langle\gamma_x^2\rangle\langle\gamma_y^2\rangle b\cos^4\theta + \langle\gamma_x^2\rangle\cos^2\theta + \langle\gamma_y^2\rangle]}. \tag{1.113}$$

In the case of a narrow antenna's pattern and $R_{xy} = 0$, for absolute value $|A_{\Delta\omega}|$ of the decorrelating factor in Eq. (1.107), we use Eq. (1.113) to find

$$|A_{\Delta\omega}| = \left\{1 + \left[\frac{\Delta kz(1 + \cos^2\theta)\theta_A^2}{8(\ln 2)\cos^4\theta}\right]^2\right\}^{-\frac{1}{2}}. \tag{1.114}$$

Now, let us consider the second case, i. e., the case where the antenna has a wide pattern ($b \ll 1$). Then, Eq. (1.109) for $R_{xy} = 1$ can be represented as

$$|A_{\Delta\omega}| = \{1 + [2\Delta kz(\langle\gamma_x^2\rangle + \langle\gamma_y^2\rangle)\cos^2\theta]^2\}^{-\frac{1}{2}} \tag{1.115}$$

if the condition

$$\theta_A \gg 4(\ln 2)\sqrt{\langle\gamma_x^2\rangle\cos^2\theta + \langle\gamma_y^2\rangle\cos^2\theta} \tag{1.116}$$

is fulfilled.

For uncorrelated slopes ($R_{xy} = 0$) and a wide antenna's pattern, Eq. (1.109) implies that

$$|A_{\Delta\omega}| = \{1 + [2\Delta kz(\langle\gamma_x^2\rangle + \langle\gamma_y^2\rangle)\cos^2\theta]^2\}^{-\frac{1}{2}} \tag{1.117}$$

Note the following interesting fact: the decorrelating factors for $R_{xy} = 1$ and $R_{xy} = 0$ are equal in the case when the antenna has a wide pattern (see Eqs. (1.115) and (1.117)).

According to Eqs. (1.109) and (1.110), the normalized FCF has the absolute value

$$|\Psi_\omega(\Delta k)| = |A_{\Delta\omega}|\exp(-2\sigma_\xi^2\Delta k^2\cos^2\theta), \tag{1.118}$$

which is usually of interest.

1.11 Calculation and analysis of the frequency coherence functions of millimeter-wave stochastic backscattered fields

It is seen from Eq. (1.118) that the FCFs of fields scattered by a terrain depend on both the terrain characteristics and the probing conditions: the flight's altitude, the width of the antenna's pattern, the angle of sight, and the spacing of signals' frequencies $\Delta f = f_2 - f_1 = \Delta\omega/2\pi = c\Delta k/2\pi$.

Let us numerically estimate the radiophysical effects. Under condition (1.99), which provides for an isotropic surface, the following formulas are suitable for practical calculations of the absolute value of the normalized FCF[5,7,32,33,71,72]:

$$|\Psi_\omega(\Delta f)| = \frac{\exp(-8.8\times 10^{-16}\Delta f^2\sigma_\xi^2\cos^2\theta)}{\left\{1 + \left[\frac{8.4\times 10^{-8}\langle\gamma^2\rangle z\Delta f\cos^2\theta}{1 + 2a(1 + \cos^2\theta)}\right]^2\right\}^{\frac{1}{2}}} \tag{1.119}$$

for $R_{xy} = 1$ and arbitrary width θ_A^2 of the antenna's pattern and

$$|\Psi_\omega(\Delta f)| = \frac{\exp(-8.8\times 10^{-16}\Delta f^2\sigma_\xi^2\cos^2\theta)}{\left\{1 + \left[\frac{8.4\times 10^{-8}\langle\gamma^2\rangle z\Delta f(a' + 1 + a'\cos^2\theta)\cos^2\theta}{1 + 2a'(1 + 2a'\cos^2\theta + \cos^2\theta)}\right]^2\right\}^{\frac{1}{2}}} \tag{1.120}$$

for $R_{xy} = 0$ and arbitrary width θ_A^2 of the antenna's pattern. Here,

$$a' = 5.55\langle\gamma^2\rangle\cos^4\theta/\theta_A^2. \tag{1.121}$$

The experimental statistical characteristics of typical terrains that are generalized in Ref. [7] and reported in Monographs [5, 6] are used to obtain the characteristic dependences of the absolute value of the FCF on difference $|\Psi_\omega|$ of the frequencies of two MMW signals reflected by a concrete surface ($\langle\gamma^2\rangle \approx 0.0016$, $\sigma_\xi \approx 0.3$ mm) and cultivated soil ($\langle\gamma^2\rangle \approx 0.25$, $\sigma_\xi \approx 2$ cm). These dependences are depicted in Figs. 1.4a (concrete surface) and 1.4b (cultivated soil)[7,71]. The aforementioned surfaces have strongly different statistical characteristics. Therefore, it is possible to estimate the range of the parameters that are of interest in specific situations. The calculations have been performed for the incidence angle $\theta = 0$; the average flight altitude $z \approx 100$ m; and the following two widths of the antenna's pattern: $\theta_A = 1°$ or $\theta_A = 0.02$ rad (Curves 1, 3) and $\theta_A = 20°$ or $\theta_A = 0.35$ rad (Curves 2, 4). Curves 1 and 2 correspond to $R_{xy} = 1$, and Curves 3 and 4 correspond to $R_{xy} = 0$.

Note the substantial effect of correlation factor R_{xy} of surface slopes on absolute value $|\Psi_\omega|$ of the FCF. It has been found that the width of the coherence band of the radar channel increases as quantity R_{xy} changes from 0 to 1. Thus, it is seen from Fig. 1.4 that, as R_{xy} grows from 0 to 1, the e^{-1}-level coherence band increases by several fractions of a gigahertz or by several gigahertz.

As the width of antenna's pattern grows, the frequency decorrelation of signals becomes stronger. When the antenna has a wide pattern, the FCF of stochastic backscattered fields is mainly determined by the characteristic slopes of surface irregularities and by the irregularities' heights. When beam antennas are used, the dependence of $|\Psi_\omega|$ on surface slopes can be disregarded. However, in this case, the effect of the width of the antenna's pattern on the absolute value of the FCF abruptly intensifies.

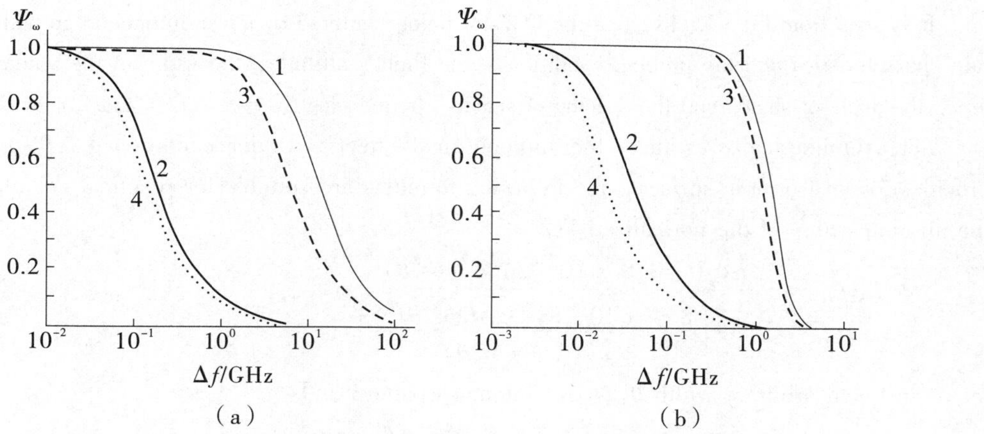

Fig. 1.4 Absolute value $|\Psi_\omega|$ of the FCF as a function of difference Δf of the frequencies of two MMW signals reflected by (a) a concrete surface and (b) cultivated soil: $\theta = 0$, $z = 100$ m, $R_{xy} =$ (Curves 1, 2) 1 and (Curves 3, 4) 0, (Curves 1, 3) $\theta_A = 1°$ or $\theta_A = 0.02$ rad, and (Curves 2, 4) $\theta_A = 20°$ or $\theta_A = 0.35$ rad

The altitude and angular dependences of the FCF have been calculated according to Eqs. (1.119) - (1.121) for the aforementioned concrete surface (Fig. 1.5) and cultivated soil

(Fig. 1.6)[7,71]. In the calculations, the parameters $\Delta f = f_2 - f_1 = 1$ GHz, $\theta = 0$, $R_{xy} = 1$ and $\theta_A = 1°$ (Curve 1), $R_{xy} = 0$ and $\theta_A = 1°$ (Curve 2), and $R_{xy} = 1$ and $\theta_A = 20°$ (Curve 3) are used in Fig. 1.5, while $\Delta f = f_2 - f_1 = 300$ MHz, $\theta_A = 1°$, $R_{xy} = 1$ and $\theta = 0$ (Curve 1), $R_{xy} = 1$ and $\theta = 30°$ (Curve 2), $R_{xy} = 0$ and $\theta = 0$ (Curve 3), and $R_{xy} = 0$ and $\theta = 30°$ (Curve 4) are used in Fig. 1.6.

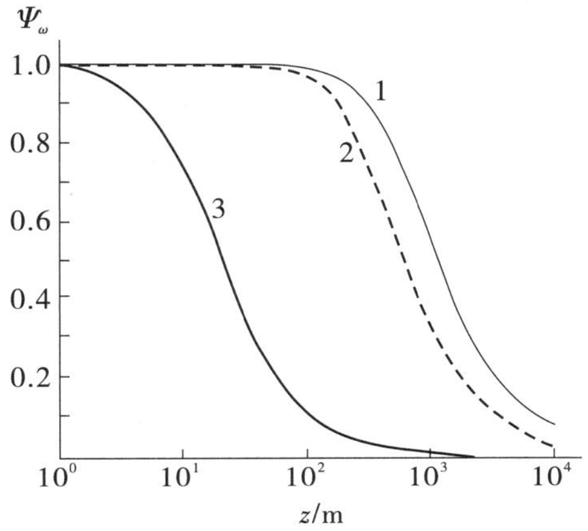

Fig. 1.5 Absolute value $|\Psi_\omega|$ of the FCF as a function of altitude z of radar probing of a concrete surface with the irregularities' characteristics $\sqrt{\langle \gamma^2 \rangle} \approx 0.04$ and $\sigma_\xi = 0.3$ mm obtained for the difference between the probing frequencies $\Delta f = f_2 - f_1 = 1$ GHz, $\theta = 0$, (Curve 1) $R_{xy} = 1$ and $\theta_A = 1°$, (Curve 2) $R_{xy} = 0$ and $\theta_A = 1°$, and (Curve 3) $R_{xy} = 1$ and $\theta_A = 20°$

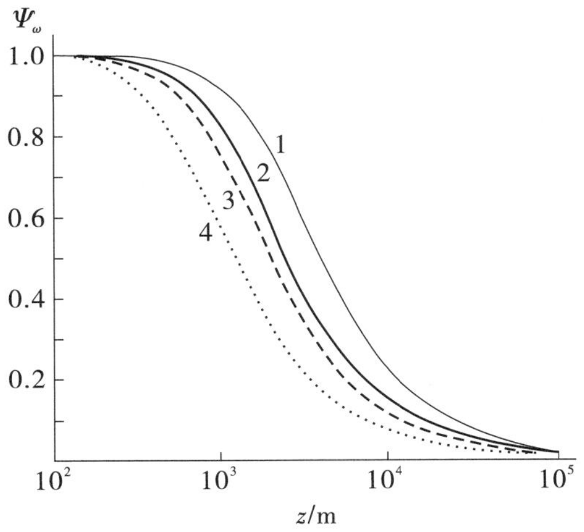

Fig. 1.6 Absolute value $|\Psi_\omega|$ of the FCF as a function of altitude z and angle θ of radar probing of cultivated soil with the irregularities' characteristics $\sqrt{\langle \gamma^2 \rangle} \approx 0.5$ and $\sigma_\xi = 2$ cm obtained for the difference between the probing frequencies $\Delta f = f_2 - f_1 = 300$ MHz, $\theta_A = 1°$, (Curve 1) $R_{xy} = 1$ and $\theta = 0$, (Curve 2) $R_{xy} = 1$ and $\theta = 30°$, (Curve 3) $R_{xy} = 0$ and $\theta = 0$, and (Curve 4) $R_{xy} = 0$ and $\theta = 30°$

It is seen from Figs. 1. 5 and 1. 6 that, as the correlation factor of surface slopes decreases or the flight altitude, the width of the antenna's pattern, or the probing angle increases, absolute value $|\Psi_{\omega}|$ of the FCF decreases.

Fig. 1. 7[7,71] shows absolute value $|\Psi_{\omega}|$ of the FCF as a function of rms height σ_{ξ} of terrain irregularities calculated for the vertically oriented antenna with $\theta_A = 1°$ when $R_{xy} = 1$ and $z \approx 100$ m. The value of θ_A and the frequency differences $\Delta f = 10$ MHz (Curve 1), 30 MHz (Curve 2), and 100 MHz (Curve 3), and 300 MHz (Curve 4) have been chosen such that, at the probing altitude specified above, the decorrelating factor in Eq. (1. 118) satisfies the condition $|A_{\Delta\omega}| \approx 1$. Then, an algorithm of remote determination of the heights of the irregularities of probed surfaces is realized[5,7,13,32,33,65-68,71]. The analytic results presented above are based on FCFs calculated for MMWs scattered by a statistically rough surface. These data were obtained earlier than the results reported in Ref. [72].

Note the good agreement of the theoretical results with laboratory data and with in situ measurements of FCFs performed (i) at the frequencies $f_1 = 8$ GHz and $f_2 = 12$[66] and 13. 9[67] GHz ($0 < \Delta f < 40$ MHz) in the case of vertical probing ($\theta = 0$) and (ii) at a frequency lying near the value $f = 100$ GHz and the incidence angle $\theta = 20°$[73].

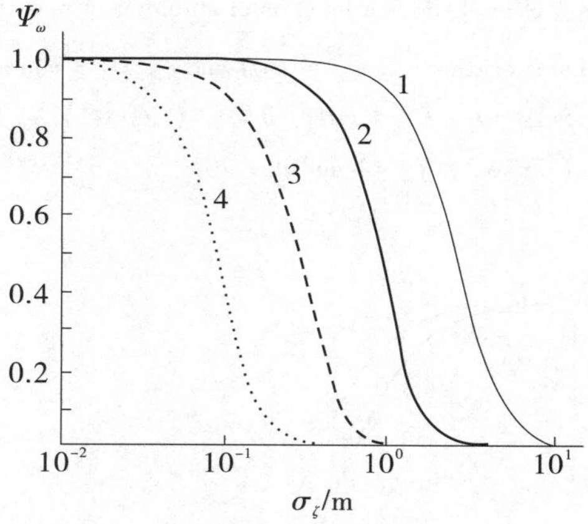

Fig. 1. 7 Absolute value $|\Psi_{\omega}|$ of the FCF as a function of the terrain irregularities' rms heights, σ_{ξ}, obtained for the vertical orientation ($\theta = 0$) of the antenna with the pattern width $\theta_A = 1°$, $R_{xy} = 1$, $z \approx$ 100 m, and the difference of the probing frequencies $\Delta f =$ (Curve 1) 10 MHz, (Curve 2) 30 MHz, (Curve 3) 100 MHz, and (Curve 4) 300 MHz

1. 12 Frequency coherence band of a spatial-temporal microwave probing radar channel

On the basis of the theoretical analysis of the stochastic functional of backscattered fields that has been proposed[7], a solution for the FCF is obtained. It should be emphasized that, with the use of FCFs, it is possible to theoretically explain and physically and illustratively interpret the process of scattering of modulated radio waves by terrains.

The main characteristic involved in this analysis is frequency coherence band Δf_c that, at any instant, determines the measure of correlation of waves at different frequencies. The reciprocal of Δf_c is the pulse spreading caused by scattering from an extended rough surface. The appropriate choice of the spectral width of simple or complex probing signals or the appropriate choice of the frequency spacing in multifrequency radar probing systems and in scattering communications channels is based on the concept of the coherence band[7,13,26,29,30,32-35,37,53-55,61,62,74-77]. In particular, it has been shown in Ref. [76] for spherical waves that the frequency decorrelation of the scattered field is observed at the mistuning such that the major semiaxis of the ellipse contributing significantly to scattering increases by a value of the effective height of irregularities.

Frequency coherence band Δf_c is determined from the relationship

$$\Delta f_k = \int_0^{\infty} \Psi_\omega(\Delta f)\,\mathrm{d}(\Delta f), \tag{1.122}$$

where Ψ_ω (Δf) is an FCF with the form of Eqs. (1.107) – (1.110).

Eq. (1.122) can be expressed through the integral representation of zero-order Macdonald function K_0 (x) or the modified Bessel function of the third kind[7,13,32,61,71]. Therefore, integration in Eq. (1.122) yields a general analytic formula for frequency coherence band Δf_c of an anisotropic surface in the case of arbitrary R_{xy}, $\langle \gamma_x^2 \rangle$ and $\langle \gamma_y^2 \rangle$:

$$\Delta f_k = \sqrt{\frac{\eta}{2}}\,\mathrm{e}^{a\eta} K_0(a\eta), \tag{1.123}$$

where

$$\begin{aligned} a &= 8\pi^2\sigma_\xi^2\cos^2\theta/c^2, \quad \eta = \eta_1/\eta_2, \\ \eta_1 &= c^2\{1 + 4b[4b\langle \gamma_x^2 \rangle\langle \gamma_y^2 \rangle(1 - R_{xy}^2)\cos^4\theta + \langle \gamma_x^2 \rangle\cos^2\theta + \langle \gamma_y^2 \rangle]\cos^2\theta\}^2, \\ \eta_2 &= 32\pi^2 z^2[4b\langle \gamma_x^2 \rangle\langle \gamma_y^2 \rangle(1 - R_{xy}^2)(1 + \cos^2\theta)\cos^2\theta + \langle \gamma_x^2 \rangle + \langle \gamma_y^2 \rangle]\cos^4\theta. \end{aligned} \tag{1.124}$$

In the case of an isotropic scattering surface and with allowance for Eq. (1.99), we obtain

$$\eta = \frac{[1 + 2a'(1 + \cos^2\theta)]^2}{141 \times 10^{-16}(\langle \gamma^2 \rangle)^2 z^2 \cos^4\theta} \tag{1.125}$$

for $R_{xy} = 1$ and arbitrary width θ_A^2 of the antenna's pattern and

$$\eta = \frac{\{1 + 2a'[1 + (1 + 2a')\cos^2\theta]\}^2}{141 \times 10^{-16}(\langle \gamma^2 \rangle)^2 z^2[a'(1 + \cos^2\theta) + \cos^2\theta]^2}, \tag{1.126}$$

for $R_{xy} = 0$, arbitrary width θ_A^2 of the antenna's pattern, and $a' = 5.55\ \langle \gamma^2 \rangle\ \cos^4\theta/\theta_A^2$.

Eqs. (1.124) – (1.126) and experimental statistical characteristics of typical terrains generalized in [7] and reported in Monographs [5, 6] are used to calculate coherence band Δf_c as a function of the width of the antenna's pattern in the case when a concrete surface ($\langle \gamma^2 \rangle \approx 0.0016$, $\sigma_\xi = 0.3$ mm) and cultivated soil ($\langle \gamma^2 \rangle \approx 0.25$, $\sigma_\xi \approx 2$ cm) are probed in the vertical direction. The results for the concrete surface (Fig. 1.8) and cultivated soil (Fig. 1.9)[7,71] are obtained for two flight altitudes: $z \approx 100$ (Curves 1, 2) and 1000 m (Curves 3, 4). Curves 1 and 3 correspond to $R_{xy} = 1$, and Curves 2 and 4 correspond to $R_{xy} = 0$.

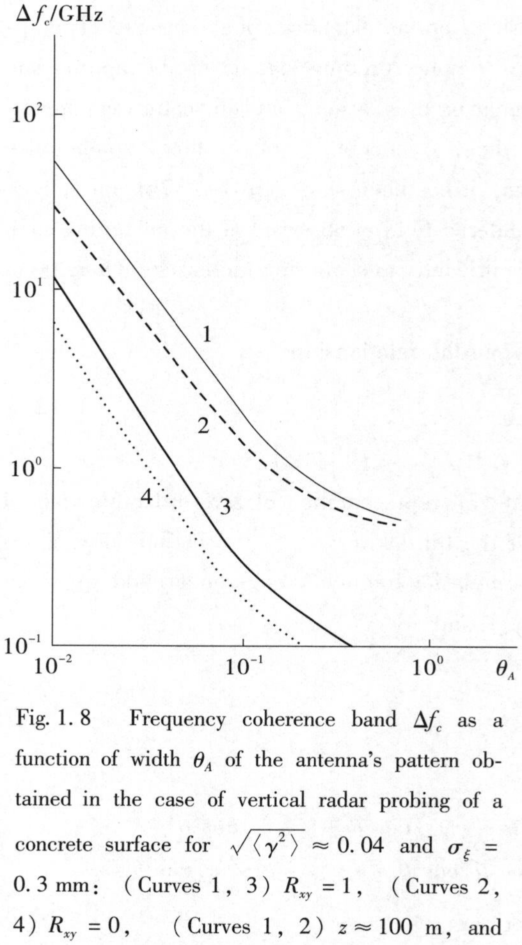

Fig. 1.8 Frequency coherence band Δf_c as a function of width θ_A of the antenna's pattern obtained in the case of vertical radar probing of a concrete surface for $\sqrt{\langle \gamma^2 \rangle} \approx 0.04$ and $\sigma_\xi = 0.3$ mm: (Curves 1, 3) $R_{xy} = 1$, (Curves 2, 4) $R_{xy} = 0$, (Curves 1, 2) $z \approx 100$ m, and (Curves 3, 4) $z \approx 1000$ m

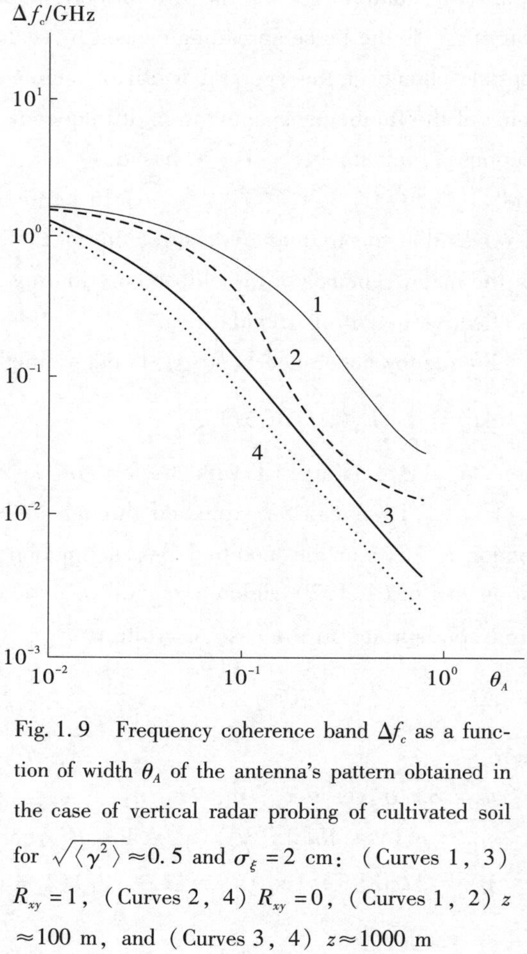

Fig. 1.9 Frequency coherence band Δf_c as a function of width θ_A of the antenna's pattern obtained in the case of vertical radar probing of cultivated soil for $\sqrt{\langle \gamma^2 \rangle} \approx 0.5$ and $\sigma_\xi = 2$ cm: (Curves 1, 3) $R_{xy} = 1$, (Curves 2, 4) $R_{xy} = 0$, (Curves 1, 2) $z \approx 100$ m, and (Curves 3, 4) $z \approx 1000$ m

As width Δf_c of the antenna's pattern increases, frequency coherence band θ_A abruptly decreases. When θ_A changes by a factor of 10 from 0.01 to 0.1 rad, quantity Δf_c decreases by two orders of magnitude for quasismooth surfaces (Fig. 1.8) and by approximately one order of magnitude for strongly rough surfaces (Fig. 1.9). Coherence bands Δf_c are always wider for surfaces with strongly correlated slopes.

Fig. 1.10[7,71] shows frequency coherence band Δf_c as a function of the rms height of irregularities for $\sqrt{\langle \gamma^2 \rangle} = 0.1$ (Curves 1, 2), $\sqrt{\langle \gamma^2 \rangle} = 0.5$ (Curves 3, 4), $R_{xy} = 1$, $\theta = 0$, and the flight altitudes $z \approx 100$ (Curves 1–3) and 1000 m (Curve 4). Curve 1 corresponds to the width of the antenna's pattern $\theta_A = 0.01$ rad, and Curves 2–4 correspond to $\theta_A = 0.35$ rad.

Coherence band Δf_c is invariant with respect to surface slopes in the case of a narrow antenna's patterns. It follows from Eq. (1.123) that this effect is observed when the condition

$$\theta_A \ll \sqrt{\langle\gamma^2\rangle}\sqrt{1+\cos^2\theta\cos^2\theta}. \qquad (1.127)$$

is satisfied.

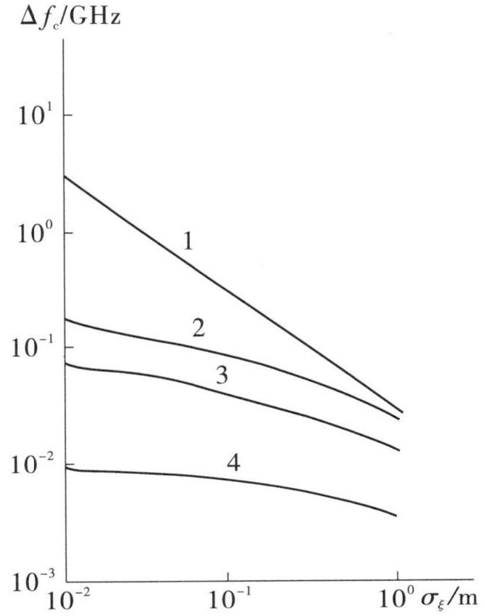

Fig. 1.10 Frequency coherence band Δf_c as a function of irregularities' rms terrain heights σ_ξ obtained for $R_{xy}=1$ and $\theta=0$: $z\approx 100$ (Curves 1 – 3) and 1000 m (Curve 4), $\sqrt{\langle\gamma^2\rangle}=0.1$ (Curves 1, 2), $\sqrt{\langle\gamma^2\rangle}=0.5$ (Curves 3, 4), and $\theta_A=0.01$ rad (Curve 1) and 0.35 rad (Curves 2 – 4)

When the relationship

$$\theta_A \gg \sqrt{\langle\gamma^2\rangle}\sqrt{1+\cos^2\theta\cos^2\theta} \qquad (1.128)$$

which is the inverse of Eq. (1.127), holds, coherence band Δf_c of a radar channel depends on the slopes ($\sqrt{\langle\gamma^2\rangle}$) of large-height irregularities.

Fig. 1.11[7,71] shows frequency coherence band Δf_c obtained as a function of the altitude of ground probing for $R_{xy}=1$, $\theta=0$, $\sigma_\xi\approx 2$ cm, $\sqrt{\langle\gamma^2\rangle}=0.5$ (Curves 1, 3), and $\sqrt{\langle\gamma^2\rangle}=0.1$ (Curve 2). The widths of the antenna's pattern are as follows: $\theta_A=0.02$ rad (Curve 1) and 0.35 rad (Curves 2, 3). It is seen from Fig. 1.11 that, as flight altitude z grows, coherence band Δf_c decreases almost linearly in the case of a wide antenna's patterns.

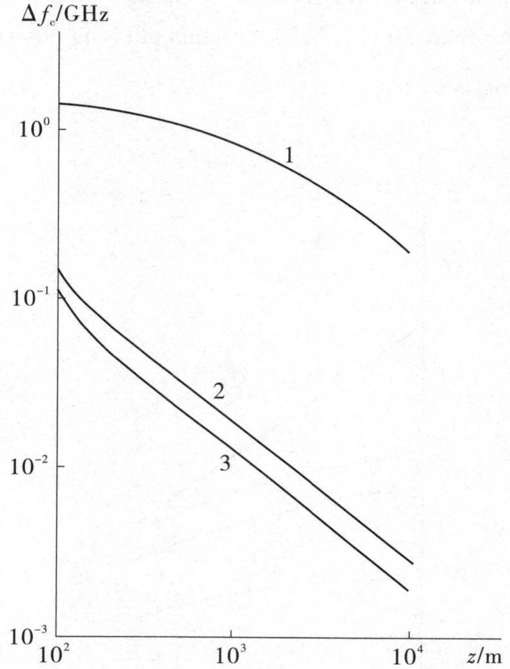

Fig. 1. 11 Frequency coherence band Δf_c as a function of altitude z of radar probing of ground obtained for $R_{xy}=1$, $\theta=0°$, and $\sigma_\xi \approx 2$ cm: $\sqrt{\langle\gamma^2\rangle}=0.5$ (Curves 1, 3) and 0. 1 and (Curve 2) $\theta_A = 0.02$ rad (Curve 1) and 0. 35 rad (Curves 2, 3)

Frequency coherence band Δf_c is shown in Fig. 1. 12[7,71] as a function of the angle of incidence of electromagnetic waves, θ, in the case when ground with $\sigma_\xi \approx 2$ cm, $\sqrt{\langle\gamma^2\rangle}=0.5$ and $R_{xy}=1$ is probed at $z \approx 100$ m and $\theta_A=0.1$ rad (Curve 1) and $\theta_A=0.35$ rad (Curve 2). When the terrain is probed in the vertical direction, frequency coherence bandwidth Δf_c reaches its maximum.

All of the results that I have obtained are universal; do not contradict theoretical predictions[73-76]; and, as has been mentioned above, coincide with laboratory and in situ measurements of Δf_c at $f_1=8$ GHz and $f_2=12.0$[66] and 13. 9 GHz $(0<\Delta f<40$ MHz)[67] in the case of vertical probing $(\theta=0)$ and at a frequency lying near the value $f=100$ GHz when the incidence angle is $\theta=20°$[73].

The method of computer-aided probabilistic simulation, specifically, the Monte Carlo method, is an alternative to the analytic approach discussed in this study. In Ref. [78], the Monte Carlo method is generalized to the 3D geometry of a scattering structure. In this case, it is possible to imitate real processes of microwave scattering. The calculation of scattering indicatrices involves construction of the trajectories of rays reflected from each discrete point of a surface that is specified by a matrix of numbers determining irregularities' heights. Practical implementation of the numerical technique allows computation of scattering indicatrices[1,2,5,7] in two cases: (i) for a synthesized (on a computer) surface with a given distribution of heights, including a fractal surface, and (ii) a real terrain. Theoretical indicatrices of scattering of horizontally polarized MMWs have been calculated by means of the Monte Carlo method and the MTP approximation at $\lambda = 8.6$ mm for synthesized and real terrains (see Ref. [5-7]).

The results of probabilistic simulation of radiowave scattering from a statistically rough surface coincide with analytic approximations[7]. Owing to this circumstance, the Monte Carlo method can be applied to calculate the characteristics of MMW scattering by real terrains, including fractal surfaces, over a wide range of probing angles.

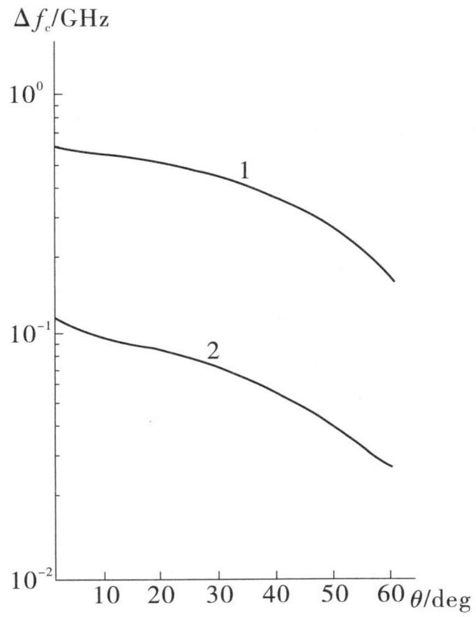

Fig. 1.12 Frequency coherence band Δf_c as a function of angle θ of ground probing obtained for $\sigma_\xi \approx 2$ cm, $\sqrt{\langle \gamma^2 \rangle} = 0.5$, $R_{xy} = 1$, and $z \approx 100$ m: $\theta_A = 0.1$ rad (Curve 1) and 0.35 rad (Curve 2)

It should be noted that the developed mathematical apparatus of FCFs is also important for application of fractal signatures in radar and remote sensing[1-3]. Combined with fractal data processing, multifrequency radio measurements are a serious alternative to available methods of increasing the signal-to-noise ratio. In one of the proposed fractal methods of radio-signal detection[79], signals are radiated at different frequencies. Probing radar frequencies should be chosen according to the values of FCF $\Psi_\omega(\Delta k)$ or coherence band Δf_c of a spatial-temporal probing radar channel. Since any target has its own characteristic dimensions, selection of probing frequencies can provide for direct determination of a new class of radar features in the form of fractal-frequency signatures[3,80,81]. Approaches to determination of generalized fractal correlators of backscattered fields have been discussed in reports [82-84].

1.13 Radar measurements of the characteristics of a rough surface and of a flight's altitude according to the frequency coherence function

The analysis of Eqs. (1.107) - (1.110) for the FCF shows that absolute value $|\Psi_\omega|$ and the phase characteristic of the FCF can be used for measuring the flight altitude and irregularities' heights. Two-frequency measurements based on the analysis of only the absolute value of the FCF have been performed for a sea surface[63,66-68]. Such measurements are successfully applied for aircraft-borne radars.

In these measurement techniques, decorrelating factor Eq. (1.109) reduces the value of $|\Psi_\omega|$ and governs the accuracy of determining irregularity heights. The range of possible irregularity heights that are determined from absolute value $|\Psi_\omega|$ can readily be estimated from the data presented in Fig. 1.7 for the necessary accuracy of measurements.

In Ref. [67], it has been shown that rms height σ_ξ is proportional to the second-order derivative of $|\Psi_\omega|$ with respect to Δk calculated at $\Delta k = 0$. In this study, it has been demonstrated that, when experimental and theoretical dependences $|\Psi_\omega(\Delta k)|$ are known, characteristic function $\xi(\vec{\rho})$ of a rough surface can be estimated.

The possibility of application of phase data contained in Ψ_ω for determination of the aforementioned quantities has not been investigated in detail. An exclusion is study[68], where particular cases are considered for a simple probing geometry and illumination in almost the vertical direction ($\theta \ll 1$). The data obtained in Refs. [7, 13, 32, 33, 65] have made it possible to analyze, on the basis of the results from Ref. [68], more general solutions to the problem of determining flight altitude and heights of irregularities of a surface and to estimate errors of measurements of these heights. The analysis of Eq. (1.107) shows that the addends corresponding to the first terms of the Taylor series and determining slant range R (see Fig. 1.3) dominate in the phase characteristic of the FCF.

Therefore, the slant range and altitude z can be determined from the phase characteristic of Ψ_ω.

For a Gaussian scattering surface, the rms deviation of phase Ψ_ω is expressed by the formula:

$$\sigma_1 \approx 2\Delta k \sigma_\xi / \cos\theta . \tag{1.129}$$

Noise component σ_2 of the phase at the output of a radar imaging system takes into account various decorrelating effects. This component can be determined from the Cramer-Rao inequality[68]:

$$\sigma_2 = [q_0^2 |\Psi_\omega|]^{-\frac{1}{2}} , \tag{1.130}$$

where q_0^2 is the signal-to-noise ratio at the receiver output in the case when $\Delta k = 0$.

Since the altitude and phase are coupled by a linear relationship, the variance of the estimated range obtained upon averaging of N independent sample values is specified as

$$\sigma_R^2 \approx \frac{\sigma_1^2 + \sigma_2^2}{4\Delta k^2 N} . \tag{1.131}$$

The substitution of Eqs. (1.129) and (1.130) into (1.131), together with the equality $\Delta k = 2\pi\Delta f / c$, yield

$$\sigma_R = \frac{\sigma_\xi}{\sqrt{N}\cos\theta}\left[1 + \frac{c^2\cos^2\theta}{16\pi^2 q_0^2 \sigma_\xi^2 (\Delta f)^1 |\Psi_\omega|}\right]^{\frac{1}{2}} . \tag{1.132}$$

Fig. 1.13[5,7,65] shows the absolute error of measured flight altitude σ_R as a function of incidence angle θ for two widths of the antenna's pattern: $\theta_A = 0.10$ rad (Curve 1) and 0.35 rad (Curve 2). In the calculations, Eq. (1.119) and the data from Figs. 1.11 and 1.12 are used;

the remaining parameters are as follows: $\Delta f = 15$ MHz, $z = 1000$ m, $\sigma_\xi = 5$ cm, $N = 1000$, and $q_0^2 = 14$ dB. It is seen that the error in the measured altitude grows at large probing angles and abruptly increases in the case of a wide antenna's patterns. The radar characteristics should be chosen such that absolute value $|\Psi_\omega(\Delta k)|$ is close to unity.

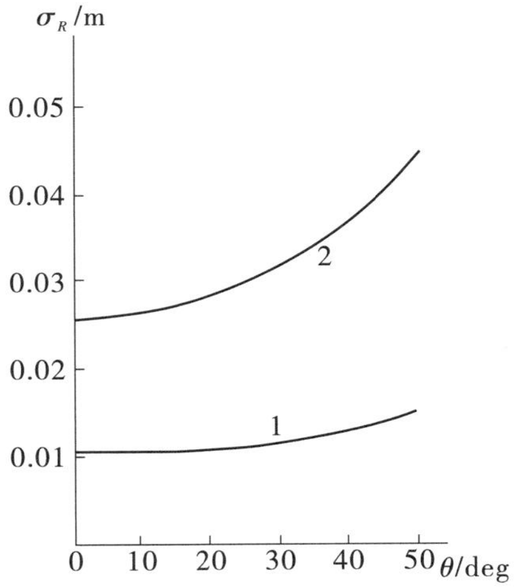

Fig. 1.13 Absolute error of measured flight altitude σ_R as a function of probing angle θ obtained for $\theta_A = 0.1$ rad (Curve 1) and 0.35 rad (Curve 2)

The above results show that FCFs can be used to measure the flight altitude with an accuracy comparable to or exceeding the accuracy ensured by the pulse method. Pulse probing necessitates power wideband radar systems, whereas two-frequency radio systems operating in the quasi-continuous mode should contain narrow-beam antennas and have a relatively narrow passband. These conclusions follow directly from similarity properties because the FCF is related with the transient impulse scattering function via the direct Fourier transform[36].

Now, consider measurements of irregularities' height σ_ξ according to the FCF[7,13,32,65]. In order to estimate the sensitivity of Ψ_ω to the height of irregularities, let us calculate partial differential of Eq. (1.107) with respect to variable σ_ξ:

$$\Delta\sigma_\xi = \frac{\Delta\Psi_\omega}{4\sigma_\xi(\Delta k)^2 |\Psi_\omega| \cos^2\theta}. \tag{1.133}$$

In the first-order approximation, we have[7]

$$\Delta\Psi_\omega \approx (2/N)^{1/2}. \tag{1.134}$$

Then, Eqs. (1.133) and (1.134) yield

$$\Delta\sigma_\xi = \frac{c^2}{8\sqrt{2}\pi^2\sigma_\xi(\Delta f)^2 |\Psi_\omega| N^{1/2}\cos^2\theta} \tag{1.135}$$

for a wide range of incidence angles θ. Note that, in Eq. (1.135), the effect of the receiver's intrinsic noise, which is involved in estimation of Ψ_ω, is disregarded.

Let us analyze the possibility of using phase relationships for determination of irregularities' height σ_ξ. A similar approach has been applied in Ref. [68, 85] for small values of θ. In this case, irregularities' height σ_ξ is determined from fluctuations of σ_R. Having determined the partial differential of σ_R^2 with respect to σ_ξ and substituted it into Eq. (1.133), we arrive at the following result:

$$2\sigma_R\Delta\sigma_R = \frac{\sigma_\xi\Delta\sigma_\xi}{N}\left(\frac{2}{\cos^2\theta} + \frac{\cos^2\theta}{q_0^2\,|\Psi_\omega|}\right). \tag{1.136}$$

For a Gaussian sample with uncorrelated values,

$$\Delta\sigma_R = \sigma_R/\sqrt{2N}. \tag{1.137}$$

Eqs. (1.132) and (1.137) yield

$$\Delta\sigma_\xi = \frac{\sigma_\xi}{\sqrt{2N}}\left(1 + \frac{c^2\cos^2\theta}{16\pi^2 q_0^2\sigma_\xi^2(\Delta f)^2\,|\Psi_\omega|}\right)\left(1 + \frac{\cos^4\theta}{2q_0^2\,|\Psi_\omega|}\right)^{-1}. \tag{1.138}$$

Relative errors $\Delta\sigma_\xi/\sigma_\xi$ in the irregularities' height that is determined by means of two methods based on FCFs are calculated according to Eqs. (1.135) and (1.138) and presented in Figs. 1.14a and 1.14b[5,7,65], respectively. The calculations have been performed for two widths of the antenna's pattern $\theta_A = 0.02$ rad (Curves 1, 3) and 0.1 rad (Curves 2, 4) and the following parameters: $\Delta f = 100$ MHz, $z = 500$ m, $\sigma_\xi = 0.1$ mm, $N = 1000$, and $q_0^2 = 14$ dB.

Although the solution has not been optimized, Fig. 1.14 shows a high resolution of the irregularities' heights. As in the case of measurements of the flight altitude, we can conclude that the error of measured σ_ξ increases in the case of a wide antenna's pattern and in the case of a large angle of incidence of a wave.

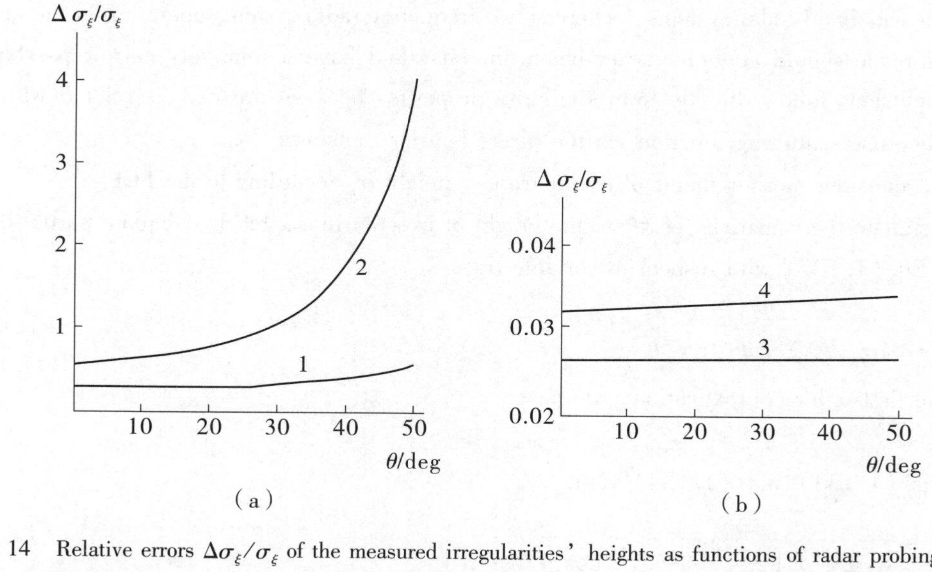

Fig. 1.14 Relative errors $\Delta\sigma_\xi/\sigma_\xi$ of the measured irregularities' heights as functions of radar probing angle θ obtained for the width of the antenna's pattern $\theta_A = 0.02$ rad (Curves 1, 3) and 0.1 rad (Curves 2, 4), the frequency difference $\Delta f = 100$ MHz, flight altitude $z = 500$ m, irregularity height $\sigma_\xi = 0.1$ m, the number of independent sample values $N = 1000$, and $q_0^2 = 14$ dB

It is obvious that the approach based on taking into account phase fluctuations depending on the flight altitude [see Eq. (1.138)] is more sensitive to irregularities' rms height σ_ξ and depends on incidence angle θ only slightly. When the signal-to-noise ratio tends toward infinity, $q_0^2 \to \infty$, Eq. (1.138) yields the following asymptotics:

$$\frac{\Delta\sigma_\xi}{\sigma_\xi} \xrightarrow[q_0^2 \to \infty]{} \frac{1}{\sqrt{2N}}. \tag{1.139}$$

The above results show that methods based on FCFs are competitive with standard short-pulse methods of measurement of flight altitudes and terrain characteristics[5,7,13,32,33,65].

1.14 The kernel of the generalized ambiguity function and a measure of noise immunity involved in radar probing of a rough surface and extended targets

Absolute value $|\chi(\Omega,\tau)|$ or $|\chi(\Omega,\tau)|^2$ of the response of an optimal filter is referred to as the ambiguity function (AF). The AF was first introduced in radar[86] for estimating the possibility of resolving point targets that are located at different ranges (the coordinate referred to as delay τ) and move at different speeds (the coordinate referred to as Doppler frequency Ω) and the possibility of developing a signal theory[8,9,29,87]. At the same time, classical deterministic AF $|\chi(\Omega,\tau)|^2$ can be applied in practice only when radio waves are reflected by a point target moving rectilinearly and uniformly. This motion determines signal delay τ and Doppler frequency shift Ω. Such a situation is very rigid, and it is violated when an electromagnetic wave is scattered by an extended target or a statistically rough surface.

Radar imaging and spatial-temporal processing of radar images are often performed when factorization condition (1.30) is fulfilled. Then, the AF represented as expression $|\chi(\Omega,\tau,\Delta,\omega_{\vec{\rho}})|^2$ becomes a product of factors, one of which characterizes only spatial signal processing, while the other characterizes only temporal signal processing[5-12]. In this case, the variables $\Delta = \{\Delta x, \Delta y\}$ and $\omega_{\vec{\rho}} = \{\omega_x, \omega_y\}$ specify the vector of the antenna's linear shift with respect to the expected position and the vector of the signal's spatial-frequency shift. The latter vector determines the angular shift of the antenna relative to a signal source. In terms of spatial frequencies, matched spatial filtering corresponds to the case when the direction in which a desired signal is received coincides with the direction of the maximum of the antenna's pattern.

When radar probing of extended chaotic surfaces is performed, the scattered field is a random function even if the probing signal is harmonic. Hence, processing of this field by a matched spatial-temporal filter is non-optimal. Enhancement of resolution and construction of an AF with allowance for the scattering properties of probed objects are usual radar problems.

Therefore, the generalized[5,7,30,88,89], mutual[90,91], and cross[92] AFs are more suitable for practice. The terminological differences result from the fact that the generalized and mutual

AFs contain random reflected signals but are constructed according to different approaches and involve different scattering characteristics, whereas the cross AF corresponds to the class of deterministic signals.

It should be emphasized that the kernel of the integral expressions for the AF has the form of various approximations in specific calculations of scattering of simple signals[89,90], frequency-modulated signals[88,91], and phase-shift keyed signals with a small bandwidth-duration product[90]. In Ref. [93], where scattering of chirp signals is studied, the kernel of transformations is determined with the use of the model of local reflection-oscillation sources that are regarded as a whole.

The results obtained can be applied to investigate realer cases within the framework of a radiophysical model of scattering of modulated radio waves and to specify the integral expressions for the AF[5,7,32,33]. Following Refs. [30, 88, 89], we represent the generalized AF in the form

$$\langle |\chi(\Omega,\tau)|^2\rangle = \frac{1}{(4\pi E_0)^2}\int_{-\infty}^{\infty}\int_{-\infty}^{\infty}\int_{-\infty}^{\infty}\int_{-\infty}^{\infty} F_0(\omega_1)F_0^*(\omega_2)\cdot$$
$$\langle K(\omega_1,t_1)K^*(\omega_2,t_2)\rangle E(t_1-\tau)E^*(t_2-\tau)\cdot$$
$$\exp[\mathrm{i}\Omega(t_1-t_2)+\mathrm{i}(\omega_1 t_1-\omega_2 t_2)]\mathrm{d}t_1\mathrm{d}t_2\mathrm{d}\omega_1\mathrm{d}\omega_2. \tag{1.140}$$

Here, $F_0(\omega)$ is the amplitude spectrum of a signal, $K(\omega, t)$ is generalized frequency response Eq. (1.50) [with Eq. (1.55)] of a radar probing channel, and $E(t)$ is the complex amplitude of a probing signal with energy E_0. In Eq. (1.140), averaging is performed to obtain a stable characteristic by a statistically rough surface. In the particular case where the process is stationary in time and frequency, Eq. (1.140) can be simplified additionally.

Second joint moment $\langle K(\omega_1,t_1)K^*(\omega_2,t_2)\rangle$ determines FCF Ψ_ω (1.107) of wave fields. Integration of Eq. (1.140) and the definition of classical AF $|\chi(\Omega,\tau)|^2$ of a signal reflected from a point target yield

$$\langle |\chi(\Omega,\tau)|^2\rangle = \frac{1}{4\pi^2}\int_{-\infty}^{\infty}\int_{-\infty}^{\infty}\Psi_\omega(-\Omega_1,\tau_1)|\chi(\Omega_1,\tau_1)|^2\cdot\exp[\mathrm{i}(\Omega_1\tau-\Omega\tau_1)]\mathrm{d}\tau_1\mathrm{d}\Omega_1, \tag{1.141}$$

where $\Psi_\omega(\Omega_1,\tau_1)$ is the FCF determined in space for two instants:

$$\Psi_\omega(\Omega_1,\tau_1) = \langle K(\omega,t)K^*(\omega+\Omega_1,t+\tau)\rangle,\ \tau_1 = t_1 - t_2,\ \Omega_1 = \omega_1 - \omega_2. \tag{1.142}$$

In the limiting case when coherence band Δf_c and correlation time τ end towards infinity, generalized AF (1.141) becomes the classical AF for a point target and, on the other hand, takes into account the scattering properties of a surface[5,7,30,88,89]. Note that, in contrast to the classical AF, the generalized AF is not invariant with respect to the 2D Fourier transform. As has been mentioned in the foregoing, the presence of fluctuations in the reflected signal strongly affects the portion of the signal energy that is used effectively in the optimal receiver.

This portion of energy is generally specified by the frequency and temporal correlation intervals of the scattered signal that are determined relative to the spectral width and the duration of a probing signal, respectively. The kernel of integral Eq. (1.141) for generalized AF $\langle |\chi(\Omega,\tau)|^2\rangle$ *coincides with the FCF specified at two spatial points* $A_1(x_{10}, y_{10}, z)$ and $A_2(x_{20}, y_{20}, z)$ at different instants t_1 and t_2 when $z_1 = z_2 = z$ (Fig. 1.3) Then, coefficients (1.86) have the form

$$p_1 = \frac{2}{z^2},\ p_2 = p_3 = \frac{\Delta k}{z},\ p_4 = \frac{b^2}{z^2} + k_1 k_2,\ p_{4,1} = \frac{b^2}{z^2} + k_1^2,\ p_{4,2} = \frac{b^2}{z^2} + k_2^2,$$

$$p_5 = \frac{b^2\cos^4\theta}{z^2} + k_1 k_2,\ p_{5,1} = \frac{b^2\cos^4\theta}{z^2} + k_1^2,\ p_{5,2} = \frac{b^2\cos^4\theta}{z^2} + k_2^2,$$

$$p_6 = \frac{b^2\cos^2\theta}{z^2} + k_1 k_2. \tag{1.143}$$

The substitution of coefficients (1.143) into Eq. (1.91) and Eq. (1.92) yields (see [7]) the explicit general expression for second joint moment $\langle K(\omega_1, t_1) K^*(\omega_2, t_2)\rangle$, which determines the FCF and the kernel of the generalized AF of signals scattered by an anisotropic rough surface. When wave parameter Eq. (1.56) is large, $p = \sqrt{z\lambda}/d \gg 1$, the very bulky relationships from [7] can be simplified substantially for the far zone by means of making allowance for the simultaneous inequalities $k \gg b/z \gg \Delta k$ and $k_1 k_2 \gg (\Delta k)^2$ [5,7]:

$$\Psi_\omega(-\Omega_1, \tau_1) = A_{\Omega,\tau} \exp B_{\Omega,\tau}, \tag{1.144}$$

where

$$A_{\Omega,\tau}^{-2} = 1 + \mathrm{i}\frac{2\Delta k z[n_1(1+\cos^2\theta) + n_2]\cos^2\theta}{1 + 4b(n_1\cos^2\theta + n_3)\cos^2\theta},$$

$$B_{\Omega,\tau} = -\mathrm{i}2z\Delta k\sec\theta - 2\sigma_\xi^2\Delta k^2\cos^2\theta - \frac{B_1}{B_2},$$

$$B_1 = \frac{k_1 k_2}{\cos^2\theta}\left\{\frac{2b}{z}(n_4 + n_5\cos^2\theta) + \mathrm{i}\Delta k[n_6 - n_5(1+\cos^2\theta)]\right\},$$

$$B_2 = 4b^2\left\{\frac{1}{z}(4n_2\cos^4\theta + 4n_3\cos^2\theta + 1) + \mathrm{i}\frac{2\Delta k}{\cos^2\theta}[n_1(1+\cos^2\theta) + n_2]\right\}, \tag{1.145}$$

and

$$n_1 = 4\langle\gamma_x^2\rangle\langle\gamma_y^2\rangle(1 - R_{xy}^2)b\cos^2\theta,\ n_2 = \langle\gamma_x^2\rangle + \langle\gamma_y^2\rangle,\ n_3 = \langle\gamma_x^2\rangle\cos^2\theta + \langle\gamma_y^2\rangle,$$
$$n_4 = \Delta x^2 + \Delta y^2 + 4b(\langle\gamma_x^2\rangle\Delta y^2\cos^2\theta + \langle\gamma_y^2\rangle\Delta x^2)\cos^4\theta,$$
$$n_5 = 8b\sqrt{\langle\gamma_x^2\rangle\langle\gamma_y^2\rangle}R_{xy}\Delta x\Delta y\cos^2\theta,$$
$$n_6 = \Delta x^2 + \Delta y^2 + 8b(\langle\gamma_x^2\rangle\Delta y^2\cos^2\theta + \langle\gamma_y^2\rangle\Delta x^2)\cos^2\theta,\ \Delta x = V_x\tau, \Delta y = V_y\tau. \tag{1.146}$$

In expressions of (1.146) for the coefficients, V_x and V_y denote the coordinate components of the horizontal velocity of an aircraft.

The quantity

$$\Delta\tau = \int_{-\infty}^{\infty}\langle|\chi(0,\tau)|^2\rangle d\tau, \tag{1.147}$$

is a measure of the delay-time (range) resolution, and the quantity

$$\Delta\Omega = \int_{-\infty}^{\infty} \langle |\chi(\Omega,0)|^2 \rangle d\Omega \tag{1.148}$$

is a measure of the Doppler-frequency (velocity) resolution.

Quantities (1.147) and (1.148) are dependent on the scattering properties of terrain and are coupled through the equations[5,30,89]

$$\langle |\chi(0,0)|^2 \rangle \Delta\Omega\Delta\tau = \text{const} , \tag{1.149}$$

which is a generalization of the well-known principle of uncertainty to the case when a statistically rough surface is probed.

The final expression for the generalized AF can be derived only for a specific signal structure because, otherwise, the integration operations in Eq. (1.141) are not determined. Analytic Eqs. (1.141) – (1.146) can be applied to establish a relationship between the generalized AF, the antenna's characteristics and orientation, and the characteristics of an anisotropic scattering surface whose slopes are correlated in the range $0 \leqslant R_{xy} \leqslant 1$ [7]. Therefore, it is possible to calculate the characteristics of signal detection in the case when waves are scattered by extended rough surfaces or targets observed against the background of such a terrain. In studies[30,89,94,95], the correlation of slopes has not been taken into account for the generalized AF.

In problems of detection of fluctuating point targets observed in the presence of a distributed interference (terrain), the measure of noise immunity[29,87,96]

$$\Delta_{11} = \frac{\langle E \rangle / N_0}{1 + \dfrac{E}{N_0}\displaystyle\int_{-\infty}^{\infty}\int_{-\infty}^{\infty} S_{\angle}(\tau_1,\Omega_1)\,|\chi_{11}(\tau - \tau_1,\Omega - \Omega_1)|^2 d\tau_1 d\Omega_1} \tag{1.150}$$

is used. Here, $\langle E \rangle$ is the average energy of a received signal; E_0 is the energy of a radiated signal; N_0 is the spectral density of noise; Ω_1 and τ_1 are the Doppler shift and time delay, respectively; $|\chi_{11}(\tau - \tau_1, \Omega - \Omega_1)|^2$ is the AF of a probing signal; and $S_{\angle}(\tau_1, \Omega_1)$ is the terrain scattering function, which characterizes the range and Doppler frequency distributions of reflections.

The second term in the denominator of Eq. (1.150) describes the deterioration of the radar noise immunity caused by reflections from terrain and is the protection ratio for the case of passive interference $\rho(\tau,\Omega)$. When $\rho(\tau,\Omega) \approx 0$, the noise immunity of a receiver is determined by its intrinsic noise. Quantity $\rho(\tau,\Omega)$ is estimated via the convolution of the scattering function and the AF of a probing signal.

Synthesis of the optimal signal for a standard receiver with a matched filter involves minimization of the total volume of the scattering function and the signal AF shifted to a target's point on the range-Doppler shift plane. If these two functions do not intersect, standard (in the presence of white noise) and optimal receivers are equivalent and their noise immunity is determined only by additive noise. When these functions intersect substantially, the optimal receiver ensures higher noise immunity but necessitates a more complicated processing scheme.

A receiver of another type[29] contains an optimal filter whose characteristic is matched with or close to a received signal arriving from a scattering surface. Then, Eq. (1.150) for the measure of noise immunity has the form

$$\Delta_{12} = \frac{\langle E\rangle |\chi_{12}(\tau,\Omega)|^2/N_0}{1+\frac{E}{N_0}\int_{-\infty}^{\infty}\int_{-\infty}^{\infty} S_{\angle}(\tau_1,\Omega_1)|\chi_{12}(\tau-\tau_1,\Omega-\Omega_1)|^2 \mathrm{d}\tau_1 \mathrm{d}\Omega_1}, \tag{1.151}$$

where $|\chi_{12}(\tau,\Omega)|^2$ is the mutual AF.

When extended targets are probed (a situation especially important if MMW signals and high-resolution radars are used), Eq. (1.150) should involve the generalized AF whose kernel is FCF (1.144) – (1.146)[5,7,32,33].

1.15 Formalism of high-order correlators and bispectra in the wave theory and multidimensional-signal precessing

So far, in statistical radiophysics and statistical radio engineering, Gaussian fields or fields that are expressed through Gaussian fields by comparatively simple formulae have been studied predominantly. It suffices to describe such fields by the second-order approximation of the correlation theory. A more detailed description of the interaction of backscattered fields can be provided by the formalism of the third-order correlators

$$R^{(3)}(t_1,t_2) = \int_{-\infty}^{\infty} E(t)E(t+t_1)E(t+t_2)\mathrm{d}t \tag{1.152}$$

and of their frequency-domain Fourier transforms referred to as a bispectrum[7,18,97,98]:

$$R^{(3)}(f_1,f_2) = \int_{-\infty}^{\infty}\int_{-\infty}^{\infty} R^{(3)}(t_1,t_2)\exp[-2\pi i(f_1t_1+f_2t_2)]\mathrm{d}t_1\mathrm{d}t_2 . \tag{1.153}$$

Then, the AF and the classical Wigner function[99] are particular cases of the third-order correlation function. The term bispectrum used for Eq. (1.153) can be explained as follows[18,19]. It is convenient to consider signal $g(x)$ of a nonlinear system as a functional that is dependent on x and specified on the space of input signal $f(x)$. A functional of the most general form can be expanded in series of regular homogeneous functionals[100]:

$$g(x) = a_0(x) + \int f(x_1)a_1(x,x_1)\mathrm{d}x_1 + \iint f(x_1)f(x_2)a_2(x,x_1,x_2)\mathrm{d}x_1\mathrm{d}x_2 + \cdots \tag{1.154}$$

The third term of series Eq. (1.154) corresponds to bilinear functionals. Thus, a system described by the equation

$$g(x) = \int_{-\infty}^{\infty}\int_{-\infty}^{\infty} f(x_1)f^*(x_2)q(x,x_1,x_2)\mathrm{d}x_1\mathrm{d}x_2 , \tag{1.155}$$

where $q(x,x_1,x_2)$ is a function of variables x_1 and x_2 that depends on x as on a parameter, is a bilinear system.

Quadratic functionals can be formed from bilinear functionals. A given quadratic functional can be represented in both the frequency and time domains with the use of suitable transformations of operators. Quadratic functionals form a normalized linear space that is isometric to the space of linear operators.

If signals under study are functions of two variables, $\vec{\rho} = \{x,y\}$, and of the corresponding spatial frequencies, $\vec{f} = \{f_x, f_y\}$, third-order correlator $R^{(3)}(t_1,t_2)$ (1.152) is a four-dimensional function of $\{x_1, y_1, x_2, y_2\}$ and spatial bispectrum (1.153) is a function of form $F^{(3)}(\vec{f}_1, \vec{f}_2)$. As soon as Gaussian process approximations become insufficient, the significance of third-order correlators abruptly increases. In particular, variable $R^{(3)}(t_1,t_2)$ turns out to be more sensitive to specific features of a real signal under study than the correlation function and more immune to interferences, such as noise and shifts[98]. The method of matched filtering that uses $R^{(3)}(t_1,t_2)$ is more informative than matched filtering based on standard correlation function $R^{(2)}(t_1)$. However, application of third-order correlators and bispectra necessitates increasing the number of computational operations and the RAM space.

1.16 The effect of irregularities of a scattering surface on the structure of reflected signals

Investigation of interaction of a probing signal and a scattering surface is one of the central tasks in direct and inverse radar problems and in active remote explorations. Below, we apply FCFs to analyze the effect of irregularities of a surface on the structure of various reflected signals[5,7,13,32,33]. The idea of the method proposed is similar to the idea of the method from Ref. [35].

As follows from functional (1.57) of stochastic backscattered fields observed at different spatial points $(\vec{\rho}_1, z_1)$ and $(\vec{\rho}_2, z_2)$ at different instants t and $t+\Delta t$, the intensity of radio waves scattered by a rough surface is expressed through FCF Ψ_ω as

$$I(t) = \int_{-\infty}^{\infty}\int_{-\infty}^{\infty} F(\omega_1)F^*(\omega_2)\Psi_\omega(\omega_1,\omega_2,t,\vec{r})\mathrm{e}^{-\mathrm{i}(\omega_2-\omega_1)t}\mathrm{d}\omega_1\mathrm{d}\omega_2 . \tag{1.156}$$

Here, $F(\omega) = \int_{-\infty}^{\infty} f(t)\mathrm{e}^{-\mathrm{i}\omega t}\mathrm{d}t$ is the spectrum of modulated probing signal $f(t)$ and $\Psi_\omega = \Psi_\omega(\omega_1,\omega_2,t,\vec{r})$ is the value of the FCF at coinciding instants $t_1 = t_2 = t$. This coincidence means that $\Delta x = \Delta y = 0$ and $z_1 = z_2 = z$.

Note that the frequency-domain description of the scattered field in form (1.156) is equivalent to the time domain description expressed by the convolution

$$I(t) = \int_{-\infty}^{\infty} G(t-t',\vec{r})I_1(t')\mathrm{d}t' \tag{1.157}$$

of the input-signal intensity

$$I_1(t) = \frac{1}{4\pi^2}\int_{-\infty}^{\infty}\int_{-\infty}^{\infty} F(\omega_1)F^*(\omega_2)\mathrm{e}^{-\mathrm{i}(\omega_2-\omega_1)t}\mathrm{d}\omega_1\mathrm{d}\omega_2 , \tag{1.158}$$

and generalized Green's function $G(t,\vec{r})$ (1.54).

For a stationary (in the broad sense of the word) radar channel with uncorrelated scattering[7,27,29,34], the FCF is a function of only the difference time $\tau = t_2 - t_1$ and the difference frequency $\Omega = \omega_2 - \omega_1$. In this case, Green's function $G(t,\vec{r})$ of the radar channel is determined from the Fourier transform of the FCF:

$$G(t,\vec{r}) = \frac{1}{2\pi}\int_{-\infty}^{\infty} \Psi_\omega(\Omega,\vec{r})\, e^{-i\Omega t} d\Omega. \tag{1.159}$$

First, let us consider reflection of a high-frequency Gaussian pulse,

$$f(t) = A_0 \exp(-i\omega_0 t - \tau_0^{-2} t) \tag{1.160}$$

with carrier frequency ω_0, e^{-1}-level duration τ_0, and the spectrum

$$F(\omega) = A_0 \tau_0 \sqrt{\pi} \exp[-\tau_0^2 (\omega - \omega_0)^2 / 4].$$

It is convenient to perform integration with respect to variables $\omega_{1,2}$ in Eq. (1.156) under the assumption that the signal is quasi-monochromatic. This means that the signal's envelope slowly changes in time and the integral is determined by the values of the integrand from the neighborhoods of the points $\omega_1 \approx \omega_0$ and $\omega_2 \approx \omega_0$ [5,7,35]. The aforementioned substitutions and integration yield

$$I(t') = A'_0 \exp\left(-\frac{\tau_0^2 \tau_1^2 \omega_0^2}{2\tau_1^2 + 0.5\tau_0^2}\right) \exp\left(-\frac{t'^2}{2\tau_1^2 + 4\tau_2^2 + 0.5\tau_0^2}\right), \tag{1.161}$$

where $t' = t - (2z/c)$ is the current time measured from the delay instant of the reflected signal, $\tau_1 = \pi\theta/(\omega_0 \theta_A)$, $\tau_2 = \sqrt{2}\sigma_\xi / c$, and A'_0 is the amplitude factor.

At initial e^{-1}-power duration $\tau_0/\sqrt{2}$ of the probing pulse, e^{-1}-power duration τ_p of the reflected pulse is

$$\tau_p = \sqrt{\frac{\tau_0^2}{2} + 8\frac{\sigma_\xi^2}{c^2} + \frac{2\pi^2\theta^2}{\omega_0^2 \theta_A^2}}. \tag{1.162}$$

In contrast to the results from [74], where the duration of the reflected pulse is described by the first two terms in Eq. (1.162), the relationship obtained above takes into account the effect of the antenna's pattern. Fig 1.15a depicts time-dependent normalized intensity $I_n(t')$ of the reflected signal with a Gaussian envelope for the wavelength $\lambda = 8.6$ mm, the width of the antenna's pattern $\theta_A = 0.01$ rad, and the irregularities' rms height $\sigma_\xi = 0.1$ m. It is seen that, with an increase in wave incidence angle θ, the maximum intensity decreases, the pulse edges become smoother, and the intensity of the reflected pulse increases.

Now, let the probing pulse be described by the function

$$f(t) = \begin{cases} A_0 \cos \omega_0 t \ (|t| \le \tau_0/2) \\ 0 \ (|t| > \tau_0/2) \end{cases} \tag{1.163}$$

with the spectrum

$$F(\omega) = \frac{A_0\tau_0}{2}\left\{\frac{\sin[(\omega - \omega_0)\tau/2]}{(\omega - \omega_0)\tau/2} + \frac{\sin[(\omega + \omega_0)\tau/2]}{(\omega + \omega_0)\tau/2}\right\} \approx \frac{A_0\tau_0}{2}\frac{\sin[(\omega - \omega_0)\tau/2]}{(\omega - \omega_0)\tau/2}.$$

In this case, integral (1.156) is calculated by means of integration with respect to the parameter and via changing variables with the Jacobian $|\Delta = 0.5|$. In the calculation, the integral identity (obtained by Prof. A. Potapov in [7, 13])

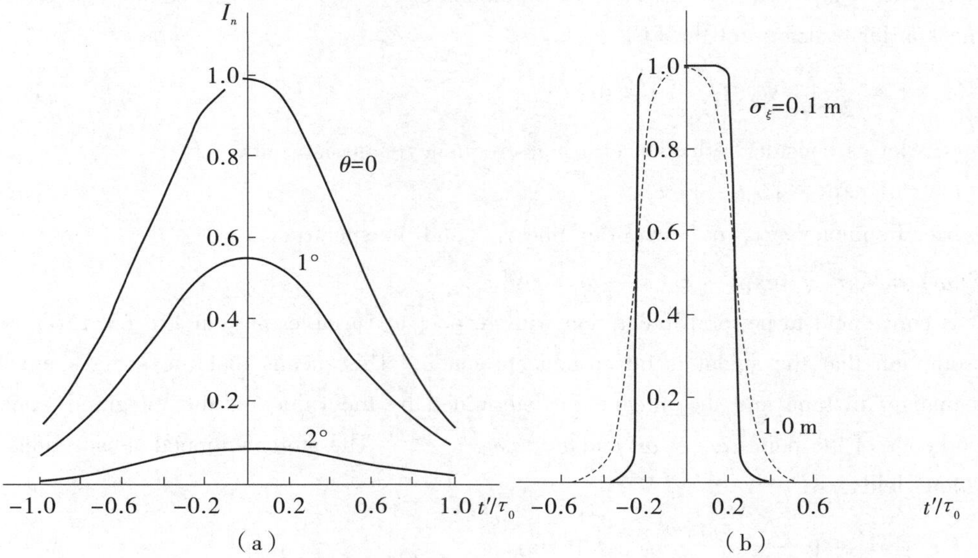

Fig. 1. 15 Normalized envelopes of the reflected signal for (a) Gaussian and (b) rectangular MMW probing radar pulses

$$\int_{-\infty}^{\infty}\{a\exp[-(ax+b)^2]\Phi(cx+d)+c\exp[-(cx+d)]^2\Phi(ax+b)\}\,dx=\sqrt{\pi}\Phi(ax+b)\Phi(cx+d)/2 \qquad (1.164)$$

which is more general than the corresponding standard tabular integral, is used.

The necessary transformations, which are bulky, yield

$$I(t') = A''_0\exp(-2\omega_0^2\tau_1^2)\left\{\Phi\left[\left(\frac{4t'+\tau_0}{4\sqrt{2\tau_1^2+4\tau_2^2}}\right)-\Phi\left(\frac{t'}{\sqrt{2\tau_1^2+4\tau_2^2}}\right)\right]\cdot\right.$$
$$\left[\Phi(\mathrm{i}\sqrt{2}\omega_0\tau_1)-\Phi\left(\frac{-\tau_0+\mathrm{i}8\omega_0\tau_1^2}{4\sqrt{2}\tau_1}\right)\right]+\Phi\left[\left(\frac{t'}{\sqrt{2\tau_1^2+4\tau_2^2}}\right)-\Phi\left(\frac{4t'-\tau_0}{4\sqrt{2\tau_1^2+4\tau_2^2}}\right)\right]\cdot$$
$$\left.\left[\Phi\left(\frac{\tau_0+\mathrm{i}8\omega_0\tau_1^2}{4\sqrt{2}\tau_1}\right)-\Phi(\mathrm{i}\sqrt{2}\omega_0\tau_1)\right]\right\}. \qquad (1.165)$$

In Eq. (1. 165), A''_0 is the amplitude factor and $\Phi(z)=erf(z)=\frac{2}{\sqrt{\pi}}\int_0^z e^{-t^2}dt$ is the probability integral or the error function. Probability integrals $erf(z)$ of complex arguments are tabulated and can be found in the literature along with the necessary relationships of symmetry[101,102].

Figure 1. 15b depicts time-dependent normalized intensity $I_n(t')$ of the reflected signal with a rectangular envelope for the wavelength $\lambda=8.6$ mm, vertical probing ($\theta=0$), and the irregularities' rms heights $\sigma_\xi=0.1$ m and 1 m. Results (1. 161), (1. 162), and (1. 165) do not contradict the data from [35, 52, 61, 103] and confirm the well-known possibility of determining σ_ξ from the duration of the edge of a received pulse.

1.17 The effect of a fractal scattering surface on the structure of reflected signals

Analysis of scattering of modulated waves by a fractal surface may be of great interest for the theory and discovery of new effects. Let us discuss this problem in brief on the basis of the results from Ref. [104]. The main aspects of this study are considered in Monographs [1, 2, 5]. Fractal Gaussian scattering surface $z(\vec{R})$ with the zero mean $\langle z(\vec{R}) \rangle = 0$, and the variance $\sigma_\xi^2 = H^2$; the structure function $\langle [z(\vec{R}+\vec{R}_0) - z(\vec{R}_0)]^2 \rangle = \Delta(\vec{R})$ is modeled by a fractal with the fractal dimension $2 < (D+1) < 3$. Hence, irregularities are so extreme that the finite region occupied by surface $z(\vec{R})$ has an infinite area but zero volume. The representation of the fractal dimension of the surface in the form $(D+1)$ means that D is the fractal dimension of the fractal nondifferentiable function[2] obtained as the section of surface $z(\vec{R})$ along a line and that $1 < D < 2$.

For corrugated $[\Delta = f(x)]$ and isotropic $[\Delta = f(R = \sqrt{x^2 + y^2})]$ scattering surfaces, structure function $\Delta(\vec{R})$ has the form

$$\Delta(\vec{R})\big|_{\vec{R}\to 0} \Rightarrow \begin{cases} L^{2D-2}|x|^{4-2D} \\ L^{2D-2}R^{4-2D} \end{cases}, \tag{1.166}$$

where L is the surface's topothesis determined from the condition[1,2,105]

$$\langle [z(\vec{R}+L) - z(\vec{R})]^2 \rangle / L^2 = 1. \tag{1.167}$$

Large-height irregularities of the surface, or its "geography", affect the asymptotics of the structure function: $\Delta(\vec{R}) \to 2\sigma_\xi^2 = 2H^2$ as $\vec{R}\to\infty$ or $|x|\to\infty$. In the isotropic case, a simple model of $\Delta(\vec{R})$ that takes into account these restrictions is the function

$$\Delta(R) = 2H^2[1 - \exp(-L^{2D-2}R^{4-2D}/2H^2)], \tag{1.168}$$

with correlation radius l that specifies the outer height of irregularities and equals the value

$$l = L(H/L)^{1/(2-D)}. \tag{1.169}$$

The effects of the fractal character and geography of a scattering surface can be distinctly discriminated under the conditions $H \gg L$ and $l \gg L$. The special case of the fractal dimension $D = 1.5$ corresponds to a Brownian fractal[1,2]. In this case, Eq. (1.166) implies that structure function $\Delta(\vec{R})$ is linear for small R or x, i.e., behaves as increments of the z coordinate of a Brownian particle.

When the fractal dimension tends toward unity (a smooth surface), $D\to 1$, topothesis L disappears from Eq. (1.166). Then, for an isotropic surface, we have

$$\Delta(\vec{R})\big|_{\vec{R}\to 0} = \langle \gamma^2 \rangle R^2. \tag{1.170}$$

In this case, Eqs. (1.168) and (1.169) can be transformed into

$$\Delta(R) = 2H^2[1 - \exp(-\langle \gamma^2 \rangle R^2/2H^2)], \tag{1.171}$$

$$l = H/\sqrt{\langle\gamma^2\rangle} \tag{1.172}$$

under the conditions $\sqrt{\langle\gamma^2\rangle} \gg 1$ and $l \gg L$, respectively.

Solution of such a fractal problem is generally nontrivial[1,2,104,105]. For the isotropic-scattering geometry displayed in Fig. 1.3, the intensity of a reflected quasi-monochromatic signal with envelope $a(t)$ can be represented in polar coordinates in the form

$$I(t) = \frac{k^2 c}{2z_0^3}\int_0^\infty dt' \left\langle a\left[t - t' + \frac{2\xi(x,y)}{c}\right]\right\rangle \cdot \int_0^\infty R\exp[-2k^2\Delta(R)]J_0(2kR\sqrt{ct'/z_0})\,dR, \tag{1.173}$$

where $t' = R^2/(cz_0)$ is the variable of integration, z_0 is the flight altitude, $\vec{R} = \vec{R}_1 - \vec{R}_2$ is the difference variable of integration, and $J_0(\cdots)$ is the zero-order Bessel function of the first kind.

For a Gaussian scattering surface, the calculated intensity of reflected signal (1.160) coincides with my results Eqs. (1.161) and (1.162).

For a corrugated scattering surface, the intensity of a monochromatic signal with envelope $a(t)$ can be represented as

$$I(t) = \frac{k\sqrt{cz_0}}{2\pi z_0^3}\int_0^\infty dt' \left\langle a\left[t - t' + \frac{2\xi(x,y)}{c}\right]\right\rangle \cdot \frac{1}{\sqrt{t'}} \cdot$$

$$\int_0^\infty R\exp[-2k^2\Delta(x)]\cos(2kx\sqrt{ct'/z_0})\,dx. \tag{1.174}$$

Having substituted the fractal structure function of Eq. (1.166) into Eqs. (1.173) and (1.174), we can show after some algebra that the tail intensity of signals reflected by a fractal surface is described by power functions (!):

$$I(t) \sim 1/(t')^{3-D}. \tag{1.175}$$

Result of (1.175) is very important because for standard cases the intensity of a reflected quasi-monochromatic signal decreases exponentially. Thus, the shape of a signal scattered by a fractal statistically rough surface substantially differs from the shape of a scattered signal obtained with allowance for classical effects of diffraction by smoothed surfaces[5,7,23].

Interest in investigations of wave scattering by rough surfaces with a non-Gaussian statistic has increased recently. It is often stated that the spatial correlation factor of a scattering surface, $R(\Delta x = x_2 - x_1, \Delta y = y_2 - y_1)$, cannot be exponential, because such a stochastic process is nondifferentiable. Sometimes, a regularizing function is applied near the zero point[106]. The possibility of using nondifferentiable functions for description of wave scattering was justified more comprehensively from the physical viewpoint after the theory of fractals, the theory of fractal measure, and scaling relationships were applied to radiophysical problems[1,2]. Note that the Gaussian model is parabolic near the incidence angle $\theta \approx 0$, while the exponential model is linear.

Earlier, irregularities of equal heights were predominantly considered in problems of wave diffraction by a statistically rough surface. Later, researchers realized that multiscale irregularities provide for more adequate results[23]. Now, it can be stated with certainty that the physical sense of the diffraction theory dealing with multiscale irregularities is clearer when the fractal approach is applied and the fractal dimension or fractal signature D is regarded as a parameter. Moreover, according to the calculation results from [1, 2], taking into account fractal properties brings the theoretical and experimental characteristics of microwave scattering indicatrices of terrain much closer. This fact has always been interpreted mainly as a result of purely instrumental errors.

1.18 The effect of hydrometeors on radar imaging

In radar probing or imaging of terrain, it is necessary to take into account distorting effects of various atmospheric elements on the characteristics of signals scattered by terrain. A substantial influence of hydrometeors (clouds and precipitation) on MMW propagation results in the fact that a resulting RI is determined by not only the reflecting properties of an object but also the parameters of meteorologic fields. This circumstance impedes interpretation of the images obtained. Below, the effect of hydrometeors on radar imaging is analyzed[7].

Attenuation of radio waves by the atmospheric mass can be predicted only probabilistically[5,6,107,108]. At the same time, experimental measurements of the rainfall characteristics have been performed only for the wavelengths $\lambda \geqslant 3$ mm[5,6,109]. The following approximation relationships between specific RCS η of rainfall and rainfall intensity I[5-7,37,110] are valid for the linear polarization:

$$\begin{aligned} &\eta = 1.04 \cdot 10^{-8} I^{1.52} \quad \text{for } \lambda = 32 \text{ mm},\\ &\eta = 2.21 \cdot 10^{-5} I^{1.05} \quad \text{for } \lambda = 8.6 \text{ mm},\\ &\eta = 1.44 \cdot 10^{-4} I^{0.59} \quad \text{for } \lambda = 4.3 \text{ mm},\\ &\eta = 8.89 \cdot 10^{-5} I^{0.57} \quad \text{for } \lambda = 3.2 \text{ mm}, \end{aligned} \tag{1.176}$$

where the dimension of $\eta \rightarrow$ is m^{-1} and $I = 0.1, \cdots, 10$ mm/h.

The presence of fractal power dependences in Eqs. (1.176) – (1.179) indicates the fractal character of radiophysical phenomena under study (see [1, 2] for details). Research in this scientific field is only now being conducted.

Only isolated synchronous measurements of attenuation factor γ and specific RCS η of rainfall have been performed[37,111]. Studies[7,112] have extended the range of applicability of Eq. (1.176) to the wavelength $\lambda = 2.2$ mm. The per-unit-length attenuation factor and the specific RCS of rainfall were measured in summertime (with the use of a relative method) through comparison signals reflected by a known amount of precipitation and by two reference corner reflectors $r_1 > r_2$.

The experimental results[7,112] have been compared to measured[113] average values of η of rainfall for the circular and linear polarizations of radiation. For circularly polarized waves, the backscattering RCS of rainfall at the wavelength $\Delta\eta = 2.2$ mm decreases by approximately 10 dB when the intensity is $I = 0.5$ mm/h and decreases to 4 dB when the intensity is 10 mm/h. At the wavelengths 3. 2 cm and 3. 2 mm, we have $\Delta\eta = 20$ dB and 18 dB, respectively[110]. As could have been expected, for the circular polarization, the signal backscattered from rainfall is compensated more poorly at higher frequencies because dimensions of drops become commensurable with the wavelength. Signal suppression is reduced also as the rainfall intensity grows, because the effect of the nonspherical shape of drops becomes more pronounced. It is usually believed that, at a drop diameter exceeding 2 mm, it is necessary to take into account its nonspherical shape. Theoretical calculations of electromagnetic-wave attenuation are based on the Mie theory for various dimension distributions of particles[6,114-116]. In the case considered, Best's distribution is used and radiation is assumed to be linearly polarized.

At the wavelength 2. 2 mm, specific RCS of rainfall η is rather accurately described by the following expression in the range $I \leqslant 10$ mm/h of the rainfall intensity:

$$\eta = 5 \cdot 10^{-6} I^{1.2}, \tag{1.177}$$

which has been obtained with the use of the least-squares method.

The approximating expression obtained via the least-squares method for measurements of attenuation factor γ [dB/km] at the wavelength 2. 2 mm has the form

$$\gamma = 1.5 I^{0.5}. \tag{1.178}$$

The relationship $\gamma = aI^b$ is widely applied to describe rainfall attenuation of radio waves. Constants a and b depend on the frequency and polarization of radio waves and on the dimension distribution and temperature of raindrops. In Ref. [117], the approximating coefficients are calculated for the frequency range 1 – 1000 GHz. For $\lambda = 2$ mm, the Laws-Parsons distribution, and the rainfall intensity $1.27 \leqslant I \leqslant 50.8$ mm/h, the following values are obtained: $a = 1.25$ and $b = 0.7$ ($t = 0$) and $a = 1.33$ and $b = 0.69$ ($t = 20$℃). The coefficients have the values $a = 1.7$ and $b = 0.51$ in Ref. [118] and $a = 1.3$ and $b = 0.635$ in Ref. [119]. The comparison of my experimental data on the average attenuation factor with the results from Refs. [118, 119] indicates that the experimental data match the measured array of coefficients γ.

Eqs. (1. 177) and (1. 178) form a closed system. Its solution yields a relationship that directly couples γ and η:

$$\eta = 1.9 \cdot 10^{-6} \gamma^{2.4}. \tag{1.179}$$

Hence, in practice, the specific RCS of rainfall can be estimated immediately if attenuation factor γ is known and vice versa: If the specific RCS of rainfall is known, the attenuation factor can be estimated.

The spectral density of the MMW power scattered by hydrometeors is of substantial practical interest. It follows from the experimental data from Ref. [7] that, at a carrier of 135 GHz, the mean spectral width of fluctuations is about 220 Hz for variations of ±20 Hz occurring over an interval of 2 min and for a rainfall intensity of 2 mm/h. The influence of snowfall, rainfall, mists, clouds, and sand-dust structures on attenuation and scattering of electromagnetic microwaves has been analyzed comprehensively and generalized in Ref. [37]. In that study, the efficiency of radar systems operating under various weather conditions has been assessed.

Fig. 1.16 illustrates measurements[7,120] of a human's RCS σ that were performed in the presence of slight drizzling rain at a wavelength of 8.6 mm by means of a radar producing a complex phase-shift keyed signal with an ultra large bandwidth-duration product[77]. The experimental data from Ref. [121] obtained at a wavelength of 2 cm are depicted in Fig. 1.16 as well. It has been shown that a human's RCS σ depends mainly on the wave polarization and the direction toward the human's body. There is a relationship between a human's RCS and mass. The data from Ref. [121] have been obtained with an error of ±1 dB for a 180-cm-tall man weighing 90 kg. The measurements described in Refs. [7, 120] correspond to a 175-cm-tall man weighing 80 kg.

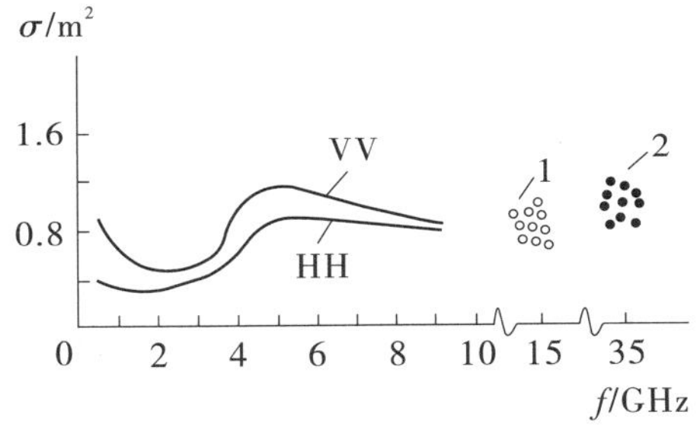

Fig. 1.16 Average RCS of a human as a function of frequency for the (HH) horizontal and (VV) vertical polarizations: (solid lines) data from Ref. [121] and (1, 2) my data for the VV polarization[7,120]

The spatial-temporal structure of meteorologic fields, which call for more comprehensive study, is very important for radar imaging. Consider the terrain-rainfall layer as a model of the environment in which an RI is formed. Let us assume that the power of the received signal is determined by two components proportional to specific surface (σ_*) and volume (η) RCSs and take into account the integral radio-wave attenuation.

Following the calculation technique applied in Ref. [122] for analyzing the influence of weather conditions on a side-looking radar, we obtain the general expression for total specific RCS σ_*^{Σ} of the terrain-inhomogeneous rainfall system in the form[7,123]:

$$\sigma_*^{\Sigma} = \sigma_* \cdot 10^{-0.2\int_0^{\infty}\gamma(r)\mathrm{d}r} + \int_0^{\infty} \eta(r) \cdot 10^{-0.2\int_0^{r}\gamma(r')\mathrm{d}r'}\mathrm{d}r . \tag{1.180}$$

In the case of surface radar probing at angle θ, we obtain from Eq. (1.180) for a homogeneous hydrometeor layer of height h

$$\sigma_*^{\Sigma} = \sigma_* \cdot 10^{-0.2\gamma h/\cos\theta} + \frac{\eta\cos\theta}{0.46\gamma}(1 - 10^{-0.2\gamma h/\cos\theta}) . \tag{1.181}$$

Eq. (1.181) yields analytic boundary conditions that make it possible to find out when rainfall zones on an RI are lighter ($\sigma_*^{\Sigma} > \sigma_*$) or darker ($\sigma_*^{\Sigma} < \sigma_*$) than the corresponding zones on the RI of the same terrain area in the absence of rainfall:

$$\sigma_*^{\Sigma} > \sigma_* \text{ when } \sigma_* < \frac{\eta\cos\theta}{0.46\gamma};$$

$$\sigma_*^{\Sigma} < \sigma_* \text{ when } \sigma_* > \frac{\eta\cos\theta}{0.46\gamma}. \qquad (1.182)$$

The boundary conditions specified by Eq. (1.182) are analyzed in detail in Ref. [7] for various spatial-temporal structures of meteorologic fields. In Ref. [7], experimental digital maps of vertical and horizontal sections of precipitation-intensity distributions and precipitation reflectance obtained at a wavelength of 3.2 cm are investigated. These measurements were performed by means of a computer-aided system designed at the Central Aerological Observatory for acquisition, processing, and representation of radar data. The system is based on an MRL-5 meteorologic radar and an SM-4 computer. Precipitation intensity I was calculated from reflectance z with the use of the known correlation relationship[124] applied in radar meteorology:

$$\log I\ [\text{mm/h}] = \{z\ [\text{d}B] - 10\log A\} / 10B, \qquad (1.183)$$

where the coefficients are as follows: $A = 250$ and $B = 1.6$ (rainfall) or $A = 1000$ and $B = 1.6$ (snowfall).

The attenuation factor and specific RCS were recalculated for a wavelength of 8.6 mm according to the technique recommended by the International Radio Consultative Committee (see, e.g., Ref. [125]). It is well known that the technique is based on the relationship between the attenuation factor and intensity of rainfall. According to Eqs. (1.176) and (1.179), spatial-temporal fields $\gamma(\vec{r})$ and $\eta(\vec{r})$ were determined at a wavelength of 8.6 mm. The calculations of spatial distributions of γ and η performed with allowance for boundary conditions (1.182) and (1.183) made it possible to determine both the total specific RCS for typical terrains in the presence of precipitation and the limit detection ranges. Such multistage calculations of space-time maps of hydrometeor distributions are exemplified in Ref. [7].

Note that application of the theory of fractals to the analysis of spatial-temporal structure of hydrometeors can yield qualitatively new concepts in the traditional theory of radio-wave propagation and scattering[1,2].

1.19 Radiophysical model of formation of reference detailed digital radar maps of terrain in the MMW band

In the theory and practice of digital correlation-extremal navigation systems (CENSs), digital terrain maps (DTMs) that are a representation of the anomalies of natural and artificial geophysical fields are used[10]. Such fields include magnetic and gravitational fields, surface-relief fields, optical and thermal fields, fields of natural gamma radiation, electrostatic and radar-contrast fields, etc. It is well known[10,126] that operation of CENSs is based on digital comparison of a current terrain image with a reference image obtained beforehand. The comparison yields an estimate of an object's location. This estimate allows formulation of a command necessary for controlling the object's motion. A generalized concept of a CENS is illustrated in Fig. 1. 17, where the following notation is employed: 1 is an aircraft, 2 is a programmed flight trajectory, 3 is an actual flight trajectory, 4 is a correction, 5 is the terrain along the flight trajectory, 6 is the digital relief of terrain, 7 is the reference terrain map, and 8 are the sections to be corrected.

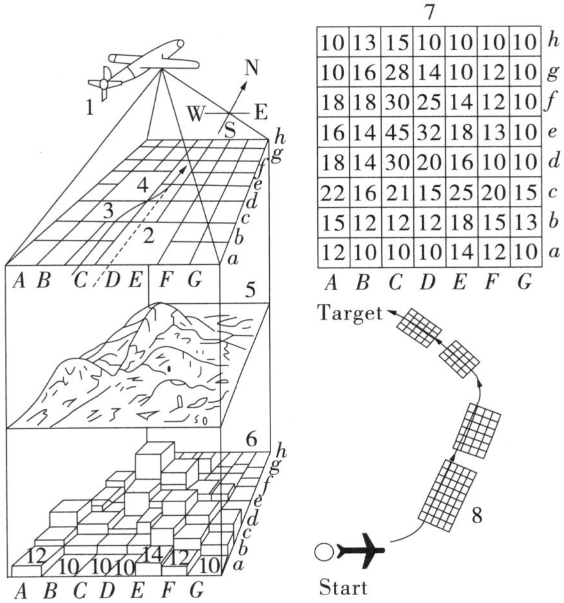

Fig. 1. 17 Operation of a CENS

The following examples of CENSs[7,127] can be enumerated. The terrain contour matching (TERCOM) system is applied when the field of the Earth's relief is used, the system of terminal homing applications of solid state imaging devices (THASSID) system is applied when the optical field is used, the microwave radiometric (MICRAD) system is applied when the thermal field is used, and the radar area guidance (RADAG) system is applied when the radarcontrast field is used. Development of such systems has resulted in the following complexes: optimal relief-measuring systems involving Kalman filtering, in particular, the sandia inertial terrain-aided navigation (SITAN) system; the automatic terrain recognition and navigation (ATRAN) system, which uses the profiles of the horizon and foreground terrain; and the combined range-only correlation (ROC) system, which operates in two modes. The ROC system uses the field of specific RCSs at large altitudes and the relief field and the field of σ_* on the final section.

The use of the field of specific RCSs σ_* of terrain is a promising direction. This technique is efficient at large altitudes over a relatively flat terrain when reliefmeasuring and thermal systems do not ensure a given accuracy[128]. At present, techniques of synthesis of reference DDRMs of an inhomogeneous terrain still have not been provided with representative data sets on the statistic and variations (including seasonal variations) of specific RCSs σ_* of various types of terrain, and the effect of spatial-temporal structures of precipitation regions has not been taken into account completely in the MMW band. The known theoretical studies are usually restricted to synthesis of Gaussian or uniformly distributed fields. At the same time, the probability densities of the brightness of real terrain images fundamentally (!) are not characterized by such distributions.

It is known that a reliable interpretation of radar and remote sensing data necessitates complexation of radar and optical images and combination of experimental and calculated data. Below, selected results that Prof. A. Potapov obtained in cooperation with various research and engineering teams are summarized[7,13,32,33,37,77,129,130]. Prof. A. Potapov began to use these results in 1987 for the development of a new method of MMW-band synthesis of nonuniform digital reference data sets[7,131,132]. A distinctive feature of this method is the use of texture signatures for classification and stochastic autoregressive synthesis of terrain images and introduction of fundamentally new (at that time) fractal methods of data processing based on fractal signatures. Model blocks were tested with the help of experimental data obtained for more than thirty types of terrain. Such tests allowed synthesis of contour and half-tone (at a wavelength of 8.6 mm) DDRMs of the region where the four-year-long experiment had been conducted[7,132].

In addition, experimental investigations performed from 1979 to 1990 made it possible not only to solve traditional scattering problems but also to reveal general processes of formation of the fine structure of MMW modulated signals and propose a new class of signatures based on the fine structure of modulated signals scattered by a statistically rough surface[7,133]. In these studies, numerical values are summarized and recommendations for practical determination and application of the aforementioned signatures are given.

The characteristics of the fine structure of reflected radar signals include intrapulse fluctuations, their statistic, correlation and spectral dependences, and the average pulse broadening specified by the reciprocal of coherence band Δf_c considered above.

Let us summarize certain results from [7, 133]. Measurements were performed with the use of transmitting-receiving devices at the wavelength $\lambda = 8.6$ mm for horizontally polarized radiation. The halfpower width of the antenna's pattern was 3° and 10° in the azimuthal and vertical planes, respectively. The probing angle was $\theta = 36°$. The duration of the probing high-frequency pulse was $\tau = 300$ ns, and the repetition rate was 4 kHz. A time-strobing circuit producing a sequence of 10 monitoring pulses with $\tau = 40$ ns and a repetition period of 150 ns was applied to directly estimate the duration of a reflected pulse. The envelope of the reflected pulse and of a strobe sequence was photographically recorded at a rate of 4 frames per second and a 150^{-1}-s-long time of exposure.

Typical instant envelopes of MMW pulses reflected by the aforementioned terrains are displayed in Fig. 1. 18. The ten most characteristic photographs were selected from numerous negatives for each terrain type. At the first stage, enlarged envelopes of reflected pulses were sampled in accordance with necessary number N_1 of values. The results served as initial data for averaging the envelopes of pulse signals reflected by various terrains (Fig. 1. 19).

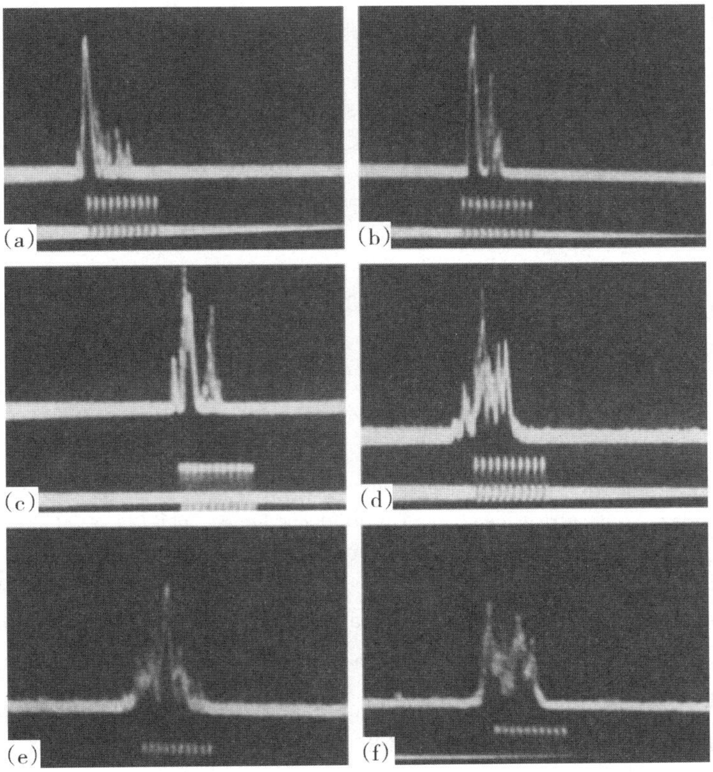

Fig. 1. 18 Experimental instant envelopes of MMW pulses reflected by (a) a lake's surface, (b) a runway strip, (c) a village, (d) a river bend, (e) a deciduous forest, and (f) a field obtained at $\lambda = 8.6$ mm and $\theta = 36°$

At the second stage of processing, representative samples of intrapulse amplitude fluctuations were considered. These samples were obtained through subtraction of discrete values of center pulses from sample values of instant envelopes. Since the considered realizations of envelopes are periodically nonstationary random processes, it was important to test intrapulse amplitude fluctuations for stationarity. Stationarity monitoring should be performed via methods of nonparametric statistics, because, owing to the discrepancy between theoretical probabilistic models of MMW scattering by terrain and the actual situation, the a priori ambiguity is high[7].

The stationarity of processes was assessed according to Wilcoxon's rank test. The asymptotic efficiency of this test is $3/\pi$ as compared to Student's test. Wilcoxon's test can be applied to short samples. Null hypothesis H_0 in Wilcoxon's test assumes that two samples are associated with identical distribution functions. If the values of each sample correspond to different intervals

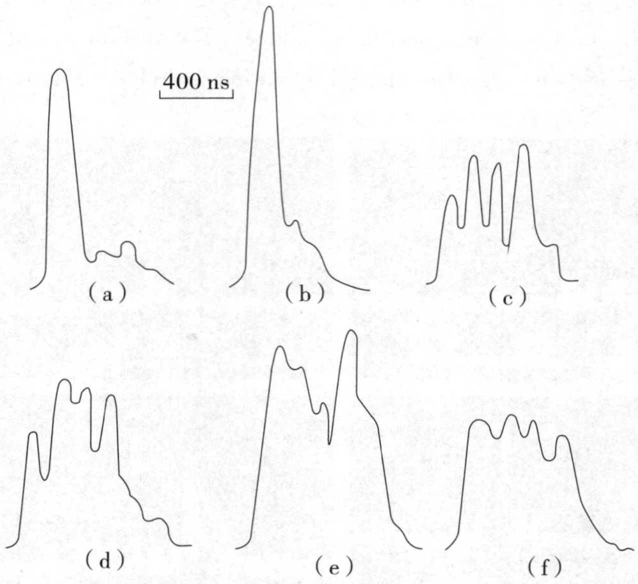

Fig. 1. 19 Average envelopes of MMW pulses reflected by (a) a lake's surface, (b) a runway strip, (c) a village, (d) a river bend, (e) a deciduous forest, and (f) a field

of a process under study, the null hypothesis means that the process is assumed stationary. It has been found that, for all types of terrain, intrapulse and intraperiod fluctuations are stationary with respect to the mean and the shape of the distribution with the confidence probability $P_{conf} = 0.95$. Most of the results on determination of a new class of signatures based on the fine structure of reflected signals are presented in Ref. [133].

Refs. [7, 134] report data on in situ modeling of a passive-active radio system that forms images and is based on the measuring complex designed earlier for a wavelength of 8.6 mm[77]. A quasi-continuous complex phase-shift keyed signal with an ultralarge bandwidth-duration product served as an illumination signal. The modulation radiometer employed exhibited a fluctuation sensitivity of 0.2 – 0.4 K for the integration constant $\tau = 1$ s.

The separate antennas of the radiometer and radar were conical horns with a pattern width of about 14°. To improve isolation, the antennas were separated by radio-wave absorbing shields. A series of experiments revealed an increase in the contrast of weakly radiating objects against a natural background. The contrasts of metal objects and concrete slabs with various RCSs and various degrees of filling of the antenna's beam against a vegetation background were examined in the presence of an open water surface in the field of view.

It has been established that the contrast of an object's image obtained through subtraction of the combined image (formed from the intrinsic radiation plus illumination) from the thermal image is improved substantially.

Experimental passive-radar data obtained with my assistance include[7] digital brightness-contrast matrices (thermal portraits) of certain types of terrain and various objects observed against the background of such terrains, the characteristics of thermal radiation produced by these terrains and objects, temperature dependences of the brightness temperature of the atmospheric mass under various weather conditions, estimates of integral attenuation in the atmosphere, and estimates of the permittivities of certain terrains. The aforementioned matrices are obtained at wavelengths of 3.5 and 8.6 mm.

Usually, the main components of the information procedure performed during synthesis of a map of an inhomogeneous terrain are topographic, geometric-metric, syntactic, semantic, structural, and other types of information[131]. The architecture of a computeraided system forming a subject map of an inhomogeneous terrain is constructed in a generalized form with allowance for the prospects of developing new information technologies (Fig. 1.20)[7,37,132].

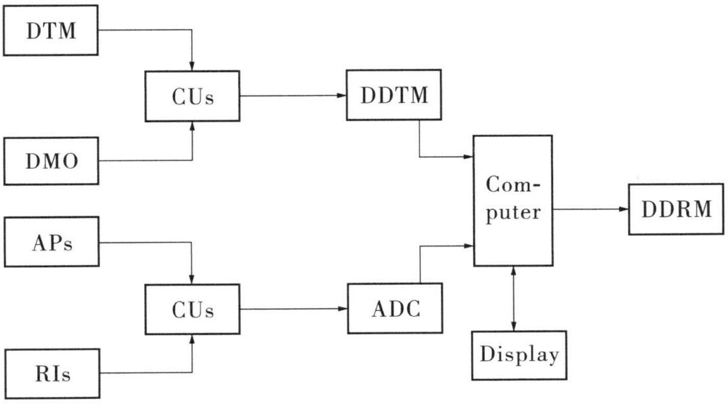

Fig. 1.20 Architecture of a system providing for a digital pattern of an inhomogeneous terrain: digital terrain map (DTM), digital map of objects (DMO), air photographs (APs), radar images (RIs), combining units or interfaces (CUs), analog-to-digital conversion (ADC), detailed digital terrain map (DDTM), and detailed digital radar map (DDRM) of terrain (reference map)

A map database is formed from (i) functionally coupled subsystems that input optical and radar information taken from aerial photographs (APs) and RIs, respectively, into a computer and (ii) digital map information obtained with allowance for the spatial location and structural details of objects on a terrain, the terrain's relief, and the steepness and exposition of slopes.

Generally, a terrain area and radar should be regarded as a spatially connected dynamic system described in external and internal representations at different abstracting levels[7].

Thus, in the external description, a DDRM is characterized by a map of the set of possible inputs (the coordinates of terrain points) onto the set of possible outputs [radiophysical characteristics of terrain: classical radar signatures (e.g. specific RCSs σ_*) and/or fractal cepstra and fractal signatures $D(\delta)$, where δ is the range of characteristic scales of secondary digital processing[1,2,79-81,135]].

When RIs are noisy and flat-contrast, synthesized images of textures should be employed. Creation of references without the use of efficient methods of digital-data compression is a challenging problem. In addition to existing techniques, efficient methods of fractal image coding compression[1,2], which are progressing rapidly today, can be applied to these problems. Radar images obtained during seasonal in situ experiments are the basis of reference synthesis.

In the theory of DDRM synthesis, the MTP approximation is used conventionally for assessing scattering properties of vegetationless terrains with allowance for the most probable angular sector of sight[7]. The conditions under which this method can be applied have been formulated above. Recently, interest in investigations of scattering from highly rough surfaces with a non-Gaussian statistics has increased substantially[1,2,5-7]. This problem is typical of exploration of certain terrain and man-made surfaces.

In the MTP approximation for matched vertical and horizontal (VV and HH) linear polarizations, the specific RCS has the following form:

$$\sigma_*^{VV} = \sigma_*^{HH} = \frac{|V(0)|^2}{2\langle\gamma^2\rangle\cos^4\theta}\exp\left(-\frac{\tan^2\theta}{2\langle\gamma^2\rangle}\right), \tag{1.184}$$

for the Gaussian probability distribution of surfacepoint heights and

$$\sigma_*^{VV} = \sigma_*^{HH} = \frac{3|V(0)|^2}{\langle\gamma^2\rangle\cos^4\theta}\exp\left(-\frac{\sqrt{6}\tan\theta}{2\langle\gamma^2\rangle}\right), \tag{1.185}$$

for the exponential probability distribution of surface point heights. Here, $|V(0)| = |(\sqrt{\varepsilon}-1)/(\sqrt{\varepsilon}+1)|$ is the absolute value of the Fresnel reflection coefficient at $\theta = 0$. The dependence of σ_* on the wavelength is specified only by coefficient $V(0)$ because the permittivity of a surface depends on frequency.

The angular dependences of specific RCS σ_* are obtained according to Eqs. (1.184) and (1.185) and depicted in Fig. 1.21 for various values of $\sqrt{\langle\gamma^2\rangle}$ [5-7]. Both of the models of backscattering show that a significant portion of power is scattered near the average normal to a surface. The Gaussian model is parabolic for the wave incidence angle $\theta \approx 0$, while the exponential model is linear. At larger incidence angles, the power decreases more rapidly in the case of the Gaussian model. As the rms value of slope ratio $\sqrt{\langle\gamma^2\rangle}$ grows, the portion of power scattered around the normal-incidence direction decreases. Simultaneously, the portion of power scattered at small grazing angles increases.

It can be shown readily[23,24] that, for irregularities whose slopes are characterized by the Gaussian distribution, effective angular half-width $\Delta\theta$ (determined at the e-decrease level) of the scattering indicatrix is $\Delta\theta = 2\sqrt{2}\sqrt{\langle\gamma_x^2\rangle}$. The effective azimuthal angular halfwidth is $\Delta\theta = 2\sqrt{2}\sqrt{\langle\gamma_y^2\rangle}/\tan\theta$, it follows from these results that the MTP approximation for radar scattering is valid in the interval of angles not exceeding the above values. The effect of the antenna's pattern on the specific RCS of the surface under study has been analyzed in [25, 136] for small incidence angles. It has been shown that, when the width of the antenna's pattern satisfies the ine-

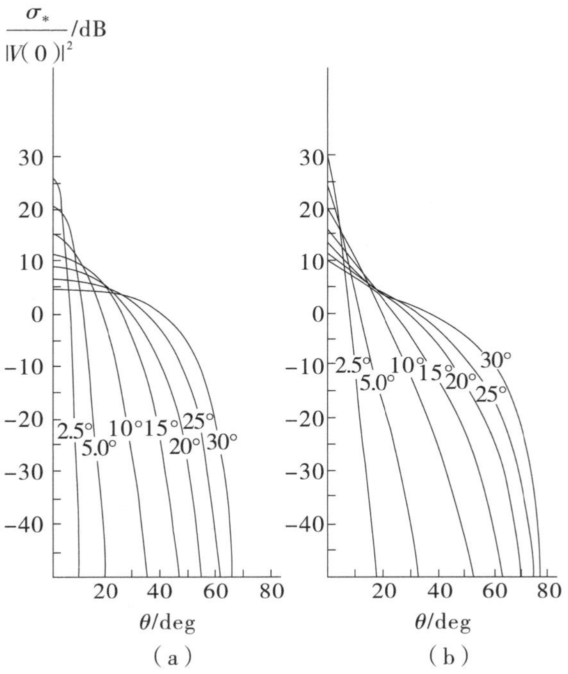

Fig. 1. 21 Specific RCS as a function of incidence angle θ obtained in the MTP approximation for surfaces with the (a) Gaussian or (b) exponential probability density of the distribution of the irregularities' heights at various irregularity slope angles: arctan ($\sqrt{\langle\gamma^2\rangle}$ = 2. 5°, ⋯, 30°)

quality $\theta_A < \sqrt{\langle\gamma^2\rangle}$, the antenna's pattern affects the received scattered-field intensity only slightly. In a system forming references, all arriving information is processed in the mode of on-line editing combined with correction of digital data by modern interactive hardware and software. Synthesized DDRMs of an inhomogeneous terrain usually use the most characteristic properties and landmarks of terrain[7].

1. 20 A radiophysical method of DDRM synthesis based on the theory of fractals

The methods of synthesis of DDRMs of inhomogeneous terrains developed by Prof. A. Potapov include the following stages[1,2,7]:

(i) Construction of topographic maps with the indicated slopes' steepnesses and expositions for an inspected region.

(ii) Photographing and radar survey of an inspected region.

(iii) Deciphering of photographs via the use of various classes of features.

(iv) Construction of maps, i. e., a diagram of terrain with coded geomorphological, soil, vegetation, and agricultural structural components.

(v) Calculation of specific RCSs of vegetationless terrains, assessment of the scattering properties of vegetation surfaces, and comparison of these properties with the surveillance databank.

(vi) Determination of the necessary number of quantization levels of the range of a terrain's specific RCSs.

(vii) Segmentation, i. e., separation of contours and homogeneous regions, each having virtually equal values of characteristics, in particular, equal specific RCSs. When large areas are formed and, as a result, a poorly informative reference is obtained, it is necessary to use additional alternative classes of features (features of the fine structure of reflected modulated MMWs[133] and/or fractal features as has been noted in Ref. [7]).

(viii) Construction of a reference contour DDRM of terrain from data obtained at the seventh stage.

(ix) Realization of stochastic autoregressive synthesis of textures of terrains with equal specific RCSs.

(x) Construction of a half-tone DDRM of terrain with allowance for the results obtained at stages 7 – 9.

Since the new interdisciplinary scientific direction in fractal radiophysics and fractal radar has been progressing rapidly at the IRE RAS during the last decade[1 – 7,79 – 84,135,137 – 161], there are spacious niches for applying fractal concepts of radiophysical investigations at almost all stages of the proposed method of synthesis of DDRMs of inhomogeneous terrains. For example, there are very interesting possibilities of establishing direct relationships between textures and fractals via the lacunarity characteristic (lacunarity features for detection of radar targets were formulated by the author in 1997)[135]. Now, we can speak about a new method of fractal synthesis of references[1,2,7,80,142,150,153-158].

Let us exemplify synthesis of contour and half-tone DDRMs of an inhomogeneous terrain at a wavelength of 8.6 mm (see [7, 131, 132]). The map analysis was combined with the results of flight experiments on MMW radar imaging of terrain[7,13,77,129,130]. The obtained data additionally were compared to photographs of a test terrain area obtained by the multispectral scanner (MSS) of a Landsat satellite in the following four bands of the visible and near infrared regions: 0.5 – 0.6 μm, 0.6 – 0.7 μm, 0.7 – 0.8 μm, and 0.8 – 1.1 μm at a resolution of 90 – 150 m. The maps of slope steepness and exposition are presented in Ref. [7].

Four ranges of specific RCSs were considered for linearly polarized radiation at a wavelength of 8.6 mm and an incidence angle of $\theta \approx 25°$. The contour DDRM of the inhomogeneous terrain displayed in Fig. 1.22 was obtained upon segmentation and combination of areas with approximately equal specific RCSs σ_*.

The stochastic autoregressive synthesis of images that is performed during construction of DDRMs makes it possible to model the textures of terrains with approximately equal specific RCSs[7]. In this case, a sequence of elements of a texture's image is modeled by realizations represented by a linear combination of weighted previous values of the elements of a noisy image. The

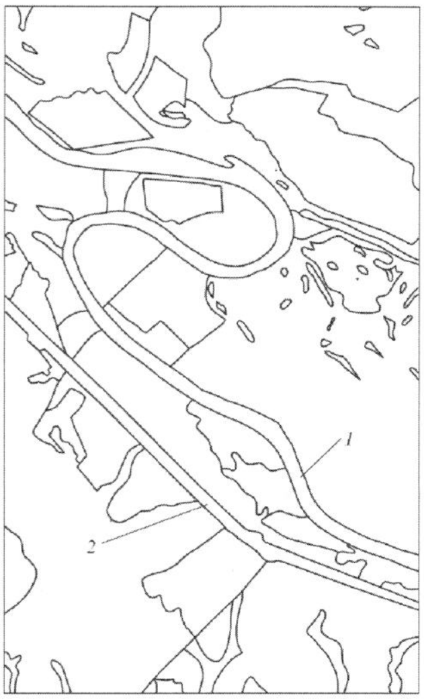

Fig. 1. 22 Synthesized contour reference map of an inhomogeneous terrain: (1) a river and (2) a railway.

characteristics necessary for synthesis have been determined from experimental RIs[7,37,162,163] with the use of the Yule-Walker system of equations and the transformation of brightness histograms.

As a result, half-tone reference DDRMs of an inhomogeneous terrain with six brightness gradations have been synthesized at a wavelength of 8. 6 mm. One of the DDRMs is displayed in Fig. 1. 23. In this case, the aforementioned six gradations correspond to six ranges of specific RCSs for an incidence angle of $\theta \approx 25°$ and a wavelength of 8. 6 mm: (1) $\sigma_* > -11$ dB (a village and seeded plots), (2) $\sigma_* = -12, \cdots, -14$ dB, (3) $\sigma_* = -15, \cdots, -17$ dB, (4) $\sigma_* = -18, \cdots, -20$ dB, (5) $\sigma_* = -21, \cdots, -23$ dB, and (6) $\sigma_* < -23$ dB (a river and a railway). The correlation and spectral characteristics of optical and radar images of certain terrains have been investigated in [7, 163]. The comparison of topographic maps with pattern synthesized contour (Fig. 1. 22) and half-tone (Fig. 1. 23) DDRMs has shown that the latter have a pronounced electrodynamic character.

As a result of experimental processing of DDRM transparencies with the help of a 2D incoherent optical correlation meter[7,163], autocorrelation coefficient R has been obtained as a function of angle φ of the transparencies' mutual rotation (Fig. 1. 24). The subsequent analysis has shown that the correlation peak is destroyed at $\varphi \approx 5°, \cdots, 7°$ for a contour DDRM (Curve 1) and at $\varphi \approx 14°, \cdots, 17°$ for a half-tone DDRM (Curve 2).

It can be concluded that both my radiophysical model of MMW formation of reference DDRMs of an inhomogeneous terrain and the method of DDRM synthesis are efficient and adequately representative of most specific features of real terrains. As has been noted above, the model can be improved via introduction of various fractal characteristics of terrains and calculated

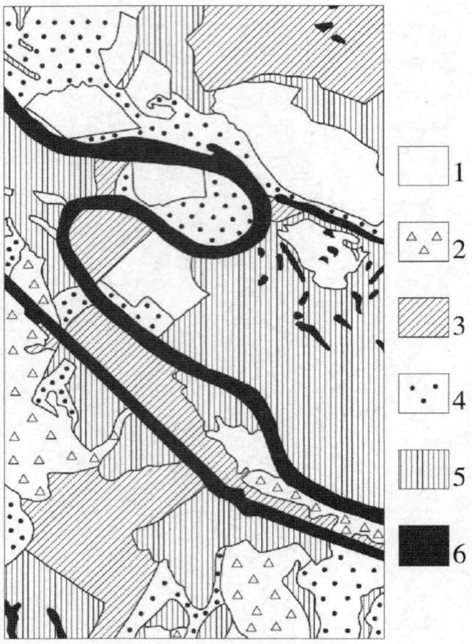

Fig. 1. 23 Synthesized half-tone reference map of an inhomogeneous terrain with six brightness gradations that correspond to six ranges of the specific RCS: (1) $\sigma_* > -11$ dB (a village and seeded plots), (2) $\sigma_* = -12, \cdots, -14$ dB, (3) $\sigma_* = -15, \cdots, -17$ dB, (4) $\sigma_* = -18, \cdots, -20$ dB, (5) $\sigma_* = -21, \cdots, -23$ dB, and (6) $\sigma_* < -23$ dB (a river and a railway)

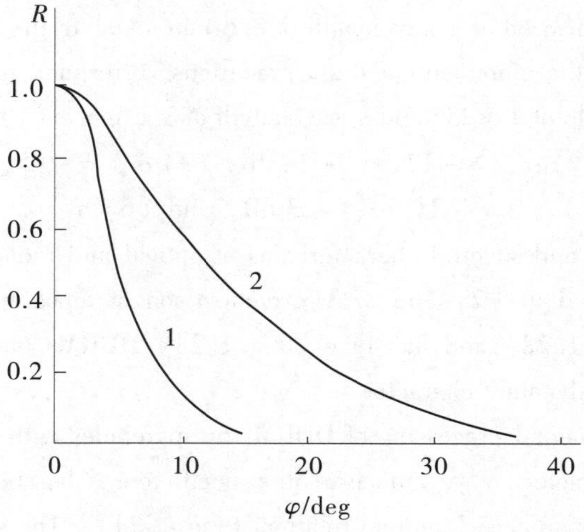

Fig. 1. 24 Autocorrelation factor R as a function of mutual rotation angle φ for the contour reference (Curve 1) and the half-tone reference (Curve 2)

2D statistics or matrices of the gradient distribution for images of fractal surfaces[1,2,7].

1.21 Indicatrices of millimeter-wave scattering by a fractal surface

Let us illustrate scattering of MMWs by fractals through simulation of this problem with the use of the 2D nondifferentiable Weierstrass function, which describes a model fractal frequency-selective scattering surface of finite dimensions[1,2,164]. These results can be applied for simulation of fractal antennas and fractal frequency-selective structures and for physical investigations of nanostructures[1-3]. Fig. 1.25 displays certain generated realizations of fractal surfaces $z(x,y)$ described by the segment-wise bounded Weierstrass function with dimension D that takes different values and finite numbers of harmonics N and M corresponding to orthogonal axes. Note that function $z(x,y)$ describes mathematical fractals only when $N\to\infty$ and $M\to\infty$ [1,2]. As fractal dimension D of a simulated surface grows, the heights of irregularities increase and their spatial correlation radius decreases.

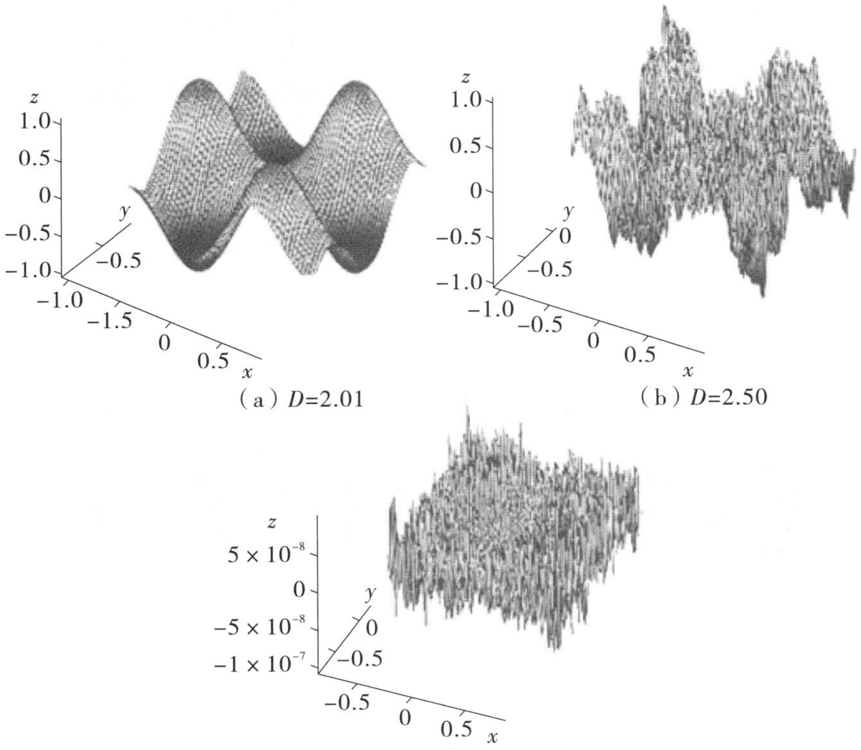

(a) D=2.01 (b) D=2.50

(c) D=2.99

Fig. 1.25 Fractal surfaces synthesized on the basis of the Weierstrass function for (a), (b) and (c)

The further theoretical estimation of the characteristics of the scattered field is based on the Kirchhoff approximation. The limits of applicability of this approximation are considered in [1, 2]. When fractal dimension D grows, it is necessary to check the accuracy of computations of scattered-field characteristics and the conditions for the validity of the Kirchhoff approximation for each spatial harmonic of a relief. Typical scattering indicatrices $g = g(\theta_2, \theta_3)$ are calculated for the wavelengths $\lambda = 1.0$ cm (Fig. 1.26) and $\lambda = 2.2$mm (Fig. 1.27) in the case of an ideal fractal rough surface ($|V| = 1$) with the fractal dimension $D = 2.5$. These results are obtained with the use of the methods described in Ref. [1, 2]. Thus, a fractal generalized Brownian surface is considered. In Fig. 1.26 and Fig. 1.27, θ_1 is the incidence angle, θ_2 is the angle of scattering, and θ_3 is the azimuthal angle.

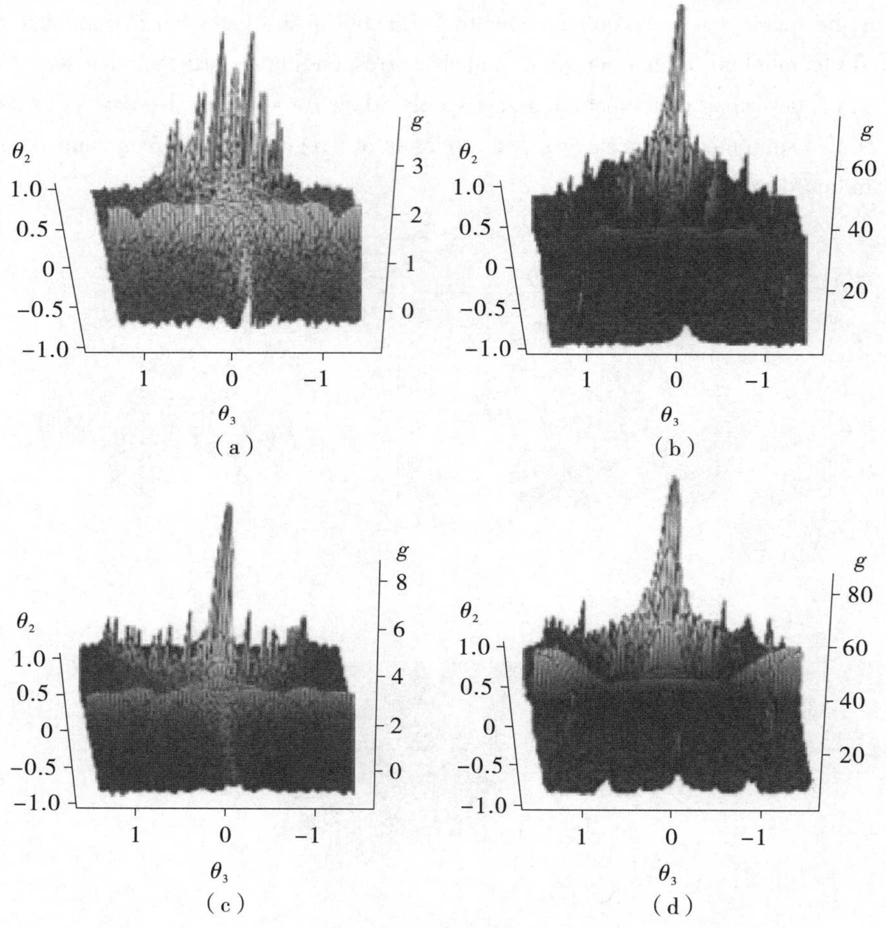

Fig. 1.26 Scattering indicatrices $g = g(\theta_2, \theta_3)$ obtained at the wavelength $\lambda = 1.0$ cm for a fractal surface of the dimension $D = 2.5$ at $\theta_1 =$ (a, b) 45° and (c, d) 22° and $N = M =$ (a, c) 1 and (b, d) 10

On the basis of a large array of similar data on the characteristics of the scattered field, the following conclusions can be drawn. When D is small, most energy is scattered in the specular direction. Sidelobes are formed owing to Bragg scattering. As fractal dimension D of a scattering surface grows, the number and intensity of sidelobes increase. As D grows, the angular range

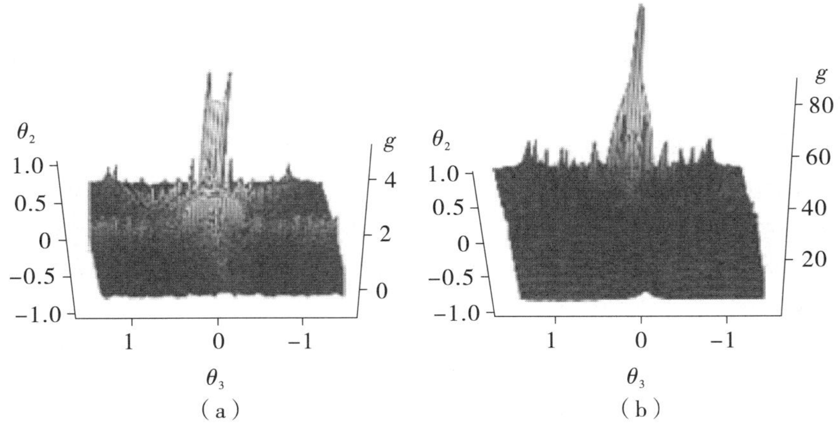

Fig. 1.27 Scattering indicatrices $g = g(\theta_2, \theta_3)$ obtained at the wavelength $\lambda = 2.2$ mm for a fractal surface of the dimension $D = 2.5$ at $\theta_1 = 22°$ and $N = M =$ (a) 1 and (b) 10

of sidelobes extends and higher-order spatial harmonics start playing a significant role.

Operating as a measuring scale, a radio wave illuminating a fractal separates spatial frequencies by means of the Bragg diffraction conditions[1,2]. When D is small, classical and fractal methods of calculation of scattered fields coincide. Thus, the fractal dimension of a rough surface can be estimated from calculated or measured characteristics of the scattered field. In practice, the dimensions of an illuminated area should be larger than the main period of the surface structure by a factor of at least 2. Then, the information on the fractal parameters of the surface is contained in the scattering characteristics.

In addition, note that the problem of determination of the radar signal reflected by arbitrary extended objects in the presence of non-Gaussian interferences becomes especially topical in the context of this study, which is devoted to formation of DDRMs and RIs[1-3,7,165,166].

1.22 Conclusions

The efficiency of radar, radio-wave imaging, remote sensing, and many other systems has been enhanced by the intense application of MMWs. Therefore, it is necessary to study the processes and characteristic features of MMW scattering by terrain. This study has adequately described the situation in this field, which, today, is again attracting the attention of numerous radiophysicists and radio engineers.

The mathematical approach that Prof. A. Potapov has been continually developing involves functionals of stochastic backscattered fields and allows qualitative estimation of the spatial-temporal and spatial-frequency characteristics of a scattered signal in the case of combined or spaced radio systems and an adequate description of the formation of observed radiophysical fields. The expressions obtained take into account the effects of irregularities' slopes and the dimensions of the antenna's aperture more accurately.

Fluctuation effects occurring during random and chaotic nonstationary reflection of waves by a medium's boundary have been investigated in the case when the boundary has an arbitrary (integer or fractal) topological dimension.

The approach proposed makes it possible to study spatial and frequency functionals of stochastic fields scattered by rough anisotropic surfaces with allowance for the mutual statistical relationship between irregularities' slopes and to specify the limits of applicability of the approximations employed. The developed generalizations and new solutions substantially extend the scope of problems of the statistical theory of wave diffraction.

It has been found that, when beam antennas are applied, the spatial correlation factors and FCFs of scattered fields are determined mainly by the width of the antennas' patterns. When antennas with wide patterns are applied, the spatial correlation factors and FCFs depend on characteristic slope ratios $\sqrt{\langle \gamma_x^2 \rangle}$ and $\sqrt{\langle \gamma_y^2 \rangle}$ of the large-height irregularities of a surface. It has been shown that, as slope correlation factor R_{xy} changes from unity to zero or as the flight altitude, the width of the antenna's pattern, or the probing angle grows, the value of the FCF decreases.

The proposed method has been applied to obtain and investigate the values of coherence band Δf_c of a radar probing channel. When width θ_A of the antenna's pattern increases, the coherence band is reduced abruptly. When θ_A changes by a factor of 10, value Δf_c decreases by two orders of magnitude for quasi-smooth surfaces and by an order of magnitude for highly rough surfaces. The value of coherence band Δf_c is always greater for a surface with strongly correlated slopes.

It has been demonstrated that Δf_c is invariant with respect to slopes of a scattering surface in the case when beam antennas are employed and the condition $\theta_A \ll \sqrt{\langle \gamma^2 \rangle} \cdot \sqrt{1 + \cos^2\theta \cos^2\theta}$ is fulfilled. If the opposite inequality holds, coherence band Δf_c depends on the slopes of large-height irregularities. As the slope ratios of irregularities decrease, coherence band Δf_c grows.

As the probing altitude increases, quantity Δf_c decreases almost linearly for antennas with wide patterns. For antennas with narrow patterns, Δf_c decreases more smoothly, while incidence angle θ increases.

The application of the FCF absolute value and phase characteristic for precision estimation of an aircraft's flight altitude and of the height of large irregularities have been investigated more accurately and justified quantitatively for various cross-correlation factors of irregularities' slopes.

The measurement errors of the aforementioned quantities increase for large probing angles and wide antenna patterns. It has been shown that measurement techniques using the absolute values and phase characteristics of FCFs are competitive with standard shortpulse methods for measuring flight altitudes and the characteristics of large-height irregularities of terrain.

Analytic expressions for the kernel of the generalized AF have been derived with allowance for the antenna's characteristics and angular orientation and the characteristics of an isotropic or anisotropic scattering surface. These expressions are valid over the entire range of the correlation factor of large-height irregularities' slopes. It has been shown that the kernel of the generalized AF can be expressed through elementary functions.

The results obtained can be used to choose the antenna's parameters, the type of modulation of the probing signal, the detection characteristics, and the measure of noise immunity for specified ranges of the statistical characteristics of a scattering surface. More general solutions that fit known results can be obtained with the use of FCFs.

The investigations have shown that the generalized complex radiophysical model and the method developed by Prof. A. Potapov[7] for formation of MMW reference DDRMs of an inhomogeneous terrain and for DDRM synthesis, respectively, are promising and highly effective.

Combination of the developed methods and fractal description of wave scattering[1-3] will undoubtedly result in the discovery of new physical laws in the wave theory. It is sure that, combined with the formalism of fractal operators, the theory of fractals and deterministic chaos applied to the problems considered above will make it possible to synthesize more adequate radiophysical and radar models that will substantially reduce discrepancies between theoretical predictions and experimental results. This study rather comprehensively covers the variety of modern problems of wave scattering that arise in theoretical fields and applications of radiophysics and radar and, generally, involves the theory of integer and fractal measures. Thus, the use of the formalism of dynamics of dissipative systems (the fractal character, fractal operators, a non-Gaussian statistic, distributions with heavy tails, the mode of deterministic chaos, the existence of strange attractors in the phase space of reflected signals, their topology, etc.) makes it possible to expect that the classical problem of wave scattering by random media will remain an area of fruitful future investigations.

The results obtained can be widely applied for designing various modern radio systems in the microwave, optical, and acoustic bands.

Chapter 2 Strong Large Deviations Principles of Non-Freidlin-Wentzell Type. Optimal Control Problem with Imperfect Information. Jumps Phenomena in Financial Markets

The chapter presents, a new large deviations principles (SLDP) of non-Freidlin-Wentzell type, corresponding to the solutions Colombeau-Ito's SDE. Using SLDP we present a new approach to construct the Bellman function $\mathbf{v}(t,x)$ and optimal control $\mathbf{u}(t,x)$ directly by way of using strong large deviations principle for the solutions Colombeau-Ito's SDE. As important application such SLDP, the generic imperfect dynamic models of air-to-surface missiles are given in addition to the related simple guidance law. A four, examples have been illustrated proposed approach and corresponding numerical simulations have been illustrated and analyzed. Using SLDP approach, Jumps phenomena, in financial markets, is also considered. Jumps phenomena, in financial markets is explained from the first principles, without any reference to Poisson jump process. In contrast with a phenomenological approach we explain such jumps phenomena from the first principles, without any reference to Poisson jump process.

2.1 Introduction

What new scalable mathematics is needed to replace the traditional Partial Differential Eqs. (PDE) approach to differential games?

Let $\mathfrak{C} = (\Omega, \Sigma, \mathbf{P})$ be a probability space. Any stochastic process on $\mathbf{R}^n$ is a Σ-measurable mapping $X: \Omega \times (0,T) \to \mathbf{R}^n$. Many stochastic optimal control problems essentially come down to constructing a function $u(t,x)$ that has the properties:

$$u(t,x) = \inf_\alpha [\bar{\mathbf{J}}(\{X^x_{s,D}(\omega)\}_{a\in[0,t]}; \{\alpha(s)\}_{a\in[0,t]})] \tag{2.1}$$

and

$$u(t,x) = \inf_\alpha [\bar{\mathbf{J}}(\{X^x_{s,D}(\omega)\}_{a\in(0,t)}; \{\alpha(s)\}_{a\in(0,t]}) + u(t, X^x_{t,D}(\omega))], \tag{2.2}$$

where $\alpha(t) \in U \nsubseteq \mathbf{R}^n$. Here $\bar{\mathbf{J}} = E_\Omega\left[\int_0^t (g(X^x_{s,D}(\omega), s))\mathrm{d}s\right]$ is the termination payoff: functional, $\alpha(t)$ is a control and $X^x_{t,D}(\omega)$ is some Markov process governed by some stochastic Ito's equation driven by a Brownian motion of the form

$$X^x_{t,D}(\omega) = x + \int_0^t f(X^x_{s,D}(\omega), \alpha(s))\mathrm{d}s + \sqrt{D}W(t,\omega). \tag{2.3}$$

Here $W(t,\omega)$ is the Brownian motion. Traditionally the function $u(t,x)$ has been computed by way of solving the associated Bellman equation, for which various numerical techniques mostly variations of the finite difference scheme have been developed. Another approach, which takes advantage of the recent developments in computing technology and allows one to construct the function $u(x,t)$ by way of backward induction governed by Bellman's principle such that described in paper [1]. In paper [1] Eq. (2.3) is approximated by an equation with affine coefficients which admits an explicit solution in terms of integrals of the exponential Brownian motion. Using Colombeau approach proposed in paper [2-4] we have replaced Eq. (2.3) by Colombeau-Ito's equation[4-6]:

$$(X^{x,\varepsilon}_{t,D,\varepsilon'}(\omega,\bar{\omega}))_{\varepsilon'} = x + \left(\int_0^t f_{\varepsilon'}(X^{x,\varepsilon}_{t,D,\varepsilon'}(\omega,\bar{\omega}),\alpha(s))\,\mathrm{d}s\right)_{\varepsilon'} +$$
$$\sqrt{D}\left(\int_0^t w_{\varepsilon'}(s,\bar{\omega})\,\mathrm{d}s\right)_{\varepsilon'} + \sqrt{\varepsilon}(W(t,\omega))_{\varepsilon'}.$$

Here $\varepsilon,\varepsilon' \in (0,1], \omega \in \Omega_1, \bar{\omega} \in \Omega_2, \Omega_1 \cap \Omega_2 = \varnothing$, where $w(t,\omega)$ is the white noise on $\mathbf{R}^n$, i. e., $w(t,\omega) = \mathrm{d}/\mathrm{d}tW(t,\omega)$ almost surely in D' and $w_{\varepsilon'}(t,\bar{\omega})$ is the smoothed white noise on $\mathbf{R}^n$, i. e., $w_{\varepsilon'}(t,\bar{\omega}) = \langle w(t,\omega), \phi_{\varepsilon'}(s-t)\rangle$, and $\phi_{\varepsilon'}$ is a model delta net[2,4]. Fortunately in contrast with Eq. (2.3) one can solve Eq. (2.4) without any approximation using strong large deviations principle of Non-Freidlin-Wentzell type[5-7].

Statement of the novelty and uniqueness of the proposed idea: a new approach, which is proposed in this paper allows one to construct the Bellman function $v(t,x)$ and optimal control $\alpha(t,x)$ directly, i. e., without any reference to the Bellman equation, by way of using strong large deviations principle for the solutions Colombeau-Ito's SDE (CISDE).

2.2 Proposed Approach

Let $\mathfrak{C}_i = (\Omega_i, \Sigma_i, \mathbf{P}_i), i = 1,2$ be a probability spaces such that: $\Omega_1 \cap \Omega_2 = \varnothing$. Let us consider m-persons Colombeau-Ito differential game $CIDG_{m;T}(\boldsymbol{f},\boldsymbol{g},\boldsymbol{y},G^n(\mathbf{R}^n),\mathfrak{C}_1,\mathfrak{C}_2)$, with the termination payoff functional for the i-th player is:

$$(\bar{\mathbf{J}}^{\varepsilon}_{\varepsilon',j})_{\varepsilon'} = \mathbf{E}_{\Omega_1}\mathbf{E}_{\Omega_2}\left[\left(\int_0^T g_{\varepsilon',i}(x^{x,\varepsilon}_{t,D,\varepsilon'}(\omega,\bar{\omega}),\boldsymbol{\alpha}(t),t,\varepsilon)\,\mathrm{d}t\right)_{\varepsilon'}\right] +$$
$$\mathbf{E}_{\Omega_1}\mathbf{E}_{\Omega_2}\left[\left(\sum_{i=1}^n [x^{x,\varepsilon}_{T,D,\varepsilon';i}(\omega,\bar{\omega}) - y_i]^2\right)_{\varepsilon'}\right] \tag{2.4}$$

and with stochastic nonlinear dynamics:

$$(\dot{\boldsymbol{x}}^{x_0,\varepsilon}_{t,D,\varepsilon'}(\omega,\bar{\omega}))_{\varepsilon'} =$$
$$(\boldsymbol{f}_{\varepsilon'}(\boldsymbol{x}^{x,\varepsilon}_{t,D,\varepsilon'}(\omega,\bar{\omega}), \sqrt{D}w_{\varepsilon'}(t,\bar{\omega}), \boldsymbol{\alpha}(t),t,\varepsilon))_{\varepsilon'} + \sqrt{\varepsilon}\,(\boldsymbol{w}(t,\omega))_{\varepsilon'}, \tag{2.5}$$
$$\varepsilon,\varepsilon' \in (0,1], \omega \in \Omega_1, \bar{\omega} \in \Omega_2.$$

Here $\forall t \in (0,T]: (x_{\varepsilon'}(t))_{\varepsilon'} \in \hat{\mathbf{R}}^n; \boldsymbol{x}^{x_0,\varepsilon}_{0,D,\varepsilon'}(\omega,\bar{\omega}) = \boldsymbol{x}_0 \in \mathbf{R}^n, \forall \varepsilon \in (0,1]: \boldsymbol{f} = [(\boldsymbol{f}_{\varepsilon'})_{\varepsilon'}], \boldsymbol{g} = [(\boldsymbol{g}_{\varepsilon'})_{\varepsilon'}]; \boldsymbol{f}(\boldsymbol{x},\circ,\circ,\circ,\circ), \boldsymbol{g}(\boldsymbol{x},\circ,\circ,\circ) \in G^n(\mathbf{R}^n), \boldsymbol{\alpha}(t) = \{\alpha_1(t),\cdots,$

$\alpha_m(t)\}; \alpha_i(t) \in U_i \not\subseteq \mathbf{R}^{k_i}, i = 1, \cdots, m,$

and m-persons Colombeau-Ito differential game

$CIDG_{m;T}(\boldsymbol{f}, \boldsymbol{g}, \boldsymbol{y}, G^n(\mathbf{R}^n), \boldsymbol{\beta}(t), \boldsymbol{\varphi}(t), \mathfrak{C}_1, \mathfrak{C}_2)$ with imperfect measurements and with imperfect information about the system[5,6]. The corresponding stochastic nonlinear dynamics is:

$$(\dot{\boldsymbol{x}}^{x_0,\varepsilon}_{t,D,\varepsilon'}(\omega,\bar{\omega}))_{\varepsilon'} = (\boldsymbol{f}_{\varepsilon'}(\boldsymbol{x}^{x_0,\varepsilon}_{t,D,\varepsilon'}(\omega,\bar{\omega}), \sqrt{D}w_{\varepsilon'}(t,\bar{\omega}), \boldsymbol{\varphi}(t), \alpha(t, x^{x_0,\varepsilon}_{t,D,\varepsilon'} + \boldsymbol{\beta}(t)), t, \varepsilon))_{\varepsilon'} +$$
$$\sqrt{\varepsilon}\,(w(t,\omega))_{\varepsilon'};\ \varepsilon, \varepsilon' \in (0,1], \omega \in \Omega_1, \bar{\omega} \in \Omega_2$$

and the playoff for the i-th player is:

$$(\bar{\mathbf{J}}^{\varepsilon}_{\varepsilon',j})_{\varepsilon'} = \mathbf{E}_{\Omega_1}\mathbf{E}_{\Omega_2}\left[\left(\int_0^T g_{\varepsilon',i}(\boldsymbol{x}^{x_0,\varepsilon}_{t,D,\varepsilon'}(\omega,\bar{\omega}), \alpha(t,\boldsymbol{\beta}(t)), t, \varepsilon)\,\mathrm{d}t\right)_{\varepsilon'}\right] +$$
$$\mathbf{E}_{\Omega_1}\mathbf{E}_{\Omega_2}\left[\left(\sum_{i=1}^{n}\left[x^{x_0,\varepsilon}_{T,D,\varepsilon';i}(\omega,\bar{\omega}) - y_i\right]^2\right)_{\varepsilon'}\right]. \tag{2.6}$$

Here

$\boldsymbol{\beta}(t) = (\beta_1(t), \cdots, \beta_n(t)), \boldsymbol{\varphi}(t) = (\varphi_1(t), \cdots, \varphi_n(t))$ and $\forall t \in [0,T]: (x_{\varepsilon'}(t))_{\varepsilon'} \in \hat{\mathbf{R}}^n;\ x^{x_0,\varepsilon}_{0,D,\varepsilon'}(\omega,\bar{\omega}) = x_0, \forall \boldsymbol{\varepsilon} \in (0,1]$:

$\boldsymbol{f} = (\boldsymbol{f}_{\varepsilon'})_{\varepsilon'}, \boldsymbol{g} = (\boldsymbol{g}_{\varepsilon'})_{\varepsilon'}; \boldsymbol{f}(x, \circ, \circ, \circ, \circ, \circ), \boldsymbol{g}(x, \circ, \circ, \circ) \in G^n(\mathbf{R}^n)$

Or

$\boldsymbol{f}(x, \circ, \circ, \circ, \circ), \boldsymbol{g}(x, \circ, \circ) \in G^n_{P,r}(E), \alpha(t) = \{\alpha_1(t), \cdots\cdots, \alpha_m(t)\}; \alpha_i(t) \in U \not\subseteq \mathbf{R}^k_i, i = 1, \cdots\cdots, m; \boldsymbol{\beta}(t) = \{\beta_1(t), \cdots\cdots, \beta_n(t)\}, \boldsymbol{\varphi}(t) = \{\varphi_1(t), \cdots\cdots, \varphi_n(t)\}$.

Here $\mathbf{R}$ is a field of the real numbers, $G(\mathbf{R}^n)$ is the algebra of Colombeau generalized functions[8,9,12], $G^n(\mathbf{R}^n) = G(\mathbf{R}^n) \times \cdots \times G(\mathbf{R}^n), G_{P,r}(E) = \dfrac{\mathcal{F}_{P,r}(E)}{K_{P,r}(E)}$ is the Colombeau type algebra[13,14], E is an appropriate algebra of functions, which is a locally convex vector space over field $\mathbb{C}$, $G^n_{P,r}(E) = G_{P,r}(E) \times \cdots \times G_{P,r}(E)$, $\hat{\hat{\mathbf{R}}}$ is the ring of Colombeau generalized numbers[11], $\hat{\mathbf{R}}^n = \hat{\mathbf{R}} \times \cdots \times \hat{\mathbf{R}}$, $t \to \alpha_i(t)$ is the control chosen by the i-th player, within a set of admissible control values U_i.

Here $t \longmapsto \left\{\left(x^{x_0,\varepsilon}_{t,D,\varepsilon';1}(\omega,\bar{\omega})\right)_{\varepsilon'}, \cdots\cdots, \left(x^{x_0,\varepsilon}_{t,D,\varepsilon';n}(\omega,\bar{\omega})\right)_{\varepsilon'}\right\}$ is the trajectory of the Eq. (2.5). Optimal control problem for the i-th player is:

$$(\bar{\bar{\mathbf{J}}}^{\varepsilon}_{\varepsilon',i})_{\varepsilon'} = \left(\min_{\alpha_i(t) \in U_i}\left(\max_{\alpha_j(t) \in U_j}\mathbf{J}^{\varepsilon}_{\varepsilon',j\neq i}\right)\right)_{\varepsilon'}. \tag{2.7}$$

We remind now some classical definitions.

Let us consider now Ito's SDE:

$$\mathrm{d}\boldsymbol{x}_t = \boldsymbol{b}(\boldsymbol{x}_t, t)\,\mathrm{d}t + \sum_{r=1}^{k}\sigma_r(\boldsymbol{x}_t, t)\,\mathrm{d}W_r(t,\omega), \tag{2.8}$$

$\boldsymbol{x}_0 = \boldsymbol{x}_0(\omega), \boldsymbol{x} \in \mathbf{R}^n$.

Theorem 1. [15,16] Let the vectors $\boldsymbol{b}(\boldsymbol{x},t), \boldsymbol{\sigma}(\boldsymbol{x},t)$ be continuous functions of $(\boldsymbol{x},t)$ such that for some constants D and C the following conditions hold:

$$\| \boldsymbol{b}(\boldsymbol{x},t) - \boldsymbol{b}(\boldsymbol{y},t) \| + \sum_{r=1}^{k} |\sigma_r(\boldsymbol{x},t) - \sigma_r(\boldsymbol{y},t)| \leqslant D \| \boldsymbol{x} - \boldsymbol{y} \| , \tag{2.9}$$

$$\| \boldsymbol{b}(\boldsymbol{x},t) \| + \sum_{r=1}^{k} |\sigma_r(\boldsymbol{x},t)| \leqslant C(1 + \| \boldsymbol{x} \|). \tag{2.10}$$

Then: ① For every random variable $\boldsymbol{x}(\omega)$ independent of the processes $\boldsymbol{W}_r(t,\omega), r = 1,2, \cdots\cdots, k$ there exists a solution $\boldsymbol{x}_t$ of the Ito's SDE (5) which is an almost surely continuous Markov process and ②Two solutions $\boldsymbol{x}_{t,1}$ and $\boldsymbol{x}_{t,2}(\omega)$ is unique up to equivalence: $\mathbf{P}[\boldsymbol{x}_{t,1}(\omega) = \boldsymbol{x}_{t,2}(\omega)] = 1$, for all $t \in [0,\infty) = I_\infty$.

Remark 1. [15,17] It is well known that the boundedness assumption on $\boldsymbol{b}(\boldsymbol{x},t)$ and $\boldsymbol{\sigma}(\boldsymbol{x},t)$ can be weakened, but somekind of restriction on the $\boldsymbol{b}(\boldsymbol{x},t)$ and $\boldsymbol{\sigma}(\boldsymbol{x},t)$ is necessary in order to guarantee the existence of a global solution, i. e. , a solution defined for all $t \in [0,\infty)$. If we remove this condition of boundedness, then a solution of Ito's SDE (5) does exist locally but, in general, blows up (or explodes) in finite time.

Definition 1. Let $\dot{\mathcal{R}}^n = \mathcal{R}^n \cup \{\Delta\}$ be the one-point compactification of $\mathbf{R}^n$ and $\dot{\boldsymbol{W}}^n = \{\boldsymbol{w} \mid [0,\infty) \ni t \longmapsto w(t) \in \dot{\mathbf{R}}^n$ is continuous and such that if $\boldsymbol{w}(t) = \Delta$, then $\boldsymbol{w}(t') = \Delta$ for all $t' \geqslant t\}$. Let $\mathfrak{H}(\mathbf{R}^n)$ be the σ-field generated by Borel cylinder sets. For $\boldsymbol{w} \in \dot{\boldsymbol{W}}^n$ we set

$$e(\boldsymbol{w}) = \inf\{t \mid \boldsymbol{w}(t) = \Delta\} \tag{2.11}$$

and call the explosion time of the trajectory $\boldsymbol{w}(t)$, $t \in [0,\infty)$.

Definition 2. [15] By a solution $\boldsymbol{x}_t(\omega)$ of the Eq. (2.8) we mean a $(\dot{\boldsymbol{W}}^n, \mathfrak{H}(\dot{\mathbf{R}}^n))$-valued random variable defined on a probability space $\boldsymbol{C} = (\boldsymbol{\Omega},\boldsymbol{\Sigma},\boldsymbol{P})$ with a reference family $(\Sigma_t)_{t\geqslant 0}$ such that:

①there exists an n-dimensional (Σ_t)-Brownian motion $\boldsymbol{W}(t,\omega) = (W_1(t,\omega),\ldots,W_n(t,\omega))$ with $\boldsymbol{W}(0,\omega) = 0$,

②for each $(t,\omega) \longmapsto x_t(\omega) \in \dot{\mathbf{R}}^n$ is Σ_t-measurable and

③if $e(\omega) = e(\boldsymbol{x}_t(\omega))$ is the explosion time of $\boldsymbol{x}_t(\omega)$ then for almost all ω,

$$④\ \boldsymbol{x}_t(\omega) - \boldsymbol{x}_0(\omega) = \int_0^t \boldsymbol{b}(\boldsymbol{x}_s(\omega),t)\mathrm{d}s + \sum_{r=1}^{k} \int_0^t \boldsymbol{\sigma}_r(\boldsymbol{x}_s(\omega),t)\mathrm{d}W_r(s,\omega), \tag{2.12}$$

for all $t \in [0,e(\omega))$.

Theorem 2. [15,16] ① Given $\mathbf{R}^n$ – continuous $\boldsymbol{b}(\boldsymbol{x},t)$ and $\boldsymbol{\sigma}(\boldsymbol{x},t)$ consider the equation (2.8). Then for any probability μ on $(\dot{\boldsymbol{W}}^n, \mathfrak{H}(\dot{\mathbf{R}}^n))$ with compact support, there exists a solution of (2.8) such that the law of $x_0(\omega)$ coincides with μ.

② Suppose $\boldsymbol{b}(\boldsymbol{x},t)$ and $\boldsymbol{\sigma}(\boldsymbol{x},t)$ are locally Lipschitz continuous, i. e. , for every $N > 0$ there exists a constant $D_N > 0$ such that

$$\| \boldsymbol{b}_n(\boldsymbol{x},t) - \boldsymbol{b}_n(\boldsymbol{y},t) \| + \sum_{r=1}^{k} |\boldsymbol{\sigma}_{r,n}(\boldsymbol{x},t) - \boldsymbol{\sigma}_{r,n}(\boldsymbol{y},t)| \leqslant D_N \| \boldsymbol{x} - \boldsymbol{y} \| \tag{2.13}$$

for every $x,y \in \boldsymbol{B}_N$, $\boldsymbol{B}_N = \{z \mid \| z \| \leqslant N\}$. Then for any probability μ on $(\dot{\boldsymbol{W}}^n, \mathfrak{H}(\dot{\mathbf{R}}^n))$ with compact support, there exists a solution of (2.8) such that the law of $\boldsymbol{x}_0(\omega)$ coincides with μ.

Theorem 3. [16] Let $\boldsymbol{x}_{t,n}(t), n = 1,2,\ldots$ be the solutions of the Ito's SDE's

$$d\boldsymbol{x}_{t,n} = \boldsymbol{b}_n(\boldsymbol{x}_{t,n},t)dt + \sum_{r=1}^{k} \sigma_{r,n}(\boldsymbol{x}_{t,n},t)dW_r(t,\omega), \tag{2.14}$$

$$\boldsymbol{x}_{0,n} = x(\omega), x \in \mathbf{R}^n.$$

Assume that: ① let the vectors $\boldsymbol{b}_n(\boldsymbol{x},t), \boldsymbol{\sigma}_n(\boldsymbol{x},t)$ be continuous functions of $(\boldsymbol{x},t)$ such that for some constants D and C the following conditions hold

$$\| \boldsymbol{b}_n(\boldsymbol{x},t) - \boldsymbol{b}_n(\boldsymbol{y},t) \| + \sum_{r=1}^{k} | \boldsymbol{\sigma}_{r,n}(\boldsymbol{x},t) - \boldsymbol{\sigma}_{r,n}(\boldsymbol{y},t) | \leqslant D \| \boldsymbol{x} - \boldsymbol{y} \|, \tag{2.15}$$

$$\| \boldsymbol{b}_n(\boldsymbol{x},t) \| + \sum_{r=1}^{k} | \boldsymbol{\sigma}_{r,n}(\boldsymbol{x},t) | \leqslant C(1 + \| \boldsymbol{x} \|), \tag{2.16}$$

② $\mathbf{E}[x^2(\omega)] < \infty$, (2.17)

③ $\forall N > 0$:

$$\lim_{n\to\infty} \sup_{\|x\| \leqslant N} [\, \| \boldsymbol{b}_n(\boldsymbol{x},t) - \boldsymbol{b}_0(\boldsymbol{y},t) \| + \sum_{r=1}^{k} | \boldsymbol{\sigma}_{r,n}(\boldsymbol{x},t) - \boldsymbol{\sigma}_{r,0}(\boldsymbol{y},t) |] = 0. \tag{2.18}$$

Then

$$\lim_{n\to\infty} \sup_{0\leqslant t\leqslant T} \mathbf{E}\, [\boldsymbol{x}_{t,n}(\omega) - \boldsymbol{x}_{t,0}(\omega)]^2 = 0. \tag{2.19}$$

Corollary 1. Let $\boldsymbol{x}_{t,n}(t), n = 1,2,...$ be the solutions of the Ito's SDE's

$$d\boldsymbol{x}_{t,n} = \boldsymbol{b}_n(\boldsymbol{x}_{t,n},t)dt + \sum_{r=1}^{k} \boldsymbol{\sigma}_{r,n} dW_r(t,\omega)\boldsymbol{x}_{0,n} = \boldsymbol{x}(\omega), x \in \mathbf{R}^n. \tag{2.20}$$

Assume that: ① Let the vectors $\boldsymbol{b}_n(\boldsymbol{x},t)$, be continuous functions of $(\boldsymbol{x},t)$ and $\sigma_n = \text{const}$ such that for some constants D and C the following conditions hold

$$\| \boldsymbol{b}_n(\boldsymbol{x},t) - \boldsymbol{b}_n(\boldsymbol{y},t) \| \leqslant D \| \boldsymbol{x} - \boldsymbol{y} \| \tag{2.21}$$

$$\| \boldsymbol{b}_n(\boldsymbol{x},t) \| + \sum_{r=1}^{k} | \sigma_{r,n} | \leqslant C(1 + \| \boldsymbol{x} \|), \tag{2.22}$$

② $\mathbf{E}[\boldsymbol{x}^2(\omega)] < \infty$, (2.23)

③ $\forall N > 0$:

$$\lim_{n\to\infty} \sup_{\|x\| \leqslant N} [\, \| \boldsymbol{b}_n(\boldsymbol{x},t) - \boldsymbol{b}_0(\boldsymbol{y},t) \| + \sum_{r=1}^{k} | \sigma_{r,n} |] = 0. \tag{2.24}$$

Then

$$\lim_{n\to\infty} \sup_{0\leqslant t\leqslant T} \mathbf{E}\, [\boldsymbol{x}_{t,n}(\omega) - \boldsymbol{x}_{t,0}(\omega)]^2 = 0. \tag{2.25}$$

Here $\boldsymbol{x}_{t,0}(\omega)$ is the solution of the ODE:

$$d\boldsymbol{x}_{t,0} = \boldsymbol{b}_0(\boldsymbol{x}_{t,0},t)dt,\ \boldsymbol{x}_{0,0} = \boldsymbol{x}(\omega), x \in \mathbf{R}^n. \tag{2.26}$$

Remark 2. Note that Theorem 3 in fact asserts that under conditions (2.15) – (2.18) any solution $\boldsymbol{x}_t(\omega)$ of the Ito's SDE (2.8) is continuously depend on functions $\boldsymbol{b}(\boldsymbol{x},t)$ and $\boldsymbol{\sigma}(\boldsymbol{x},t)$.

Note that the assumptions of the Lipschitz continuously (2.15) and boundedness (2.16) on $\boldsymbol{b}(\boldsymbol{x},t)$ and $\boldsymbol{\sigma}(\boldsymbol{x},t)$ in the Theorem 3 cannot be weakened.

Theorem. Assume that: ①Let $\boldsymbol{x}_{t,n}(t), n = 1,2,...$ be the solutions of the Ito's SDE's

$$\boldsymbol{x}_{t,n} = \boldsymbol{b}_n(\boldsymbol{x}_{t,n},t)dt + \boldsymbol{\sigma}_n(\boldsymbol{x}_{t,n},t)dW(t,\omega),$$

$\boldsymbol{x}_{0,n} = x(\omega), x \in \mathbf{R}^n$.

And let $\tilde{\boldsymbol{x}}_{t,n}(t), n = 1,2,\ldots$ be the solutions of the Ito's SDE's

$\tilde{\boldsymbol{x}}_{t,n} = \tilde{\boldsymbol{b}}_n(\tilde{\boldsymbol{x}}_{t,n}, t)\mathrm{d}t + \tilde{\boldsymbol{\sigma}}_n(\tilde{\boldsymbol{x}}_{t,n}, t)\mathrm{d}\boldsymbol{W}(t,\omega)$,

$\tilde{\boldsymbol{x}}_{0,n} = \boldsymbol{x}(\omega), x \in \mathbf{R}^n$.

Here

$$\boldsymbol{\sigma}_n(\boldsymbol{x}_{t,n}, t)\mathrm{d}\boldsymbol{W}(t,\omega) = \sum_{r=1}^{k} \boldsymbol{\sigma}_{r,n}(\boldsymbol{x}_{t,n}, t)\mathrm{d}W_r(t,\omega),$$

$$\tilde{\boldsymbol{\sigma}}_n(\tilde{\boldsymbol{x}}_{t,n}, t)\mathrm{d}\boldsymbol{W}(t,\omega) = \sum_{r=1}^{k} \tilde{\boldsymbol{\sigma}}_{r,n}(\boldsymbol{x}_{t,n}, t)\mathrm{d}W_r(t,\omega).$$

② The inequalities

$\|\boldsymbol{b}_n(\boldsymbol{x},t)\| + \|\boldsymbol{\sigma}_n(\boldsymbol{x},t)\| \leqslant K_n(1 + \|\boldsymbol{x}\|)$,

$\|\boldsymbol{b}_n(\boldsymbol{x},t) - \boldsymbol{b}_n(\boldsymbol{y},t)\| + \|\boldsymbol{\sigma}_n(\boldsymbol{x},t) - \boldsymbol{\sigma}_n(\boldsymbol{y},t)\| \leqslant K_n\|\boldsymbol{x} - \boldsymbol{y}\|$,

$\|\tilde{\boldsymbol{b}}_n(\boldsymbol{x},t)\| + \|\tilde{\boldsymbol{\sigma}}_n(\boldsymbol{x},t)\| \leqslant K_n(1 + \|\boldsymbol{x}\|)$,

$\|\tilde{\boldsymbol{b}}_n(\boldsymbol{x},t) - \tilde{\boldsymbol{b}}_n(\boldsymbol{y},t)\| + \|\tilde{\boldsymbol{\sigma}}_n(\boldsymbol{x},t) - \tilde{\boldsymbol{\sigma}}_n(\boldsymbol{y},t)\| \leqslant K_n\|\boldsymbol{x} - \boldsymbol{y}\|$,

$\|\boldsymbol{b}_n(\boldsymbol{x},t) - \tilde{\boldsymbol{b}}_n(\boldsymbol{x},t)\| \leqslant \delta_{1,n}\|\boldsymbol{x}\|$,

$\|\tilde{\boldsymbol{\sigma}}_n(\boldsymbol{x},t) - \tilde{\boldsymbol{\sigma}}_n(\boldsymbol{x},t)\| \leqslant \delta_{2,n}\|\boldsymbol{x}\|$

where $0 \leqslant t \leqslant T$, is satisfied. Then the inequality

$$\sup_{0\leqslant t\leqslant T}\mathbf{E}[\|\boldsymbol{x}_{t,n} - \tilde{\boldsymbol{x}}_{t,n}\|^2] \leqslant \mathrm{e}^{L_n}(\delta_{1,n}^2 + \delta_{2,n}^2)E\left[\int_0^T \|\tilde{\boldsymbol{x}}_{t,n}\|^2 \mathrm{d}t\right]$$

is satisfied.

Proof. See Appendix A.

Remark 3. [17] If conditions (2.9), (2.10) are valid only in every cylinder $U_R \times I_\infty$, with $C = C(R), D = D(R)$, one can construct a sequence of functions $\boldsymbol{b}_n(\boldsymbol{x},t)$ and $\boldsymbol{\sigma}_n(\boldsymbol{x},t)$ such that for $\|\boldsymbol{x}\| < n$

$$\boldsymbol{b}_n(\boldsymbol{x},t) = \boldsymbol{b}(\boldsymbol{x},t), \boldsymbol{\sigma}_n(\boldsymbol{x},t) = \boldsymbol{\sigma}(\boldsymbol{x},t), \tag{2.27}$$

and therefore for each $\boldsymbol{b}_n(\boldsymbol{x},t), \boldsymbol{\sigma}_n(\boldsymbol{x},t)$ satisfy conditions (2.9), (2.10) everywhere in $\mathbf{R}^n$.

By Theorem 1, there exists a sequence of Markov processes $\boldsymbol{x}_{t,n}(\omega)$ corresponding to the functions $\boldsymbol{b}_n(\boldsymbol{x},t)$ and $\boldsymbol{\sigma}_n(\boldsymbol{x},t)$.

Assumption 1. Suppose now that the distribution of $\boldsymbol{x}_0(\omega)$ has compact support in $\mathbf{R}^n$. Then as, well known, that the first exit random times $\boldsymbol{\tau}_m(\omega)$ of the processes $\boldsymbol{x}_{t,m}(\omega)$ from the set $\|\boldsymbol{x}\| < n$ are identical for $m \geqslant n$ [15,18,19]. Let this common value be $\boldsymbol{\tau}_n(\omega)$. It is also clear that the processes themselves coincide up to time $\boldsymbol{\tau}_n(\omega)$, i. e.

$$\mathbf{P}[\sup_{0\leqslant\tau(\omega)\leqslant\tau_n(\omega)} \|\boldsymbol{x}_{\tau(\omega),n}(\omega) - \boldsymbol{x}_{\tau(\omega),m}(\omega)\| > 0] = 0, m > n. \tag{2.28}$$

Or in the equivalent form

$$\mathbf{P}[\sup_{0\leqslant t\leqslant\tau_n(\omega)} \|\boldsymbol{x}_{t,n}(\omega) - \boldsymbol{x}_{t,m}(\omega)\| > 0] = 0, m > n. \tag{2.29}$$

Definition 1. ①Let $\boldsymbol{\tau}_\infty(\omega)$ denote the (finite or infinite) limit of the monotone increasing sequence $\tau_n(\omega)$ as $n \to \infty$. We call the random variable $\tau_\infty(\omega)$ the first exit time from every bounded domain, or briefly the explosion time.

②We now define a new stochastic process $x_t(\omega)$ by setting[17]:

$$x_t(\omega) = x_{t,n}(\omega) \text{ for } t < \tau_n(\omega). \tag{2.30}$$

It well known that this is always a Markov process for $t < \tau_n(\omega)$[18,19].

We also can define a new stochastic process $x_t(\omega)$ by setting

$$x_t(\omega) = \mathbf{P} - \lim_{n\to\infty} x_{t,n}(\omega) \tag{2.31}$$

If finite or infinite limit in RHS of Eq. (2.31) exist.

③In general case we set

$$(x_{t,\varepsilon'}(\omega))_{\varepsilon'} = (x_{t,n}(\omega))_n, n = \frac{1}{\varepsilon'}. \tag{2.32}$$

We note that the Colombeau-Ito's equation

$$(x_{t,\varepsilon'}(\omega))_{\varepsilon'} - (x_{0,\varepsilon}(\omega))_{\varepsilon'} = \left(\int_0^t b(x_{s,\varepsilon'}(\omega), s)\,\mathrm{d}s\right)_{\varepsilon'} + \sum_{r=1}^{k}\left(\int_0^t \sigma_r(x_{s,\varepsilon'}(\omega), t)\,\mathrm{d}W_r(s,\omega)\right)_{\varepsilon'} \tag{2.33}$$

is satisfied for all $t \in [0, \tau_\infty(\omega))$.

④Markov process $x_t(\omega)$ is regular if for all $s < \infty$, $x \in \mathbf{R}^n$,

$$\mathbf{P}^{s,x}\{\tau_\infty(\omega) = \infty\} = 1. \tag{2.34}$$

Assumption 2. We assume now that: ① $\forall \epsilon \in (0,1]$: $(b_{\varepsilon'}(x,t,\epsilon))_{\varepsilon'} \triangleq (b_{1,\varepsilon'}(x,t,\epsilon),\ldots,b_{n,\varepsilon'}(x,t,\epsilon))_{\varepsilon'} \in G^n(\mathbf{R}^n)$ (or $G^n_{P,r}(E)$), $\epsilon = (\epsilon_1,\ldots,\epsilon_n)$, $\epsilon \in (0,1]$: n for all $t \in [0,\infty)$ and ② $\forall \epsilon \in (0,1]^n$ there exist infinite Colombeau constants $(C^\varepsilon_{\varepsilon'})_{\varepsilon'}$ and $(D^\varepsilon_{\varepsilon'})_{\varepsilon'}$ such that $\forall \epsilon \in (0,1]$:

$$\text{(i)}\ (\|b_{\varepsilon'}(x,t,\epsilon)\|^2)_{\varepsilon'} \leqslant ((C^\varepsilon_{\varepsilon'})_{\varepsilon'})(1 + \|x\|^2), \varepsilon' \in (0,1], \tag{2.35}$$

$$\text{(ii)}\ (\|b_{i,\varepsilon'}(x,t,\epsilon) - b_{i,\varepsilon'}(y,t,\epsilon)\|)_{\varepsilon'_i} \leqslant ((D^\varepsilon_{\varepsilon'})_{\varepsilon'})\|x - y\| \tag{2.36}$$

for all $t \in [0,\infty)$ and for all $x \in \mathbf{R}^n$ and for all $y \in \mathbf{R}^n$.

Definition 2.[4] (1) Let $C = (\Omega, \Sigma, \mathbf{P})$ be a probability space. Let εR be the space of nets $(X_\varepsilon(\omega))_\varepsilon$ of measurable functions on Ω.

Let εR_M be the space of nets $(X_\varepsilon)_\varepsilon \in \varepsilon R$, $\varepsilon \in (0,1]$, with the property that for almost all $\omega \in \Omega$ there exist constants $r, C > 0$.

And $\varepsilon_0 \in (0,1]$ such that $|(X_\varepsilon)_\varepsilon| \leqslant C\varepsilon^{-r}$, $\varepsilon \leqslant \varepsilon_0$.

(2) Let NR is the space of nets $|X_\varepsilon|_\varepsilon \in \varepsilon R$, $\varepsilon \in (0,1]$, with the property that for almost all $\omega \in \Omega$ and all $b \in \mathbf{R}_+$ there exist constants $C > 0$ and $\varepsilon_0 \in (0,1]$ such that $|(X_\varepsilon)_\varepsilon| \leqslant C\varepsilon^b$, $\varepsilon \leqslant \varepsilon_0$. The differential algebra GR of Colombeau generalized random variables is the factor algebra GR $= \varepsilon R/\text{NR}$.

Let us consider now a family $(x^{x_0,\varepsilon}_{t,\epsilon,\varepsilon'})_{\varepsilon'}$ of the solutions Colombeau-Ito's SDE:

$$(\mathrm{d}x^{x_0,\varepsilon}_{t,\epsilon,\varepsilon'}(\omega))_{\varepsilon'} = (b_{\varepsilon'}(x^{x_0,\varepsilon}_{t,\epsilon,\varepsilon'}(\omega), t, \epsilon))_{\varepsilon'} + \sqrt{\varepsilon}(\mathrm{d}W(t,\omega))_{\varepsilon'}, \tag{2.37}$$

$$(\boldsymbol{x}_{0,,\epsilon,\varepsilon'}^{x_0,\varepsilon})_{\varepsilon'} = (\boldsymbol{x}_{\varepsilon'}^{x_0}(\omega))_{\varepsilon'}, \in GR, (\mathbf{E}[\boldsymbol{x}_{0,\epsilon,\varepsilon'}^{x_0,\varepsilon}])_{\varepsilon'} = x_0 \in \hat{\mathbf{R}}^n, \tag{2.38}$$

$t \in [0,T], \varepsilon, \varepsilon' \in (0,1]$.

Here ① $\boldsymbol{W}(t,\omega) = (W_1(t,\omega),\ldots,W_n(t,\omega))$ is n-dimensional Brownian motion,

② $\forall t \in [0,T] : (\boldsymbol{b}_{\varepsilon'}(\boldsymbol{x},t,\epsilon))_{\varepsilon'} \in G^n(\mathbf{R}^n)$ (or $G^n_{P,r}(E)$), $\boldsymbol{b}_0(\boldsymbol{x},t,\epsilon) \equiv \boldsymbol{b}_{\varepsilon'=0}(\boldsymbol{x},t,\epsilon)$: $\mathbf{R}^n \to \mathbf{R}^n$ is a polynomial on variable $\boldsymbol{x} = (x_1,\cdots,x_n)$, i. e.

$$b_{0,i}(\boldsymbol{x},t,\epsilon) = \sum_{\alpha,\,|\alpha|\leqslant r} b_{0,i}^{\alpha}(t,\epsilon)x^{\alpha}, \alpha = (i_1,\cdots,i_n), |\alpha| = \sum_{j=1}^{n} i_j, 0 \leqslant i_j \leqslant p, \tag{2.39}$$

or

③ $\forall t \in [0,T] : (\boldsymbol{b}_{\varepsilon'}(\boldsymbol{x},t,\epsilon))_{\varepsilon'} \in G^n_{P,r}(E), \boldsymbol{b}_0(\boldsymbol{x},t,\epsilon) \equiv \boldsymbol{b}_{\varepsilon'=0}(\boldsymbol{x},t,\epsilon) : \mathbf{R}^n \to \mathbf{R}^n$ is $\mathbf{R}$-analytic function on variable $\boldsymbol{x} = (x_1,\cdots,x_n)$, i. e.

$$b_{0,i}(\boldsymbol{x},t,\epsilon) = \sum_{r=1}^{\infty}\sum_{\alpha,\,|\alpha|\leqslant r} b_{0,i}^{\alpha}(t,\epsilon)x^{\alpha}, \alpha = (i_1,\cdots,i_n), |\alpha| = \sum_{j=1}^{n} i_j, \tag{2.40}$$

$0 \leqslant i_j \leqslant p$ and

④ $\lim_{\|x\|\to\infty} \|\boldsymbol{b}_0(\boldsymbol{x},t,\epsilon)\| / \|\boldsymbol{x}\| = \infty$,

⑤ $b_{i,\varepsilon'}(\boldsymbol{x}(t),t,\epsilon) = b_{i,0}(x_{\varepsilon'}(t),t,\epsilon)$. (2.41)

Here $\boldsymbol{x}_{\varepsilon'}(t) = (x_{1,\varepsilon'}(t),\cdots,x_{n,\varepsilon'}(t))$ and

$$x_{i,\varepsilon'}(t) = \begin{cases} x_{i,\varepsilon'}(t) = \dfrac{x_i(t)}{1+(\varepsilon')^{2l}x_i^{2l}(t)}, l \geqslant 1 \\ \text{or} \\ x_{i,\varepsilon'}(t) = x_i(t)\theta_{\epsilon_i}[x_i(t)]. \end{cases} \quad i = 1,\cdots,n. \tag{2.42}$$

Here $\theta_{\epsilon_i}[z] \in C^{\infty}(\mathbf{R})$, supp $(\theta_{\epsilon_i}[z]) \subseteq [-\nu(\epsilon_i),\nu(\epsilon_i)]$

$$\begin{cases} \theta_{\epsilon_i}[z] = 1 \Leftrightarrow z \in (-\nu_1(\epsilon_i),\nu_1(\epsilon_i)] \subsetneqq [-\nu(\epsilon_i),\nu(\epsilon_i)], \\ \theta_{\epsilon_i}[z] = 0 \Leftrightarrow z \in \mathbf{R} \setminus [-\nu(\epsilon_i),\nu(\epsilon_i)], \\ 0 \leqslant \theta_{\epsilon_i}[z] \leqslant 1 \Leftrightarrow z \in [-\nu(\epsilon_i),\nu(\epsilon_i)] \setminus [-\nu_1(\epsilon_i),\nu_1(\epsilon_i)]. \end{cases}$$

Remark 5. By Theorem 1 for every Colombeau generalized random variable $(\boldsymbol{x}_{\varepsilon'}^{x_0}(\omega))_{\varepsilon'}$, $\in GR$ such that

$(\mathbf{E}[\boldsymbol{x}_{0,\varepsilon'}^{x_0,\varepsilon}])_{\varepsilon'} = \boldsymbol{x}_0 \in \hat{\mathbf{R}}^n$, and independent of the processes $W_1(t,\omega),\cdots,W_n(t,\omega)$ there exist Colombeau generalized stochastic process

$(\boldsymbol{x}_{t,\epsilon,\varepsilon'}^{x_0,\varepsilon}(\omega))_{\varepsilon'}, \varepsilon' \in (0,1]$, such that $(\boldsymbol{x}_{0,\epsilon,\varepsilon'}^{x_0,\varepsilon}(\omega))_{\varepsilon'} = (\boldsymbol{x}_{\varepsilon'}^{x_0}(\omega))_{\varepsilon'}$, and $(\boldsymbol{x}_{t,,\epsilon,\varepsilon'}^{x_0,\varepsilon}(\omega))_{\varepsilon'}$ is the solution of the Colombeau-Ito's SDE (2.37), (2.38), which is an almost surely continuous Colombeau generalized stochastic process and is unique up to equivalence

$(\mathbf{P}[\|\boldsymbol{x}_{t,\epsilon,\varepsilon',1}^{x_0,\varepsilon}(\omega) - \boldsymbol{x}_{t,\epsilon,\varepsilon',2}^{x_0,\varepsilon}(\omega)\| > 0])_{\varepsilon'} = 0$, for all $t \in [0,\infty)$.

Remark 6. One can construct a sequence of Colombeau generalized functions $(\boldsymbol{b}_{\varepsilon',n}(\boldsymbol{x},t,\epsilon))_{\varepsilon'}$ such that for $\|x\| < n$:

$$\boldsymbol{b}_{\varepsilon',n}(\boldsymbol{x},t,\boldsymbol{\epsilon}) = \boldsymbol{b}_{\varepsilon'}(\boldsymbol{x},t,\boldsymbol{\epsilon}),\varepsilon' \in (0,1],\boldsymbol{\epsilon} \in (0,1]^n,$$

and therefore for each $\boldsymbol{b}_{\varepsilon',n}(\boldsymbol{x},t,\boldsymbol{\epsilon})$, satisfy conditions (2.21), (2.22) everywhere in $\mathbf{R}^n$. By Theorem 1, there exists a sequence of Colombeau generalized stochastic processes $(\boldsymbol{x}^{x_0,\varepsilon}_{t,\varepsilon',n}(\omega))_{\varepsilon'}$ corresponding to Colombeau generalized functions $(\boldsymbol{b}_{\varepsilon',n}(\boldsymbol{x},t,\boldsymbol{\epsilon}))_{\varepsilon'}$. Suppose now that for each $\varepsilon' \in (0,1],\boldsymbol{\epsilon} \in (0,1]^n$ the distribution of $\boldsymbol{x}^{x_0}_{0,\varepsilon'}(\omega)$ has compact support in $\mathbf{R}^n$. Then there exit times of the processes $\boldsymbol{x}^{x_0,\varepsilon}_{t,\epsilon,\varepsilon',m}(\omega),\varepsilon',\varepsilon \in (0,1]$, from the set $\|x\| < n$ are identical for $m \geqslant n$. Let this common value be $\boldsymbol{\tau}_{\varepsilon',n}(\omega,\boldsymbol{\epsilon})$. It is also clear that the processes $(\boldsymbol{x}^{x_0,\varepsilon}_{t,\epsilon,\varepsilon',n}(\omega))_{\varepsilon'}$ and $(\boldsymbol{x}^{x_0,\varepsilon}_{t,\epsilon,\varepsilon',m}(\omega))_{\varepsilon'}$ themselves coincide up to time $(\boldsymbol{\tau}_{\varepsilon',n}(\omega,\boldsymbol{\epsilon}))_{\varepsilon'}$ i. e.,

$$(\mathbf{P}[\sup_{0\leqslant t\leqslant \tau_{\varepsilon',n}(\omega,\epsilon)} \|\boldsymbol{x}^{x_0,\varepsilon}_{t,\epsilon,\varepsilon',m}(\omega) - \boldsymbol{x}^{x_0,\varepsilon}_{t,\epsilon,\varepsilon',n}(\omega)\| > 0])_{\varepsilon'} = 0, \text{ for all } m \geqslant n. \tag{2.43}$$

Definition 3. ① Let $\boldsymbol{\tau}_{\varepsilon'}(\omega,\varepsilon,\boldsymbol{\epsilon}),\varepsilon',\boldsymbol{\epsilon} \in (0,1]^n$ denote the (finite or infinite) limit of the monotone increasing sequence $\boldsymbol{\tau}_{\varepsilon',n}(\omega,\varepsilon,\boldsymbol{\epsilon})$ as $n\to\infty$. We call the generalized random variable $(\boldsymbol{\tau}_{\varepsilon'}(\omega,\varepsilon,\boldsymbol{\epsilon}))_{\varepsilon'},\varepsilon' \in (0,1]$ the first exit time of the sample function from every bounded domain, or briefly the generalized explosion time.

②We now define Colombeau generalized stochastic process $(\boldsymbol{x}^{x_0,\varepsilon}_{t,,\epsilon,\varepsilon'}(\omega))_{\varepsilon'}$ by setting

$$\boldsymbol{x}^{x_0,\varepsilon}_{t,\epsilon,\varepsilon'}(\omega) = \boldsymbol{x}^{x_0,\varepsilon}_{t,\epsilon,\varepsilon',n}(\omega) \text{ for } t = t(\omega) < \boldsymbol{\tau}_{\varepsilon',n}(\omega,\varepsilon,\boldsymbol{\epsilon}). \tag{2.44}$$

③That this is always a Markov process for $t = t(\omega) < (\boldsymbol{\tau}_{\varepsilon',n}(\omega,\varepsilon,\boldsymbol{\epsilon}))_{\varepsilon'}$.

④Colombeau generalized stochastic process $(\boldsymbol{x}^{x_0,\varepsilon}_{t,\epsilon,\varepsilon'}(\omega))_{\varepsilon'}$ defined by setting Eq. (2.42) on the random generalized interval $[0,(\boldsymbol{\tau}_{\varepsilon',n}(\omega,\varepsilon,\boldsymbol{\epsilon}))_{\varepsilon'}]$ is regular, if for any $s < \infty,\boldsymbol{x} \in \mathbf{R}^n,\boldsymbol{\epsilon} \in (0,1]^n$:

$$(\mathbf{P}^{s,x}\{\boldsymbol{\tau}_{\varepsilon'}(\omega,\varepsilon,\boldsymbol{\epsilon}) = \infty\})_{\varepsilon'} = 1,\varepsilon' \in (0,1]. \tag{2.45}$$

⑤Colombeau generalized stochastic process $(\boldsymbol{x}^{x_0,\varepsilon}_{t,\epsilon,\varepsilon'}(\omega))_{\varepsilon'}$, defined by setting Eq. (2.44) is a strongly regular if for any $s < \infty,\boldsymbol{x} \in \mathbf{R}^n$, $\varepsilon' \in [0,1],\boldsymbol{\epsilon} \in (0,1]^n$:

$$(\mathbf{P}^{s,x}\{\boldsymbol{\tau}_{\varepsilon'}(\omega,\varepsilon,\boldsymbol{\epsilon}) = \infty\})_{\varepsilon'} = 1. \tag{2.46}$$

Remark 7. We note that: ③ does not imply ④.

Proposition 1. Assume that Colombeau generalized stochastic process $(\boldsymbol{x}^{x_0,\varepsilon}_{t,\varepsilon'}(\omega))_{\varepsilon'}$ defined by setting Eq. (2.44) is a strongly regular.

Then ① $\forall \boldsymbol{\epsilon},\boldsymbol{\epsilon} \in (0,1]^n, \forall \delta,\delta > 0$:

$$\lim_{\varepsilon'\to 0}\mathbf{E}[\|\boldsymbol{x}^{x_0,\varepsilon}_{t,\epsilon,\varepsilon'}(\omega) - \boldsymbol{x}^{x_0,\varepsilon}_{t,\epsilon,\varepsilon'=0}(\omega)\|^2]0. \tag{2.47. a}$$

$$\lim_{\varepsilon'\to 0}\mathbf{P}[\|\boldsymbol{x}^{x_0,\varepsilon}_{t,\epsilon,\varepsilon'}(\omega) - \boldsymbol{x}^{x_0,\varepsilon}_{t,\epsilon,\varepsilon'=0}(\omega)\| > \delta] = 0. \tag{2.47. b}$$

② $\forall \delta,\delta > 0$:

$$\lim_{\varepsilon'\to 0,\epsilon\to 0}\mathbf{E}[\|\boldsymbol{x}^{x_0,\varepsilon}_{t,\epsilon,\varepsilon'}(\omega) - \boldsymbol{x}^{x_0,\varepsilon}_{t,\epsilon=0,\varepsilon'=0}(\omega)\|^2] = 0. \tag{2.47. c}$$

$$\lim_{\varepsilon'\to 0,\epsilon\to 0}\mathbf{P}[\|\boldsymbol{x}^{x_0,\varepsilon}_{t,\epsilon,\varepsilon'}(\omega) - \boldsymbol{x}^{x_0,\varepsilon}_{t,\epsilon=0,\varepsilon'=0}(\omega)\| > \delta] = 0. \tag{2.47. d}$$

Proof. Immediately follows from Theorem A1. (I) (see Appendix A) and Definitions 1, 3.

Let us consider now a family $(\boldsymbol{x}^{x_0,\varepsilon}_{t,\epsilon,\varepsilon'}(\omega))_{\varepsilon'}$ of the solutions of the Colombeau SDE:

$$(\mathrm{d}\boldsymbol{x}^{x_0,\varepsilon}_{t,\epsilon,\varepsilon'}(\omega))_{\varepsilon',\epsilon} = (\boldsymbol{b}_{\varepsilon',\epsilon}(\boldsymbol{x}^{x_0,\varepsilon}_{t,\epsilon,\varepsilon'}(\omega),t,\omega))_{\varepsilon'} + \sqrt{\varepsilon}\mathrm{d}\boldsymbol{W}(t,\omega), \tag{2.48}$$

$(\boldsymbol{x}_{0,\varepsilon'}^{x_0,\varepsilon})_{\varepsilon'} = x_0 \in \hat{R}^n, t \in [0,T], \varepsilon, \varepsilon' \in (0,1], \boldsymbol{\epsilon} \in (0,1]^n$.

Here $\boldsymbol{W}(t)$ is n-dimensional Brownian motion, and $\forall \boldsymbol{\epsilon} \in (0,1]^n, \forall t \in [0,T]$ and for almost all $\omega \in \Omega$: $(b_{\varepsilon',\epsilon}(x,t,\omega))_{\varepsilon'} \in G^n(\mathbf{R}^n)$, $b_{0,0}(\cdot,t) \equiv b_{\varepsilon'=0,\epsilon=0}(\cdot,t,\omega): \mathbf{R}^n \to \mathbf{R}^n$ is a polynomialvector-function on a variable $\boldsymbol{x} = (x_1,\cdots,x_n)$, i. e. , $b_{i,0,0}(x,t) = \sum_{\alpha,|\alpha| \leqslant r} b_{i,0,0}^{\alpha}(t)x^{\alpha}$, $\alpha = (i_1,\cdots,i_n)$, $|\alpha| = \sum_{j=1}^{n} i_j, 0 \leqslant i_j \leqslant p$, and

$$b_{i,\varepsilon',\epsilon}(\boldsymbol{x}(t),t,\omega) = b_{i,0,0}(\boldsymbol{x}_{\varepsilon',\epsilon}(t,\omega),t). \tag{2.49}$$

Here $\boldsymbol{x}_{\varepsilon',\epsilon}(t,\omega) = (x_{1,\varepsilon',\varepsilon,\epsilon}(t,\omega),\cdots,x_{n,\varepsilon',\varepsilon,\epsilon}(t,\omega))$,

$$x_{i,\varepsilon',\epsilon}(t,\omega) = \frac{x_i(t)}{1 + \varepsilon' x_i^{2l}(t) + \varepsilon' \left\{ \epsilon_i \int_0^t \theta_{\epsilon_i}[x_i(\tau)] x_i^{2l}(\tau)\mathrm{d}\tau + \sqrt{\delta}W_i(t) \right\}^2}, \tag{2.50}$$

$i = 1,\cdots,n$. Now we let

$$u_i(t) = \epsilon_i \int_0^t \theta_{\epsilon_i}[x_i(\tau)] x_i^{2l}(\tau)\mathrm{d}\tau + \sqrt{\delta}W_i(t) \tag{2.51}$$

and rewrite Eq. (2.48) of the canonical Colombeau-Ito form:

$$(\mathrm{d}\boldsymbol{x}_{t,\varepsilon',\epsilon}^{x_0,\varepsilon}(\omega))_{\varepsilon',\epsilon} = (\boldsymbol{b}_{\varepsilon',\epsilon}(\boldsymbol{x}_{t,\varepsilon',\epsilon}^{x_0,\varepsilon}(\omega), \boldsymbol{u}_{t,,\varepsilon',\epsilon}^{\delta}(\omega),t))_{\varepsilon'} + \sqrt{\varepsilon}\mathrm{d}\boldsymbol{W}(t,\omega),$$

$$\boldsymbol{u}_{t,\varepsilon',\varepsilon,\epsilon}^{\delta}(\omega) = (u_{1,t,\varepsilon',\epsilon}^{\delta}(\omega),\cdots,u_{n,t,\varepsilon',\epsilon}^{\delta}(\omega)), \tag{2.52}$$

$$(\mathrm{d}u_{i,t,\varepsilon',\epsilon}^{\delta}(\omega))_{\varepsilon'} = \epsilon_i (\theta_{\epsilon_i}[x_{i,t,\varepsilon',\epsilon}^{x_0,\delta}(\omega)] [x_{i,t,\varepsilon',\epsilon}^{x_0,\delta}(\omega)]^{2l})_{\varepsilon'} + \sqrt{\delta}\mathrm{d}W_i(t), \tag{2.53}$$

$i = 1,\cdots,n, (\boldsymbol{x}_{0,\varepsilon'}^{x_0,\varepsilon})_{\varepsilon'} = x_0 \in \hat{\mathbf{R}}^n, t \in [0,T], \varepsilon,\varepsilon',\epsilon,\delta \in (0,1]$.

Theorem 3. Let us consider a pair of the Colombeau-Ito's SDE:

$$(\mathrm{d}\boldsymbol{x}_{t,\varepsilon',\mu}^{x_0,\varepsilon}(\omega))_{\varepsilon'} = (\boldsymbol{g}_{\varepsilon'}^{\mu}(\boldsymbol{x}_{t,\varepsilon'}^{x_0,\varepsilon}(\omega),t))_{\varepsilon'} + \sqrt{\varepsilon}(\mathrm{d}\boldsymbol{W}(t,\omega))_{\varepsilon'}, \tag{2.54}$$

$$(\boldsymbol{x}_{0,\varepsilon',\mu}^{x_0,\varepsilon})_{\varepsilon'} = \boldsymbol{x}_0 \in \hat{\mathbf{R}}^m, t \in [0,T], \varepsilon,\varepsilon' \in (0,1], \mu = 1,2. \tag{2.55}$$

Assume now that: ① Conditions (2.35) and (2.36) is satisfied.

② For a given $N > 0$, $\forall \boldsymbol{x} \in \mathbf{R}^n$ such that $\|\boldsymbol{x}\| \leqslant N: \boldsymbol{g}_{\varepsilon'}^1(\boldsymbol{x},t) = \boldsymbol{g}_{\varepsilon'}^2(\boldsymbol{x},t)$.

Let $\boldsymbol{x}_{t,\varepsilon',\mu}^{x_0,\varepsilon}(\omega), \mu = 1,2$ be a pair of the solutions of the Colombeau-Ito's SDE (2.54), (2.55) and let $\mathcal{F}_{\varepsilon',\mu}^{N,t}(\omega), \mu = 1,2$ be a set $\mathcal{F}_{\varepsilon',\mu}^{N}(\omega) = \{t \mid \sup_{0\leqslant s\leqslant t} \|\boldsymbol{x}_{s,\varepsilon',\mu}^{x_0,\varepsilon}(\omega)\| \leqslant N\}$. We let now $\boldsymbol{\tau}_{\varepsilon',\mu}^{N}(\omega) = \sup\{t \mid t \in \mathcal{F}_{\varepsilon,\varepsilon',\mu}^{N}(\omega)\}$.

Then $\forall \varepsilon' \in (0,1]$:

(i) $\mathbf{P}\{\boldsymbol{\tau}_{\varepsilon,\varepsilon',1}^{N}(\omega) = \boldsymbol{\tau}_{\varepsilon,\varepsilon',2}^{N}(\omega)\} = 1$ and

(ii) $\mathbf{P}\{\sup_{0\leqslant s\leqslant \tau_1} \|\boldsymbol{x}_{t,\varepsilon',1}^{x_0,\varepsilon}(\omega) - \boldsymbol{x}_{t,\varepsilon',2}^{x_0,\varepsilon}(\omega)\| = 0\} = 1$.

Proof. A proof of this statement complete similarly to a classical case. For example, see Ref. [15], Chapt. 2, Subsect. 6, Theorem 2.

Let us rewrite now Eq. (2.52) – Eq. (2.53) in the next form (with $\theta_{\epsilon_i} \equiv 1$)

$$(\boldsymbol{x}_{t,\varepsilon',\epsilon}^{x_0,\varepsilon}(\omega,\delta))_{\varepsilon',\epsilon} = \boldsymbol{x}_0 + \left(\int_0^t \boldsymbol{b}_{\varepsilon',\epsilon}(\boldsymbol{x}_{\tau,\varepsilon',\epsilon}^{x_0,\varepsilon}(\omega,\delta), \boldsymbol{u}_{\tau,\varepsilon',\epsilon}^{\delta}(\omega),\tau)\mathrm{d}\tau\right)_{\varepsilon'} + \sqrt{\varepsilon}\boldsymbol{W}(t,\omega), \tag{2.56}$$

$$\boldsymbol{u}^{\delta}_{t,\varepsilon',\epsilon}(\omega) = (u^{\delta}_{1,t,\varepsilon',\epsilon}(\omega),\cdots,u^{\delta}_{n,t,\varepsilon',\epsilon}(\omega)),$$

$$(u^{\delta}_{i,t,\varepsilon',\epsilon}(\omega))_{\varepsilon'} = \epsilon_i\left(\int_0^t [x^{x_0,\varepsilon}_{i,\tau,\varepsilon',\epsilon}(\omega,\delta)]^{2l}\mathrm{d}\tau\right)_{\varepsilon'} + \sqrt{\delta}W_i(t), i = 1,\cdots,n, \tag{2.57}$$

Let $G_N(\boldsymbol{y}),\boldsymbol{y} \in \mathbf{R}^n$ be a function: (i) $G_N(\boldsymbol{y}) = \boldsymbol{y}$ if $\|(\boldsymbol{y})\| \leqslant N$, (ii) $G_N(\boldsymbol{y}) = 0$ if $\|\boldsymbol{y}\| > N$.

We set now $\boldsymbol{b}^N_{\varepsilon',\epsilon}(\boldsymbol{x},\boldsymbol{u},t) = \boldsymbol{b}_{\varepsilon',\epsilon}(G_N(\boldsymbol{x}),G_N(\boldsymbol{u}),t)$.

Let $(\boldsymbol{y}^{x_0,\varepsilon}_{t,\varepsilon',\epsilon}(\omega,\delta,N))_{\varepsilon'} = \{(\boldsymbol{x}^{x_0,\varepsilon}_{t,\varepsilon',\epsilon}(\omega,\delta,N))_{\varepsilon'},\boldsymbol{u}^{\delta}_{t,\varepsilon',\epsilon}(\omega,N)\}$

be a family of the solution of the Colombeau-Ito's SDE:

$$(\boldsymbol{x}^{x_0,\varepsilon}_{t,\varepsilon',\epsilon}(\omega,\delta,N))_{\varepsilon'} = \boldsymbol{x}_0 + \left(\int_0^t \boldsymbol{b}^N_{\varepsilon',\epsilon}[\boldsymbol{x}^{x_0,\varepsilon}_{\tau,\varepsilon',\epsilon}(\omega,\delta,N),\boldsymbol{u}^{\delta}_{\tau,\varepsilon',\epsilon}(\omega,\tau,N)]\mathrm{d}\tau\right)_{\varepsilon'} +$$

$$\sqrt{\varepsilon}\boldsymbol{W}(t,\omega), \tag{2.58}$$

$$\boldsymbol{u}^{\delta}_{t,\varepsilon',\epsilon}(\omega,N) = (u^{\delta}_{1,t,\varepsilon',\epsilon}(\omega,N),\cdots,u^{\delta}_{n,t,\varepsilon',\epsilon}(\omega,N)),$$

$$(u^{\delta}_{i,t,\varepsilon',\epsilon}(\omega,N))_{\varepsilon'} = \epsilon_i\left(\int_0^t \left[G_N(x^{x_0,\varepsilon}_{i,\tau,\varepsilon',\epsilon}(\omega,\delta,N))\right]^{2l}\mathrm{d}\tau\right)_{\varepsilon'} + \sqrt{\delta}W_i(t), \tag{2.59}$$

$i = 1,\cdots,n$,

Definition 4. (1) Let $\left\{\left(\boldsymbol{y}^{x_0,\varepsilon}_{t,\varepsilon',\epsilon}(\omega,\delta,N)\right)_{\varepsilon',\epsilon}\right\}_{N=1}^{\infty}$ be a sequence of the solution of the Colombeau-Ito's SDE (2.56) – (2.57). Let $\mathcal{F}^N_{\varepsilon,\varepsilon',\epsilon}(\omega,\delta)$ be a set

$$\mathcal{F}^N_{\varepsilon,\varepsilon',\epsilon}(\omega,\delta) = \left\{t \mid \sup_{0\leqslant s\leqslant t}\|\boldsymbol{y}^{x_0,\varepsilon}_{t,\varepsilon',\epsilon}(\omega,\delta,N)\| \leqslant N\right\}. \tag{2.60}$$

(2) We let now

$$\boldsymbol{\tau}^N_{\varepsilon,\varepsilon',\epsilon}(\omega,\delta) = \sup\left\{t \mid t \in \mathcal{F}^N_{\varepsilon,\varepsilon',\epsilon}(\omega)\right\}, \tag{2.61}$$

$$\boldsymbol{\tau}^{\infty}_{\varepsilon,\varepsilon',\epsilon}(\omega,\delta) = \lim_{N\to\infty}\boldsymbol{\tau}^N_{\varepsilon,\varepsilon',\epsilon}(\omega,\delta). \tag{2.62}$$

(3) Let $\tilde{\boldsymbol{y}}^{x_0,\varepsilon}_{t,\varepsilon',\epsilon}(\omega,\delta)$ be a net of the stochastic processes defined by setting

$$\tilde{\boldsymbol{y}}^{x_0,\varepsilon}_{t,\varepsilon',\epsilon}(\omega,\delta) = \boldsymbol{y}^{x_0,\varepsilon}_{t,\varepsilon',\epsilon}(\omega,\delta,N) \text{ iff } t < \boldsymbol{\tau}^N_{\varepsilon,\varepsilon',\epsilon}(\omega,\delta). \tag{2.63}$$

(4) Let $\left(\boldsymbol{y}^{x_0,\varepsilon}_{t,\varepsilon',\epsilon}(\omega,\delta)\right)_{\varepsilon'}$ be Colombeau generalized stochastic process defined by setting

$$\left(\boldsymbol{y}^{x_0,\varepsilon}_{t,\varepsilon',\epsilon}(\omega,\delta)\right)_{\varepsilon'} = \left(\tilde{\boldsymbol{y}}^{x_0,\varepsilon}_{t,\varepsilon',\epsilon}(\omega,\delta)\right)_{\varepsilon'}. \tag{2.64}$$

Remark 5. We note that according to the Theorem 3 $\forall M(M \geqslant N)$ one obtain

$$\mathbf{P}\left[\sup_{0\leqslant t\leqslant \tau^N_{\varepsilon,\varepsilon',\epsilon}(\omega)}\|\boldsymbol{y}^{x_0,\varepsilon}_{t,\varepsilon',\epsilon}(\omega,\delta,N) - \boldsymbol{y}^{x_0,\varepsilon}_{t,\varepsilon',\epsilon}(\omega,\delta,M)\| > 0\right] = 0,$$

Therefore definitions (2.63) – (2.64) is correct.

Definition 5. Let $(\boldsymbol{y}^{x_0,\varepsilon}_{t,\varepsilon',\epsilon})_{\varepsilon'}$ be a family of the solutions Colombeau-Ito's SDE (2.58) – (2.59).

(1) A family $(\boldsymbol{y}^{x_0,\varepsilon}_{t,\varepsilon',\epsilon})_{\varepsilon'},\boldsymbol{\varepsilon},\boldsymbol{\varepsilon}' \in (0,1],\boldsymbol{\epsilon} \in (0,1]^n$ is regular if

$$\left(\lim_{c\to\infty}\mathbf{P}\left[\|\boldsymbol{y}^{x_0,\varepsilon}_{t,\varepsilon',\epsilon}(\omega,\delta)\| > c\right]\right)_{\varepsilon'} = 0. \tag{2.65}$$

Or in the next equivalent form

$$\left(\mathbf{P}\left[\boldsymbol{\tau}_{\varepsilon,\varepsilon',\epsilon}^{\infty}(\omega,\delta)=\infty\right]\right)_{\varepsilon'}=1. \tag{2.66}$$

(2) A family $(\boldsymbol{y}_{t,\varepsilon',\epsilon}^{x_0,\varepsilon})_{\varepsilon'}$ is a strongly regular if $\forall\varepsilon',\varepsilon'\in[0,1]$, $\forall\epsilon,\epsilon\in(0,1]^n$:

$$\left(\lim_{c\to\infty}\mathbf{P}\left[\left\|\boldsymbol{x}_{t,\varepsilon',\epsilon}^{x_0,\varepsilon}(\omega,\delta)\right\|>c\right]\right)_{\varepsilon'}=0. \tag{2.67}$$

or in the next equivalent form: $\forall\varepsilon',\varepsilon'\in[0,1],\forall\epsilon,\epsilon\in(0,1]^n$:

$$\left(\mathbf{P}\left[\boldsymbol{\tau}_{\varepsilon,\varepsilon',\epsilon}^{\infty}(\omega,\delta)=\infty\right]\right)_{\varepsilon'}=1. \tag{2.68}$$

Definition 6. Let $(\boldsymbol{y}_{t,\varepsilon',\epsilon}^{x_0,\varepsilon})_{\varepsilon'}$ be a family of the solutions Colombeau-Ito's SDE (2.58) – (2.59). A family $(\boldsymbol{y}_{t,\varepsilon',\epsilon}^{x_0,\varepsilon})_{\varepsilon'},\varepsilon,\varepsilon'\in(0,1],\epsilon\in(0,1]^n$ is a non-regular if

$$\exists t'\forall t\geq t':\left(\lim_{c\to\infty}\mathbf{P}\left[\left\|\boldsymbol{x}_{t,\varepsilon',\epsilon}^{x_0,\varepsilon}(\omega,\delta)\right\|>c\right]\right)_{\varepsilon'}\neq 0. \tag{2.69}$$

Or in the next equivalent form

$$\left(\mathbf{P}\left[\boldsymbol{\tau}_{\varepsilon,\varepsilon',\epsilon}^{\infty}(\omega,\delta)<\infty\right]\right)_{\varepsilon'}=1. \tag{2.70}$$

Proposition 2. Assume that Colombeau generalized stochastic process $\left(\boldsymbol{y}_{t,\varepsilon',\epsilon}^{x_0,\varepsilon}(\omega,\delta)\right)_{\varepsilon'}$ defined by setting (2.64) is a strongly regular. Then

① $$\lim_{\substack{\varepsilon'\to 0\\ \epsilon\to 0,}}\lim_{\delta\to 0}\mathbf{E}\left[\left\|\boldsymbol{y}_{t,\varepsilon',\epsilon}^{x_0,\varepsilon}(\omega,\delta)-\boldsymbol{y}_t^{x_0,\varepsilon}(\omega)\right\|^2\right]=0 \tag{2.71.a}$$

② $\forall\sigma>0$:

$$\lim_{\substack{\varepsilon'\to 0\\ \epsilon\to 0,}}\lim_{\delta\to 0}\mathbf{P}\left[\left\|\boldsymbol{y}_{t,\varepsilon',\epsilon}^{x_0,\varepsilon}(\omega,\delta)-\boldsymbol{y}_t^{x_0,\varepsilon}(\omega)\right\|>\sigma\right]=0. \tag{2.71.b}$$

Here, $\boldsymbol{y}_t^{x_0,\varepsilon}(\omega)=\boldsymbol{y}_{t,\varepsilon'=0,\epsilon=0}^{x_0,\varepsilon}(\omega,\delta=0)$.

Proof. Immediately follows from Theorem 3 and Theorem A. 1 (see Appendix A).

Proposition 3. Let $\left[\boldsymbol{y}_{t,\varepsilon',\epsilon}^{x_0,\varepsilon}(\omega,\delta)\right]_{\varepsilon'}=\left\{\left(\boldsymbol{x}_{t,\varepsilon',\epsilon}^{x_0,\varepsilon}\right)_{\varepsilon'},\left[\boldsymbol{u}_{t,\varepsilon',\epsilon}^{\delta}(\omega)\right]_{\varepsilon'}\right\}$ be a family of the solutions Colombeau-Ito's SDE (2.58) – (2.59) with $\theta_\epsilon[z]\equiv 1$. A family $\left(\boldsymbol{y}_{t,\varepsilon',\epsilon}^{x_0,\varepsilon}\right)_{\varepsilon'},\varepsilon,\varepsilon'\in(0,1],\epsilon\in(0,1]^n$ is regular.

Proof. Assume that: process $\left(\boldsymbol{y}_{t,\varepsilon',\epsilon}^{x_0,\varepsilon}(\omega,\delta)\right)_{\varepsilon'}$ is a non-regular. Therefore $\left(\mathbf{P}^{s,x}\left\{\boldsymbol{\tau}_{\varepsilon'}(\omega,\varepsilon)<\infty\right\}\right)_{\varepsilon'}=1$ and consequently

$$\left(\mathbf{P}^{s,x}\left[\boldsymbol{y}_{\tau_{\varepsilon'}(\omega,\varepsilon),\varepsilon',\epsilon}^{x_0,\varepsilon}(\omega,\delta)=\infty\right]\right)_{\varepsilon'}>0. \tag{2.72}$$

But the other hand from Eqs. (2.58) – (2.59) we obtain

$$\left(\boldsymbol{x}_{\tau_{\varepsilon'}(\omega,\varepsilon),\varepsilon',\epsilon}^{x_0,\varepsilon}(\omega,\delta)\right)_{\varepsilon'}=\boldsymbol{x}_0+\left(\int_0^{\tau_{\varepsilon'}(\omega,\varepsilon)}\boldsymbol{b}_{\varepsilon',\epsilon}[\boldsymbol{x}_{v,\varepsilon',\epsilon}^{x_0,\varepsilon}(\omega,\delta),\boldsymbol{u}_{v,\varepsilon',\epsilon}^{\delta}(\omega),v,\varepsilon]\mathrm{d}v\right)_{\varepsilon'}+$$
$$\left(\sqrt{\varepsilon}\boldsymbol{W}[\boldsymbol{\tau}_{\varepsilon'}(\omega,\varepsilon),\omega]\right)_{\varepsilon'}, \tag{2.73}$$

$$\boldsymbol{u}^{\delta}_{v,\varepsilon',\epsilon}(\omega) = \left(u^{\delta}_{1,v,\varepsilon',\epsilon}(\omega),\cdots,u^{\delta}_{n,v,\varepsilon',\epsilon}(\omega)\right),$$

$$\left(u^{\delta}_{i,\tau_{\varepsilon'}(\omega,\varepsilon),\varepsilon',\epsilon}(\omega)\right)_{\varepsilon'} = \epsilon\left(\int_0^{\tau_{\varepsilon'}(\omega,\varepsilon)}\left[x^{x_0,\varepsilon}_{i,v,\varepsilon',\epsilon}(\omega,\delta)\right]^{2l}\mathrm{d}v\right)_{\varepsilon'} + \left(\sqrt{\delta}W_i[\boldsymbol{\tau}_{\varepsilon'}(\omega,\varepsilon)]\right)_{\varepsilon'},$$

$$i = 1,\cdots,n, \tag{2.74}$$

From (2.72) and Eqs. (2.73) and (2.74) we obtain

$$\left(\mathbf{P}^{0,x_0}\left[\int_0^{\tau_{\varepsilon'}(\omega,\varepsilon)}\boldsymbol{b}_{\varepsilon',\epsilon}\left(\boldsymbol{x}^{x_0,\varepsilon}_{v,\varepsilon',\epsilon}(\omega,\delta),\boldsymbol{u}^{\delta}_{v,\varepsilon',\epsilon}(\omega),v,\varepsilon\right)\mathrm{d}v = \infty\right]\right)_{\varepsilon'} = 0,$$

and therefore

$$\left(\mathbf{P}^{0,x_0}\left[\boldsymbol{x}^{x_0,\varepsilon}_{\tau_{\varepsilon'}(\omega,\varepsilon),\varepsilon',\epsilon}(\omega,\delta) = \infty\right]\right)_{\varepsilon'} =$$

$$\left(\mathbf{P}^{0,x_0}\left\{x_0 + \int_0^{\tau_{\varepsilon'}(\omega,\varepsilon)}\boldsymbol{b}_{\varepsilon',\epsilon}\left[\boldsymbol{x}^{x_0,\varepsilon}_{v,\varepsilon',\epsilon}(\omega,\delta),\boldsymbol{u}^{\delta}_{v,\varepsilon',\epsilon}(\omega),v,\varepsilon\right]\mathrm{d}v + \sqrt{\varepsilon}\boldsymbol{W}\left[\boldsymbol{\tau}_{\varepsilon'}(\omega,\varepsilon),\omega\right]\right.\right.$$

$$\left.\left. = \infty\right\}\right)_{\varepsilon'} = 0,$$

$$\left(\mathbf{P}^{0,0}\left[u^{\delta}_{i,\tau_{\varepsilon'}(\omega,\varepsilon),\varepsilon',\epsilon}(\omega) = \infty\right]\right)_{\varepsilon'} =$$

$$\left(\mathbf{P}^{s,x}\left\{\int_0^{\tau_{\varepsilon'}(\omega,\varepsilon)}\left[x^{x_0,\varepsilon}_{i,v,\varepsilon',\epsilon}(\omega,\delta)\right]^{2l}\mathrm{d}v + \sqrt{\delta}W_i\left[\boldsymbol{\tau}_{\varepsilon'}(\omega,\varepsilon)\right] = \infty\right\}\right)_{\varepsilon'} = 0.$$

Thus

$$\left(\mathbf{P}^{s,x}\left[\boldsymbol{y}^{x_0,\varepsilon}_{\tau_{\varepsilon'}(\omega,\varepsilon),\varepsilon',\epsilon}(\omega,\delta) = \infty\right]\right)_{\varepsilon'} = 0.$$

But this is the contradiction. This contradiction completed the proof.

Definition 7. CISDE (2.37), (2.38) is $\hat{\mathbf{R}}$-dissipative if there exist Lyapunov candidate function $\left(V_{\varepsilon'}(\boldsymbol{x},t)\right)_{\varepsilon'}:\hat{\mathbf{R}}^n \times [0,T] \to \hat{\mathbf{R}}$ and positive infinite Colombeau constants $\tilde{C} = (C_{\varepsilon'})_{\varepsilon'} \in \hat{\mathbf{R}}_+$, $\tilde{r} = (r_{\varepsilon'})_{\varepsilon'} \in \hat{\mathbf{R}}_+$, such that:

①$\forall \varepsilon' \in (0,1]: V_{*,\varepsilon'} = \lim_{R\to\infty}(\inf_{\|x\|>R}V_{\varepsilon'}(\boldsymbol{x},t)) = \infty$, and

②$\forall[(x_{\varepsilon'})_{\varepsilon'}]((\|x_{\varepsilon'}\|)_{\varepsilon'} \geqslant \tilde{r})$. the inequality

$$(\dot{V}_{\varepsilon'}(\boldsymbol{x}_{\varepsilon'},t;\boldsymbol{b}_{\varepsilon'}))_{\varepsilon'} \leqslant \tilde{C}\left[\left(V_{\varepsilon'}(\boldsymbol{x}_{\varepsilon'},t)\right)_{\varepsilon'}\right] \tag{2.75}$$

is satisfied. Here

$$\left(\dot{V}_{\varepsilon'}(\boldsymbol{x}_{\varepsilon'},t;\boldsymbol{b}_{\varepsilon'})\right)_{\varepsilon'} \equiv \left(\frac{\partial V_{\varepsilon'}(\boldsymbol{x}_{\varepsilon'},t)}{\partial t}\right)_{\varepsilon'} + \sum_{i=1}^{n}\left[\left(\frac{\partial V_{\varepsilon'}(\boldsymbol{x}_{\varepsilon'},t)}{\partial x'_{i,\varepsilon}}b'_{i,\varepsilon}(\boldsymbol{x}_{\varepsilon'},t)\right)_{\varepsilon'}\right]. \tag{2.76}$$

Or in the next equivalent form:

CISDE (2.37), (2.38) is $\hat{\mathbf{R}}$-dissipative if there exist Lyapunov candidate function $(V_{\varepsilon'}(\boldsymbol{x},t))_{\varepsilon'}:\hat{\mathbf{R}}^n \times [0,T] \to \hat{\mathbf{R}}$ and positive infinite Colombeau constants $\tilde{C} = [(C_{\varepsilon'})_{\varepsilon'}] \in \hat{\mathbf{R}}_+$,

$\tilde{r} = [(r_{\varepsilon'})_{\varepsilon'}] \in \hat{\mathbf{R}}_+$, such that:

① $\forall \varepsilon' \in (0,1]: V_{*,\varepsilon'} = \lim_{R\to\infty}\left(\inf_{\|x\|>R} V_{\varepsilon'}(\boldsymbol{x},t)\right) = \infty$, and

② $\forall \varepsilon' \in (0,1]\, \forall x_{\varepsilon'}\left[\left(x_{\varepsilon'} \in \mathbf{R}^n\right) \wedge \left(\|\boldsymbol{x}_{\varepsilon'}\| \geqslant r_{\varepsilon'}\right)\right]$ the inequality

$$\left(\dot{V}_{\varepsilon'}\left(\boldsymbol{x}_{\varepsilon'},t;\boldsymbol{b}_{\varepsilon'}\right)\right)_{\varepsilon'} \leqslant \left((C_{\varepsilon'})_{\varepsilon'}\right)\left(V_{\varepsilon'}(\boldsymbol{x}_{\varepsilon'},t)\right)_{\varepsilon'} \text{ is satisfied.} \tag{2.77}$$

Here

$$\left(\dot{V}_{\varepsilon'}(\boldsymbol{x}_{\varepsilon'},t;\boldsymbol{b}_{\varepsilon'})\right)_{\varepsilon'} \equiv \left(\frac{\partial V_{\varepsilon'}(\boldsymbol{x}_{\varepsilon'},t)}{\partial t}\right)_{\varepsilon'} + \left(\sum_{i=1}^{n}\frac{\partial V_{\varepsilon'}(\boldsymbol{x}_{\varepsilon'},t)}{\partial x_{\varepsilon'}} b_{i,\varepsilon'}(\boldsymbol{x}_{\varepsilon'},t)\right)_{\varepsilon'}. \tag{2.78}$$

Definition 8. CISDE (2.37), (2.38) is a strongly $\hat{\mathbf{R}}$-dissipative if Lyapunov candidate function $\left(V_{\varepsilon'}(\boldsymbol{x},t)\right)_{\varepsilon'}: \hat{\mathbf{R}}^n \times [0,T] \to \hat{\mathbf{R}}$,

$\varepsilon' \in [0,1]$ and positive finite Colombeau constants

$\tilde{C} = (C_{\varepsilon'})_{\varepsilon'} \in \hat{\mathbf{R}}_+$, $\tilde{r} = (r_{\varepsilon'})_{\varepsilon'} \in \hat{\mathbf{R}}_+$, such that:

① $\forall \varepsilon' \in (0,1]: V_{*,\varepsilon'} = \lim_{r\to\infty}(\inf_{\|x\|>r} V_{\varepsilon'}(\boldsymbol{x},t)) = \infty$, and

② $\forall [(\boldsymbol{x}_{\varepsilon'})_{\varepsilon'}]([(\|\boldsymbol{x}_{\varepsilon'}\|)_{\varepsilon'}] \geqslant \tilde{r})$ the inequality

$$(\dot{V}_{\varepsilon'}(\boldsymbol{x}_{\varepsilon'},t;\boldsymbol{b}_{\varepsilon'}))_{\varepsilon'} \leqslant \tilde{C}[(V_{\varepsilon'}(\boldsymbol{x}_{\varepsilon'},t))_{\varepsilon'}] \text{ is satisfied.} \tag{2.79}$$

Here

$$(\dot{V}_{\varepsilon'}(\boldsymbol{x}_{\varepsilon'},t;\boldsymbol{b}_{\varepsilon'}))_{\varepsilon'} \equiv \left(\frac{\partial V_{\varepsilon'}(\boldsymbol{x}_{\varepsilon'},t)}{\partial t}\right)_{\varepsilon'} + \sum_{i=1}^{n}\left[\left(\frac{\partial V_{\varepsilon'}(\boldsymbol{x}_{\varepsilon'},t)}{\partial x_{i,\varepsilon'}} b_{i,\varepsilon'}(\boldsymbol{x}_{\varepsilon'},t)\right)_{\varepsilon'}\right]. \tag{2.80}$$

Proposition 4. Let $\left(\boldsymbol{x}^{x_0,\varepsilon}_{t,\varepsilon',\epsilon}(\omega)\right)_{\varepsilon'}$ be generalized stochastic process satisfying C-I SDE (2.37), (2.38) on the time interval $[s,T]$ and $\left(\boldsymbol{\tau}_{\varepsilon',U}(\omega,\epsilon)\right)_{\varepsilon'}$ – is a generalized random variable equal to the time at which the sample function of the generalized process $\left(\boldsymbol{x}^{x_0,\varepsilon}_{t,\varepsilon',\epsilon}(\omega)\right)_{\varepsilon'}$ first leaves the bounded neighborhood U, and let $\left(\boldsymbol{\tau}_{\varepsilon',U}(\omega,t,\epsilon)\right)_{\varepsilon'} = \left(\min\left\{\boldsymbol{\tau}_{\varepsilon',U}(\omega,\epsilon), t\right\}\right)_{\varepsilon'}$. Suppose moreover that $\forall \varepsilon' \in (0,1]: \mathbf{P}\left[\boldsymbol{x}^{x_0,\varepsilon}_{s,\varepsilon',\epsilon}(\omega) \in U\right] = 1$. Then

$$\left(\mathbf{E}\left\{V_{\varepsilon'}\left[\boldsymbol{x}^{x_0,\varepsilon}_{\tau_{\varepsilon',U}(\omega,t,\epsilon),\varepsilon',\epsilon}(\omega),\boldsymbol{\tau}_{\varepsilon',U}(\omega,t,\epsilon)\right] - V_{\varepsilon'}\left[\boldsymbol{x}^{x_0,\varepsilon}_{s,\varepsilon',\epsilon}(\omega),s\right]\right\}\right)_{\varepsilon'} =$$

$$\left(\mathbf{E}\left\{\int_s^{\tau_{\varepsilon',U}(\omega,t,\epsilon)} \dot{V}_{\varepsilon'}\left[\boldsymbol{x}^{x_0,\varepsilon}_{u,\varepsilon',\epsilon}(\omega),u\right]\mathrm{d}u\right\}\right)_{\varepsilon'}.$$

Proof. Similarly as the proof of the corresponding classical result, see Ref. [17] Lemma 3.2.

Theorem 4. (1) Assume that: (i) for CISDE (2.37) – (2.38) the inequalities (2.35) and (2.36) is satisfied and (ii) CISDE (2.37) – (2.38) is $\hat{\mathbf{R}}$-dissipative.

Then ① Colombeau generalized stochastic process $\left(x_{t,\varepsilon',\epsilon}^{x_0,\varepsilon}(\omega)\right)_{\varepsilon'}$, $\varepsilon' \in (0,1], \epsilon \in (0,1]^n$ defined by setting (2.44) is regular, and ② the inequality

$$\left(\mathbf{E}\left[V_{\varepsilon'}\left(x_{t,\varepsilon',\epsilon}^{x_0,\varepsilon}(\omega),t\right)\right]\right)_{\varepsilon'} \leqslant \left(\mathbf{E}\left[V_{\varepsilon'}\left(x_{t_0,\varepsilon',\epsilon}^{x_0,\varepsilon}(\omega),t_0\right)\right]\right)_{\varepsilon'} \exp\left[\left(C_{\varepsilon'}\right)_{\varepsilon'}(t-t_0)\right] \tag{2.81}$$

is satisfied.

Proof. (1) From (2.78) it follows that the Colombeau generalized function $\left(W_{\varepsilon'}(x_{\varepsilon'},t)\right)_{\varepsilon'} = \left(V_{\varepsilon'}(x_{\varepsilon'},t)\right)_{\varepsilon'} \exp\left[-(C_{\varepsilon'})_{\varepsilon'}(t-t_0)\right]$ is satisfies the inequality: $\left(\dot{W}_{\varepsilon'}(x_{\varepsilon'},t)\right)_{\varepsilon'} \leqslant 0$. Hence, by Proposition 4, for $\left(\tau_{\varepsilon',n}(\omega,t,\epsilon)\right)_{\varepsilon'} = (\min\{\tau_{\varepsilon',n}(\omega,\epsilon),t\})_{\varepsilon'}$ we have

$$\left(\mathbf{E}\left[V_{\varepsilon'}\left[x_{\tau_{\varepsilon',\epsilon,n}(\omega,t,\epsilon),\varepsilon'}^{x_0,\varepsilon}(\omega),\tau_{\varepsilon',n}(\omega,t,\epsilon)\right]\exp\left[-(C_{\varepsilon'})_{\varepsilon'}[\tau_{\varepsilon',n}(\omega,t,\epsilon)-t_0]\right]\right]\right)_{\varepsilon'} -$$

$$\left(\mathbf{E}\left[V_{\varepsilon'}\left(x_{t_0,\varepsilon',\epsilon}^{x_0,\varepsilon}(\omega),t_0\right)\right]\right)_{\varepsilon'} = \left(\mathbf{E}\left[\int_{t_0}^{\tau_{\varepsilon',n}(\omega,t,\epsilon)} \dot{W}_{\varepsilon'}\left(x_{u,\varepsilon',\epsilon}^{x_0,\varepsilon}(\omega),u\right)du\right]\right)_{\varepsilon'} \leqslant 0. \tag{2.82}$$

This, together with the inequalities $\left(\tau_{\varepsilon',n}(\omega,t,\epsilon)\right)_{\varepsilon'} \leqslant t$, $\left(V_{\varepsilon'}(x_{\varepsilon'},t)\right)_{\varepsilon'} \geqslant 0$, implies

$$\left(\mathbf{E}\left[V_{\varepsilon'}\left(x_{\tau_{\varepsilon',\epsilon,n}(\omega,t),\varepsilon'}^{x_0,\varepsilon}(\omega),\tau_{\varepsilon',U}(\omega,t,\epsilon)\right)\right]\right)_{\varepsilon'} \leqslant$$

$$\left(\mathbf{E}\left[V_{\varepsilon'}\left(\tilde{x}_{t_0,\epsilon,\varepsilon'}^{x_0,\varepsilon}(\omega),t_0\right)\right]\right)_{\varepsilon'} \exp\left[(C_{\varepsilon'})_{\varepsilon'}(t-t_0)\right] \tag{2.83}$$

From (2.83) one derive the estimate

$$\left(\mathbf{P}\left\{\tau_{\varepsilon',n}(\omega,\epsilon) < t\right\}\right)_{\varepsilon'} \leqslant \frac{\exp\left[(C_{\varepsilon'})_{\varepsilon'}(t-t_0)\right]\left(\mathbf{E}\left[V_{\varepsilon'}\left(\tilde{x}_{t_0,\varepsilon'}^{x_0,\varepsilon}(\omega),t_0\right)\right]\right)_{\varepsilon'}}{\left(\inf_{\|x\|\geqslant n,u>t_0} V_{\varepsilon'}(x,u)\right)_{\varepsilon'}}$$

Letting $n \to \infty$ and making use of the Definition 7 we now get (2.66).

(2) Assume that CISDE (2.37) – (2.38) is a strongly $\hat{\mathbf{R}}$-dissipative. Then ① Colombeau generalized stochastic process $= \left(x_{t,\varepsilon'}^{x_0,\varepsilon}(\omega)\right)_{\varepsilon'}$, $\varepsilon' \in [0,1], \epsilon \in [0,1]^n$, defined by setting Eq. (2.44) is a strongly regular and ② the inequality

$$\left(\mathbf{E}\left[V_{\varepsilon'}\left(x_{t,\varepsilon'}^{x_0,\varepsilon}(\omega),t\right)\right]\right)_{\varepsilon'} \leqslant \left(\mathbf{E}\left[V_{\varepsilon'}\left(x_{t_0,\varepsilon'}^{x_0,\varepsilon}(\omega),t_0\right)\right]\right)_{\varepsilon'} \exp\left[(C_{\varepsilon'})_{\varepsilon'}(t-t_0)\right] \tag{2.84}$$

is satisfied.

Proof. (2) Similarly as the proof of the Theorem 4.

Theorem 5. We set now $\theta_{\epsilon_i}[z] \equiv 1, i = 1,\cdots,n$. For any solution

$$\left(x_{t,\varepsilon',\epsilon}^{x_0,\varepsilon}(\omega,\delta)\right)_{\varepsilon'} = \left(x_{1,t,\varepsilon',\epsilon}^{x_0,\varepsilon},\cdots,x_{n,t,\varepsilon',\epsilon}^{x_0,\varepsilon}\right)_{\varepsilon'}$$

of a strongly $\hat{\mathbf{R}}$-dissipative CISDE (2.48) – (2.50) and any $\mathbf{R}$-valued parameters $\lambda_1,\cdots,\lambda_n$, *there exist finite Colombeau constant* $\tilde{C}' = (C'_{\varepsilon'})_{\varepsilon'} > 0$, such that $\forall \boldsymbol{\lambda}[\boldsymbol{\lambda} = (\lambda_1,\cdots,\lambda_n)]$, the inequality

$$\lim_{\substack{\varepsilon\to 0\\ \varepsilon'\to 0\\ \epsilon\to 0\\ \left(\frac{\varepsilon'}{\varepsilon}\right)\to 0}} \lim_{\delta\to 0}\mathbf{E}_{\Omega}\left[\ \| \boldsymbol{x}^{x_0,\varepsilon}_{t,\varepsilon',\epsilon}(\omega,\delta) - \lambda \|^2\right] \leqslant \tilde{C}' \ \| \boldsymbol{U}(t,\boldsymbol{\lambda}) \|^2 \tag{2.85}$$

is satisfies. Or in the next equivalent form: for a sufficiently small $\epsilon \approx 0$ and for a sufficiently small $\varepsilon \approx 0, \varepsilon' \approx 0$ such that

$\dfrac{\varepsilon'}{\varepsilon} \approx 0$, the inequality

$$\left(\lim_{\delta\to 0}\mathbf{E}_{\Omega}\left[\ \| \boldsymbol{x}^{x_0,\varepsilon}_{t,\varepsilon',\epsilon}(\omega,\delta) - \boldsymbol{\lambda} \|^2\right]\right)_{\varepsilon'} \leqslant \tilde{C}' \ \| \boldsymbol{U}(t,\boldsymbol{\lambda}) \|^2 \tag{2.86}$$

is satisfied.

Here the vector-function $\boldsymbol{U}(t,\boldsymbol{\lambda}) = (U_1(t,\lambda),\cdots,U_n(t,\lambda))$ is the solution of the differential master equation:

$$\dot{\boldsymbol{U}}(t,\boldsymbol{\lambda}) = \mathbf{J}[\boldsymbol{b}_0(\boldsymbol{\lambda},t)]\boldsymbol{U}(t,\lambda) + \boldsymbol{b}_0(\lambda,t),\ \boldsymbol{U}(0,\lambda) = \boldsymbol{x}_0 - \boldsymbol{\lambda}, \tag{2.87}$$

Here $\mathbf{J} = \mathbf{J}[\boldsymbol{b}_0(\lambda,t)]$ is a Jacobian i. e., J is $n \times n$ – matrix:

$$\mathbf{J}[\boldsymbol{b}_0(\lambda,t)] = \mathbf{J}[\partial \boldsymbol{b}_{0,i}(x,t)/\partial \boldsymbol{x}_j]_{x=\lambda}. \tag{2.88}$$

Proof. We let now

$$\boldsymbol{x}^{x_0,\varepsilon}_{t,\varepsilon',\epsilon} - \boldsymbol{\lambda} = \boldsymbol{y}^{x_0,\varepsilon}_{t,\varepsilon',\epsilon}. \tag{2.88a}$$

Replacement $\boldsymbol{x}^{x_0,\varepsilon}_{t,\varepsilon',\epsilon} = \boldsymbol{y}^{x_0,\varepsilon}_{t,\varepsilon',\epsilon} + \boldsymbol{\lambda}$ into Eq. (2.52) – Eq. (2.53) gives

$$\left(\mathrm{d}\boldsymbol{y}^{x_0,\varepsilon}_{t,\varepsilon',\epsilon}(\omega,\delta)\right)_{\varepsilon'} = \left(\boldsymbol{b}_{\varepsilon',\epsilon}\left[\boldsymbol{y}^{x_0,\varepsilon}_{t,\varepsilon',\epsilon}(\omega,\delta) + \lambda, \boldsymbol{u}^{\delta}_{t,\varepsilon',\varepsilon,\epsilon}(\omega), t, \varepsilon\right]\right)_{\varepsilon'} + \sqrt{\varepsilon}\mathrm{d}\boldsymbol{W}(t,\omega),$$

$$\boldsymbol{u}^{\delta}_{t,\varepsilon',\varepsilon,\epsilon}(\omega) = \left(u^{\delta}_{1,t,\varepsilon',\varepsilon,\epsilon}(\omega),\cdots,u^{\delta}_{n,t,\varepsilon',\varepsilon,\epsilon}(\omega)\right), \tag{2.89}$$

$$\left(\mathrm{d}u^{\delta}_{i,t,\varepsilon',\varepsilon,\epsilon}(\omega)\right)_{\varepsilon'} = \epsilon\left(\left[x^{x_0,\varepsilon}_{i,t,\varepsilon',\varepsilon,\epsilon}(\omega,\delta)\right]^{2l}\right)_{\varepsilon'} + \sqrt{\delta}\mathrm{d}W_i(t),$$

$$i = 1,\cdots,n, \left(x^{x_0,\varepsilon}_{0,\varepsilon',\epsilon}\right)_{\varepsilon'} = \boldsymbol{x}_0 \in \hat{\mathbf{R}}^n, t \in [0,T], \varepsilon, \varepsilon', \epsilon, \delta \in (0,1].$$

Thus we need to estimate the quantity

$$\lim_{\substack{\varepsilon\to 0\\ \varepsilon'\to 0\\ \epsilon\to 0\\ \left(\frac{\varepsilon'}{\varepsilon}\right)\to 0}} \lim_{\delta\to 0}\mathbf{E}_{\Omega}\left[\ \| \boldsymbol{y}^{x_0,\varepsilon}_{t,\varepsilon',\epsilon}(\omega,\delta) \|^2\right].$$

Application of the Theorem B.4 (see Appendix B) to Eq. (2.89) gives the inequality (2.85) directly.

Theorem 6. (Strong large deviations principle)[5,7]. Assume that CISDE (2.37) – (2.38) is a strongly $\hat{\mathbf{R}}$-dissipative. Then:

①For any solution

$$\left(\boldsymbol{x}_{t,\varepsilon',\epsilon}^{x_0,\varepsilon}(\omega)\right)_{\varepsilon'} = \left(x_{1,t,\varepsilon',\epsilon}^{x_0,\varepsilon},\cdots,x_{n,t,\varepsilon',\epsilon}^{x_0,\varepsilon}\right)_{\varepsilon'}$$

of a strongly $\hat{\mathbf{R}}$ -dissipative CISDE (2.37) – (2.42) and any $\mathbf{R}$- valued parameters $\lambda_1,\cdots,\lambda_n$, there exist finite Colombeau constant $\tilde{C}' = \left(C'_{\varepsilon'}\right)_{\varepsilon'} > 0$, such that $\forall\boldsymbol{\lambda}[\boldsymbol{\lambda} = (\lambda_1,\cdots,\lambda_n)]$ the inequality

$$\lim_{\substack{\varepsilon\to 0\\ \varepsilon'\to 0\\ \epsilon\to 0\\ \left(\frac{\varepsilon'}{\varepsilon}\right)\to 0}} \mathbf{E}_\Omega\left[\ \| x_{t,\varepsilon',\epsilon}^{x_0,\varepsilon}(\omega) - \boldsymbol{\lambda}\|^2\right] \leqslant \tilde{C}'\ \|\boldsymbol{U}(t,\boldsymbol{\lambda})\|^2 \tag{2.90}$$

is satisfied. Or in the next equivalent form: for a sufficiently small $\epsilon \approx 0$ and for a sufficiently small $\varepsilon \approx 0, \varepsilon' \approx 0$ such that $\varepsilon'/\varepsilon \approx 0$, the inequality

$$\left(\mathbf{E}_\Omega\left[\ \|\boldsymbol{x}_{t,\varepsilon',\epsilon}^{x_0,\varepsilon}(\omega) - \boldsymbol{\lambda}\|^2\right]\right)_{\varepsilon'} \leqslant \tilde{C}'\ \|\boldsymbol{U}(t,\lambda)\|^2.$$

is satisfied.

②For any solution

$$\left(\boldsymbol{x}_{t,\varepsilon',\epsilon}^{x_0,\varepsilon}(\omega)\right)_{\varepsilon'} = \left(x_{1,t,\varepsilon',\epsilon}^{x_0,\varepsilon},\cdots,x_{n,t,\varepsilon',\epsilon}^{x_0,\varepsilon}\right)_{\varepsilon'}$$

of a strongly $\hat{\mathbf{R}}$ -dissipative CISDE (2.37) – (2.42) and any $\mathbf{R}$- valued parameters $\lambda_1,\cdots,\lambda_n$, there exist finite Colombeau constant $\tilde{C}' = (C'_{\varepsilon'})_{\varepsilon'} > 0$, such that $\forall\lambda[\lambda = (\lambda_1,\cdots,\lambda_n)]$ the inequality

$$\lim_{\varepsilon\to 0}\mathbf{E}_\Omega\left[\ \|\boldsymbol{x}_{t,\varepsilon'=0,\epsilon=0}^{x_0,\varepsilon}(\omega) - \boldsymbol{\lambda}\|^2\right] \leqslant \tilde{C}'\ \|\boldsymbol{U}(t,\boldsymbol{\lambda})\|^2 \tag{2.91}$$

is satisfies. Here the vector-function $\boldsymbol{U}(t,\boldsymbol{\lambda}) = (U_1(t,\lambda),\cdots,U_n(t,\lambda))$ is the solution of the differential master equation:

$$\dot{\boldsymbol{U}}(t,\lambda) = \mathbf{J}[\boldsymbol{b}_0(\lambda,t)]\boldsymbol{U}(t,\lambda) + \boldsymbol{b}_0(\lambda,t),\boldsymbol{U}(0,\lambda) = \boldsymbol{x}_0 - \lambda, \tag{2.92}$$

Where $\mathbf{J} = \mathbf{J}[\boldsymbol{b}_0(\lambda,t)]$ is a Jacobian i. e., $\mathbf{J}$ is $n \times n$ -matrix:

$\mathbf{J}[\boldsymbol{b}_0(\lambda,t)] = \mathbf{J}[\partial\boldsymbol{b}_{0,i}(x,t)/\partial\boldsymbol{x}_j]_{x=\boldsymbol{\lambda}}$.

Proof 1. From the equality

$$\mathbf{E}_\Omega\left[\ \|\boldsymbol{x}_{t,\varepsilon',\epsilon}^{x_0,\varepsilon}(\omega) - \lambda\|^2\right] =$$

$$\mathbf{E}_\Omega\left\{\ \|[\boldsymbol{x}_{t,\varepsilon',\epsilon}^{x_0,\varepsilon}(\omega) - \boldsymbol{x}_{t,\varepsilon',\epsilon}^{x_0,\varepsilon}(\omega,\delta)] + [\boldsymbol{x}_{t,\varepsilon',\epsilon}^{x_0,\varepsilon}(\omega,\delta) - \lambda]\|^2\right\},$$

by using the triangle inequality, one obtain

$$\sqrt{\mathbf{E}_\Omega[\ \|\boldsymbol{x}_{t,\varepsilon',\epsilon}^{x_0,\varepsilon}(\omega) - \lambda\|^2]} \leqslant \sqrt{\mathbf{E}_\Omega[\ \|\boldsymbol{x}_{t,\varepsilon',\epsilon}^{x_0,\varepsilon}(\omega) - \boldsymbol{x}_{t,\varepsilon',\epsilon}^{x_0,\varepsilon}(\omega,\delta)\|^2]} +$$
$$\sqrt{\mathbf{E}_\Omega[\ \|\boldsymbol{x}_{t,\varepsilon',\epsilon}^{x_0,\varepsilon}(\omega,\delta) - \lambda\|^2]}.$$

Therefore statement ① immediately follows from Theorem A. 1 (see appendix A), Proposition 2 and Theorem 5.

Proof 2. From the equality

$$\mathbf{E}_{\Omega}\left[\ \| \boldsymbol{x}_{t,\varepsilon'=0,\epsilon=0}^{x_0,\varepsilon}(\omega) - \lambda \|^2 \right] =$$

$$\mathbf{E}_{\Omega}\left\{\ \| [\boldsymbol{x}_{t,\varepsilon'=0,\epsilon=0}^{x_0,\varepsilon}(\omega) - \boldsymbol{x}_{t,\varepsilon',\epsilon}^{x_0,\varepsilon}(\omega)] + [\boldsymbol{x}_{t,\varepsilon',\epsilon}^{x_0,\varepsilon}(\omega) - \lambda] \|^2 \right\},$$

by using the triangle inequality, one obtain

$$\sqrt{\mathbf{E}_{\Omega}[\ \| \boldsymbol{x}_{t,\varepsilon'=0,\epsilon=0}^{x_0,\varepsilon}(\omega) - \lambda \|^2]} \leqslant \sqrt{\mathbf{E}_{\Omega}[\ \| \boldsymbol{x}_{t,\varepsilon'=0,\epsilon=0}^{x_0,\varepsilon}(\omega) - \boldsymbol{x}_{t,\varepsilon',\epsilon}^{x_0,\varepsilon}(\omega) \|^2]} +$$

$$\sqrt{\mathbf{E}_{\Omega}[\ \| \boldsymbol{x}_{t,\varepsilon',\epsilon}^{x_0,\varepsilon}(\omega) - \lambda \|^2]}.$$

Therefore statement ② immediately follows from Theorem A. 1 (see appendix A), Proposition 1 and statement ①.

Remark 5. We note that in general case the inequality

$$\left(\delta_{\varepsilon'}(t)\right)_{\varepsilon'} \equiv \left(\lim_{\varepsilon\to 0}\mathbf{E}_{\Omega}\left[\ \| \boldsymbol{x}_{t,\varepsilon'}^{x_0,\varepsilon}(\omega) - \boldsymbol{x}_{t,\varepsilon'}^{x_0,\varepsilon=0} \|^2 \right]\right)_{\varepsilon'} \neq 0$$

is satisfied, see Example 1.

Example 1. Figures 2. 1 – 2. 2.

$$\dot{x}_t^{x_0,\varepsilon} = -a\cdot\left(x_t^{x_0,\varepsilon}\right)^3 - b\cdot\left(x_t^{x_0,\varepsilon}\right)^2 - c\cdot x_t^{x_0,\varepsilon} - \sigma\cdot t^n -$$

$$\chi\cdot t^m\cdot\sin(\Omega\cdot t^k) + \sqrt{\varepsilon}w(t), x_0^{x_0,\varepsilon} = x_0. \tag{2.93}$$

From Eq. (2. 93) and general differential master equation (2. 92) one obtain the next linear differential master equation:

$$\dot{u}(t) = -(3a\lambda^2 + 2b\lambda + c)u(t) - (a\cdot\lambda^3 + b\cdot\lambda^2 + c\cdot\lambda) -$$

$$\sigma\cdot t^n - \chi\cdot t^m\sin(\Omega\cdot t^k), u(0) = x_0 - \lambda. \tag{2.94}$$

From the differential master equation (2. 94) one obtain the transcendental master equation:

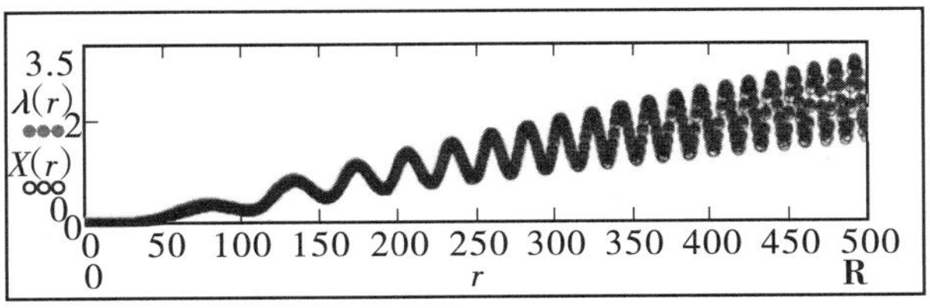

Fig. 2. 1 The solution of the Eq. (2. 8) in a comparison with a corresponding solution $x(t)$ of the ODE (10)

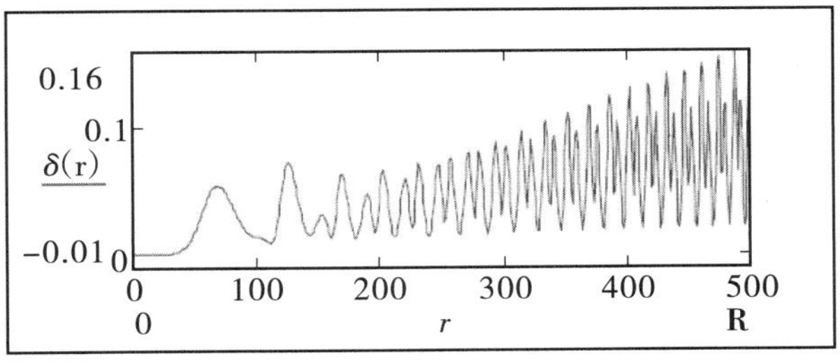

Fig. 2. 2 $\delta(r)$ versus **R**

$$[x_0 - \lambda(t)]\exp\{-[3a\cdot\lambda^2(t) + 2b\cdot\lambda(t)]t\} - \int_0^t[\sigma\cdot\tau^n + \chi\cdot\tau^m\sin(\Omega\cdot\tau^k) +$$
$$a\cdot\lambda^3(t) + b\cdot\lambda^2(t)]\times\exp\{-[3a\cdot\lambda^2(t) + 2b\cdot\lambda(t)]\cdot(t-\tau)\}d\tau$$
$$= 0. \tag{2.95}$$

Example 1. Numerical simulation: Figures 2. 1 and 2. 2.

$a = 1, b = 5, c = 1, \sigma = \chi = -2, m = n = k = 2, \Omega = 5,$

$x_0 = 0, T = 5, R = T/0.001.$

$$\delta(r) = \lim_{\varepsilon\to 0}\mathbf{E}\left(\|\boldsymbol{x}_t^{x_0,\varepsilon} - x_t^{x_0,\varepsilon=0}\|^2\right). \tag{2.96}$$

$$\dot{x}_t^0 = -a(x_t^0)^3 - b(x_t^0)^2 - cx_t^0 - \sigma\cdot t^n - \chi\cdot t^m\cdot\sin(\Omega\cdot t^k). \tag{2.97}$$

Let $\mathfrak{C} = (\Omega, \Sigma, P)$ be a probability space. Let us consider now m-persons Colombeau-Ito's stochastic differential game $\mathrm{CIDG}_{m;T}(\boldsymbol{f}, \boldsymbol{g}, \boldsymbol{y}, G^n(\mathbf{R}^n), \mathfrak{C})$ with nonlinear dynamics:

$$\left(\dot{\boldsymbol{x}}_{t,\varepsilon'}^{x_0,\varepsilon}(\omega)\right)_{\varepsilon'} = \left(\boldsymbol{f}_{\varepsilon'}\left[\boldsymbol{x}_{t,\varepsilon'}^{x_0,\varepsilon}(\omega), \boldsymbol{\alpha}(t), t\right]\right)_{\varepsilon'} + \sqrt{\varepsilon}\,(\boldsymbol{w}(t,\omega))_{\varepsilon'}. \tag{2.98}$$

Here $\varepsilon, \varepsilon' \in (0,1], \varepsilon \ll 1; \forall t \in [0,T]: (\boldsymbol{x}_{\varepsilon'}(t))_{\varepsilon'} \in \hat{\mathbf{R}}^n; \boldsymbol{x}_{0,\varepsilon'}^{x_0,\varepsilon}(\omega) = \boldsymbol{x}_0 \in \mathbf{R}^n, \boldsymbol{f}_{\varepsilon'} = (f_{\varepsilon',1}, \cdots, f_{\varepsilon',n}), \boldsymbol{f} = (\boldsymbol{f}_{\varepsilon'})_{\varepsilon'}, \boldsymbol{g} = (\boldsymbol{g}_{\varepsilon'})_{\varepsilon'};$

$\boldsymbol{f}(x,\circ,\circ), \boldsymbol{g}(x,\circ,\circ) \in G^n(\mathbf{R}^n), \boldsymbol{\alpha}(t) = \{\alpha_1(t), \cdots, \alpha_m(t)\}; \alpha_i(t) \in U_i \subsetneqq \mathbf{R}^{k_i}, i = 1, \cdots, m.$

Here $t \mapsto \alpha_i(t)$, is the control chosen by the i-th player, within a set of admissible control values $U_i \nsubseteq \mathbf{R}^{k_i}$ and the playoff of the i-th player is

$$\left(\bar{\mathbf{J}}_{\varepsilon',i}^{\varepsilon}\right)_{\varepsilon'} = \mathbf{E}\left[\left(\int_0^T g_{\varepsilon',i}^2\left[x_{t,\varepsilon'}^{x_0,\varepsilon}(\omega), \boldsymbol{\alpha}(t), t\right]dt\right)_{\varepsilon'}\right] +$$
$$\mathbf{E}\left[\left(\sum_{i=1}^n\left[x_{T,\varepsilon';i}^{x,\varepsilon}(\omega) - y_i\right]^2\right)_{\varepsilon'}\right]. \tag{2.99}$$

Definition 9. $\mathrm{CIDG}_{m;T}(\boldsymbol{f}, \boldsymbol{g}, \boldsymbol{y}, G^n(\mathbf{R}^n), \boldsymbol{C})$ (2.98) – (2.99) is a strongly $\hat{\mathbf{R}}$-dissipative if CISDE (2.98) is a strongly $\hat{\mathbf{R}}$-dissipative.

Theorem 7. Suppose that:

① $\mathrm{CIDG}_{m;T}(\boldsymbol{f}, \boldsymbol{g}, \boldsymbol{y}, G^n(\mathbf{R}^n), \boldsymbol{C})$ (2.98) – (2.99) is a strongly $\hat{\mathbf{R}}$-dissipative, ② $f_0(\circ, \boldsymbol{\alpha}, t) \equiv \boldsymbol{f}_{\varepsilon'=0}(\circ, \boldsymbol{\alpha}, t): \mathbf{R}^n \to \mathbf{R}^n$ is a polynomial on a variable $\boldsymbol{x} = (x_1, \cdots, x_n)$ and a linear function on a variable $\boldsymbol{\alpha}(t) = \{\alpha_1(t), \cdots, \alpha_m(t)\}$ i. e., $f_{0,i}(\boldsymbol{x}, \boldsymbol{\alpha}, t) = \sum_{\mu, |\mu| \leq r} f_{0,i}^{\mu}(t)x^{\mu} + \sum_{l=1}^m c_{l,i}(t)\alpha_l(t), \boldsymbol{\mu} = (i_1, \cdots, i_n), |\boldsymbol{\mu}| = \sum_{j=1}^n i_j, 0 \leq i_j \leq p$, ③ $\boldsymbol{g}_0(\circ, \boldsymbol{\alpha}, t) \equiv \boldsymbol{g}_{\varepsilon'=0}(\circ, \boldsymbol{\alpha}, t): \mathbf{R}^n \to \mathbf{R}^n$ is a polynomial on a variable $\boldsymbol{x} = (x_1, \cdots, x_n)$ and a linear function on a variable $\boldsymbol{\alpha}(t) = \{\alpha_1(t), \cdots, \alpha_m(t)\}$ i. e., $g_{0,i}(\boldsymbol{x}, \boldsymbol{\alpha}, t) = \sum_{\mu, |\mu| \leq r} g_{0,i}^{\mu}(t)\boldsymbol{x}^{\mu} + \sum_{l=1}^m d_{l,i}(t)\alpha_l(t), \boldsymbol{\mu} = (i_1, \cdots, i_n), |\boldsymbol{\mu}| = \sum_{j=1}^n i_j, 0 \leq i_j \leq p.$

Then For any solution

$$\{\boldsymbol{x}_{t,\varepsilon'}^{x_0,\varepsilon};\breve{\boldsymbol{\alpha}}(t)\} = \left(\left\{x_{1,t,\varepsilon'}^{x_0,\varepsilon},\cdots,x_{n,t,\varepsilon'}^{x_0,\varepsilon}\right\};\left\{\alpha_1(t),\cdots,\alpha_m(t)\right\}\right) \tag{2.100}$$

of the $\mathrm{CIDG}_{m;T}(\boldsymbol{f},\boldsymbol{g},\boldsymbol{y},G^n(\mathbf{R}^n),\boldsymbol{C})$ (2.98) – (2.99) and any **R**- valued parameter $s\{\boldsymbol{\lambda}^{(1)},\boldsymbol{\lambda}^{(2)}\} = \left\{\left(\lambda_1^{(1)},\cdots,\lambda_n^{(1)}\right),\left(\lambda_1^{(2)},\cdots,\lambda_m^{(2)}\right)\right\}$ there exist finite Colombeau constant $\tilde{C}' = \left[\left(C'_{\varepsilon'}\right)_{\varepsilon'}\right] > 0$, such that $\forall\boldsymbol{\lambda}[\boldsymbol{\lambda} = \{\boldsymbol{\lambda}^{(1)},\boldsymbol{\lambda}^{(2)}\}]$ the inequalities

$$\lim_{\substack{\varepsilon\to0\\ \varepsilon'\to0\\ \left(\frac{\varepsilon'}{\varepsilon}\right)\to0}} \mathbf{E}_\Omega\left[\|\boldsymbol{x}_{t,\varepsilon'}^{x_0,\varepsilon}(\omega)-\boldsymbol{\lambda}^{(1)}\|^2\right] \leqslant \tilde{C}'\|\boldsymbol{U}(t,\boldsymbol{\lambda}^{(1)})\|^2,$$

$$\lim_{\substack{\varepsilon\to0\\ \varepsilon'\to0\\ \left(\frac{\varepsilon'}{\varepsilon}\right)\to0}} \mathbf{E}_\Omega\left[\|\boldsymbol{x}_{T,\varepsilon'}^{x_0,\varepsilon}(\omega)-\boldsymbol{\lambda}^{(2)}\|^2\right] \leqslant \tilde{C}'\|\boldsymbol{V}(T,\boldsymbol{\lambda}^{(1)})\|^2,$$

$$\lim_{\substack{\varepsilon\to0\\ \varepsilon'\to0\\ \left(\frac{\varepsilon'}{\varepsilon}\right)\to0}} \mathbf{E}_\Omega\left[\|\boldsymbol{x}_{T,,\varepsilon'}^{x_0,\varepsilon}(\omega)-\boldsymbol{y}\|^2\right] \leqslant \tilde{C}'\|\boldsymbol{V}(T,\boldsymbol{y})\|^2,$$

$$\lim_{\substack{\varepsilon\to0\\ \varepsilon'\to0\\ \left(\frac{\varepsilon'}{\varepsilon}\right)\to0}} \left(\bar{\mathbf{J}}_{\varepsilon',j}^{\varepsilon}\right) \leqslant |V_i(T,\boldsymbol{y},0)| + \|\boldsymbol{U}(T,\boldsymbol{y})\|^2,\ i=1,\cdots,m, \tag{2.101.a}$$

$$\lim_{\varepsilon\to0}\mathbf{E}_\Omega\left[\|\boldsymbol{x}_{t,\varepsilon'=0}^{x_0,\varepsilon}(\omega)-\boldsymbol{\lambda}^{(1)}\|^2\right] \leqslant \tilde{C}'\|\boldsymbol{U}(t,\boldsymbol{\lambda}^{(1)})\|^2,$$

$$\lim_{\varepsilon\to0}\mathbf{E}_\Omega\left[\|\boldsymbol{x}_{T,\varepsilon'=0}^{x_0,\varepsilon}(\omega)-\boldsymbol{\lambda}^{(2)}\|^2\right] \leqslant \tilde{C}'\|\boldsymbol{V}(T,\boldsymbol{\lambda}^{(1)})\|^2,$$

$$\lim_{\varepsilon\to0}\mathbf{E}_\Omega\left[\|\boldsymbol{x}_{T,\varepsilon'=0}^{x_0,\varepsilon}(\omega)-\boldsymbol{y}\|^2\right] \leqslant \tilde{C}'\|\boldsymbol{V}(T,\boldsymbol{y})\|^2,$$

$$\lim_{\varepsilon\to0}\left(\bar{\mathbf{J}}_{\varepsilon'=0,j}^{\varepsilon}\right) \leqslant |V_i(T,\boldsymbol{y},0)| + \|\boldsymbol{U}(T,\boldsymbol{y})\|^2,\ i=1,\cdots,m$$

is satisfied. Or in the next equivalent form: for a sufficiently small $\varepsilon\approx0,\varepsilon'\approx0$ such that $\frac{\varepsilon'}{\varepsilon}\approx0$, the inequalities

$$\left[\left(\liminf_{\varepsilon\to0}\mathbf{E}_\Omega\left[\|\boldsymbol{x}_{t,\varepsilon'}^{x_0,\varepsilon}(\omega)-\boldsymbol{\lambda}^{(1)}\|^2\right]\right)_{\varepsilon'}\right] \leqslant \tilde{C}'\|\boldsymbol{U}(t,\boldsymbol{\lambda}^{(1)})\|^2,$$

$$\left[\left(\liminf_{\varepsilon\to0}\mathbf{E}_\Omega\left[\|\boldsymbol{z}_{T,\varepsilon'}^{z_0,\varepsilon}(\omega)-\boldsymbol{\lambda}^{(2)}\|^2\right]\right)_{\varepsilon'}\right] \leqslant \tilde{C}'\|\boldsymbol{V}(T,\boldsymbol{\lambda})\|^2, \tag{2.101.b}$$

$$\left[\left(\liminf_{\varepsilon\to0}\mathbf{E}_\Omega\left[\|\boldsymbol{x}_{T,\varepsilon'}^{x,\varepsilon}(\omega)-\boldsymbol{y}\|^2\right]\right)_{\varepsilon'}\right] \leqslant \tilde{C}'\|\boldsymbol{U}(T,\boldsymbol{y})\|^2.$$

$$\left(\bar{\mathbf{J}}_{\varepsilon',j}^{\varepsilon}\right)_{\varepsilon'} \leqslant |V_i(T,\boldsymbol{y},0)| + \|\boldsymbol{U}(T,\boldsymbol{y})\|^2,\ i=1,\cdots,m$$

is satisfied. Here $\boldsymbol{z}_{T,\varepsilon'}^{z_0,\varepsilon}(\omega) = \int_0^T \boldsymbol{g}_{\varepsilon'}\left(\boldsymbol{x}_{T,\varepsilon'}^{x_0,\varepsilon}(\omega),\boldsymbol{\alpha}(t),t\right)\mathrm{d}t$.

Here a function $\boldsymbol{W}(t,\boldsymbol{\lambda}) = \{\boldsymbol{U}(t,\boldsymbol{\lambda}),\boldsymbol{V}(t,\boldsymbol{\lambda})\}^t = \{(U_1(t,\boldsymbol{\lambda}),\cdots,U_n(t,\boldsymbol{\lambda}));(V_1(t,\boldsymbol{\lambda}),\cdots,V_m(t,\boldsymbol{\lambda}))\}^t$ is the solution of the differential master game with linear dynamics:

$$\dot{\boldsymbol{W}}(t,\lambda) = \bar{\mathbf{J}}[\hat{\boldsymbol{b}}_0(\lambda,t)]\boldsymbol{W}(t,\lambda) + \hat{\boldsymbol{b}}_0(\lambda,t) + \langle \boldsymbol{d}(t), \breve{\boldsymbol{\alpha}}_i(t)\rangle, \tag{2.102}$$
$$\boldsymbol{U}(0,\lambda) = \boldsymbol{x}_0 - \lambda, \boldsymbol{V}(0,\lambda) = 0.$$

And with the playoff of the i-th player is:

$$\breve{\mathbf{J}}_i = |V_i(T,\boldsymbol{y},0)| + \|\boldsymbol{U}(T,\boldsymbol{y})\|^2. \tag{2.103}$$

Here

$$\hat{\boldsymbol{b}}_0(\lambda,t) = \{\boldsymbol{f}(\lambda,0,t);\boldsymbol{g}(\lambda,0,t)\}^t \tag{2.104}$$

and

$$\mathbf{J} = \mathbf{J}[\hat{\boldsymbol{b}}_0(\lambda,t)] \tag{2.105}$$

is Jacobian i. e., $\mathbf{J}$ is $(n+m)\times(n+m)$-matrix:

$$\mathbf{J}[\hat{\boldsymbol{b}}_0(\lambda,t)] = \mathbf{J}[\partial\hat{\boldsymbol{b}}_{0,i}(x,t)/\partial x_j]_{x=\lambda}. \tag{2.106}$$

Proof. Let us rewrite Eq. (2.98) – Eq. (2.99) of the next equivalent form

$$(\mathrm{i})\ \left(\dot{\boldsymbol{x}}_{t,\varepsilon'}^{x_0,\varepsilon}(\omega)\right)_{\varepsilon'} = \left(\boldsymbol{f}_{\varepsilon'}\left[\boldsymbol{x}_{t,\varepsilon'}^{x_0,\varepsilon}(\omega),\boldsymbol{\alpha}(t),t\right]\right)_{\varepsilon'} + \sqrt{\varepsilon}\left(w(t,\omega)\right)_{\varepsilon'} \tag{2.107}$$

$$(\mathrm{ii})\ \dot{z}_{t,\varepsilon'}^{z_0,\varepsilon}(\omega) = \boldsymbol{g}_{\varepsilon'}^2\left[\boldsymbol{x}_{T,\varepsilon'}^{x_0,\varepsilon}(\omega),\boldsymbol{\alpha}(t),t\right] + \sqrt{\varepsilon}\left(w(t,\omega)\right)_{\varepsilon'}. \tag{2.108}$$

Then the playoff of the i-th player is

$$\left(\bar{\mathbf{J}}_{\varepsilon',i}^{\varepsilon}\right)_{\varepsilon'} = \mathbf{E}\left[\left(z_{T,\varepsilon',i}^{z_0,\varepsilon}(\omega)\right)_{\varepsilon'}\right] + \mathbf{E}\left[\left(\sum_{i=1}^{n}\left[x_{T,\varepsilon';i}^{x,\varepsilon}(\omega) - y_i\right]^2\right)_{\varepsilon'}\right].$$

Here

$$z_{t,\varepsilon',i}^{z_0,\varepsilon}(\omega) = \int_0^t g_{\varepsilon',i}\left(x_{T,\varepsilon'}^{x_0,\varepsilon}(\omega),\boldsymbol{\alpha}(t),t\right)\mathrm{d}t,\ z_0 = 0.$$

The inequalities (2.101) immediately follow from Eq. (2.107) – Eq. (2.108), Theorem 6 and definitions.

Example 2. 2-Persons Ito's stochastic differential game, with a small white noise.

$$(1)\ \dot{x}_1 = x_2, \dot{x}_2 = -kx_2^3 + \alpha_1(t) + \alpha_2(t) + \sqrt{\varepsilon}w(t,\omega); k > 0,$$
$$t \in [0,T], x_1(0) = x_{10}, x_2(0) = x_{20}; \varepsilon \ll 1; \tag{2.109}$$
$$(2)\ \alpha_1(t) \in [-\rho_1,\rho_1], \alpha_2(t) \in [-\rho_2,\rho_2];$$
$$(3)\ J_i = x_1^2(T), i = 1,2.$$

Optimal control problem for the first player is:

$$\min\alpha_1(t) \in [-\rho_1,\rho_1]\left(\max\alpha_2(t) \in [-\rho_2,\rho_2]^{[x_1^2(T)]}\right) \tag{2.110}$$

and optimal control problem for the second player is:

$$\max\alpha_2(t) \in [-\rho_2,\rho_2]\left(\min\alpha_1(t) \in [-\rho_1,\rho_1]^{[x_1^2(T)]}\right) \tag{2.111}$$

Using Eq. (2.102) one obtained the corresponding linear master game:

$$(1)\ \dot{u}_1 = u_2, \dot{u}_2 = -3k\lambda_2^2 u_2 - k\lambda_2^3 + \breve{\alpha}_1(t) + \breve{\alpha}_2(t),\ u_1(0) = x_{10} - \lambda_1,$$
$$x_2(0) = x_{20} - \lambda_2; \tag{2.112}$$

(2) $\breve{\alpha}_1(t) \in (-\rho_1, \rho_1), \breve{\alpha}_2(t) \in (-\rho_2, \rho_2)$;

(3) $\bar{\mathbf{J}}_i = u_1^2(T), i = 1, 2.$

Optimal control problem for the first player is:

$$\min \breve{\alpha}_1(t) \in [-\rho_1, \rho_1] (\max \breve{\alpha}_2(t) \in [-\rho_2, \rho_2]^{[u_1^2(T)]}), \tag{2.113}$$

and optimal control problem for the second player is:

$$\max \breve{\alpha}_2(t) \in [-\rho_2, \rho_2] (\min \breve{\alpha}_1(t) \in [-\rho_1, \rho_1]^{[u_1^2(T)]}), \tag{2.114}$$

Having solved by standard way [20 – 22] linear master game Eqs. (2.112) – (2.114), one obtain optimal feedback control for the first player [5 – 6]:

$$\alpha_1(t) = \breve{\alpha}_1(t, x_1(t), x_2(t)) = \rho_1 \mathrm{sign}[x_1(t) + [\Theta_\tau(t)] x_2(t)] \tag{2.115}$$

and optimal feedback control for the second player:

$$\alpha_2(t) = \breve{\alpha}_2(t, x_1(t), x_2(t)) = \rho_2 \mathrm{sign}[x_1(t) + [\Theta_\tau(t)] x_2(t)]. \tag{2.116}$$

Here

$$\Theta_\tau(t) = \theta_\tau(\eta_\tau(t)), \theta_\tau = \boldsymbol{\tau} - t, \eta_\tau(t) = t - \left(\mathrm{ceil}\left(\frac{t}{\tau}\right) - 1\right)\boldsymbol{\tau}, \tag{2.117}$$

And where ceil (x) is a part-whole of a number $x \in \mathbf{R}$. Thus, for numerical simulation we obtain nonlinear ODE:

$$\dot{x}_1 = x_2, \dot{x}_2 = -k x_2^3 + \alpha_2(t) - \rho_1 \mathrm{sign}[x_1(t) + [\Theta_\tau(t)] x_2(t)]. \tag{2.118}$$

Numerical simulation: Figures 2.3 – 2.6.

$k = 1, |\rho_1| \leqslant 400, x_1(0) = 300\ \mathrm{m}, x_2(0) = 30\ \mathrm{m/s}, T = 80\ \mathrm{s},$

$\alpha_2(t) = A \sin^2(\omega t), A = 100, \omega = 5.$

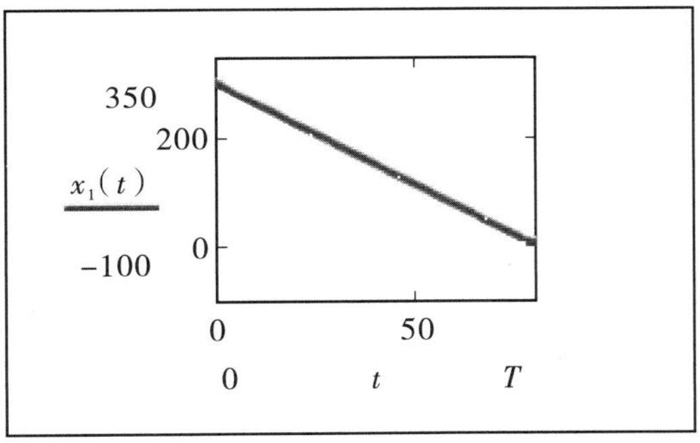

Fig. 2.3 Optimal trajectory: $x_1(t)$. $x_1(T) = 0.4$ m

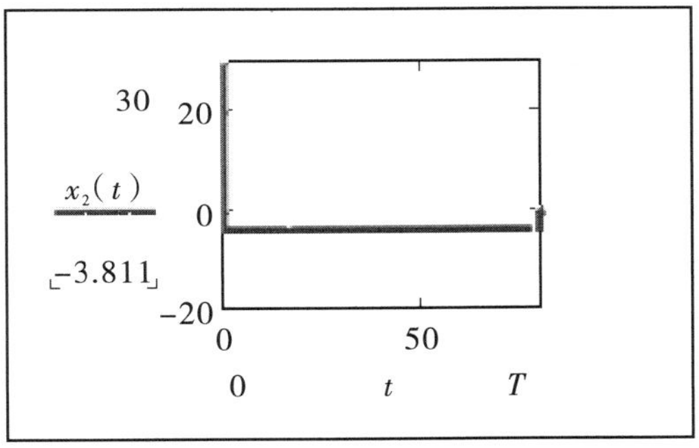

Fig. 2.4 Optimal velocity: $x_2(t)$. $x_2(T) = 0.4$ m/s

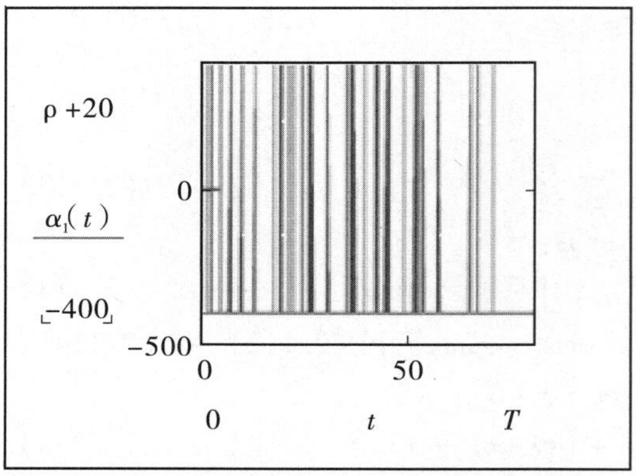

Fig. 2.5 Optimal control of the first player: $\alpha_2(t) = \rho_2 \mathrm{sign}(x_1(t) + (\Theta_\tau(t)) \cdot x_2(t))$

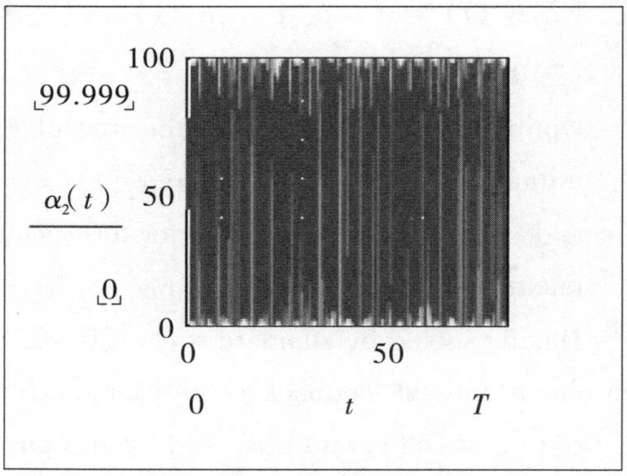

Fig. 2.6 Control of the second player: $\alpha_2(t)$.

Let $\mathfrak{C} = (\Omega, \Sigma, \mathbf{P})$ be a probability space. Let us consider now m-persons Colombeau-Ito's differential game $\mathrm{CIDG}_{m;T}(\boldsymbol{f}, \boldsymbol{g}, \boldsymbol{y}, \boldsymbol{G}^n(\mathbf{R}^n), \boldsymbol{\beta}(t), \boldsymbol{\varphi}(t), \mathfrak{C})$ with imperfect measurements and with imperfect information about the system [5], [6]. The corresponding stochastic nonlinear dynamics is:

$$\left(\dot{\boldsymbol{x}}_{t,\varepsilon'}^{x_0,\varepsilon}(\omega)\right)_{\varepsilon'} = \left(\boldsymbol{f}_{\varepsilon'}\left(\boldsymbol{x}_{t,\varepsilon'}^{x_0,\varepsilon}(\omega), \boldsymbol{\varphi}(t), \boldsymbol{\alpha}(t, x_{t,\varepsilon'}^{x_0,\varepsilon} + \boldsymbol{\beta}(t)), t\right)\right)_{\varepsilon'} + \sqrt{\varepsilon}\left(w(t,\omega)\right)_{\varepsilon'};$$

$$\varepsilon, \varepsilon' \in (0,1], \omega \in \Omega, \qquad (2.119)$$

and the playoff of the i-th player is:

$$\left(\bar{\mathbf{J}}_{\varepsilon',j}^{\varepsilon}\right)_{\varepsilon'} = \mathbf{E}_\Omega\left(\left(\int_0^T \boldsymbol{g}_{\varepsilon',i}(\boldsymbol{x}_{t,\varepsilon'}^{x_0,\varepsilon}(\omega), \boldsymbol{\alpha}(t, \boldsymbol{\beta}(t)), t)\,\mathrm{d}t\right)_{\varepsilon'}\right) +$$

$$\mathbf{E}_\Omega\left(\left(\sum_{i=1}^{n} (\boldsymbol{x}_{T,D,\varepsilon';i}^{x_0,\varepsilon}(\omega) - y_i)^2\right)_{\varepsilon'}\right). \qquad (2.120)$$

Here

$\boldsymbol{\beta}(t) = [\beta_1(t), \cdots, \beta_n(t)]$, $\boldsymbol{\varphi}(t) = [\varphi_1(t), \cdots, \varphi_n(t)]$ and $\forall t(0,T)$: $\left(\boldsymbol{x}_{\varepsilon'}(t)\right)_{\varepsilon'} \in \hat{\mathbf{R}}$; $\boldsymbol{x}_{0,\varepsilon'}^{x_0,\varepsilon}(\omega) = x_0$, $\boldsymbol{f} = \left(\boldsymbol{f}_{\varepsilon'}\right)_{\varepsilon'}$, $\boldsymbol{g} = \left(\boldsymbol{g}_{\varepsilon'}\right)_{\varepsilon'}$; $\boldsymbol{f}(x, \circ, \circ, \circ), \boldsymbol{g}(x, \circ, \circ) \in \boldsymbol{G}^n(\mathbf{R}^n)$, $\boldsymbol{\alpha}(t) = [\alpha_1(t), \cdots, \alpha_m(t)]$; $\boldsymbol{\alpha}_i(t) \in U_i \subsetneq \mathbf{R}^{k_i}$, $i = 1, \cdots, m$; $\boldsymbol{\beta}(t) = [\beta_1(t), \cdots, \beta_n(t)]$, $\boldsymbol{\varphi}(t) = [\varphi_1(t), \cdots, \varphi_n(t)]$.

Definition 10. $\mathrm{CIDG}_{m;T}(\boldsymbol{f}, \boldsymbol{g}, \boldsymbol{y}, \boldsymbol{G}^n(\mathbf{R}^n), \mathfrak{C})$ is a strongly $\hat{\mathbf{R}}$- dissipative if corresponding CISDE (2.119) is a strongly $\hat{\mathbf{R}}$- dissipative.

Theorem 8. Suppose that:

① $\mathrm{CIDG}_{m;T}(\boldsymbol{f}, \boldsymbol{g}, \boldsymbol{y}, \boldsymbol{G}^n(\mathbf{R}^n), \boldsymbol{\beta}(t), \boldsymbol{\varphi}(t), \boldsymbol{C})$ (2.119) – (2.120) is a strongly $\hat{\mathbf{R}}$- dissipative, ② $\boldsymbol{f}_0(\circ, \boldsymbol{\varphi}(t), \boldsymbol{\alpha}, t) \equiv \boldsymbol{f}_{\varepsilon'=0}(\circ, \boldsymbol{\varphi}(t), \boldsymbol{\alpha}, t) : \mathbf{R}^n \to \mathbf{R}^n$ is a polynomial on a variable $\boldsymbol{x} = (x_1, \cdots, x_n)$ and a linear function on a variable $\boldsymbol{\alpha}(t) = [\alpha_1(t), \cdots, \alpha_m(t)]$, i. e., $\boldsymbol{f}_{0,i}(\boldsymbol{x},$

$\varphi(t),\boldsymbol{\alpha},t) = \sum_{\mu,|\mu|\leq r} f^{\mu}_{0,i}(t)\boldsymbol{x}^{\mu} + \sum_{l=1}^{m} c_{l,i}(t)\alpha_l(t), \boldsymbol{\mu} = (i_1,\cdots,i_n), |\boldsymbol{\mu}| = \sum_{j=1}^{n} i_j, 0 \leq i_j \leq p,$

③$\boldsymbol{g}_0(\circ,\boldsymbol{\alpha},t) \equiv \boldsymbol{g}_{\varepsilon'=0}(\circ,\boldsymbol{\alpha},t):\mathbf{R}^n \to \mathbf{R}^n$ is a polynomial on a variable $\boldsymbol{x} = (x_1,\cdots,x_n)$ and a linear function on a variable $\boldsymbol{\alpha}(t) = [\alpha_1(t),\cdots,\alpha_m(t)]$, i. e., $g_{0,i}(\boldsymbol{x},\boldsymbol{\alpha},t) = \sum_{\mu,|\mu|\leq r} g^{\mu}_{0,i}(t)\boldsymbol{x}^{\mu} + \sum_{l=1}^{m} d_{l,i}(t)\alpha_l(t), \boldsymbol{\mu} = (i_1,\cdots,i_n), |\boldsymbol{\mu}| = \sum_{j=1}^{n} i_j, 0 \leq i_j \leq p.$

Then for any solution

$(\boldsymbol{x}^{x_0,\varepsilon}_{t,\varepsilon'};\breve{\boldsymbol{\alpha}}(t)) = \left[\left(\boldsymbol{x}^{x_0,\varepsilon}_{1,t,\varepsilon'},\cdots,\boldsymbol{x}^{x_0,\varepsilon}_{n,t,\varepsilon'}\right);(\boldsymbol{\alpha}_1(t),\cdots,\boldsymbol{\alpha}_m(t))\right]$ of

The $\mathrm{CIDG}_{m;T}(\boldsymbol{f},\boldsymbol{g},\boldsymbol{y},\boldsymbol{G}^n(\mathbf{R}^n),\boldsymbol{\beta}(t),\boldsymbol{\varphi}(t),\mathfrak{C})$ (2.119) – (2.120) and for any $\mathbf{R}$-valued parameters $(\boldsymbol{\lambda}^{(1)},\boldsymbol{\lambda}^{(2)}) = [(\boldsymbol{\lambda}^{(1)}_1,\cdots,\boldsymbol{\lambda}^{(1)}_n),(\lambda^{(2)}_1,\cdots,\lambda^{(2)}_m)]$ there exist finite Colombeau constant $\tilde{C}' = (C'_{\varepsilon'})_{\varepsilon'} > 0$, such that $\forall\boldsymbol{\lambda}(\boldsymbol{\lambda} = (\boldsymbol{\lambda}^{(1)},\boldsymbol{\lambda}^{(2)}))$ the inequalities

$$\begin{aligned}
&(1)\ \left(\liminf_{\varepsilon\to 0}\mathbf{E}_{\Omega}\left[\|\boldsymbol{x}^{x_0,\varepsilon}_{t,,\varepsilon'=0}(\omega) - \boldsymbol{\lambda}^{(1)}\|^2\right]\right)_{\varepsilon'} \leq \tilde{C}'\|\boldsymbol{U}(t,\boldsymbol{\lambda}^{(1)})\|^2,\\
&(2)\ \left(\liminf_{\varepsilon\to 0}\mathbf{E}_{\Omega}\left[\|\boldsymbol{z}^{z_0,\varepsilon}_{T,,\varepsilon'=0}(\omega) - \boldsymbol{\lambda}^{(2)}\|^2\right]\right)_{\varepsilon'} \leq \tilde{C}'\|\boldsymbol{V}(T,\boldsymbol{\lambda})\|^2,\\
&(3)\ \left(\liminf_{\varepsilon\to 0}\mathbf{E}_{\Omega}\left[\|\boldsymbol{x}^{x,\varepsilon}_{T,\varepsilon'=0}(\omega) - \boldsymbol{y}\|^2\right]\right)_{\varepsilon'} \leq \tilde{C}'\|\boldsymbol{U}(T,\boldsymbol{y})\|^2.\\
&(4)\ \left(\breve{\mathbf{J}}^{\varepsilon}_{\varepsilon',j}\right)_{\varepsilon'} \leq |V_i(T,\boldsymbol{y},0)| + \|\boldsymbol{U}(T,\boldsymbol{y})\|^2, i = 1,\cdots,m,
\end{aligned} \tag{2.121}$$

is satisfied.

Here $\boldsymbol{z}^{z_0,\varepsilon}_{T,\varepsilon'}(\omega) = \int_0^T \boldsymbol{g}_{\varepsilon'}(\boldsymbol{x}^{x_0,\varepsilon}_{T,\varepsilon'}(\omega),\boldsymbol{\alpha}(t),t)\,dt.$

Here a function

$\boldsymbol{W}(t,\boldsymbol{\lambda}) = \{\boldsymbol{U}(t,\boldsymbol{\lambda}),\boldsymbol{V}(t,\boldsymbol{\lambda})\}^t = \{(\boldsymbol{U}_1(t,\boldsymbol{\lambda}),\cdots,\boldsymbol{U}_n(t,\boldsymbol{\lambda}));(\boldsymbol{V}_1(t,\boldsymbol{\lambda}),\cdots,\boldsymbol{V}_m(t,\boldsymbol{\lambda}))\}^t$

is the solution of the differential master game with linear imperfect dynamics and with imperfect measurements:

$$\begin{aligned}
&\dot{\boldsymbol{W}}(t,\boldsymbol{\lambda}) = \mathbf{J}[\boldsymbol{b}_0(\boldsymbol{\lambda},\boldsymbol{\varphi}(t),t)]\boldsymbol{W}(t,\boldsymbol{\lambda}) + \boldsymbol{b}_0[\boldsymbol{\lambda},\boldsymbol{\varphi}(t),t] + [\boldsymbol{d}(t),\breve{\boldsymbol{\alpha}}_l(t,\boldsymbol{\beta}(t))],\\
&\boldsymbol{U}(0,\boldsymbol{\lambda}) = \boldsymbol{x}_0 - \boldsymbol{\lambda}, \boldsymbol{V}(0,\boldsymbol{\lambda}) = 0
\end{aligned} \tag{2.122}$$

and the play off of the i-th player is:

$$\breve{\mathbf{J}}_i = |V_i(T,\boldsymbol{y},0)| + \|\boldsymbol{U}(T,\boldsymbol{y})\|^2. \tag{2.123}$$

Here

$$\hat{\boldsymbol{b}}_0[\boldsymbol{\lambda},\boldsymbol{\varphi}(t),t] = [\boldsymbol{f}(\boldsymbol{\lambda},\boldsymbol{\varphi}(t),0,t);\boldsymbol{g}(\boldsymbol{\lambda},0,t)]^t, \tag{2.124}$$

And here $\mathbf{J} = \mathbf{J}[\hat{\boldsymbol{b}}_0(\boldsymbol{\lambda},t)]$ is Jacobian, i. e., $\mathbf{J}$ is $(n+m)\times(n+m)$-matrix:

$$\mathbf{J}[\hat{\boldsymbol{b}}_0(\boldsymbol{\lambda},\boldsymbol{\varphi}(t),t)] = \mathbf{J}[\partial\hat{\boldsymbol{b}}_{0,i}(\boldsymbol{x},\boldsymbol{\varphi}(t),t)/\partial x_j]_{x=\lambda}. \tag{2.125}$$

Proof. The proof is completely to similarly a proof of the Theorem 7.

Example 3.2 m-Persons Ito's stochastic differential game with a small white noise, and with imperfect measurements.

(1) $\dot{x}_1 = x_2, \dot{x}_2 = -k_1 x_2^3 + k_2 x_2^2 + \alpha_1[t, x_1(t), x_2(t) + \beta(t)] + \alpha_2(t) + \sqrt{\varepsilon} w(t,\omega); k_1 > 0, t \in [0,T], x_1(0) = x_{10}, x_2(0) = x_{20}; \varepsilon \ll 1;$ (2.126)

(2) $\alpha_1(t) \in [-\rho_1, \rho_1], \alpha_2(t) \in [-\rho_2, \rho_2];$

(3) $J_i = x_1^2(T), i = 1,2.$

(4) $\beta(t) = A\sin^2(\omega t).$

From Eq. (2.122) one obtained corresponding linear master game:

(1) $\dot{u}_1 = u_2, \dot{u}_2 = -(3k_1\lambda_2^2 - 2k_2\lambda_2)u_2 - k\lambda_2^3 + \breve{\alpha}_1[t, u_1(t), u_2(t) + \beta(t)] + \breve{\alpha}_2(t),$
$u_1(0) = x_{10} - \lambda_1, x_2(0) = x_{20} - \lambda_2;$

(2) $\breve{\alpha}_1(t) \in [-\rho_1, \rho_1], \breve{\alpha}_2(t) \in [-\rho_2, \rho_2];$ (2.127)

(3) $\bar{\mathbf{J}}_i = u_1^2(T), i = 1,2.$

Having solved by standard way linear master game (2.126) one obtain local optimal feedback control of the first player[5]:

$$\alpha_1(t_{n+1}) = -\rho_1 \text{sign}\{x_1(t_n) + (t_{n+1} - t_n)[x_2(t_n) + \beta(t_n)]\}, \quad (2.128)$$

and local optimal feedback control of the second player:

$$\alpha_2(t_{n+1}) = \rho_2 \text{sign}\{x_1(t_n) + (t_{n+1} - t_n)[x_2(t_n) + \beta(t_n)]\}. \quad (2.129)$$

Thus, finally we obtain global optimal control of the next form[5]:

$$\alpha_1(t) = -\rho_1 \text{sign}\{x_1(t) + [\Theta_\tau(t)] \cdot [x_2(t) + \beta(t)]\}, \quad (2.130)$$

$$\alpha_2(t) = \rho_2 \text{sign}\{x_1(t) + [\Theta_\tau(t)] \cdot [x_2(t) + \beta(t)]\}. \quad (2.131)$$

Here $\Theta_\tau(t) = \theta_\tau[\eta_\tau(t)], \boldsymbol{\theta}_\tau = \tau - t, \eta_\tau(t) = t - [\text{ceil}(t/\tau) - 1]\boldsymbol{\tau}$, where ceil (x) is a part-whole of a number $x \in \mathbf{R}$. Thus, for numerical simulation we obtain ODE:

$$\begin{aligned} &\dot{x}_1 = x_2, \dot{x}_2 = -k_1 x_2^3 + k_2 x_2^2 - \rho_1 \text{sign}\{x_1(t)[\Theta_\tau(t)][x_2(t) + \beta(t)]\} + \\ &\rho_2 \text{sign}\{x_1(t) + [\Theta_\tau(t)][x_2(t) + \beta(t)]\}. \end{aligned} \quad (2.132)$$

Numerical simulation: Figures 2.7 – 2.12.

Game with imperfect measurements (red curves $x_1(t)$ and $x_2(t)$) in comparison with a classical game with perfect measurements: blue curves $y_1(t)$ and $y_2(t)$, $\beta(t) = A\sin^2(\omega t)$, $A = 100, \omega = 5$.

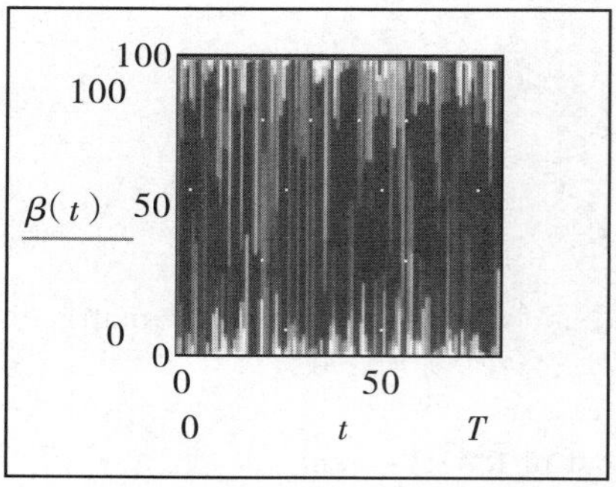

Fig. 2.7 Uncertainty of speed measurements $\beta(t)$

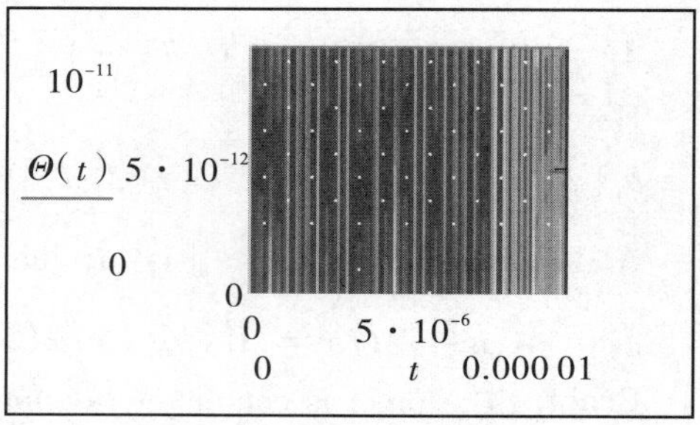

Fig. 2.8 Cutting function: $\Theta_\tau(t), \tau = 10^{11}$

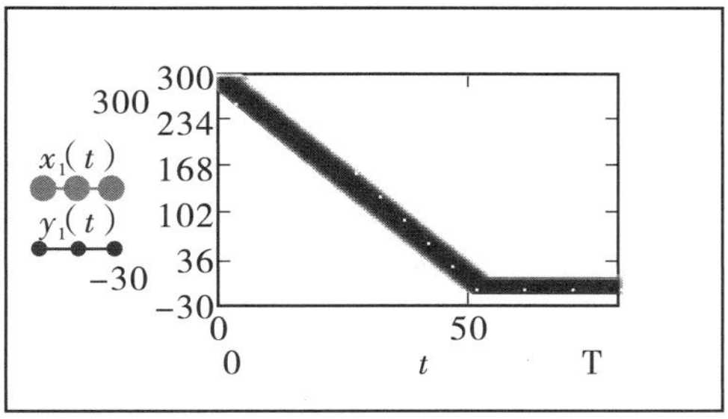

Fig. 2. 9 Optimal trajectory

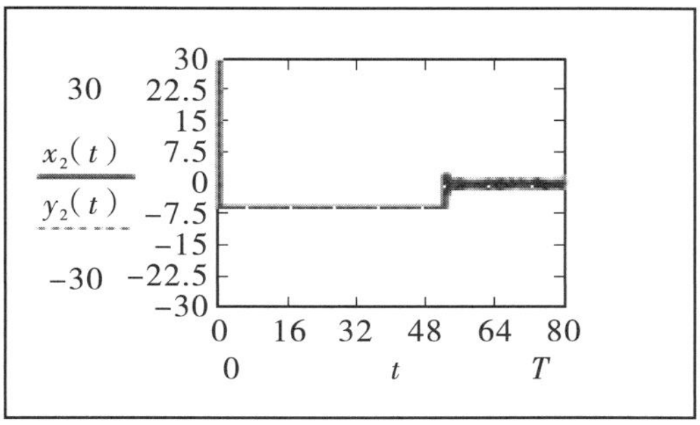

Fig. 2. 10 Optimal velocity

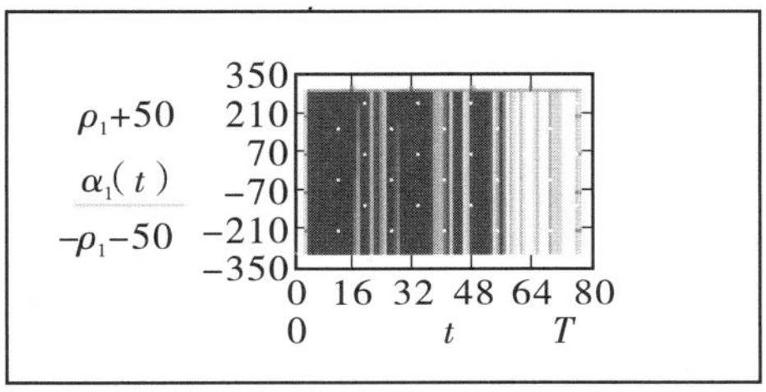

Fig. 2. 11 Optimal control of the first player

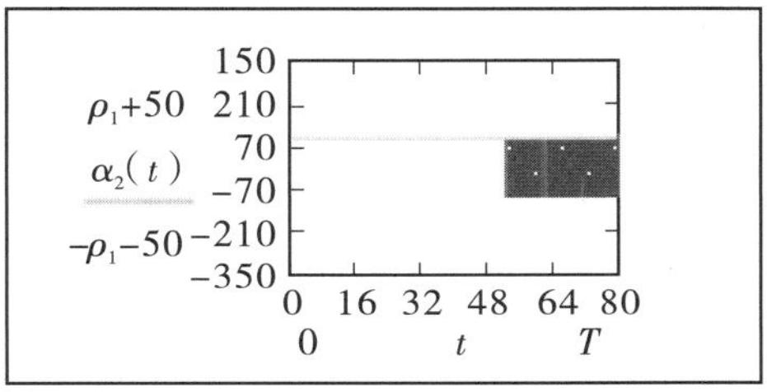

Fig. 2. 12 Optimal control of the second player

2. 3 Homing missile guidance with imperfect measurements capable to defeat in conditions of hostile active radio-electronic jamming

Homing missile guidance strategies (guidance laws) dictate the manner in which the missile will guide to intercept, orrendezvous with, the target[20-21]. The feedback nature of homing guidance allows the guided missile (or, more generally, the pursuer) to tolerate some level of (sensor) measurement uncertainties, errors in the assumptions used to model the engagement (e. g. , unanticipated target maneuver), and errors in modeling missile capability (e. g. , deviation of actual missile speed of response to guidance commands from the design assumptions). Nevertheless, the selection of a guidance strategy and its subsequent mechanization are crucial design factors that can have substantial impact on guided missile performance. Key drivers to guidance law design include the type of targeting sensor to be used (passive IR, active or semi-active RF, etc.), accuracy of the targeting and inertial measurement unit (IMU) sensors, missile maneuverability, and, finally yet important, the types of targets to be engaged and their associated maneuverability levels.

Fig. 2. 13 shows the intercept geometry of a missile in planar pursuit of a target. Taking the origin of the reference frame to be the instantaneous position of the missile, the equation of motion in polar form are[22]:

$$\ddot{R} = R\dot{\sigma}^2 + a_M^r[t,\tilde{R}(t),\tilde{\dot{R}}(t)] + a_T^r(t), \tag{2.133}$$

$$R\ddot{\sigma} + 2\dot{R}\dot{\sigma} = a_M^n[t,\tilde{\sigma}(t),\tilde{\dot{\sigma}}(t)] + a_T^n(t), \tag{2.134}$$

$a_M^r(t) \in [-\bar{a}_M^r, \bar{a}_M^r], a_T^r(t) \in [-\bar{a}_T^r, \bar{a}_T^r]$,

$a_M^n(t) \in [-\bar{a}_M^n, \bar{a}_M^n], a_T^n(t) \in [-\bar{a}_T^n, \bar{a}_T^n]$.

(1) The variable $R = R(t)$ denotes a true target-to-missile range $R_{TM}(t)$.

(2) The variable $\tilde{R} = \tilde{R}(t)$ denotes it is measured target-to-missile range :$R_{TM}(t)$.

(3) The variable $\sigma = \sigma(t)$ denotes a true line-of-sight angle (LOST), i. e., it is true angle between the constant reference direction and target-to-missile direction.

(4) The variable $\tilde{\sigma} = \tilde{\sigma}(t)$ denotes it's really measured line-of-sight angle (LOSM), i. e., it is measured angle between the constant reference direction and target-to-missile direction.

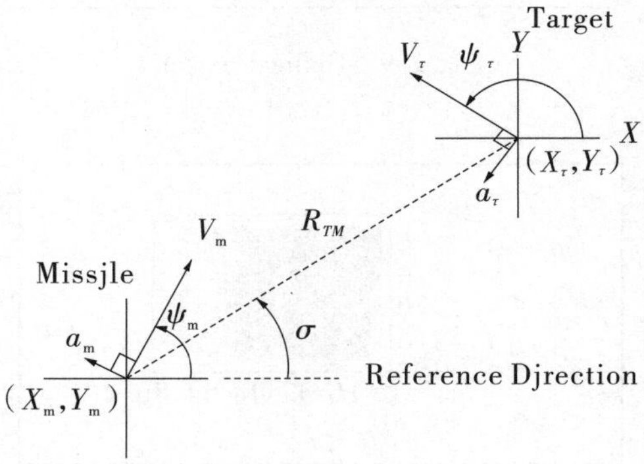

Fig. 2. 13 Planar intercept geometry

(5) The variable $a_M^n(t) = a_M^n[t,\sigma(t),\dot{\sigma}(t)]$ denotes the missiles acceleration along direction which perpendicularly to line-of-sight direction.

(6) The variable $a_M^r(t) = a_M^r[t,R(t),\dot{R}(t)]$ denotes the missile acceleration along target-to-missile direction.

(7) The variable $a_T^n(t)$ denotes the target acceleration along direction which perpendicularly to line-of-sight direction.

(8) The variable $a_T^r(t)$ denotes the target acceleration along target-to-missile direction.

Using replacement $z = R\sigma$ into Eqs. (2.133) – (2.134) one obtain:

$$\ddot{R} = \frac{\dot{z}^2}{R} + a_M^r[t,\tilde{R}(t),\tilde{\dot{R}}(t)] + a_T^r(t), \tag{2.135}$$

$$\ddot{z} = -\frac{\ddot{R}z}{R} + a_M^n[t,\tilde{z}(t),\tilde{\dot{z}}(t)] + a_T^n(t), \tag{2.136}$$

$$\tilde{z}(t) = \tilde{R}(t)\,\tilde{\sigma}(t), \tag{2.137}$$

$$\tilde{\dot{z}}(t) = \tilde{\dot{R}}(t)\,\tilde{\sigma}(t) + \tilde{R}(t)\,\tilde{\dot{\sigma}}(t). \tag{2.138}$$

Here we denoted:

$$\tilde{\dot{R}}(t) \triangleq \tilde{\dot{R}}, \tilde{\dot{\sigma}}(t) \triangleq \tilde{\dot{\sigma}}, \tilde{\dot{z}}(t) \triangleq \tilde{\dot{z}}(t), \tag{2.139}$$

$$\tilde{\ddot{\sigma}}(t) \triangleq \tilde{\ddot{\sigma}}(t), \tilde{\ddot{z}}(t) \triangleq \tilde{\ddot{z}}(t). \tag{2.140}$$

Suppose that:

$$\tilde{R}(t) = R(t) + \beta_1(t), \tilde{\sigma}(t) = \sigma(t) + \beta_2(t). \tag{2.141}$$

Therefore:

$$\tilde{\dot{R}}(t) = \dot{R}(t) + \dot{\beta}_1(t) = \dot{R}(t) + \bar{\beta}_1(t), \bar{\beta}_1(t) = \dot{\beta}_1(t). \quad (2.142)$$

$$\tilde{\dot{\sigma}}(t) = \dot{\sigma}(t) + \dot{\beta}_2(t), \tilde{\ddot{\sigma}}(t) = \ddot{\sigma}(t) + \ddot{\beta}_2(t). \quad (2.143)$$

$$\tilde{\dot{z}}(t) = \tilde{R}(t)\tilde{\dot{\sigma}}(t) = [R(t) + \beta_1(t)][\dot{\sigma}(t) + \dot{\beta}_2(t)] = \quad (2.144)$$

$$R(t)\dot{\sigma}(t) + \beta_1(t)[\dot{\sigma}(t) + \dot{\beta}_2(t)] + R(t)\dot{\beta}_2(t) \approx$$

$$\dot{z}(t) + \beta_1(t)[\tilde{\dot{\sigma}}(t) + \dot{\beta}_2(t)] + \tilde{R}(t)\dot{\beta}_2(t) = \dot{z}(t) + \bar{\beta}_3(t), \quad (2.145)$$

$$\bar{\beta}_3(t) = \beta_1(t)[\tilde{\dot{\sigma}}(t) + \dot{\beta}_2(t)] + \tilde{R}(t)\dot{\beta}_2(t), \quad (2.146)$$

$$\tilde{z}(t) = z(t) + \int_0^t \bar{\beta}_3(\tau)\mathrm{d}\tau = z(t) + \bar{\beta}_4(t). \quad (2.147)$$

Let us consider antagonistic Colombeau differential game $\mathrm{CIDG}_{2;T}(\boldsymbol{f},\boldsymbol{g},\boldsymbol{y},\boldsymbol{G}^n(\mathbf{R}^n),\boldsymbol{\beta}(t),\mathfrak{C}),\boldsymbol{\beta}(t) = (\beta_1(t),\beta_2(t))$ with non-linear dynamics given by Eq. (2.135) – Eq. (2.136), and imperfect measurements [5–6]. Using replacement $\dot{R}(t) = V_r(t)$, $\dot{z}(t) = \eta(t)$, from Eq. (2.135) – Eq. (2.147) one obtain:

$$\dot{R} = V_r, \quad (2.148)$$

$$\dot{V}_r = \frac{\dot{z}^2}{R} + a_M^r(t) + a_T^r(t), \quad (2.149)$$

$$a_M^r(t) = \breve{a}_M^r[t,\tilde{R}(t),\tilde{V}_r(t)] - k_1\tilde{V}_r^3(t), \quad (2.150)$$

$$\tilde{R}(t) = R(t) + \beta_1(t), \tilde{V}_r(t) = V_r + \bar{\beta}_1(t), \quad (2.151)$$

$$\dot{z} = \eta, \quad (2.152)$$

$$\dot{\eta} = -\frac{V_r\eta}{R} + a_M^n(t) + a_T^n(t), \quad (2.153)$$

$$a_M^n(t) = \breve{a}_M^n[t,\tilde{z}(t),\tilde{\eta}(t)] - k_2\tilde{\eta}^3(t), \quad (2.154)$$

$$\tilde{\eta}(t) = \eta(t) + \bar{\beta}_3(t), \tilde{\dot{\eta}}(t) = \dot{\eta}(t) + \bar{\beta}_4(t), \quad (2.155)$$

$$\breve{a}_M^r(t) \in [-\bar{a}_M^r,\bar{a}_M^r], \breve{a}_T^r(t) \in [-\bar{a}_T^r,\bar{a}_T^r], \quad (2.156)$$

$$\breve{a}_M^n(t) \in [-\bar{a}_M^n,\bar{a}_M^n], \breve{a}_T^n(t) \in [-\bar{a}_T^n,\bar{a}_T^n], \quad (2.157)$$

$$\mathbf{J}_i = R^2(t_1), i = 1,2. \quad (2.158)$$

Optimal control problem of the first player is:

$$\bar{\mathbf{J}}_1 = \left\{ \begin{array}{l} \min['\gamma_1(t_1)] \\ \breve{a}_M^r(t) \in [-\bar{a}_M^r,\bar{a}_M^r], \breve{a}_M^n(t) \in [-\bar{a}_M^n,\bar{a}_M^n] \end{array} \right\}, \quad (2.159)$$

where

$$'\gamma_1(t_1) = \left\{ \begin{array}{l} \max[R^2(t_1)] \\ \breve{a}_T^r(t) \in [-\bar{a}_T^r,\bar{a}_T^r], \breve{a}_T^n(t) \in [-\bar{a}_T^n,\bar{a}_M^n] \end{array} \right\}. \quad (2.160)$$

Optimal control problem of the second player is:

$$\bar{\mathbf{J}}_2 = \left\{\begin{array}{l}\max[\,'\gamma_2(t_1)] \\ \breve{a}_T^r(t) \in [-\bar{a}_T^r, \bar{a}_M^r], \breve{a}_T^n(t) \in [-\bar{a}_T^n, \bar{a}_M^n]\end{array}\right\}, \quad (2.161)$$

where

$$'\gamma_2(t_1) = \left\{\begin{array}{l}\min[R^2(t_1)] \\ \breve{a}_M^r(t) \in [-\bar{a}_M^r, \bar{a}_M^r], \breve{a}_M^n(t) \in [-\bar{a}_M^n, \bar{a}_M^n]\end{array}\right\}. \quad (2.162)$$

From Eqs. (2.148) – (2.162) one obtain corresponding linear master game:

$$\dot{r}(t) = v_r(t) + \lambda_2, \quad (2.163)$$

$$\dot{v}_r(t) = -3k_1\lambda_2^2\tilde{v}_r(t) - k_1\lambda_2^3 + \breve{a}_M^r(t) + \breve{a}_T^t(t), \quad (2.164)$$

$$\breve{a}_M^r(t) = \breve{a}_M^r[t, \tilde{r}(t), \tilde{v}_r(t)], \quad (2.165)$$

$$\tilde{r}(t) = \lambda_1 + r(t) + \beta_1(t), \quad (2.166)$$

$$\tilde{v}_r(t) = \lambda_2 + v_r(t) + \bar{\beta}_1(t), \quad (2.167)$$

$$\dot{z}_1(t) = \eta_1(t) + \lambda_3, \quad (2.168)$$

$$\dot{\eta}_1(t) = -3k_2\lambda_3^2\tilde{\eta}_1(t) - k_2\lambda_3^3 + \breve{a}_M^n(t) + \breve{a}_T^n(t), \quad (2.169)$$

$$\tilde{\eta}_1(t) = \lambda_3 + \eta_1(t) + \bar{\beta}_3(t), \quad (2.170)$$

$$\breve{a}_M^r(t) \in [-\bar{a}_M^r, \bar{a}_M^r], \breve{a}_T^r(t) \in [-\bar{a}_T^r, \bar{a}_T^r], \quad (2.171)$$

$$\breve{a}_M^n(t) \in [-\bar{a}_M^n, \bar{a}_M^n], \breve{a}_T^n(t) \in [-\bar{a}_T^n, \bar{a}_T^n], \quad (2.172)$$

$$\mathbf{J}_i = r^2(t_1), i = 1, 2. \quad (2.173)$$

Optimal control problem of the first player is:

$$\mathbf{J}_1 = \left\{\begin{array}{l}\min[\,'\gamma_1(t_1)] \\ \breve{a}_M^r(t) \in [-\bar{a}_M^r, \bar{a}_M^r], \breve{a}_M^n(t) \in [-\bar{a}_M^n, \bar{a}_M^n]\end{array}\right\}, \quad (2.174)$$

where

$$'\gamma_1(t_1) = \left\{\begin{array}{l}\max[J_1(t_1)] \\ \breve{a}_T^r(t) \in [-\bar{a}_T^r, \bar{a}_T^r], \breve{a}_T^n(t) \in [-\bar{a}_T^n, \bar{a}_T^n]\end{array}\right\}. \quad (2.175)$$

Optimal control problem of the second player is:

$$\bar{\mathbf{J}}_2 = \left\{\begin{array}{l}\max[\,'\gamma_2(t_1)] \\ \breve{a}_T^r(t) \in [-\bar{a}_T^r, \bar{a}_T^r], \breve{a}_T^n(t) \in [-\bar{a}_T^n, \bar{a}_M^n]\end{array}\right\}, \quad (2.176)$$

where

$$'\gamma_2(t_1) = \left\{\begin{array}{l}\min[J_2(t_1)] \\ \breve{a}_M^r(t) \in [-\bar{a}_M^r, \bar{a}_M^r], \breve{a}_M^n(t) \in [-\bar{a}_M^n, \bar{a}_M^n]\end{array}\right\}. \quad (2.177)$$

From Eq. (2.26) we obtain quasi optimal solution for the antagonistic differential game $\mathrm{CIDG}_{2;T}(\boldsymbol{f}, \boldsymbol{g}, \boldsymbol{y}, \boldsymbol{G}^n(\mathbf{R}^n), \boldsymbol{\beta}(t), \mathfrak{C})$ given by Eqs. (2.21) – (2.23). Optimal control $[a_M^r(t), a_M^n(t)]$ of the first player are [5–6]:

$$\breve{a}_M^n[t, \tilde{R}(t), \tilde{V}_r(t)] = -\bar{a}_M^r \mathrm{sign}\{[R(t) + \beta_1(t)] + \Theta_\tau(t)[V_r(t) + \bar{\beta}_1(t)]\}, \quad (2.178)$$

$$\breve{a}^n_M[t,\tilde{z}(t),\tilde{\dot{z}}(t)] = -\bar{a}^n_M \mathrm{sign}\{[z(t)+\bar{\beta}_4(t)] + \Theta_\tau(t)[\dot{z}(t)+\bar{\beta}_3(t)]\}. \tag{2.179}$$

Thus, for numerical simulation we obtain ODE's:

$$\dot{R} = V_r, \tag{2.180}$$

$$\dot{V}_r = \frac{\dot{z}^2}{R} - \bar{a}^r_M \mathrm{sign}\{[R(t)+\bar{\beta}_1(t)] + \Theta_\tau(t)[V_r(t)+\bar{\beta}_1(t)]\} - k_1[V_r(t)+\bar{\beta}_1(t)]^3 + a^r_T(t), \tag{2.181}$$

$$\dot{z} = \eta, \tag{2.182}$$

$$\dot{\eta} = -\frac{V_r\eta}{R} - \bar{a}^n_M \mathrm{sign}\{[z(t)+\bar{\beta}_4(t)] + \Theta_\tau(t)[\dot{z}(t)+\bar{\beta}_3(t)]\} - k_2[z(t)+\bar{\beta}_4(t)]^3. \tag{2.183}$$

Example 4. Figures 2.14 – 2.24. $\tau = 10^{-3}$, $k_1 = k_2 = 10^{-3}$, $\bar{a}^r_T = 20\ \mathrm{m/s^2}$, $\bar{a}^\tau_T = 20\ \mathrm{m/s^2}$, $R(0) = 200$ m, $V_r(0) = 10$ m/s, $z(0) = 60$, $\dot{z}(0) = 40$, $\bar{a}^r_T(t) = \bar{a}^r_T[\sin(\omega t)]^p$, $a^\tau_T(t) = \bar{a}^\tau_T \cdot [\sin(\omega t)]^q$, $\beta_i(t) = \bar{\beta}[\sin(\omega t)]^p$, $i = 1,2,3,4$, $\omega = 50$, $\bar{\beta} = 20$, $p = 2$, $q = 1$.

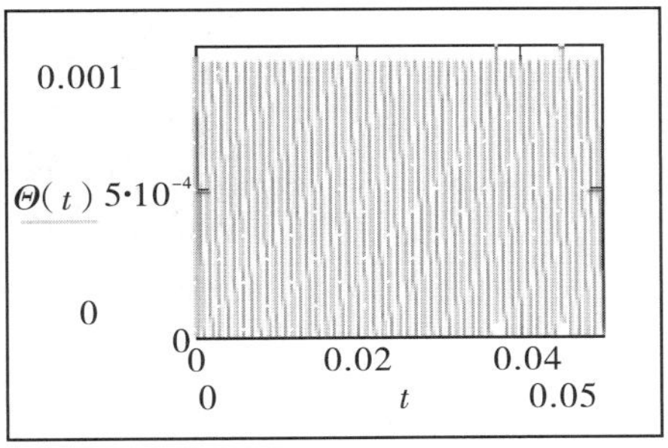

Fig. 2.14 Cutting function: $\Theta_\tau(t)$

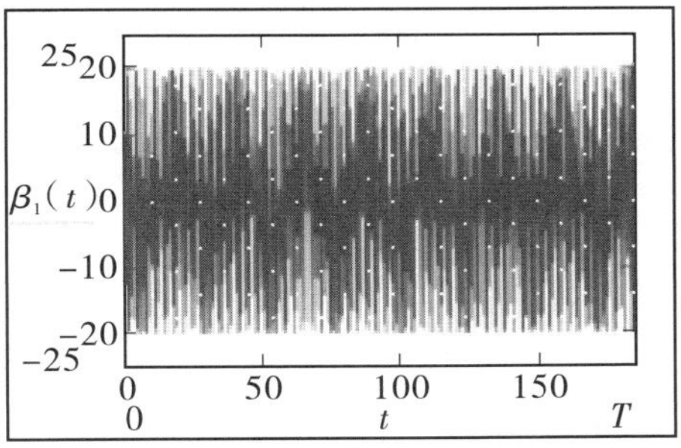

Fig. 2.15 Uncertainty $\beta_1(t)$ of measurements of a variable $R(t)$.

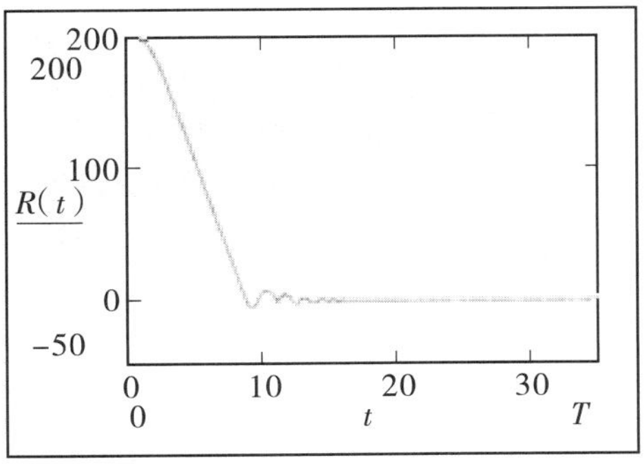

Fig. 2.16 Target-to-missile range $R(t)$. $R(30) = 7.20^3$ m

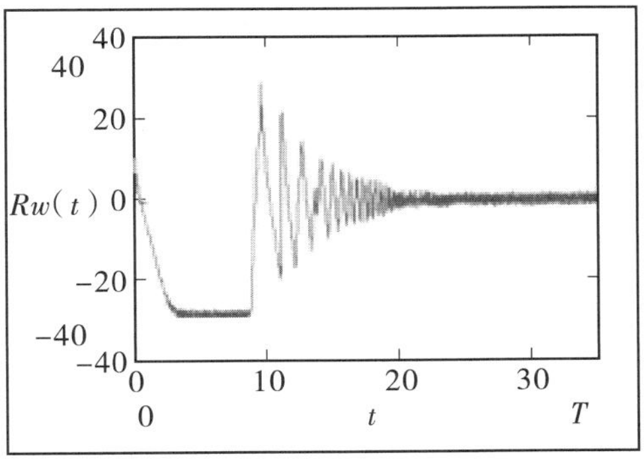

Fig. 2.17 Speed of rapprochement missile-to-target: $\dot{R}(t)$

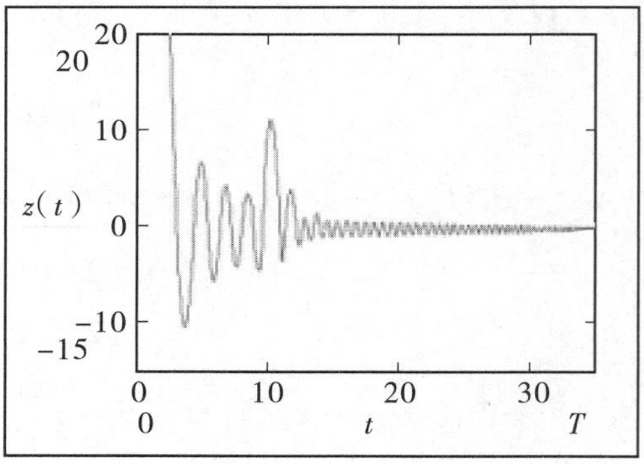

Fig. 2. 18　Variable $\dot{z}(t) = R(t)\dot{\sigma}(t)$

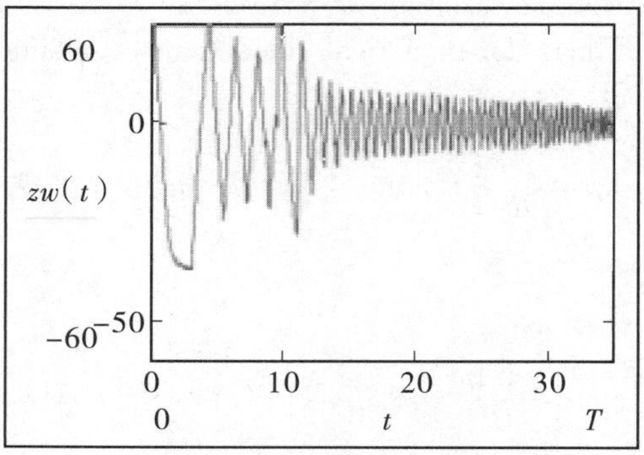

Fig. 2. 19　Variable $\ddot{z}(t), \ddot{z}(T) = 2.172.$

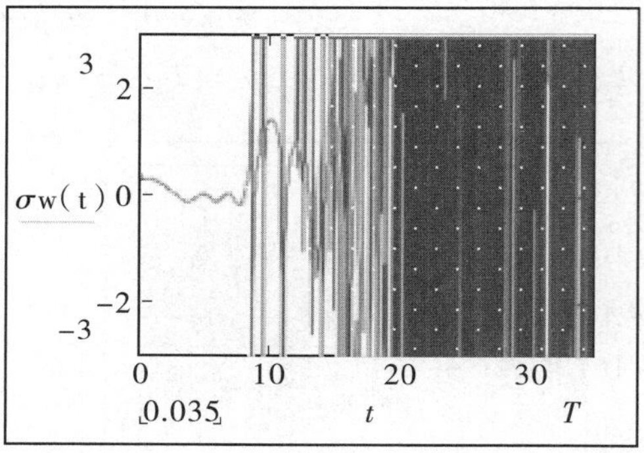

Fig. 2. 20　Variable $\dot{\sigma}(t), \dot{\sigma}(0) = 0.3$

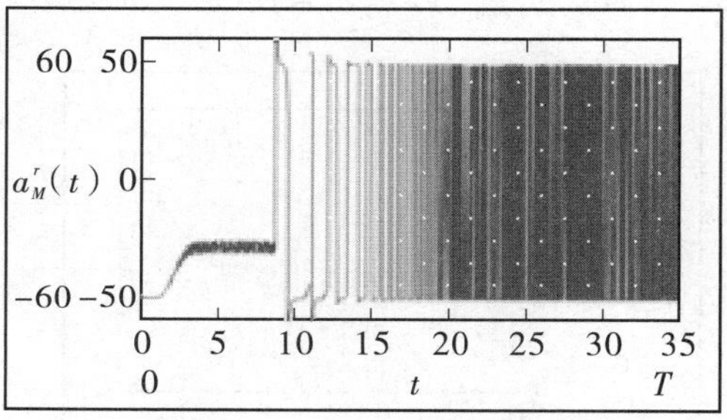

Fig. 2. 21　Missile acceleration along target-to-missile direction: $a_M^r(t)$

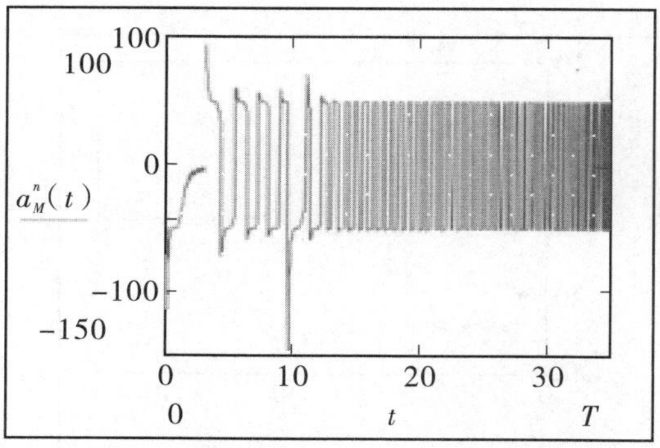

Fig. 2. 22　Missile acceleration along direction which perpendicularly to line-of-sight direction: $a_M^n(t)$

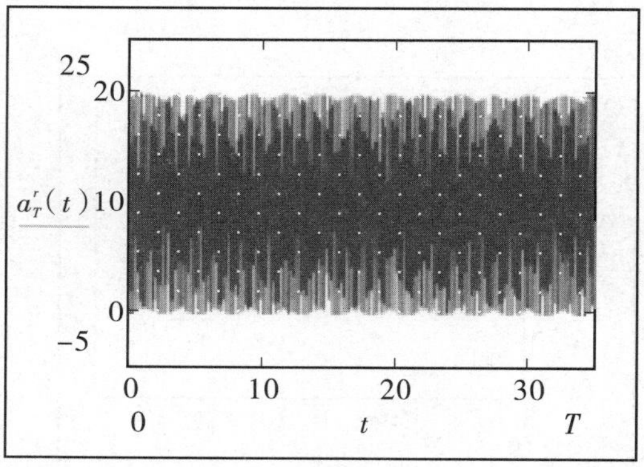

Fig. 2. 23　Target acceleration along target-to-missile direction: $a_T^r(t)$

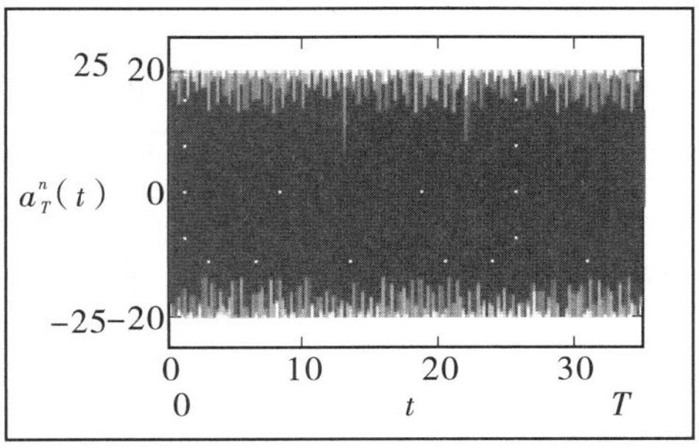

Fig. 2. 24 Target acceleration along direction which perpendicularly to line-of-sight direction: $a_T^n(t)$

2. 4 Jumps in financial markets

A classical model of financial market return process, such as the Black-Scholes[23], is the lognormal diffusion process, such that the log-return process has a normal distribution. However, real markets exhibit several deviations from this ideal, although useful, model. The market distribution, say for stocks, should have several realistic properties not found in the ideal lognormal model: ① the model must permit large random fluctuations such as crashes or sudden upsurges, ②the log-return distribution should be skew since large downward outliers are larger than upward outliers, and ③ the distribution should be leptokurtic since the mode is usually higher and the tails thicker than for a normal distribution. For modeling these extra properties phenomenologically, a jump-diffusion process with log-uniform jump-amplitude Poisson process it is usually applied. Let $S(t)$ be the price of a stock or stock fund satisfies a Markov, continuous-time, jump-diffusion stochastic differential equation

$$dS(t) = S(t)[\mu dt + \sigma dZ(t) + J(Q)dP(t)], S(0) = S_0, \tag{2.184}$$

where μ is the mean return rate, σ is the diffusive volatility, $Z(t)$ is a one-dimensional stochastic diffusion process, $J(Q)$ is a log-return mean μ_j and variance σ_j^2 random jump-amplitude and $P(t)$ is a simple Poisson jump process with jump rate λ. The processes $Z(t)$ and $P(t)$ are pairwise independent, while $J(Q)$ is also independent except that it is conditioned on the existence of a jump in $dP(t)$ it is conditioned on the existence of a jump in $dP(t)$. The numerical simulation a jumps in financial markets based on, jump-diffusion stochastic differential equation (2. 184), was considered in many papers, see for examples Ref. [23] – [25].

In contrast with a phenomenological approach we explain jumps phenomena in financial markets from the first principles, without any reference to Poisson jump process.

We claim that jumps phenomena in financial markets completely induced by nonlinearity and additive "small" white noise in corresponding Colombeau-Ito's stochastic equations.

Let $\boldsymbol{C}_i = (\Omega_i, \boldsymbol{\Sigma}_i, \mathbf{P}_i), i = 1,2$ be a probability spaces such that: $\Omega_1 \cap \Omega_2 = \varnothing$. Let $\boldsymbol{W}(t, \omega)$ be a Wiener process on $\boldsymbol{C}_1$ and let $\boldsymbol{W}(t, \bar{\omega})$ be a Wiener process on $\boldsymbol{C}_2$. Let us consider now a family $\left(\dot{\boldsymbol{x}}_{t,D,\varepsilon'}^{x_0,\varepsilon}(\omega,\bar{\omega})\right)_{\varepsilon'}$ of the Colombeau generalized stochastic processes which is a solution of the Colombeau-Ito's stochastic equation with stochastic coefficients:

$$\left(\dot{\boldsymbol{x}}_{t,D,\varepsilon'}^{x_0,\varepsilon}(\omega,\bar{\omega})\right)_{\varepsilon'} = \left(\boldsymbol{b}_{\varepsilon'}(\bar{\omega},\boldsymbol{x}_{t,D,\varepsilon'}^{x,\varepsilon}(\omega,\bar{\omega}),t)\right)_{\varepsilon'} +$$

$$\sqrt{D}\boldsymbol{w}_{\varepsilon'}(t,\bar{\omega}) + \sqrt{\varepsilon}\boldsymbol{w}(t,\omega);\varepsilon,\varepsilon' \in (0,1],\omega \in \Omega_1,\bar{\omega} \in \Omega_2. \tag{2.185}$$

Here $\boldsymbol{w}(t,\bar{\omega})$ and $\boldsymbol{w}(t,\omega)$ is a white noise on $\mathbf{R}^n$, i. e.,

$$\boldsymbol{w}(t,\bar{\omega}) = \frac{\mathrm{d}}{\mathrm{d}t}\boldsymbol{W}(t,\bar{\omega}),\boldsymbol{w}(t,\omega) = \frac{\mathrm{d}}{\mathrm{d}t}\boldsymbol{W}(t,\omega) \tag{2.186}$$

is almost surely in D', and $\boldsymbol{w}_{\varepsilon'}(t,\bar{\omega})$ is the smoothed white noise on $\mathbf{R}^n$, i. e.,

$$\boldsymbol{w}_{\varepsilon'}(t,\bar{\omega}) = \langle \boldsymbol{w}(t,\bar{\omega}),\varphi_{\varepsilon'}(s-t)\rangle, \tag{2.187}$$

where $\varphi_{\varepsilon'}$ is a model delta net[2],[4].

Definition 11. (1) CISDE (2.185) is $\hat{\mathbf{R}}$- dissipative if there exist Lyapunov candidate function $\left(V_{\varepsilon'}(\bar{\omega},x,t)\right)_{\varepsilon'}:\hat{\mathbf{R}}^n \times [0,T] \to \hat{\mathbf{R}}$ and positive infinite Colombeau constants $\tilde{C} = [(C_{\varepsilon'})_{\varepsilon'}] \in \hat{\mathbf{R}}_+,\tilde{r} = [(r_{\varepsilon'})_{\varepsilon'}] \in \hat{\mathbf{R}}_+$, such that:

(1) $\forall \varepsilon' \in (0,1]:\mathbf{P}_2\{V_{*,\varepsilon'}(\bar{\omega}) = \infty\} = 1$, where $V_{*,\varepsilon'}(\bar{\omega}) = \lim\limits_{R\to\infty}[\inf_{\|x\|>R}V_{\varepsilon'}(\bar{\omega},\boldsymbol{x},t)]$, and

(2) $\forall[(x_{\varepsilon'})_{\varepsilon'}]\{[(\|x_{\varepsilon'}\|)_{\varepsilon'}] \geqslant \tilde{r}\}$ the inequality

$$\left[\left(\dot{V}_{\varepsilon'}(\bar{\omega},\boldsymbol{x}_{\varepsilon'},t;\boldsymbol{b}_{\varepsilon'})\right)_{\varepsilon'}\right] \leqslant \tilde{C}\left[\left(V_{\varepsilon'}(\bar{\omega},\boldsymbol{x}_{\varepsilon'},t)\right)_{\varepsilon'}\right]\mathbf{P}_2 - \text{o. s.} \tag{2.188}$$

is satisfied. Here

$$\left[\left(\dot{V}_{\varepsilon'}(\bar{\omega},\boldsymbol{x}_{\varepsilon'},t;\boldsymbol{b}_{\varepsilon'})\right)_{\varepsilon'}\right] \equiv$$

$$\left[\left(\frac{\partial V_{\varepsilon'}(\bar{\omega},\boldsymbol{x}_{\varepsilon'},t)}{\partial t}\right)_{\varepsilon'}\right] + \sum_{i=1}^{n}\left[\left(\frac{\partial V_{\varepsilon'}(\bar{\omega},\boldsymbol{x}_{\varepsilon'},t)}{\partial x_{i,\varepsilon'}}b_{i,\varepsilon'}(\bar{\omega},\boldsymbol{x}_{\varepsilon'},t)\right)_{\varepsilon'}\right]. \tag{2.189}$$

Or in the next equivalent form:

(1) CISDE (2.185) is $\hat{\mathbf{R}}$-dissipative if there exist Lyapunov candidate function $(V_{\varepsilon'}(\bar{\omega},\boldsymbol{x},t))_{\varepsilon'}:\hat{\mathbf{R}}^n \times [0,T] \to \hat{\mathbf{R}}$ and positive infinite Colombeau constants $\tilde{C} = [(C_{\varepsilon'})_{\varepsilon'}] \in \hat{\mathbf{R}}_+,\tilde{r} = [(r_{\varepsilon'})_{\varepsilon'}] \in \hat{\mathbf{R}}_+$, such that:

(i) $\forall \varepsilon' \in (0,1]:\mathbf{P}_2\{V_{*,\varepsilon'}(\bar{\omega}) = \infty\} = 1$, where $V_{*,\varepsilon'}(\bar{\omega}) = \lim\limits_{R\to\infty}(\inf_{\|x\|>R}V_{\varepsilon'}(\bar{\omega},\boldsymbol{x},t))$, and

(ii) $\forall \varepsilon' \in (0,1]\forall x_{\varepsilon'}[(\boldsymbol{x}_{\varepsilon'} \in \mathbf{R}^n) \wedge (\|\boldsymbol{x}_{\varepsilon'}\| \geqslant r_{\varepsilon'})]$ the inequality

$$\left(\dot{V}_{\varepsilon'}(\bar{\omega},\boldsymbol{x}_{\varepsilon'},t;\boldsymbol{b}_{\varepsilon'})\right)_{\varepsilon'} \leqslant (C_{\varepsilon'})_{\varepsilon'}(V_{\varepsilon'}(\bar{\omega},\boldsymbol{x}_{\varepsilon'},t))_{\varepsilon'}\mathbf{P}_2 - \text{a. s.} \tag{2.190}$$

is satisfied. Here

$$\left(\dot{V}_{\varepsilon'}(\bar{\omega},\boldsymbol{x}_{\varepsilon'},t;\boldsymbol{b}_{\varepsilon'})\right)_{\varepsilon'}\equiv$$

$$\left(\frac{\partial V_{\varepsilon'}(\bar{\omega},\boldsymbol{x}_{\varepsilon'},t)}{\partial t}\right)_{\varepsilon'}+\left(\sum_{i=1}^{n}\frac{\partial V_{\varepsilon'}(\bar{\omega},\boldsymbol{x}_{\varepsilon'},t)}{\partial x_{i,\varepsilon'}}b_{i,\varepsilon'}(\bar{\omega},\boldsymbol{x}_{\varepsilon'},t)\right)_{\varepsilon'}. \tag{2.191}$$

(2) CISDE (2.185) is a strongly $\hat{\mathbf{R}}$-dissipative if there exist Lyapunov candidate function $(V_{\varepsilon'}(\bar{\omega},\boldsymbol{x},t))_{\varepsilon'}:\hat{\mathbf{R}}^n\times[0,T]\to\hat{\mathbf{R}},\varepsilon'\in[0,1]$ and positive finite Colombeau constants $\tilde{C}=[(C_{\varepsilon'})_{\varepsilon'}]\in\hat{\mathbf{R}}_+,\tilde{r}=[(r_{\varepsilon'})_{\varepsilon'}]\in\hat{\mathbf{R}}_+$, such that:

(i) $\forall\varepsilon'\in(0,1]:V_{*,\varepsilon'}=\lim_{r\to\infty}(\inf_{\|x\|>r}V_{\varepsilon'}(x,t))=\infty$, and (ii) $\forall[(\boldsymbol{x}_{\varepsilon'})_{\varepsilon'}]\{[(\|\boldsymbol{x}_{\varepsilon'}\|)_{\varepsilon'}]\geqslant\tilde{r}\}$ the inequality

$$\left(\dot{V}_{\varepsilon'}(\bar{\omega},\boldsymbol{x}_{\varepsilon'},t;\boldsymbol{b}_{\varepsilon'})\right)_{\varepsilon'}\leqslant\tilde{C}\left[\left(V_{\varepsilon'}(\bar{\omega},\boldsymbol{x}_{\varepsilon'},t)\right)_{\varepsilon'}\right]\mathbf{P}_2-\text{a. s.}$$

is satisfied. Here

$$\left(\dot{V}_{\varepsilon'}(\bar{\omega},\boldsymbol{x}_{\varepsilon'},t;\boldsymbol{b}_{\varepsilon'})\right)_{\varepsilon'}\equiv$$

$$\left(\frac{\partial V_{\varepsilon'}(\bar{\omega},\boldsymbol{x}_{\varepsilon'},t)}{\partial t}\right)_{\varepsilon'}+\sum_{i=1}^{n}\left[\left(\frac{\partial V_{\varepsilon'}(\bar{\omega},\boldsymbol{x}_{\varepsilon'},t)}{\partial x_{i,\varepsilon'}}b_{i,\varepsilon'}(\bar{\omega},\boldsymbol{x}_{\varepsilon'},t)\right)_{\varepsilon'}\right].$$

Let us now consider a family $\left(\boldsymbol{x}^{x_0,\varepsilon}_{t,\epsilon,\varepsilon'}(\bar{\omega})\right)_{\varepsilon'}$ of the solutions of the Colombeau SDE:

$$\left(\mathrm{d}\boldsymbol{x}^{x_0,\varepsilon}_{t,\epsilon,\varepsilon'}(\omega,\bar{\omega},\delta)\right)_{\varepsilon'}=\left(\boldsymbol{b}_{\varepsilon',\epsilon}\left(\boldsymbol{x}^{x_0,\varepsilon}_{t,\epsilon,\varepsilon'}(\omega,\bar{\omega},\delta),t,\bar{\omega}\right)\right)_{\varepsilon'}+\sqrt{D}\mathrm{d}\boldsymbol{W}(t,\bar{\omega})+\sqrt{\varepsilon}\boldsymbol{w}(t,\omega), \tag{2.192}$$

$$\left(\boldsymbol{x}^{x_0,\varepsilon}_{0,\varepsilon'}\right)_{\varepsilon'}=\boldsymbol{x}_0\in\hat{\mathbf{R}}^n,t\in[0,T],\varepsilon,\varepsilon'\in(0,1],\boldsymbol{\epsilon}\in(0,1]^n.$$

Here $\boldsymbol{W}(t)$ is n-dimensional Brownian motion, and $\forall\boldsymbol{\epsilon}\in(0,1]^n$, $\forall t\in[0,T]$ and for almost all $\bar{\omega}\in\Omega_2$: $\left(\boldsymbol{b}_{\varepsilon',\epsilon}(x,t,\bar{\omega})\right)_{\varepsilon'}\in G^n(\mathbf{R}^n)$, $\boldsymbol{b}_{0,0}(\cdot,t,\bar{\omega})\equiv\boldsymbol{b}_{\varepsilon'=0,\epsilon=0}(\cdot,t,\bar{\omega}):\mathbf{R}^n\to\mathbf{R}^n$ is a polynomial vector-function on a variable $\boldsymbol{x}=(x_1,\cdots,x_n)$, i. e. ,

$$b_{i,0,0}(\boldsymbol{x},t,\bar{\omega})=\sum_{\alpha,|\alpha|\leqslant r}b^{\alpha}_{i,0,0}(t,\bar{\omega})x^{\alpha},\alpha=(i_1,\cdots,i_n),|\alpha|=\sum_{j=1}^{n}i_j,0\leqslant i_j\leqslant p,$$

and

$$b_{i,\varepsilon',\epsilon}(\boldsymbol{x}(t),t,\bar{\omega})=b_{i,0,0}(\boldsymbol{x}_{\varepsilon',\epsilon}(t,\bar{\omega}),t,\bar{\omega}).$$

Here

$$\boldsymbol{x}_{\varepsilon',\epsilon}(t,\omega)=(x_{1,\varepsilon',\varepsilon,\epsilon}(t,\bar{\omega}),\cdots,x_{n,\varepsilon',\varepsilon,\epsilon}(t,\bar{\omega})),$$

$$x_{i,\varepsilon',\epsilon}(t,\bar{\omega})=\frac{x_i(t)}{1+\varepsilon'x_i^{2l}(t)+\varepsilon'\left\{\epsilon_i\int_0^t\theta_{\epsilon_i}[x_i(\tau)]x_i^{2l}(\tau)\mathrm{d}\tau+\sqrt{\delta}W_i(t,\bar{\omega})\right\}^2},$$

$i=1,\cdots,n.$

Now we let

$$u_i(t)=\epsilon_i\int_0^t\theta_{\epsilon_i}[x_i(\tau)]x_i^{2l}(\tau)\mathrm{d}\tau+\sqrt{\delta}W_i(t,\bar{\omega}),$$

and rewrite Eq. (2.192) in the canonical Colombeau-Ito form:

$$\left(\mathrm{d}\boldsymbol{x}^{x_0,\varepsilon}_{t,\varepsilon',\epsilon}(\omega,\bar{\omega},\delta)\right)_{\varepsilon'} = \left(\boldsymbol{b}_{\varepsilon',\epsilon}(\boldsymbol{x}^{x_0,\varepsilon}_{t,\varepsilon',\epsilon}(\omega,\bar{\omega},\delta),\boldsymbol{u}^{\delta}_{t,\varepsilon',\epsilon}(\omega,\bar{\omega}),t)\right)_{\varepsilon'} +$$

$$\sqrt{D}\boldsymbol{w}_{\varepsilon'}(t,\bar{\omega}) + \sqrt{\varepsilon}\boldsymbol{w}(t,\omega),\boldsymbol{u}^{\delta}_{t,\varepsilon',\varepsilon,\epsilon}(\omega,\bar{\omega}) = [u^{\delta}_{1,t,\varepsilon',\epsilon}(\omega,\bar{\omega}),\cdots,u^{\delta}_{n,t,\varepsilon',\epsilon}(\omega,\bar{\omega})],$$

$$\left(\mathrm{d}u^{\delta}_{i,t,\varepsilon',\epsilon}(\omega,\bar{\omega})\right)_{\varepsilon'} = \epsilon_i\left(\theta_{\epsilon_i}\left[x^{x_0,\delta}_{i,t,\varepsilon',\epsilon}(\omega,\bar{\omega})\right]\left[x^{x_0,\delta}_{i,t,\varepsilon',\epsilon}(\omega,\bar{\omega})\right]^{2l}\right)_{\varepsilon'} + \sqrt{\delta}\mathrm{d}W_i(t,\bar{\omega}),$$

(2.193)

$$i = 1,\cdots,n,\left(x^{x_0,\varepsilon}_{0,\varepsilon'}\right)_{\varepsilon'} = \boldsymbol{x}_0 \in \hat{\mathbf{R}}^n,t \in [0,T],\varepsilon,\varepsilon',\epsilon,\delta \in (0,1].$$

Theorem 9. (Strong large deviations principle SLDP)

We set now $\theta_{\epsilon_i}[z] \equiv 1, i = 1,\cdots,n$.

(1) Assume that CISDE (2.193) is a strongly $\hat{\mathbf{R}}$-dissipative. Then for any solution

$$\left(\boldsymbol{x}^{x_0,\varepsilon}_{t,\varepsilon',\epsilon}(\omega,\bar{\omega},\delta)\right)_{\varepsilon'} = \left(x^{x_0,\varepsilon}_{1,t,\varepsilon',\epsilon}(\omega,\bar{\omega},\delta),\cdots,x^{x_0,\varepsilon}_{n,t,\varepsilon',\epsilon}(\omega,\bar{\omega},\delta)\right)_{\varepsilon'}$$

of a strongly $\hat{\mathbf{R}}$-dissipative CISDE (2.193) and any $\mathbf{R}$-valued parameters $\lambda_1,\cdots,\lambda_n$, there exist finite Colombeau constant $\tilde{C}' = (C'_{\varepsilon'})_{\varepsilon'} > 0$, such that $\forall\boldsymbol{\lambda}[\boldsymbol{\lambda} = (\lambda_1,\cdots,\lambda_n)]$, the inequality

$$\lim_{\substack{\varepsilon\to 0\\ \varepsilon'\to 0\\ \epsilon\to 0\\ (\frac{\varepsilon'}{\varepsilon})\to 0}} \lim_{\delta\to 0}\mathbf{E}_{\Omega_1}\left[\|\boldsymbol{x}^{x_0,\varepsilon}_{t,\varepsilon',\epsilon}(\omega,\bar{\omega},\delta) - \boldsymbol{\lambda}\|^2\right] \leqslant \tilde{C}'\|\boldsymbol{U}(\bar{\omega},t,\boldsymbol{\lambda})\|^2\ \mathbf{P}_2 - \text{a. s.} \quad (2.194)$$

is satisfied. Or in the next equivalent form: for a sufficiently small $\epsilon \approx 0$ and for a sufficiently small $\varepsilon \approx 0,\varepsilon' \approx 0$ such that $\frac{\varepsilon'}{\varepsilon} \approx 0$, the inequality

$$\left[\left(\lim_{\delta\to 0}\mathbf{E}_{\Omega_1}\left[\|\boldsymbol{x}^{x_0,\varepsilon}_{t,\varepsilon',\epsilon}(\omega,\bar{\omega},\delta) - \boldsymbol{\lambda}\|^2\right]\right)_{\varepsilon'}\right] \leqslant \tilde{C}'\|\boldsymbol{U}(\bar{\omega},t,\boldsymbol{\lambda})\|^2\mathbf{P}_2 - \text{a. s.}$$

is satisfied.

Here the vector-function is the solution of the differential master equation:

$\dot{\boldsymbol{U}}(\bar{\omega},t,\boldsymbol{\lambda}) = \mathbf{J}[\boldsymbol{b}_0(\boldsymbol{\lambda},t,\bar{\omega})]\boldsymbol{U}(\bar{\omega},t,\boldsymbol{\lambda}) + \boldsymbol{b}_0(\boldsymbol{\lambda},t,\bar{\omega}),\ \boldsymbol{U}(0,\boldsymbol{\lambda},\bar{\omega}) = \boldsymbol{x}_0 - \boldsymbol{\lambda}.$

Here $\mathbf{J} = \mathbf{J}[\boldsymbol{b}_0(\boldsymbol{\lambda},t,\bar{\omega})]$ is a Jacobian, i.e., $\mathbf{J}$ is $n\times n$-matrix:

$\mathbf{J}[\boldsymbol{b}_0(\boldsymbol{\lambda},t,\bar{\omega})] = \mathbf{J}[\partial\boldsymbol{b}_{0,i}(\boldsymbol{x},t,\bar{\omega})/\partial x_j]_{x=\lambda}$.

(2) Assume that CISDE (2.185) is a strongly $\hat{\mathbf{R}}$-dissipative. Then for any solution

$$\left(\boldsymbol{x}^{x_0,\varepsilon}_{t,\varepsilon',\epsilon}(\omega,\bar{\omega})\right)_{\varepsilon'} = \left(x^{x_0,\varepsilon}_{1,t,\varepsilon',\epsilon}(\omega,\bar{\omega}),\cdots,x^{x_0,\varepsilon}_{n,t,\varepsilon',\epsilon}(\omega,\bar{\omega})\right)_{\varepsilon'}$$

of a strongly $\hat{\mathbf{R}}$-dissipative CISDE (2.187) and any $\mathbf{R}$-valued parameters $\lambda_1,\cdots,\lambda_n$, there exist finite Colombeau constant $\tilde{C}' = (C'_{\varepsilon'})_{\varepsilon'} > 0$, such that $\forall\boldsymbol{\lambda}[\boldsymbol{\lambda} = (\lambda_1,\cdots,\lambda_n)]$, the inequality

$$\left(\mathbf{E}_{\Omega_1}\left[\|\boldsymbol{x}^{x_0,\varepsilon}_{t,\varepsilon',\epsilon}(\omega,\bar{\omega}) - \boldsymbol{\lambda}\|^2\right]\right)_{\varepsilon'} \leqslant \tilde{C}'\|\boldsymbol{U}(\bar{\omega},t,\boldsymbol{\lambda})\|^2\ \mathbf{P}_2 - \text{a. s.}$$

is satisfied.

(3) For any solution

$$\left(\boldsymbol{x}^{x_0,\varepsilon}_{t,\varepsilon',\epsilon}(\omega,\bar{\omega})\right)_{\varepsilon'} = \left(x^{x_0,\varepsilon}_{1,t,\varepsilon',\epsilon},\cdots,x^{x_0,\varepsilon}_{n,t,\varepsilon',\epsilon}\right)_{\varepsilon'}$$

of a strongly $\hat{\mathbf{R}}$-dissipative CISDE (2.187) and any $\mathbf{R}$-valued parameters $\lambda_1,\cdots,\lambda_n$, there exist finite Colombeau constant $\tilde{C}' = (C'_{\varepsilon'})_{\varepsilon'} > 0$, such that $\forall \boldsymbol{\lambda}[\boldsymbol{\lambda} = (\lambda_1,\cdots,\lambda_n)]$, the inequality

$$\lim_{\varepsilon\to 0}\mathbf{E}_{\Omega_1}\left[\|\boldsymbol{x}^{x_0,\varepsilon}_{t,\varepsilon'=0,\epsilon=0}(\omega,\bar{\omega}) - \boldsymbol{\lambda}\|^2\right] \leqslant \tilde{C}' \|\boldsymbol{U}(\bar{\omega},t,\boldsymbol{\lambda})\|^2 \mathbf{P}_2 - \text{a. s.} \tag{2.195}$$

is satisfied. Here the vector-function $\boldsymbol{U}(\bar{\omega},t,\boldsymbol{\lambda}) = (U_1(\bar{\omega},t,\boldsymbol{\lambda}),\cdots,U_n(\bar{\omega},t,\boldsymbol{\lambda}))$ is the solution of the differential master equation:

$$\dot{\boldsymbol{U}}(\bar{\omega},t,\boldsymbol{\lambda}) = \mathbf{J}[\boldsymbol{b}_0(\boldsymbol{\lambda},t,\bar{\omega})]\boldsymbol{U}(\bar{\omega},t,\boldsymbol{\lambda}) + \boldsymbol{b}_0(\boldsymbol{\lambda},t,\bar{\omega}),\ \boldsymbol{U}(0,\boldsymbol{\lambda},\bar{\omega}) = \boldsymbol{x}_0 - \boldsymbol{\lambda},$$

where $\mathbf{J} = \mathbf{J}[\boldsymbol{b}_0(\lambda,t,\bar{\omega})]$ is a Jacobian, i. e., $\mathbf{J}$ is $n\times n$-matrix:

$$\mathbf{J}[\boldsymbol{b}_0(\boldsymbol{\lambda},t,\bar{\omega})] = \mathbf{J}[\partial \boldsymbol{b}_{0,i}(\boldsymbol{\lambda},t,\bar{\omega})/\partial x_j]_{x=\lambda}.$$

Proof (1) we now let $\boldsymbol{x}^{x_0,\varepsilon}_{t,\varepsilon',\epsilon}(\omega,\bar{\omega}) - \boldsymbol{\lambda} = \boldsymbol{y}^{x_0,\varepsilon}_{t,\varepsilon',\epsilon}(\omega,\bar{\omega})$.

Replacement $\boldsymbol{x}^{x_0,\varepsilon}_{t,\varepsilon',\epsilon}(\omega,\bar{\omega}) = \boldsymbol{y}^{x_0,\varepsilon}_{t,\varepsilon',\epsilon}(\omega,\bar{\omega}) + \boldsymbol{\lambda}$ into Eq. (2.193) gives

$$\left(\mathrm{d}\boldsymbol{y}^{x_0,\varepsilon}_{t,\varepsilon',\epsilon}(\omega,\bar{\omega},\delta)\right)_{\varepsilon'} = \left(\boldsymbol{b}_{\varepsilon',\epsilon}(\boldsymbol{y}^{x_0,\varepsilon}_{t,\varepsilon',\epsilon}(\omega,\bar{\omega},\delta) + \boldsymbol{\lambda},\boldsymbol{u}^{\delta}_{t,\varepsilon',\varepsilon,\epsilon}(\omega,\bar{\omega}),t,\varepsilon)\right)_{\varepsilon'} +$$

$$\sqrt{\varepsilon}\mathrm{d}\boldsymbol{W}(t,\omega),\boldsymbol{u}^{\delta}_{t,\varepsilon',\varepsilon,\epsilon}(\omega) = (u^{\delta}_{1,t,\varepsilon',\varepsilon,\epsilon}(\omega,\bar{\omega}),\cdots,u^{\delta}_{n,t,\varepsilon',\varepsilon,\epsilon}(\omega,\bar{\omega})),$$

$$(\mathrm{d}u^{\delta}_{i,t,\varepsilon',\varepsilon,\epsilon}(\omega,\bar{\omega}))_{\varepsilon'} = \epsilon([\boldsymbol{x}^{x_0,\varepsilon}_{i,t,\varepsilon',\varepsilon,\epsilon}(\omega,\bar{\omega},\delta)]^{2l})_{\varepsilon'} + \sqrt{\delta}\mathrm{d}W_i(\omega,t),$$

$$i = 1,\cdots,n,(\boldsymbol{x}^{x_0,\varepsilon}_{0,\varepsilon',\epsilon})_{\varepsilon'} = \boldsymbol{x}_0 \in \hat{\mathbf{R}}^n, t\in[0,T],\varepsilon,\varepsilon',\epsilon,\delta\in(0,1].$$

Thus we need to estimate the quantity

$$\lim_{\substack{\varepsilon\to 0\\ \varepsilon'\to 0\\ \epsilon\to 0\\ \left(\frac{\varepsilon'}{\varepsilon}\right)\to 0}}\lim_{\delta\to 0}\mathbf{E}_{\Omega_1}\left[\|\boldsymbol{y}^{x_0,\varepsilon}_{t,\varepsilon',\epsilon}(\omega,\bar{\omega},\delta)\|^2\right]$$

$\omega\in\Omega_1,\bar{\omega}\in\Omega_2$.

Application of the Theorem B.4 (see Appendix B) to Eq. (2.193) gives the inequality (2.194) directly.

(2) From the equality

$$\mathbf{E}_{\Omega_1}\left[\|\boldsymbol{x}^{x_0,\varepsilon}_{t,\varepsilon',\epsilon}(\omega,\bar{\omega}) - \boldsymbol{\lambda}\|^2\right] =$$

$$\mathbf{E}_{\Omega_1}\left[\|[\boldsymbol{x}^{x_0,\varepsilon}_{t,\varepsilon',\epsilon}(\omega,\bar{\omega}) - \boldsymbol{x}^{x_0,\varepsilon}_{t,\varepsilon',\epsilon}(\omega,\bar{\omega},\delta)] + [\boldsymbol{x}^{x_0,\varepsilon}_{t,\varepsilon',\epsilon}(\omega,\bar{\omega},\delta) - \boldsymbol{\lambda}]\|^2\right],$$

by using the triangle inequality, one obtain

$$\sqrt{\mathbf{E}_{\Omega_1}[\|\boldsymbol{x}^{x_0,\varepsilon}_{t,\varepsilon',\epsilon}(\omega,\bar{\omega}) - \boldsymbol{\lambda}\|^2]} \leqslant \sqrt{\mathbf{E}_{\Omega_1}[\|\boldsymbol{x}^{x_0,\varepsilon}_{t,\varepsilon',\epsilon}(\omega,\bar{\omega}) - \boldsymbol{x}^{x_0,\varepsilon}_{t,\varepsilon',\epsilon}(\omega,\bar{\omega},\delta)\|^2]} +$$

$$\sqrt{\mathbf{E}_{\Omega_1}[\|\boldsymbol{x}^{x_0,\varepsilon}_{t,\varepsilon',\epsilon}(\omega,\delta) - \boldsymbol{\lambda}\|^2]}.$$

Therefore statement (2) immediately follows from Theorem A1 (see appendix A), Proposition 2 and Theorem 5.

(3) From the equality

$$\mathbf{E}_{\Omega_1}[\,\|\boldsymbol{x}^{x_0,\varepsilon}_{t,\varepsilon'=0,\epsilon=0}(\omega,\bar{\omega})-\boldsymbol{\lambda}\|^2]=$$

$$\mathbf{E}_{\Omega_1}[\,\|(\boldsymbol{x}^{x_0,\varepsilon}_{t,\varepsilon'=0,\epsilon=0}(\omega,\bar{\omega})-\boldsymbol{x}^{x_0,\varepsilon}_{t,\varepsilon',\epsilon}(\omega,\bar{\omega})]+[\boldsymbol{x}^{x_0,\varepsilon}_{t,\varepsilon',\epsilon}(\omega,\bar{\omega})-\boldsymbol{\lambda}]\|^2],$$

by using the triangle inequality, one obtain

$$\sqrt{\mathbf{E}_{\Omega_1}[\,\|\boldsymbol{x}^{x_0,\varepsilon}_{t,\varepsilon'=0,\epsilon=0}(\omega,\bar{\omega})-\boldsymbol{\lambda}\|^2]}\leqslant\sqrt{\mathbf{E}_{\Omega_1}[\,\|\boldsymbol{x}^{x_0,\varepsilon}_{t,\varepsilon'=0,\epsilon=0}(\omega,\bar{\omega})-\boldsymbol{x}^{x_0,\varepsilon}_{t,\varepsilon',\epsilon}(\omega,\bar{\omega})\|^2]}+$$

$$\sqrt{\mathbf{E}_{\Omega_1}[\,\|\boldsymbol{x}^{x_0,\varepsilon}_{t,\varepsilon',\epsilon}(\omega,\bar{\omega})-\boldsymbol{\lambda}\|^2]}.$$

Therefore statement (3) immediately follows from Theorem A2 (see appendix A), Proposition 1 and statement (2).

The stochastic dynamics (2.185) we take now in the following form

$$\left(\dot{\boldsymbol{x}}^{x_0,\varepsilon}_{t,D,\varepsilon'}(\omega,\bar{\omega})\right)_{\varepsilon'}=\left(F_{\varepsilon'}(\boldsymbol{x}^{x,\varepsilon}_{t,D,\varepsilon'}(\omega,\bar{\omega}),t)\right)_{\varepsilon'}+\sqrt{D}w_{\varepsilon'}(t,\bar{\omega})+\sqrt{\varepsilon}w(t,\omega);$$

$$\varepsilon,\varepsilon'\in(0,1],\omega\in\Omega_1,\bar{\omega}\in\Omega_2.$$

The force field $F_{\varepsilon'}$ is assumed to derive from a metastable potential which undergoes an arbitrary periodic modulation in time with period τ, i. e, $F_{\varepsilon'}(x,t+\tau)=F_{\varepsilon'}(x,t)$.

The random time-dependent force field $F_{\varepsilon'}(x,t)$ takes the following form

$$F_{\varepsilon'}(x,t)=-\dot{V}(x)+A\sin(\Omega\cdot t)+B\cos(\Theta\cdot t)+\sqrt{D}w_{\varepsilon'}(t,\bar{\omega})+\sqrt{\varepsilon}w(t,\omega);$$

$$\varepsilon,\varepsilon'\in(0,1],\omega\in\Omega_1,\bar{\omega}\in\Omega_2.$$

As an example we consider a force field with a double well potential $V(x)$ as cartooned in Fig. 2.25.

The stochastic dynamics (2.185) takes the following form:

$$\dot{\boldsymbol{x}}^{x_0}_{t,D}(\omega,\bar{\omega})=-a[\boldsymbol{x}^{x_0}_{t,D}(\omega,\bar{\omega})]^3+b\boldsymbol{x}^{x_0}_{t,D}(\omega,\bar{\omega})+A\sin(\Omega\cdot t)+B\cos(\Theta\cdot t)+$$

$$c+\sqrt{D}w_{\varepsilon'}(t,\bar{\omega})+\sqrt{\varepsilon}w(t,\omega),x^{x_0}_{0,D}(\omega,\bar{\omega})=x_0. \tag{2.196}$$

Using Theorem 9. (3) one obtain the next differential linear master equation

$$\dot{U}(t,\lambda)=-(3a\lambda^2-b)U(t,\lambda)+A\sin(\Omega\cdot t)+B\cos(\Theta\cdot t)-$$

$$(a\lambda^3-b\lambda)+c+\sqrt{D}\dot{W}(t,\bar{\omega})+\sqrt{\varepsilon}w(t,\omega),U(t,\lambda)=x_0-\lambda.$$

Solving this differential linear master equation, we obtain the next transcendental master equation

$$(x_0-\lambda(t))\exp[-(3a\lambda^2-b)t]-(a\lambda^3-b\lambda-c)\times\int_0^t d\tau\exp[(3a\lambda^2-b)(t-\tau)]+$$

$$\int_0^t d\tau[A\sin(\Omega\cdot t)+B\cos(\Theta\cdot t)]\exp[(3a\lambda^2-b)(t-\tau)]+$$

$$\sqrt{D}\int_0^t\exp[(3a\lambda^2-b)(t-\tau)]d\tau W(\tau,\bar{\omega}).$$

Note that

$$\int_0^t\exp[(3a\lambda^2-b)(t-\tau)]d\tau W(\tau,\bar{\omega})=W(t,\bar{\omega})-$$

$$(3a\lambda^2 - b)\int_0^t W(\tau,\bar{\omega})\exp[(3a\lambda^2 - b)(t-\tau)]\mathrm{d}\tau.$$

Finally we obtain the next transcendental master equation

$$(x_0 - \lambda(t))\exp[-(3a\lambda^2 - b)t] - (a\lambda^3 - b\lambda - c) \times \int_0^t \mathrm{d}\tau\exp[(3a\lambda^2 - b)(t-\tau)] + \int_0^t \mathrm{d}\tau[A\sin(\Omega\cdot t) + B\cos(\Theta\cdot t)]\exp[(3a\lambda^2 - b)(t-\tau)] + \sqrt{D}W(t,\bar{\omega}) - \sqrt{D}(3a\lambda^2 - b)\cdot \int_0^t W(\tau,\bar{\omega})\exp[(3a\lambda^2 - b)(t-\tau)]\mathrm{d}\tau. \tag{2.197}$$

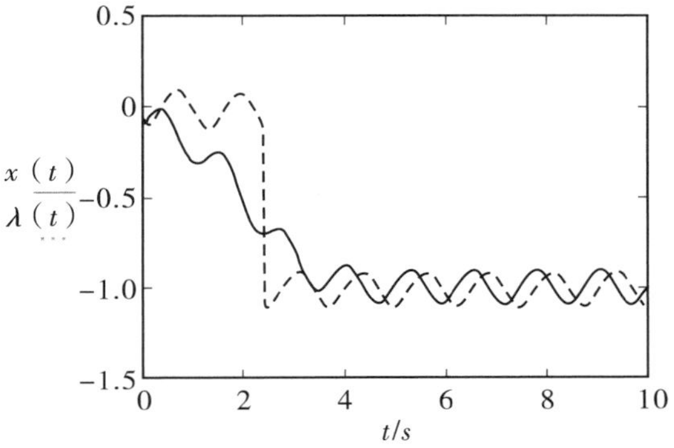

Fig. 2. 25 Comparison of: ① dynamics (2. 196) with $D = 0, \varepsilon = 0$ (solid curve) and ② quasi classical (dotted curve) dynamics in the limit $\varepsilon \to 0$, calculated by using SLDP. $a = 1, b = 1, c = 0, A = 0.5, B = 0, \Omega = 5, \Theta = 0, D = 0, x_0 = -0.1$

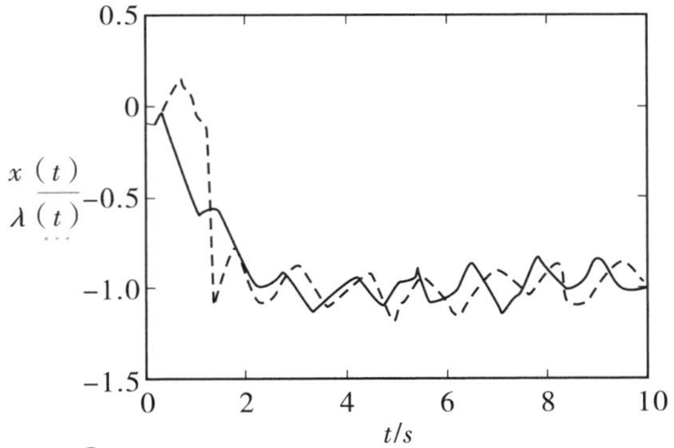

Fig. 2. 26 Comparison of: ① stochastic dynamics (2. 194) with $\varepsilon = 0, D \neq 0$ (solid curve) and ② quasi-classical (dotted curve) stochastic dynamics in the limit $\varepsilon \to 0$, calculated by using SLDP. $a = 1, b = 1, c = 0, A = 0.5, B = 0, \Omega = 5, \Theta = 0, D = 0.01, x_0 = -0.1$

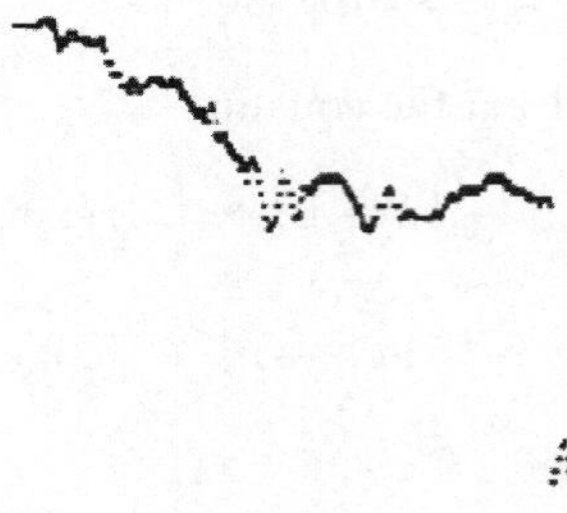

Fig. 2.27 Evolution of SLM (NYSE) February 1993[26].

Figure 2.27 shows the evolution of SLM over a one month period (February 1993)[26]. The price behavior over this period is clearly dominated by a large downward jump, which accounts for half of the monthly return. If we go down to an intraday scale shown in Figure 2.29, we see that the price moves essentially through jumps.

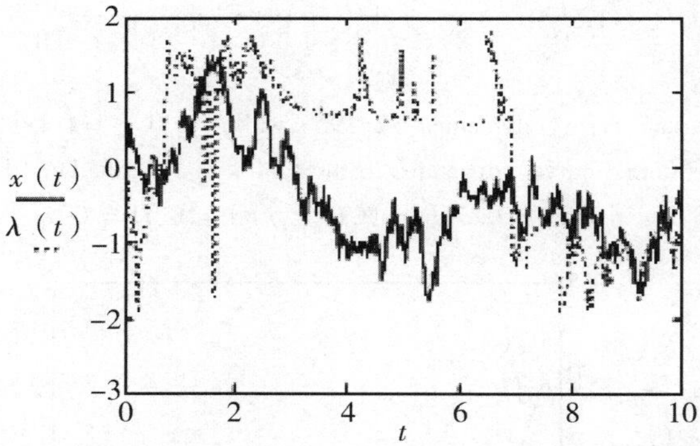

Fig. 2.28 Comparison of: ① stochastic dynamics (2.196) with $\varepsilon=0, D\neq 0$ (solid curve) and ② quasi-classical (dotted curve) stochastic dynamics in the limit $\varepsilon\to 0$, calculated by using SLDP. $a=1, b=1, c=0, A=0.5, B=0, \Omega=5, \Theta=0, D=1, x_0=-1$

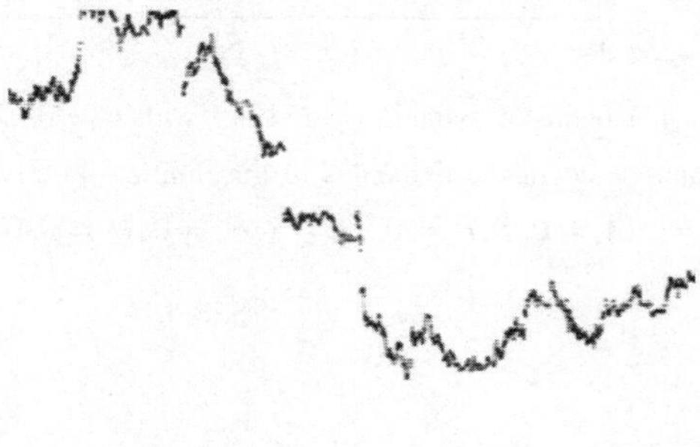

Fig. 2.29 Evolution of SLM (NYSE) January-March 1993[26].

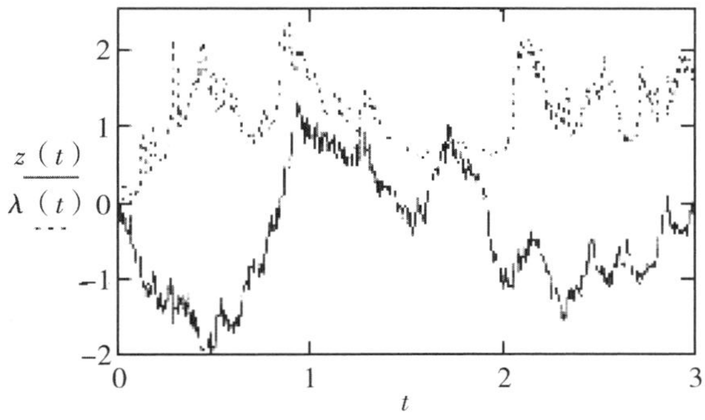

Fig. 2. 30 Comparison of stochastic dynamics (2. 196) with $\varepsilon = 0, D \neq 0$ (solid curve) and quasi-classical (dotted curve) stochastic dynamics in the limit $\varepsilon \to 0$, calculated by using SLDP. $a = 1, b = 1, c = 0, A = 0.5, B = 0, \Omega = 5, \Theta = 0, D = 2, x_0 = -1$

2. 5 Comparison of the quasi-classical stochastic dynamics obtained by using saddle-point approximation with a non-perturbative quasi-classical stochastic dynamics obtained by using SLDP

The double stochastic dynamics we take of the next form[7,27]:

$\dot{x}(t) = F[x(t), t] + \sqrt{D}w(t, \omega) + \sqrt{\varepsilon}w(t, \bar{\omega}).$

The force field $F[x(t), t]$ is assumed to derive from a metastable potential which undergoes an arbitrary periodic modulation in time with period τ, i. e., $F[x, t+\tau] = F[x, t]$. An example, is a static potential $V(x)$ supplemented by an additive sinusoidal and more general driving. The time-dependent force field $F[x, t]$ takes the following form:

$F[x, t] = -\dot{V}(x) + A\sin(\Omega \cdot t) + B\cos(\Theta \cdot t).$

We have compared now by quantity

$$\delta(\bar{\omega}, t) = \| x_{t,D}^{x_0}(\bar{\omega}) - \lambda(\bar{\omega}, t) \|. \tag{2. 198}$$

The above analytical predictions for the ε-limit (2. 195) given by master equation (2. 197) with very accurate numerical results for stochastic dynamics given by Ito's equation

$$\mathbf{x}_{t,D}^{x_0}(\bar{\omega}) = x_0 + \int_0^t F(x_{t,D}^{x_0}(\bar{\omega}), s)\,ds + \sqrt{D}W(t, \bar{\omega}). \tag{2. 199}$$

And we have compared now by quantity

$$\sigma(\bar{\omega}, t) = \| \mathbf{x}_{t,D}^{x_0}(\bar{\omega}) - E_p(\bar{\omega}, t) \|. \tag{2. 200}$$

The quasi-classical analytical predictions for the ε-limit

$$\lim_{\varepsilon \to 0} \mathbf{E}_{\Omega_1}[\, \| \mathbf{x}_{t,\varepsilon'=0,\epsilon=0}^{x_0,\varepsilon}(\omega, \bar{\omega}) \|^2], \tag{2. 201}$$

given by saddle-point approximation[7,27], denoted by $E_p(\bar{\omega}, t)$, with very accurate numerical results for stochastic dynamics given by Ito's equation (2. 199).

Example 5. Double well potential. As an example we consider the force field with a double well potential $V(x)$ as cartooned in Fig. 2. 31.

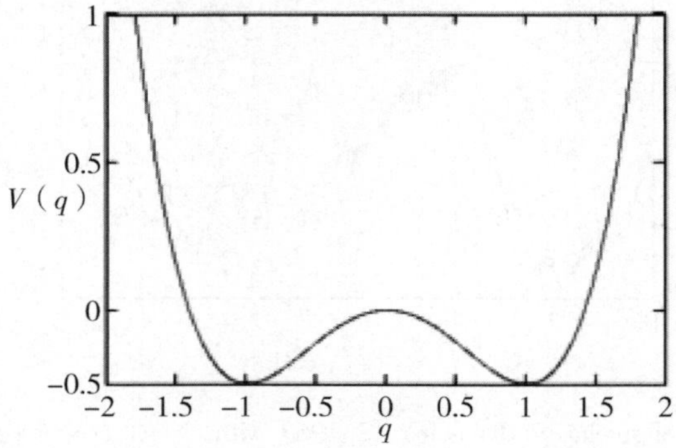

Fig. 2. 31 Double well potential, $a = 1, b = 1, c = 0$

$$V(x) = -\frac{a}{4}x^4 - \frac{b}{2}x^2 - cx, a > 0, b > 0.$$

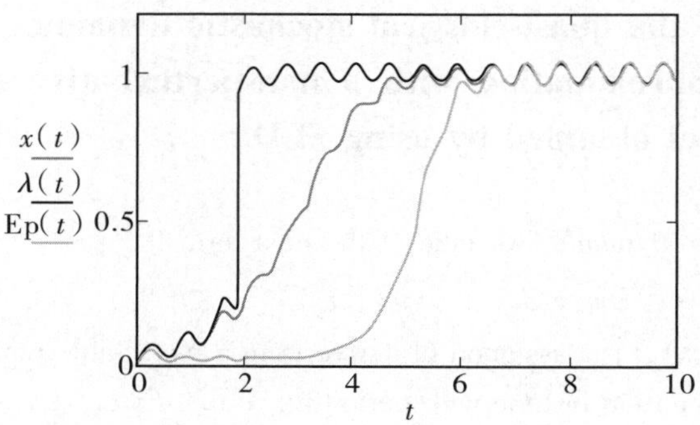

Fig. 2. 32 Comparison of: ① classical dynamics (2. 196) with $D = 0, \varepsilon = 0$ (dark grey curve), ② corresponding quasi-classical dynamics with $D = 0, \varepsilon \neq 0$ in the limit $\varepsilon \to 0$, calculated using SLDP (black curve) and ③ quasi-classical dynamics in the limit $\varepsilon \to 0$, calculated using saddle-point approximation[7,22] (light grey curve).

$a = 1, b = 1, c = 0, A = 1, B = 0, \Omega = 5, \Theta = 0, D = 0, x_0 = 0$

The time-dependent force field (2. 33) takes the following form:

$$F[x,t] = -ax^3 + bx + c + A\sin(\Omega \cdot t) + B\cos(\Theta \cdot t).$$

The stochastic dynamics (2. 199) takes the following form:

$$\dot{\boldsymbol{x}}_{t,D}^{x_0}(\omega) = -a[\boldsymbol{x}_{t,D}^{x_0}(\omega)]^3 + b\boldsymbol{x}_{t,D}^{x_0}(\omega) + c + A\sin(\Omega \cdot t) + B\cos(\Theta \cdot t), \boldsymbol{x}_{0,D}^{x_0}(\omega) = \boldsymbol{x}_0.$$

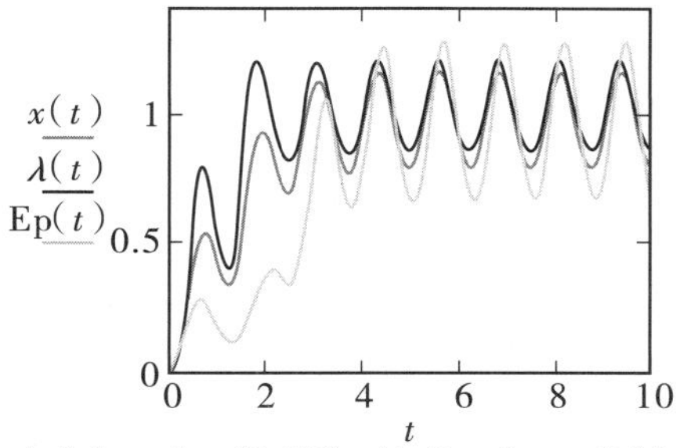

Fig. 2.33 Comparison of: ① classical dynamics (2.196) with $D=0,\varepsilon=0$ (dark grey curve), ② corresponding quasi-classical dynamics with $D=0,\varepsilon\neq 0$ in the limit $\varepsilon\to 0$, calculated using SLDP (black curve) and ③ quasi-classical dynamics in the limit $\varepsilon\to 0$, calculated using saddle-point approximation (light grey curve): $a=1,b=1,c=0,A=0.3,B=0,\Omega=10,\Theta=0,D=0,x_0=0$

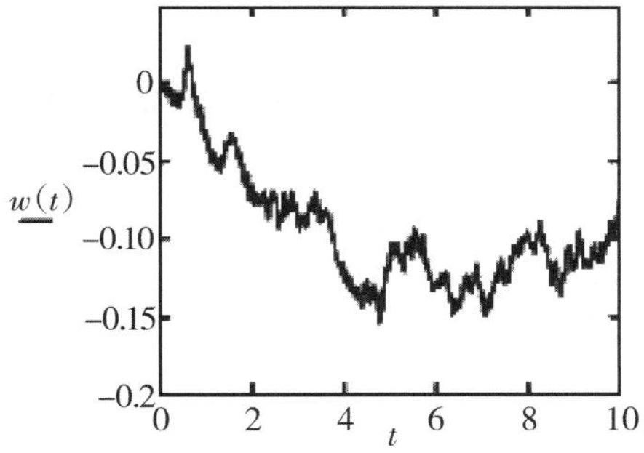

Fig. 2.34 The realization of a Wiener process $w(t)=\sqrt{D}W(t)$, where $W(t)$ is standard Wiener process, $D=10^{-3}$

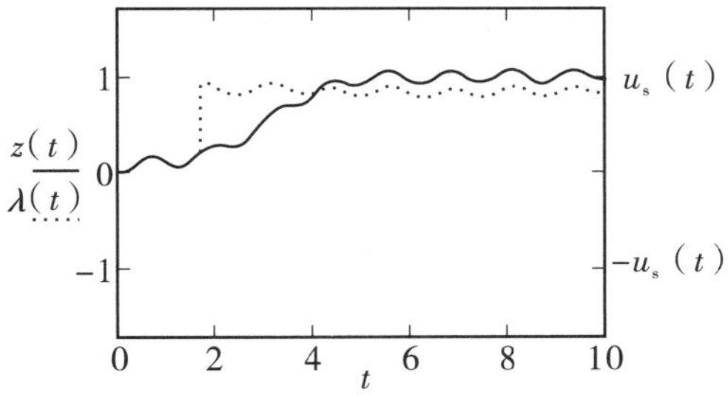

Fig. 2.35 Comparison of the quasi-classical stochastic dynamics (solid curve) obtained by using Saddle-point approximation[7,22] and quasi-classical stochastic dynamics (dotted curve) obtained by using SLDP: $a=1,b=1,c=0,A=0.3,B=0,\Omega=5,\Theta=0,D=10^{-3},x_0=0$

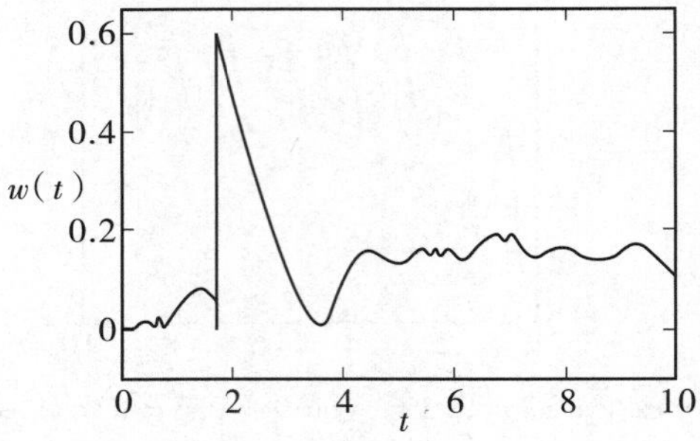

Fig. 2. 36 Comparison of the quasi classical stochastic dynamics, obtained by using Saddle-point approximation[7] and quasi classical stochastic dynamics obtained by using SLDP:
$a=1, b=1, c=0, A=0.3, B=0, \Omega=5, \Theta=0, D=10^{-3}, x_0=0$

2.6 Strong large deviations principles of Non-Freidlin-Wentzell Type.

Definition 12. [4] Let $\boldsymbol{C}=(\Omega,\Sigma,P)$ be a probability space. Let εR be the space of nets $(X_{\varepsilon'}(\omega))_{\varepsilon'}$ of measurable functions on Ω. Let εR_M be the space of nets $(X_{\varepsilon'})_{\varepsilon'} \in \varepsilon R, \varepsilon' \in (0,1]$, with the property that for almost all $\omega \in \Omega$ there exist constants, $C>0$ and $\varepsilon_0 \in (0,1]$ such that $|(X_{\varepsilon'})_{\varepsilon'}| \leqslant C(\varepsilon')^{-r}, \varepsilon' \leqslant \varepsilon_0$.

Definition 13. Let ξ be a distribution $\xi \in D'$. Distribution ξ is the generalized probability density of net $(X_{\varepsilon'})_{\varepsilon'} \in \varepsilon R, \varepsilon' \in (0,1]$ if $\forall f \in D: (\mathbf{E}[X_{\varepsilon'}(\omega)])_{\varepsilon'}=\xi(f)$.

Let us consider now Colombeau-Ito's SDE:

$$(\mathrm{d}\boldsymbol{x}_{t'',\varepsilon'}^{x_0,\varepsilon}(\omega))_{\varepsilon'}=(\boldsymbol{b}_{\varepsilon'}(\boldsymbol{x}_{t'',\varepsilon'}^{x_0,\varepsilon}(\omega),t''))_{\varepsilon'}+\sqrt{\varepsilon}\mathrm{d}\boldsymbol{W}(t'',\omega) \quad (2.202)$$

$$(\mathbf{E}[f(\boldsymbol{x}_{t',\varepsilon'}^{x_0,\varepsilon}(\omega))])_{\varepsilon'}=f(\boldsymbol{x}_0), \boldsymbol{x}_0=\boldsymbol{q}' \in \mathbf{R}^n, t'' \in [t',T], \quad (2.203)$$

$\varepsilon, \varepsilon' \in (0,1], f \in C_0^\infty(\mathbf{R}^n), \boldsymbol{x}_0 \in \operatorname{supp}(f)$.

Assumption 1. We assume now that there exist Colombeau constants $(C_{\varepsilon'})_{\varepsilon'}$ and $(D_{\varepsilon'})_{\varepsilon'}$ such that

$$(1)\ (\|\boldsymbol{b}_{\varepsilon'}(\boldsymbol{x},t)\|)_{\varepsilon'} \leqslant (C_{\varepsilon'})_{\varepsilon'}(1+\|\boldsymbol{x}\|),$$

$$(2)\ (\|\boldsymbol{b}_{\varepsilon'}(\boldsymbol{x},t)-\boldsymbol{b}_{\varepsilon'}(\boldsymbol{y},t)\|)_{\varepsilon'} \leqslant (D_{\varepsilon'})_{\varepsilon'}\|\boldsymbol{x}-\boldsymbol{y}\| \quad (2.204)$$

for all $t \in [0,\infty)$ and all $\boldsymbol{x}$ and $\boldsymbol{y} \in \mathbf{R}^n$.

Here $\boldsymbol{W}(t)$ is n-dimensional Brownian motion, and

$$\forall t \in [0,T]: (\boldsymbol{b}_{\varepsilon'}(\boldsymbol{x},t))_{\varepsilon'} \in G^n(\mathbf{R}^n), \boldsymbol{b}_0(\cdot,t) \equiv \boldsymbol{b}_{\varepsilon'=0}(\cdot,t): \mathbf{R}^n \to \mathbf{R}^n$$

is a polynomial on variable $x=(x_1,\cdots,x_n)$, i. e., $b_{0,i}(x,t)=\sum_{\alpha,|\alpha|\leqslant r} b_{0,i}^{\alpha}(t)x^{\alpha}, \alpha=(i_1,\cdots,i_n), |\alpha|=\sum_{j=1}^{n} i_j, 0 \leqslant i_j \leqslant p$.

Assumption 2. We assume now without loss of generality that $\boldsymbol{x}_0 \neq 0$.

Colombeau-Ito's SDE (2.202) – (2.203) is well-known to be equivalent to the Colombeau-Fokker-Planck equation

$$\left(\frac{\partial p_{\varepsilon'}^{\varepsilon}(\boldsymbol{q}',t' \mid \boldsymbol{q}'',t'')}{\partial t''}\right)_{\varepsilon'} = \varepsilon \sum_{i=1}^{n}\left(\frac{\partial^2 p_{\varepsilon'}^{\varepsilon}(\boldsymbol{q}',t' \mid \boldsymbol{q}'',t'')}{\partial q_i''\partial q_i''}\right)_{\varepsilon'} - \sum_{i=1}^{n}\left(\frac{\partial}{\partial q_i''}(b_{i,\varepsilon'}(\boldsymbol{q}'',t'')p_{\varepsilon'}^{\varepsilon}(\boldsymbol{q}',t' \mid \boldsymbol{q}'',t''))\right)_{\varepsilon'}. \tag{2.205}$$

With initial condition of the form

$$(p_{\varepsilon'}^{\varepsilon}(\boldsymbol{q}',t' \mid \boldsymbol{q}'',t'))_{\varepsilon'} = (p_{\varepsilon'}(\boldsymbol{q}''-\boldsymbol{q}'))_{\varepsilon'} = \delta(\boldsymbol{q}''-\boldsymbol{q}'). \tag{2.206}$$

Colombeau PDE (2.205) – (2.206) can be solved formally in terms of Feynman path integral of the forms [27] – [28]:

$$(p_{\varepsilon'}^{\varepsilon}(\boldsymbol{q}',t' \mid \boldsymbol{q}'',t'))_{\varepsilon'} = \lim_{\Delta t\to 0}\boldsymbol{I}_N(\boldsymbol{q}',t' \mid \boldsymbol{q}'',t''). \tag{2.207}$$

Here

$$\boldsymbol{I}_N(\boldsymbol{q}',t' \mid \boldsymbol{q}'',t'') = \check{\boldsymbol{N}}_N\int_{-\infty}^{\infty}\mathrm{d}\boldsymbol{q}_0\int_{-\infty}^{\infty}\mathrm{d}\boldsymbol{q}_1\cdots\int_{-\infty}^{\infty}\mathrm{d}\boldsymbol{q}_m\cdots\int_{-\infty}^{\infty}\mathrm{d}\boldsymbol{q}_{N-1}\times(\boldsymbol{p}_{\varepsilon'}(\boldsymbol{q}_0-\boldsymbol{q}'))_{\varepsilon'}\cdot$$

$$\exp\left[-\frac{1}{2\varepsilon}(\boldsymbol{S}_{\varepsilon'}(\boldsymbol{q}_0,\boldsymbol{q}_1,\cdots,\boldsymbol{q}_{N-1},\boldsymbol{q}_N,\varepsilon))_{\varepsilon'}\right] \tag{2.208}$$

$$\boldsymbol{q}_N = \boldsymbol{q}'',\mathrm{d}\boldsymbol{q}_m = \prod_{j=1}^{n}\mathrm{d}q_{j,m},m = 0,\cdots,N,\Delta t = (t''-t')/N,t_m = m\Delta t,$$

$$(\boldsymbol{S}_{\varepsilon'}(\boldsymbol{q}_0,\boldsymbol{q}_1,\cdots,\boldsymbol{q}_{N-1},\boldsymbol{q}_N,\varepsilon))_{\varepsilon'} = \Delta t\sum_{m=1}^{N}\left(\boldsymbol{L}_{\varepsilon'}\left(\frac{q_m-q_{m-1}}{\Delta t},\frac{q_m-q_{m-1}}{2},t_m\right)\right)_{\varepsilon'} \tag{2.209}$$

and

$$\boldsymbol{L}_{\varepsilon'}\left(\frac{\boldsymbol{q}_m-\boldsymbol{q}_{m-1}}{\Delta t},\frac{\boldsymbol{q}_m-\boldsymbol{q}_{m-1}}{2},t_m\right) = \left\|\frac{\boldsymbol{q}_m-\boldsymbol{q}_{m-1}}{\Delta t}+\boldsymbol{b}_{\varepsilon'}\left(\frac{\boldsymbol{q}_m+\boldsymbol{q}_{m-1}}{2},t_m\right)\right\|^2 - \varepsilon\sum_{i=1}^{n}b_{i,i,\varepsilon'}\left(\frac{q_m+q_{m-1}}{2},t_m\right), \tag{2.210}$$

$$b_{i,i,\varepsilon'}(\boldsymbol{q},t) = \frac{\partial b_{i,\varepsilon'}(\boldsymbol{q},t)}{\partial q_i};\varepsilon,\varepsilon' \in (0,1].$$

Here $(p_{\varepsilon'}^{\varepsilon}(\boldsymbol{q}',t' \mid \boldsymbol{q}'',t''))_{\varepsilon'}$ is the generalized probability density that the system (2.202) will end up at q'' at time t'' if it started at q' at time t' and here $\check{\boldsymbol{N}}_N$ is the usual overall normalization of the path integral

$$\check{N}_N = (2\pi\varepsilon\Delta t)^{-nN/2}. \tag{2.211}$$

Remark 1. Note that Colombeau-Fokker-Planck Eqs. (2.205) – (2.206) is just a Euclidean Colombeau-Schrodinger equation, and is well-known that one can transform Colombeau-Schrodinger equation

$$\left(\frac{\partial u_{\varepsilon'}}{\partial t}\right)_{\varepsilon'} = \frac{\mathrm{i}\varepsilon}{2}(\Delta u_{\varepsilon'})_{\varepsilon'} - \mathrm{i}(V_{\varepsilon'}u_{\varepsilon'})_{\varepsilon'},(u_{\varepsilon'}(0))_{\varepsilon'} = (\varphi_{\varepsilon'})_{\varepsilon'},\varepsilon' \in (0,1] \tag{2.212}$$

Into mathematically rigorous path integral by standard method using Trotter's Product Formula[29]. Here Δ is the Laplace operator $\partial^2/\partial x_1^2+\cdots+\partial^2/\partial x_n^2$, $V_{\varepsilon'}$ is a realmeasurable function on $\mathbf{R}^n$, $\varphi_{\varepsilon'}$ and each $u_{\varepsilon'}$ are elements of $L_2(\mathbf{R}^n)$ and ε is a constant. Let $\mathcal{F}$ denote the Fourier transformation, $\mathcal{F}^{-1}$ its inverse. We define now as usual[29]:

$$(\Delta\varphi_{\varepsilon'})_{\varepsilon'} = (\mathcal{F}^{-1}[(-\|\lambda\|^2)\mathcal{F}(\varphi_{\varepsilon'})])_{\varepsilon'}, \varepsilon' \in (0,1] \tag{2.213}$$

on the domain $D(\Delta)$ of all square-integrable $(\varphi_{\varepsilon'})_{\varepsilon'}$ such that $(\mathcal{F}^{-1}[(-\|\lambda\|^2)\mathcal{F}(\varphi_{\varepsilon'})])_{\varepsilon'}$ is also square-integrable. (Here λ denotes the variable in momentum space and $\|\lambda\|^2 = \lambda_1^2 + \cdots + \lambda_n^2$). Then Δ is self-adjoint, and

$$(u_{\varepsilon'}(t))_{\varepsilon'} = (K_{\varepsilon}^t \varphi_{\varepsilon'})_{\varepsilon'}, K_{\varepsilon}^t = \exp\left[\frac{it\varepsilon}{2}\Delta\right] \tag{2.214}$$

is the solution of the Eq. (2.212) for $(V_{\varepsilon'})_{\varepsilon'} = 0$. The operator $V_{\varepsilon'}$ of multiplication by the function $V_{\varepsilon'}$, on the domain $D(V_{\varepsilon'})$ of all $\varphi_{\varepsilon'}$ in $L_2(\mathbf{R}^n)$ such that $V_{\varepsilon'}\varphi_{\varepsilon'}$ is also in $L_2(\mathbf{R}^n)$, is self-adjoint, and

$$(u_{\varepsilon'}(t))_{\varepsilon'} = (M_{V_{\varepsilon'}}^t \varphi_{\varepsilon'})_{\varepsilon'}, M_{V_{\varepsilon'}}^t = \exp[-itV_{\varepsilon'}] \tag{2.215}$$

is the solution of the Eq. (2.212) with $\varepsilon = 0$. Kato has found conditions under which the operator $\mathcal{R}_{\varepsilon,\varepsilon'}$ is self-adjoint[29]

$$(\mathcal{R}_{\varepsilon,\varepsilon'}u_{\varepsilon'})_{\varepsilon'} = \frac{i\varepsilon}{2}(\Delta u_{\varepsilon'})_{\varepsilon'} - i(V_{\varepsilon'}u_{\varepsilon'})_{\varepsilon'}, \tag{2.216}$$

Under these conditions if we let

$$U_{\varepsilon,V_{\varepsilon'}}^t = \exp[t\mathcal{R}_{\varepsilon,\varepsilon'}]. \tag{2.217}$$

Then a theorem of Trotter[29] asserts that for all $\varphi_{\varepsilon'}$ in $L_2(\mathbf{R}^n)$

$$(U_{\varepsilon,V_{\varepsilon'}}^t \varphi_{\varepsilon'})_{\varepsilon'} = \left(\lim_{N\to\infty}\left[\left(K_{\varepsilon}^{\frac{t}{N}} M_{V_{\varepsilon'}}^{\frac{t}{N}}\right)^N \varphi_{\varepsilon'}\right]\right)_{\varepsilon'} \tag{2.218}$$

This is discussed in detail in [29] (See [29] Appendix B). Using now Eq. (2.213) – Eq. (2.217) by simple calculation one obtain[29]

$$\left(\left(K_{\tau}^{\frac{t}{N}} M_{V_{\varepsilon'}}^{\frac{t}{N}}\right)^N \varphi_{\varepsilon'}(x)\right)_{\varepsilon'} = \left(\frac{2\pi it\varepsilon}{N}\right)^{-\frac{1}{2}nN} \times$$

$$\left(\int\cdots\int d^n\boldsymbol{x}_0\cdots d^n\boldsymbol{x}_{N-1}\exp[iS_{\varepsilon'}(\boldsymbol{x}_0,\cdots,\boldsymbol{x}_N;t)]\right)_{\varepsilon'}. \tag{2.219}$$

Here $S_{\varepsilon'}(x_0,\cdots,x_n;t) = \sum_{i=1}^{N}\left[\frac{1}{2\varepsilon}\frac{\|x_i - x_{i-1}\|^2}{(t/N)^2} - V_{\varepsilon'}(x_i)\right]\frac{t}{N}$, where we set $x_N = x$.

Theorem 10. (Suzuki-Trotter Formula)[31-33]. Let $\{A_j\}_{j=1}^p$ be an family of any bounded operator in a Banach algebra $\boldsymbol{C}$ with a norm $\|\circ\|_C$. Let $\Phi_n(\{A_j\})$ be a function

$$\Phi_n(\{A_j\}) = \left(\exp\left(\frac{A_1}{n}\right)\cdots\exp\left(\frac{A_p}{n}\right)\right)^n. \tag{2.220}$$

For any bounded operators $\{A_j\}_{j=1}^p$ in a Banach algebra $\boldsymbol{C}$:

$$\lim_{n\to\infty}\|\Phi_n(\{A_j\}) - \exp(\sum_{j=1}^p A_j)\|_C = 0. \tag{2.221}$$

Remark 2. Note that one can transform Colombeau-Fokker-Planck Eq. (2.205) – Eq. (2.206) into mathematically rigorous path integral by standard method using Suzuki-Trotter's Product Formula (2.221) (see, e.g., Ref. [33]). However path integral representation of the solutions of the Colombeau-Fokker- Planck Eq. (2.205) – Eq. (2.206), given by canonical Eq. (2.207) – Eq. (2.211), does not valid under canonical assumptions which is discussed above in Remark 1.

Remark 3. Note that formal Colombeau pseudo-differential operator[30 – 35], given by formula (see also [36 – 39])

$$(P^t_{\varepsilon'})_{\varepsilon'} = \left(\exp\left(\sum_{i=1}^{n}\int_0^t b_{i,\varepsilon'}\left(\overset{2}{x},\tau\right)\mathrm{d}\tau\,\overset{1}{\frac{\partial}{\partial x_i}}\right)\right)_{\varepsilon'},\tag{2.222}$$

evidently does not define any contraction Colombeau semi-group on $L_2(\mathbf{R}^n)$. Even in the case when a functions $(b_{i,\varepsilon'}(x,t))_{\varepsilon'}, i = 1,\cdots,n$ is the Colombeau constants $(b_{i,\varepsilon'})_{\varepsilon'} \in \tilde{\mathbf{R}}, i = 1,\cdots, n$ formal Colombeau pseudo-differential operator (see [30 – 35]), given by formula

$$(P^t_{\varepsilon'})_{\varepsilon'} = \left(\exp\left(t\sum_{i=1}^{n} b_{i,\varepsilon'}\frac{\partial}{\partial x_i}\right)\right)_{\varepsilon'},\tag{2.223}$$

does not define any contraction Colombeau semi-group on $L_2(\mathbf{R}^n)$. Nevertheless formal Colombeau pseudo-differential operator (2.223) define an contraction Colombeau semi-group: ① on a test space $H^\infty(S_R), R = (R_1,\cdots,R_n)$, with members $\varphi(\boldsymbol{x}), \boldsymbol{x} \in \mathbf{R}^n$ such that $\mathcal{F}[\varphi](\boldsymbol{\xi})$ is supported inside region $S_R = \{\boldsymbol{\xi} \mid |\xi_i| < R_i; i = 1,\cdots,n\}$ [40] and ② on a corresponding dual space $H^{-\infty}(S_R)$. Pseudo-differential calculus on a test space $H^\infty(S_R)$ and on dual space $H^{-\infty}(S_R)$ is discussed in detail in Ref. [40].

Theorem 11.[40]. Let

$$D = (D_1,\cdots,D_n), D_i = \frac{\partial}{\partial x_i} \triangleq \partial x_i\tag{2.224}$$

and

$$A(D) = \sum_{|\alpha|=0}^{M} a_\alpha D^\alpha, a_\alpha \in \boldsymbol{C}, M \leqslant \infty.$$

Then (1) $A(D)\varphi(\boldsymbol{x}) \in H^\infty(S_R)$ if $\varphi(\boldsymbol{x}) \in H^\infty(S_R)$,

(2) $A(D)'\gamma(\boldsymbol{x}) \in H^{-\infty}(S_R)$ if $'\gamma(\boldsymbol{x}) \in H^{-\infty}(S_R)$,

(3) Operator $A(D)$ is bounded on $H^\infty(S_R)$,

(4) Operator $A(D)$ is bounded on $H^{-\infty}(S_R)$.

Remark 4. (1) From Theorem 10 – 11 one obtain, that: ① operator $\sum_{i=1}^{n} b_{i,\varepsilon'}\partial q_i$, $\varepsilon \in (0,1]$ is bounded on $H^\infty(S_R)$, ② the generalized function $(u_{\varepsilon'}(\boldsymbol{q}'',t''))_{\varepsilon'}$ given by formula

$$(u_{\varepsilon'}(\boldsymbol{q}'',t''))_{\varepsilon'} = (p^{\varepsilon}_{\varepsilon'}(\boldsymbol{q}',t' \mid \boldsymbol{q}'',t''))_{\varepsilon'} = \left(P^{t''-t'}_{\varepsilon'}\varphi(\boldsymbol{q}''-\boldsymbol{q}')\right)_{\varepsilon'} =$$

$$\left(\exp\left(t''\sum_{i=1}^{n} b_{i,\varepsilon'}\partial q''_i\right)\varphi(\boldsymbol{q}''-\boldsymbol{q}')\right)_{\varepsilon'} =$$

$$\left(\mathcal{F}^{-1}\left\{\exp\left[-(t''-t')\sum_{i=1}^{n} b_{i,\varepsilon'}(i\xi_i)\right]\mathcal{F}[\varphi(\boldsymbol{q}''-\boldsymbol{q}')](\boldsymbol{\xi})\right\}\right)_{\varepsilon'} =$$

$$\frac{1}{(2\pi)^n}\left(\int_{-\infty}^{\infty} d^n\boldsymbol{\xi}\exp\left[i\langle \boldsymbol{q}'',\boldsymbol{\xi}\rangle - (t''-t')\sum_{i=1}^{n} b_{i,\varepsilon'}(i\boldsymbol{\xi}_i)\right]\mathcal{F}[\varphi(\boldsymbol{q}''-\boldsymbol{q}')](\boldsymbol{\xi})\right)_{\varepsilon'}$$

(2.225)

is the solution of the Colombeau-Fokker-Planck Eqs. (2.205) – (2.206) with initial condition $\varphi(\boldsymbol{q}''-\boldsymbol{q}') \in H^{\infty}(S_R)$, for the case: $\varepsilon = 0$ and $(b_{i,\varepsilon'}(\boldsymbol{q}'',t))_{\varepsilon'} \equiv (b_{i,\varepsilon'})_{\varepsilon'}, i = 1,\cdots,n$, $(b_{i,\varepsilon'})_{\varepsilon'} \in \tilde{\mathbf{R}}, i = 1,\cdots,n$ is the Colombeau constants, and ③ $\forall t'': u_{\varepsilon'}(\boldsymbol{q}'',t'') \in H^{\infty}(S_R)$.

(2) From Theorem 10 – 11 one obtain that:

①Operator Δ is bounded on $H^{\infty}(S_R)$,

②The Colombeau generalized function $(u_{\varepsilon'}(\boldsymbol{q}'',t''))_{\varepsilon'}$ given by formula

$$(u_{\varepsilon'}(\boldsymbol{q}'',t''))_{\varepsilon'} = (K_{\varepsilon}^{t''-t'}\varphi_{\varepsilon'}(\boldsymbol{q}''-\boldsymbol{q}'))_{\varepsilon'}, \tag{2.226}$$

$$K_{\varepsilon}^{t''-t'} = \exp\left[\frac{\varepsilon}{2}(t''-t')\Delta\right] \tag{2.227}$$

$$(u_{\varepsilon'}(\boldsymbol{q},t''))_{\varepsilon'} =$$

$$\left(\mathcal{F}^{-1}\left\{\exp\left[-\frac{\varepsilon}{2}(t''-t')\,\|\boldsymbol{\xi}\|^2\right]\mathcal{F}[\varphi_{\varepsilon'}(\boldsymbol{q}''-\boldsymbol{q}')](\boldsymbol{\xi})\right\}\right)_{\varepsilon'} \tag{2.228}$$

is the solution of the Colombeau-Fokker-Planck Eqs. (2.205) – (2.206) with initial condition $\varphi_{\varepsilon'}(\boldsymbol{q}''-\boldsymbol{q}') \in H^{\infty}(S_R)$, for the case: $(b_{i,\varepsilon'}(\boldsymbol{q}'',t))_{\varepsilon'} \equiv 0, i = 1,\cdots,n$, and ③ $\forall t'': u_{\varepsilon'}(\boldsymbol{q}'', t'') \in H^{\infty}(S_R)$.

(3) From Theorem 10 – 11 one obtain that: operator

$R_{\varepsilon,\varepsilon'} = \sum_{i=1}^{n} b_{i,\varepsilon'}\dfrac{\partial}{\partial q_i''} + \varepsilon\Delta, \varepsilon \in (0,1]$, is bounded on $H^{\infty}(S_R)$.

If we let now

$$U_{\varepsilon,\varepsilon'}^{t''-t'} = \exp[(t''-t')\mathcal{R}_{\varepsilon,\varepsilon'}], \tag{2.229}$$

Then Theorem 10 – 11 asserts that for all $\varphi_{\varepsilon'}$ in $H^{\infty}(S_R)$

$$(U_{\varepsilon,\varepsilon'}^{t''-t'}\varphi_{\varepsilon'})_{\varepsilon'} = \left(\lim_{N\to\infty}\left[\left(K_{\varepsilon}^{\frac{t''-t'}{N}}P_{\varepsilon'}^{\frac{t''-t'}{N}}\right)^N\varphi_{\varepsilon'}(\boldsymbol{q}''-\boldsymbol{q}')\right]\right)_{\varepsilon'}, \tag{2.230}$$

Where the limit is calculated by norm in $H^{\infty}(S_R)$.

From Eq. (2.226) – Eq. (2.230) by simple calculation one obtain

$$\left(K_{\varepsilon}^{\frac{t''-t'}{N}}P_{\varepsilon'}^{\frac{t''-t'}{N}}\right)^N\varphi_{\varepsilon'}(\boldsymbol{q}''-\boldsymbol{q}') =$$

$$\frac{1}{(2\pi)^{\frac{Nn}{2}}}\int_{-\infty}^{\infty} d\boldsymbol{q}\int_{-\infty}^{\infty} d\boldsymbol{\xi}\exp\left\{i\sum_{m=0}^{N}\left[(\boldsymbol{q}_{m+1}-\boldsymbol{q}_m)\boldsymbol{\xi}_m + \frac{t''-t'}{N}\sum_{i=1}^{n} b_{i,\varepsilon'}\boldsymbol{\xi}_{i,m}\right] - \right.$$

$$\left.\frac{t''-t'}{N}\frac{\varepsilon}{2}\sum_{m=0}^{N}\|\boldsymbol{\xi}_m\|^2\right\}\varphi_{\varepsilon'}(\boldsymbol{q}_0-\boldsymbol{q}'). \tag{2.231}$$

Here

$$dq = d\boldsymbol{q}_0\cdots d\boldsymbol{q}_m\cdots d\boldsymbol{q}_N, d\boldsymbol{\xi} = d\boldsymbol{\xi}_0\cdots d\boldsymbol{\xi}_m\cdots d\boldsymbol{\xi}_N, \boldsymbol{q}_N = \boldsymbol{q}'',$$

$$d\boldsymbol{q}_m = \prod_{j=1}^{n} d\boldsymbol{q}_{j,m}, d\boldsymbol{\xi}_m = \prod_{j=1}^{n} d\boldsymbol{\xi}_{j,m}, m = 0,\cdots,N.$$

Integrating on variable ξ gives

$$\left(K_{\varepsilon}^{\frac{t''-t'}{N}} P_{\varepsilon'}^{\frac{t''-t'}{N}}\right)^N \varphi_{\varepsilon'}(\boldsymbol{q}''-\boldsymbol{q}') = \boldsymbol{I}_{N,\varepsilon'}(\boldsymbol{q}',t' \mid \boldsymbol{q}'',t') =$$

$$\frac{1}{(2\pi)^{\frac{Nn}{2}}}\int_{-\infty}^{\infty} \mathrm{d}\boldsymbol{q}\exp\left\{\frac{1}{2\boldsymbol{\varepsilon}}\frac{t''-t'}{N}\sum_{m=0}^{N}\sum_{i=1}^{n}\left[\frac{(\boldsymbol{q}_{i,m+1}-\boldsymbol{q}_{i,m})}{\frac{t''-t'}{N}}-b_{i,\varepsilon'}\right]^2\right\}\varphi(\boldsymbol{q}_0-\boldsymbol{q}').$$

Finally we obtain

$$(U_{\varepsilon,\varepsilon'}^{t''-t'}\varphi_{\varepsilon'})_{\varepsilon'} = \lim_{N\to\infty}(\boldsymbol{I}_{N,\varepsilon'}(\boldsymbol{q}',t' \mid \boldsymbol{q}'',t'))_{\varepsilon'}. \tag{2.232}$$

Remark 5. Note that $\delta_{\varepsilon'}(x) \in H^{\infty}(S_R), R = 1/\varepsilon', x \in \mathbf{R}, \varepsilon' \in (0,1]$, where

$$\delta_{\varepsilon'}(x) = \frac{1}{\pi\varepsilon' x}\sin\left(\frac{x}{\varepsilon'}\right). \tag{2.233}$$

By simple calculation one obtain

$$\mathcal{F}[\delta_{\varepsilon'}(x)] = \begin{cases}\frac{1}{2\pi}, x \in (-r,r)\\ 0, x \notin (-r,r)\end{cases}, \tag{2.234}$$

$r = 1/\varepsilon'$ [30].

Assumption 3. We assume now that

$$(p_{\varepsilon'}(\boldsymbol{q}''-\boldsymbol{q}'))_{\varepsilon'} = \prod_{i=1}^{n}(\delta_{\varepsilon'}(q_i''-q'_i))_{\varepsilon'}. \tag{2.235}$$

Remark 6. Note that

$$(p_{\varepsilon'}(\boldsymbol{q}_0-\boldsymbol{q}'))_{\varepsilon'} = \delta(\boldsymbol{q}_0-\boldsymbol{q}'). \tag{2.236}$$

Definition 14. We let now $n = 1$. A tagged partition of the real line $\mathbf{R} = (-\infty, +\infty)$ is a finite sequence $-\infty = x_0 < x_1 < x_2 < \cdots < x_{P-1} < x_P = +\infty$. This partitions the open interval $(-\infty, +\infty)$ into n sub-intervals $\boldsymbol{J}_r = [x_{r-1}, x_r], r = 1,\cdots,p, \boldsymbol{J}_0 = (-\infty, x_1], \boldsymbol{J}_P = [x_{P-1}, +\infty)$ indexed by $r = 1,\cdots,p$. Let $b_{\varepsilon'}(q'',t,r)$ be a quantity

$$b_{\varepsilon'}(t,r) = \sup_{q''\in J_r} b_{\varepsilon'}(q'',t) \tag{2.237}$$

and let $\partial q_i'' b_{\varepsilon'}(t,r)$ be a quantity

$$\partial q'' b_{\varepsilon'}(t,r) = \sup_{q''\in J_r}\frac{\partial b_{\varepsilon'}(q'',t)}{\partial q''}. \tag{2.238}$$

Let $\dot{b}_{\varepsilon'}(q'',t)$ be a function

$$\dot{b}_{\varepsilon'}(q'',t) = \sum_{r=1}^{P} 1_{J_r}(q'')b_{\varepsilon'}(t,r). \tag{2.239}$$

Let $\partial\dot{b}_{\varepsilon'}(q'',t)$ be a function

$$\partial\dot{b}_{\varepsilon'}(q'',t) = \sum_{r=0}^{P} 1_{J_r}(q'')\partial b_{\varepsilon'}(t,r). \tag{2.240}$$

Here $1_{J_r}(\boldsymbol{q}'')$ is indicator function of a subset :$\boldsymbol{J}_r = [x_{r-1}, x_r]$.

Definition 15. Let $H^{\infty}(S_R;\check{S}_U), R = (R_1,\cdots,R_n), U = (U_1,\cdots,U_n)$ be a test space with a member $\varphi(\boldsymbol{x},\boldsymbol{p}), \boldsymbol{x} \in \mathbf{R}_x^n, p \in \mathbf{R}_p^n$, such that $\forall \check{\boldsymbol{x}}, \check{\boldsymbol{x}} \in \check{S}_U, \check{S}_U \subseteq S_U = \{\boldsymbol{x} \mid |x_i| < U_i \leqslant \infty; i = 1,\cdots,n\}$ function $\varphi(\check{\boldsymbol{x}},\boldsymbol{p})$ is supported inside region $S_R = \{\boldsymbol{p} \mid |p_i| < R_i; i = 1,\cdots,n\}$, i. e.

$\forall \check{x},\check{x} \in S_U : \mathcal{F}^{-1}[\varphi(\check{x},p)](\check{x},x) \in H^\infty(S_R)$.

Definition 16. (1) We let $\mathcal{F}^{\#}[\psi(x)](p) = \varphi(x,p)$ if there exist a function $\varphi(x,p) \in H^\infty(S_R;S_U)$ such that

$$\mathcal{F}^{-1}[\varphi(x,p)](x,x) = \psi(x). \tag{2.241}$$

(2) We let now $H^\infty(S_R;\check{S}_{U,p})$ if $n = 1$ and $\check{S}_U = S_U \setminus \{x_0,\cdots,x_p\}$.

Remark 6. ① From Theorem 11 – 12 one obtain, that: operator $b_{\varepsilon'}(q'',t)\partial q = \sum_{r=1}^{p} \mathbf{1}_{J_r}(q'')b_{\varepsilon'}(t,r)\partial q$, $\varepsilon \in (0,1]$ is bounded on $H^\infty(S_R;\check{S}_U)$.

② The Colombeau generalized function $(u_{\varepsilon',p}(q'',t''))_{\varepsilon'}$ given by formula

$$(u_{\varepsilon',p}(q'',t''))_{\varepsilon'} = (P_{\varepsilon',p}^{t''-t'}\varphi_{\varepsilon'}(q''-q'))_{\varepsilon'} =$$
$$(\exp[-(t''-t')\grave{b}_{\varepsilon'}(q'',t)\partial q'']\varphi_{\varepsilon'}(q''-q'))_{\varepsilon'} =$$
$$\left(\exp\left[-(t''-t')\sum_{r=1}^{P}\mathbf{1}_{J_r}(q'')b_{\varepsilon'}(t,r)\partial q''\right]\varphi_{\varepsilon'}(q''-q')\right)_{\varepsilon'} =$$
$$\left(\prod_{r=1}^{p}\exp[-(t''-t')\mathbf{1}_{J_r}(q'')b_{\varepsilon'}(t,r)\partial q'']\varphi_{\varepsilon'}(q''-q')\right)_{\varepsilon'} \tag{2.242}$$

is the solution (except points $\{x_0,\cdots,x_p\}$) of the Colombeau-Fokker-Planck Eqs. (2.205) – (2.206) with initial condition $\varphi_{\varepsilon'}(q''-q') \in H^\infty(S_R;\check{S}_{U,p})$, for the case:

$$(b_{\varepsilon'}(q'',t))_{\varepsilon'} = \grave{b}_{\varepsilon'}(q'',t) = \sum_{r=1}^{P}\mathbf{1}_{J_r}(q'')b_{\varepsilon'}(t,r). \tag{2.243}$$

③ We note that:

$$\exp[-(t''-t')\mathbf{1}_{J_r}(q'')b_{\varepsilon'}(t,r)\partial q'']\varphi_{\varepsilon'}(q''-q') =$$
$$\mathcal{F}^{-1}\left\{\exp[-(t''-t')\mathbf{1}_{J_r}(q'')b_{\varepsilon'}(t,r)(i\xi)]\mathcal{F}^{\#}[\varphi_{\varepsilon'}(q''-q')]\right\}. \tag{2.244}$$

④ $\forall t'' : u_{\varepsilon',p}(q'',t'') \in H^\infty(S_R;\check{S}_{U,p})$.

(2) From Theorem 11 – 12 one obtain, that:

①Operator Δ is bounded on $H^\infty(S_R;\check{S}_{U,p})$,

②The Colombeau generalized function $(u_{\varepsilon',p}(\boldsymbol{q}'',t''))_{\varepsilon'}$ given by formula

$$(u_{\varepsilon',p}(q'',t''))_{\varepsilon'} = (K_{\varepsilon}^{t''-t'}\varphi_{\varepsilon'}(q''-q'))_{\varepsilon'}, \tag{2.245}$$

$$K_{\varepsilon,p}^{t''-t'} = \exp\left[\frac{\varepsilon}{2}(t''-t')\Delta\right]. \tag{2.246}$$

$$(u_{\varepsilon',p}(q,t''))_{\varepsilon'} =$$
$$\left(\mathcal{F}^{-1}\left[\exp\left(-\frac{\varepsilon}{2}(t''-t')\|\boldsymbol{\xi}\|^2\right)\mathcal{F}^{\#}[\varphi_{\varepsilon'}(q''-q')](\boldsymbol{\xi})\right]\right)_{\varepsilon'} \tag{2.247}$$

is the solution of the Colombeau-Fokker-Planck Eqs. (2.205) – (2.206) with initial condition $\varphi_{\varepsilon'}(q''-q') \in H^\infty(S_R;\check{S}_{U,p})$, for the case: $(b_{\varepsilon'}(q'',t))_{\varepsilon'} \equiv 0$, and

③ $\forall t'' : u_{\varepsilon',p}(q'',t'') \in H^\infty(S_R;\check{S}_{U,p})$.

(3) From Theorem 11 – 12 one obtain, that: operator

$\mathcal{R}_{\varepsilon,\varepsilon',p} = \sum_{r=1}^{p}\mathbf{1}_{J_r}(q'')b_{\varepsilon'}(t,r)\dfrac{\partial}{\partial q''} + \varepsilon\Delta, \varepsilon' \in (0,1]$, is bounded on $H^\infty(S_R;\check{S}_{U,p})$,

If we let now

$$U^{t''-t'}_{\varepsilon,\varepsilon',p} = \exp[(t''-t')\mathcal{R}_{\varepsilon,\varepsilon',p}], \tag{2.248}$$

then Theorem 10 – 11 asserts that for all $\varphi_{\varepsilon'}$ in $H^{\infty}(S_R;\check{S}_{U,p})$

$$\left(U^{t''-t'}_{\varepsilon,\varepsilon',p}\varphi_{\varepsilon'}\right)_{\varepsilon'} = \left(\lim_{N\to\infty}\left[\left(K^{\frac{t''-t'}{N}}_{\varepsilon,p}P^{\frac{t''-t'}{N}}_{\varepsilon',p}\right)^N\varphi_{\varepsilon'}(q''-q')\right]\right)_{\varepsilon'}, \tag{2.249}$$

where the limit is calculated by norm in $H^{\infty}(S_R;\check{S}_{U,p})$.

From Eq. (2.246) – Eq. (2.249) by simple calculation one obtain

$$\left(K^{\frac{t''-t'}{N}}_{\varepsilon,p}P^{\frac{t''-t'}{N}}_{\varepsilon',p}\right)^N\varphi_{\varepsilon'}(\boldsymbol{q}''-\boldsymbol{q}') = \frac{1}{(2\pi)^{\frac{Nn}{2}}}\int_{-\infty}^{\infty}\mathrm{d}\boldsymbol{q}\int_{-\infty}^{\infty}\mathrm{d}\boldsymbol{\xi}\exp\left\{\boldsymbol{i}\sum_{m=0}^{N}\left[(\boldsymbol{q}_{m+1}-\boldsymbol{q}_m)\xi_m + \frac{t''-t'}{N}\sum_{r=1}^{p}\boldsymbol{I}_{J_r}(q_m)b_{\varepsilon'}(t,r)\xi_m\right] - \frac{t''-t'}{N}\frac{\varepsilon}{2}\sum_{m=0}^{N}\xi_m^2\right\}\varphi(q_0-q'). \tag{2.250}$$

Here $\mathrm{d}q = \mathrm{d}q_0\cdots\mathrm{d}q_m\cdots\mathrm{d}q_{N-1}$, $\mathrm{d}\xi = \mathrm{d}\xi_0\cdots\mathrm{d}\xi_m\cdots\mathrm{d}\xi_N$, $q_N = q''$, $m = 0,\cdots,N$.

Integrating on variable ξ gives

$$\left(K^{\frac{t''-t'}{N}}_{\varepsilon,p}P^{\frac{t''-t'}{N}}_{\varepsilon',p}\right)^N\varphi_{\varepsilon'}(\boldsymbol{q}''-\boldsymbol{q}') = \boldsymbol{I}_{N,\varepsilon',p}(\boldsymbol{q}',t'\mid\boldsymbol{q}'',t'') = \frac{1}{(2\pi)^{\frac{Nn}{2}}}\int_{-\infty}^{\infty}\mathrm{d}\boldsymbol{q}\exp\left\{\frac{1}{2\varepsilon}\frac{t''-t'}{N}\sum_{m=0}^{N}\left[\frac{(\boldsymbol{q}_{m+1}-\boldsymbol{q}_m)}{\frac{t''-t'}{N}} - \sum_{r=1}^{p}\boldsymbol{I}_{J_r}(q_m)b_{\varepsilon'}(t,r)\right]^2\right\}\varphi_{\varepsilon'}(q_0-q') \tag{2.251}$$

From Eq. (2.251) we obtain

$$(u_{\varepsilon',p}(q'',t''))_{\varepsilon'} = \left(\lim_{N\to\infty}\boldsymbol{I}_{N,\varepsilon',p}(q',t'\mid q'',t')\right)_{\varepsilon'}. \tag{2.252}$$

Here the limit is calculated by norm in $H^{\infty}(S_R;\check{S}_{U,p})$.

Let δ_p be the quantity $\delta_p = \max_{1\leqslant r\leqslant p}\{\delta_r \mid r = 1,\cdots,p\}$, $\delta_r = |x_{r-1},x_r|$. We assume that: ① $\delta_p \to 0$ if $p\to\infty$, ② $x_0\to\infty$ if $p\to\infty$, ③ $x_p\to\infty$ if $p\to\infty$.

Finally we obtain

$$(u_{\varepsilon'}(q'',t''))_{\varepsilon'} = \left(\lim_{p\to\infty}u_{\varepsilon',p}(q'',t'')\right)_{\varepsilon'} = \left(\lim_{p\to\infty}U^{t''-t'}_{\varepsilon,\varepsilon',p}\varphi_{\varepsilon'}(q''-q')\right)_{\varepsilon'}. \tag{2.253}$$

Here the limit is calculated by norm $\|\cdot\|_{W^{2,2}(\mathbf{R},\omega)}$ of the weighted Sobolev space $W^{2,2}(\mathbf{R},\omega)$ [41].

Theorem 13. Assume that $\varphi_{\varepsilon'}(q''-q') \in H^{\infty}(S_R)\cap W^{2,2}(\mathbf{R},\omega)$, $\varepsilon'\in(0,1]$, $\omega = \omega(q'')$. Then:

① $\forall\varepsilon'\in(0,1]$ there exist p_0 such that $\forall p_1\forall p_2[(p_1\geqslant p_0)\wedge(p_2\geqslant p_0)]$ the inequality $\|u_{\varepsilon',p_1}(q',t'\mid q'',t'') - u_{\varepsilon',p_1p_2}(q',t'\mid q'',t'')\|_{W^{2,2}(\mathbf{R},\omega)} \leqslant [(t''-t')/(p_1)]C_1\exp[C_2(t''-t')]\|\varphi_{\varepsilon'}\|_{W^{2,2}(\mathbf{R},\omega)}$

holds for each $t''\in[t',\infty)$.

②The Colombeau generalized function $(u_{\varepsilon'}(q'',t''))_{\varepsilon'}$ given by formula (2.253) is the solution of the Colombeau-Fokker-Planck Eq. (2.205), Eq. (2.206) (except a set Lebesgue measure zero) with initial condition $(\varphi_{\varepsilon'}(q''-q'))_{\varepsilon'}$ such that

$$\forall \varepsilon' \in (0,1]: \varphi_{\varepsilon'}(q''-q') \in H^{\infty}(S_R).$$

Let us consider now n-dimensional case. We shall be working with rectangular parallelograms in $\mathbf{R}^n$, those parallelograms whose edges are mutually orthogonal. Actually, we shall be even more restrictive, and consider only those whose edges are in the directions of the coordinate axes.

Definition 17. We call them special rectangles. Each of these may be expressed as a Cartesian product of intervals in $\mathbf{R}$:

$$\boldsymbol{I} = [a_1, b_1] \times \cdots \times [a_n, b_n] = \{q'' \mid q'' \in \mathbf{R}^n, a_i \leqslant q''_i \leqslant b_i, i = 1, \cdots, n\}$$

and we let $\boldsymbol{I}_{\infty} = \mathbf{R}^n \setminus \boldsymbol{I}$.

Definition 18. We define a partition of $\boldsymbol{I}$ to be a collection of non-overlapping special rectangles $\boldsymbol{I}_1, \boldsymbol{I}_2, \cdots, \boldsymbol{I}_r, \cdots, \boldsymbol{I}_P$ whose union is $\boldsymbol{I}$. "Non-overlapping" requires that the interiors of these rectangles are mutually disjoint.

Definition 19. Let $b_{i,\varepsilon'}(\boldsymbol{q}'', t, r), i = 1, \cdots, n$ be a quantity

$$b_{i,\varepsilon'}(t,r) = \sup_{q'' \in \boldsymbol{I}_r} b_{i,\varepsilon'}(\boldsymbol{q}'', t) \tag{2.254}$$

and let $\partial q_j b_{i,\varepsilon'}(t,r), i = 1, \cdots, n, j = 1, \cdots, n$ be a quantity

$$\partial q_j b_{i,\varepsilon'}(t,r) = \sup_{\boldsymbol{q}'' \in \boldsymbol{I}_r} \frac{\partial b_{i,\varepsilon'}(\boldsymbol{q}'', t)}{\partial q_j}. \tag{2.255}$$

Definition 20. Let $\check{b}_{i,\varepsilon'}(\boldsymbol{q}'', t, r), i = 1, \cdots, n$ be a function

$$\grave{b}_{i,\varepsilon'}(\boldsymbol{q}'', t) = \sum_{r=1}^{P} \boldsymbol{1}_{I_r}(\boldsymbol{q}'') b_{i,\varepsilon'}(t,r). \tag{2.256}$$

Let $\partial q_j \grave{b}_{i,\varepsilon'}(\boldsymbol{q}'', t)$ be a function

$$\partial q_j \grave{b}_{i,\varepsilon'}(\boldsymbol{q}'', t) = \sum_{r=0}^{P} \boldsymbol{1}_{I_r}(\boldsymbol{q}'') \partial q_j b_{i,\varepsilon'}(t,r). \tag{2.257}$$

Here $\boldsymbol{1}_{I_r}(\boldsymbol{q}'')$ is indicator function of a subset $\boldsymbol{I}_r \subset \boldsymbol{I}$.

Definition 21. We let now $H_n^{\infty}(S_R; \check{S}_{U,p})$ if $\check{S}_U = S_U \setminus \{\partial I_0, \cdots, \partial I_p\}$.

Remark 7. (1) From Theorem 11 – 12 one obtain, that:

① Operator $\grave{b}_{\varepsilon'}(\boldsymbol{q}'', t)\partial \boldsymbol{q}'' = \sum_{i=1}^{n} \tilde{b}_{i,\varepsilon'}(\boldsymbol{q}'', t)\partial q''_i = \sum_{i=1}^{n} \sum_{r=1}^{p} \boldsymbol{1}_{I_r}(\boldsymbol{q}'') b_{i,\varepsilon'}(t,r)\partial q''_i$, $\boldsymbol{\varepsilon}' \in (0,1]$ is bounded on $H_n^{\infty}(S_R; \check{S}_{U,p})$.

② The Colombeau generalized function $(u_{\varepsilon',p}(\boldsymbol{q}'', t''))_{\varepsilon'}$ given by formula

$$(u_{\boldsymbol{\varepsilon}',p}(\boldsymbol{q}'', t''))_{\varepsilon'} = \left(P_{\varepsilon',p}^{t''-t'} \varphi_{\varepsilon'}(\boldsymbol{q}'' - \boldsymbol{q}') \right)_{\varepsilon'} =$$

$$\left(\exp[-(t''-t')\grave{b}_{\varepsilon'}(\boldsymbol{q}'', t)\partial \boldsymbol{q}''] \varphi_{\varepsilon'}(\boldsymbol{q}'' - \boldsymbol{q}') \right)_{\varepsilon'} =$$

$$\left(\exp\left[-(t''-t') \sum_{i=1}^{n} \sum_{r=1}^{P} \boldsymbol{1}_{I_r}(\boldsymbol{q}'') b_{i,\varepsilon'}(t,r)\partial q''_i \right] \varphi_{\varepsilon'}(\boldsymbol{q}'' - \boldsymbol{q}') \right)_{\varepsilon'} =$$

$$\left(\prod_{i=1}^{n}\prod_{r=1}^{p}\exp[-(t''-t')\mathbf{1}_{I_r}(\mathbf{q}'')b_{i,\varepsilon'}(t,r)\partial q''_i]\varphi_{\varepsilon'}(\mathbf{q}''-\mathbf{q}')\right)_{\varepsilon'}. \tag{2.258}$$

is the solution (except points $\mathbf{q}'' \in \bigcup_{r=1}^{p} \partial I_r$) of the Colombeau-Fokker-Planck Eqs. (2.205), (2.206) with initial condition $(\varphi_{\varepsilon'}(\mathbf{q}''-\mathbf{q}'))_{\varepsilon'}$ such that

$\forall \varepsilon' \in (0,1]: \varphi_{\varepsilon'}(\mathbf{q}''-\mathbf{q}') \in H_n^{\infty}(S_R;\check{S}_{U,p})$, for the case: $\varepsilon = 0$ and

$$(b_{i,\varepsilon'}(\mathbf{q}'',t))_{\varepsilon'} = \dot{b}_{\varepsilon'}(\mathbf{q}'',t) = \sum_{r=1}^{P}\mathbf{1}_{I_r}(\mathbf{q}'')b_{i,\varepsilon'}(t,r), i = 1,\cdots,n. \tag{2.259}$$

③ We note that:

$$\exp[-(t''-t')\mathbf{1}_{I_r}(\mathbf{q}'')b_{i,\varepsilon'}(t,r)\partial q''_i]\varphi_{\varepsilon'}(\mathbf{q}''-\mathbf{q}') =$$

$$\mathcal{F}^{-1}\{\exp[-(t''-t')\mathbf{1}_{I_r}(\mathbf{q}'')b_{i,\varepsilon'}(t,r)(i\xi_i)]\mathcal{F}^{\#}[\varphi_{\varepsilon'}(\mathbf{q}''-\mathbf{q}')]\} \tag{2.260}$$

④ $\forall t'': u_{\varepsilon',p}(\mathbf{q}'',t'') \in H_n^{\infty}(S_R;\check{S}_{U,p})$.

(2) From Theorem 11 − 12 one obtain, that:

①Operator Δ is bounded on $H_n^{\infty}(S_R;\check{S}_{U,p})$,

②The Colombeau generalized function $(u_{\varepsilon',p}(\mathbf{q}'',t''))_{\varepsilon'}$ given by formula

$$(u_{\varepsilon',p}(\mathbf{q}'',t''))_{\varepsilon'} = \left(K_{\varepsilon,p}^{t''-t'}\varphi_{\varepsilon'}(\mathbf{q}''-\mathbf{q}')\right)_{\varepsilon''} \tag{2.261}$$

$$K_{\varepsilon,p}^{t''-t'} = \exp\left[\frac{\varepsilon}{2}(t''-t')\Delta\right] \tag{2.262}$$

$$(u_{\varepsilon',p}(\mathbf{q}'',t''))_{\varepsilon'} =$$

$$\left(\mathcal{F}^{-1}\left\{\exp\left(-\frac{\varepsilon}{2}(t''-t')\ \|\boldsymbol{\xi}\|^2\right)\mathcal{F}^{\#}[\varphi_{\varepsilon'}(\mathbf{q}''-\mathbf{q}')](\boldsymbol{\xi})\right\}\right)_{\varepsilon'} \tag{2.263}$$

is the solution of the Colombeau-Fokker-Planck Eq. (2.205) – Eq. (2.206) with initial condition $\left(\varphi_{\varepsilon'}(\mathbf{q}''-\mathbf{q}')\right)_{\varepsilon'}$ such that

$\forall \varepsilon' \in (0,1]: \varphi_{\varepsilon'}(\mathbf{q}''-\mathbf{q}') \in H_n^{\infty}(S_R;\check{S}_{U,p})$, for the case: $\left(b_{i,\varepsilon'}(\mathbf{q}'',t)\right)_{\varepsilon'} \equiv 0, i = 1, \cdots, n$ and ③ $\forall t'': u_{\varepsilon',p}(\mathbf{q}'',t'') \in H_n^{\infty}(S_R;\check{S}_{U,p})$.

(3) From Theorem 11 − 12 one obtain, that: operator

$\mathcal{R}_{\varepsilon,\varepsilon',p} = \sum_{i=1}^{n}\sum_{r=1}^{p} b_{i,\varepsilon'}\dfrac{\partial}{\partial \mathbf{q}''_i} + \varepsilon\Delta, \varepsilon \in (0,1]$, is bounded on $H_n^{\infty}(S_R;\check{S}_{U,p})$.

If we let now

$$U_{\varepsilon,\varepsilon',p}^{t''-t'} = \exp[(t''-t')\mathcal{R}_{\varepsilon,\varepsilon',p}], \tag{2.264}$$

then Theorem 10 − 11 asserts that for all $\varphi_{\varepsilon'}$ in $H_n^{\infty}(S_R;\check{S}_{U,p})$

$$\left(U_{\varepsilon,\varepsilon',p}^{t''-t'}\varphi_{\varepsilon'}\right)_{\varepsilon'} = \left(\lim_{N\to\infty}\left[\left(K_{\varepsilon,p}^{\frac{t''-t'}{N}}P_{\varepsilon',p}^{\frac{t''-t'}{N}}\right)^{N}\varphi_{\varepsilon'}(\mathbf{q}''-\mathbf{q}')\right]\right)_{\varepsilon'} \tag{2.265}$$

Where the limit is calculated by norm in $H_n^{\infty}(S_R;\check{S}_{U,p})$.

From Eqs. (2.258) – (2.265) by simple calculation one obtain

$$\left(K_{\varepsilon,p}^{\frac{t''-t'}{N}}P_{\varepsilon',p}^{\frac{t''-t'}{N}}\right)^{N}\varphi_{\varepsilon'}(\mathbf{q}''-\mathbf{q}') =$$

$$\frac{1}{(2\pi)^{\frac{Nn}{2}}}\int_{-\infty}^{\infty}\mathrm{d}\boldsymbol{q}\int_{-\infty}^{\infty}\mathrm{d}\boldsymbol{\xi}\exp\left\{\boldsymbol{i}\sum_{m=0}^{N}\left[(\boldsymbol{q}_{m+1}-\boldsymbol{q}_m)\boldsymbol{\xi}_m+\frac{t''-t'}{N}\sum_{i=1}^{n}\sum_{r=1}^{p}\boldsymbol{1}_{J_r}(\boldsymbol{q}_m)b_{i,\varepsilon'}(t,r)\boldsymbol{\xi}_{i,m}\right]-\right.$$
$$\left.\frac{t''-t'}{N}\frac{\varepsilon}{2}\sum_{m=0}^{N}\boldsymbol{\xi}_m^2\right\}\varphi_{\varepsilon'}(\boldsymbol{q}_0-\boldsymbol{q}'). \tag{2.266}$$

Here $\mathrm{d}\boldsymbol{q}=\mathrm{d}\boldsymbol{q}_0,\cdots,\mathrm{d}\boldsymbol{q}_m,\cdots,\mathrm{d}\boldsymbol{q}_{N-1},\mathrm{d}\boldsymbol{\xi}=\mathrm{d}\boldsymbol{\xi}_0,\cdots,\mathrm{d}\boldsymbol{\xi}_m,\cdots,\mathrm{d}\boldsymbol{\xi}_N,\boldsymbol{q}_N=\boldsymbol{q}''$,

$$\mathrm{d}\boldsymbol{q}_m=\prod_{i=1}^{n}\mathrm{d}q_{i,m},\mathrm{d}\boldsymbol{\xi}_m=\prod_{i=1}^{n}\mathrm{d}\xi_{i,m},m=0,\cdots,N.$$

Integrating on variable $\boldsymbol{\xi}$ gives

$$\left(K_{\varepsilon,p}^{\frac{t''-t'}{N}}P_{\varepsilon',p}^{\frac{t''-t'}{N}}\right)^N\varphi_{\varepsilon'}(q''-q')=\boldsymbol{I}_{N,\varepsilon',p}(q',t'\mid q'',t')=$$
$$\frac{1}{(2\pi)^{\frac{Nn}{2}}}\int_{-\infty}^{\infty}\mathrm{d}\boldsymbol{q}\exp\left\{\frac{1}{2\varepsilon}\frac{t''-t'}{N}\sum_{m=0}^{N}\sum_{i=1}^{n}\left[\frac{(q_{i,m+1}-q_{i,m})}{\frac{t''-t'}{N}}-\sum_{r=1}^{p}\boldsymbol{1}_{I_r}(\boldsymbol{q}_m)b_{i,\varepsilon'}(t,r)\right]^2\right\}\cdot$$
$$\varphi_{\varepsilon'}(\boldsymbol{q}_0-\boldsymbol{q}') \tag{2.267}$$

From Eq. (2.267) we obtain

$$(u_{\varepsilon',p}(\boldsymbol{q}'',t''))_{\varepsilon'}=\lim_{N\to\infty}\boldsymbol{I}_{N,\varepsilon',p}(\boldsymbol{q}',t'\mid\boldsymbol{q}'',t'').$$

Here the limit is calculated by norm in $H_n^{\infty}(S_R;\check{S}_{U,p})$.

Let $\delta_p=\max\limits_{1\leqslant r\leqslant p}\{\delta_r\mid r=1,\cdots,p\}$, $\delta_r=\mathrm{diam}(\boldsymbol{I}_r)$, where $\mathrm{diam}(I_r)=\sup\limits_{x,y\in I_r}\|x-y\|$.

We assume that: ① $\delta_p\to 0$ if $p\to\infty$, ② $a_i\to-\infty, i=1,\cdots,n$ if $p\to\infty$, ③ $b_i\to\infty, i=1,\cdots,n$ if $p\to\infty$.

Finally we obtain

$$(u_{\varepsilon'}(\boldsymbol{q}'',t''))_{\varepsilon'}=\left(\lim_{p\to\infty}u_{\varepsilon',p}(\boldsymbol{q}'',t'')\right)_{\varepsilon'}=\left(\lim_{p\to\infty}U_{\varepsilon,\varepsilon',p}^{t''-t'}\varphi_{\varepsilon'}(\boldsymbol{q}''-\boldsymbol{q}')\right)_{\varepsilon'}. \tag{2.268}$$

Here the limit is calculated by norm $\|\cdot\|_{W^{2,2}(\mathbf{R}^n,\omega)}$ of the weighted Sobolev space $(\mathbf{R}^n,\omega)$ [41].

Theorem 14. Assume that $\varphi_{\varepsilon'}(\boldsymbol{q}''-\boldsymbol{q}')\in H_n^{\infty}(S_R)\cap W^{2,2}(\mathbf{R}^n,\omega),\varepsilon'\in(0,1],\omega=\omega(\boldsymbol{q}'')$. Then:

① $\forall\varepsilon'\in(0,1]$ there exist p_0 such that $\forall p_1\forall p_2[(p_1\geqslant p_0)\wedge(p_2\geqslant p_0)]$ the inequality $\|u_{\varepsilon',p_1}(\boldsymbol{q}',t'\mid\boldsymbol{q}'',t'')-u_{\varepsilon',p_1p_2}(\boldsymbol{q}',t'\mid\boldsymbol{q}'',t'')\|_{W^{2,2}(\mathbf{R}^n,\omega)}\leqslant[(t''-t')/(p_1)]C_1\exp[C_2(t''-t')]\|\varphi_{\varepsilon'}\|_{W^{2,2}(\mathbf{R}^n,\omega)}$ holds for each $t''\in[t',\infty)$.

② The Colombeau generalized function $(u_{\varepsilon'}(\boldsymbol{q}'',t''))_{\varepsilon'}$ given by formula (2.268) is the solution of the Colombeau-Fokker-Planck Eqs. (2.205) – (2.206) (except a set Lebesgue measure zero) with initial condition $(\varphi_{\varepsilon'}(\boldsymbol{q}''-\boldsymbol{q}'))_{\varepsilon'}$ such that

$\forall\varepsilon'\in(0,1]:\varphi_{\varepsilon'}(\boldsymbol{q}''-\boldsymbol{q}')\in H_n^{\infty}(S_R)$.

Remark 8. The continuous-space-time conditional probability when $p\to\infty$ in Eqs. (2.268) is symbolically indicated by the path-integral expression [22-23]:

$$\left(p_{\varepsilon'}^{\varepsilon}(\boldsymbol{q}',t'\mid\boldsymbol{q}'',t'')\right)_{\varepsilon'}=$$
$$\int_{q(t')=\boldsymbol{q}'}^{q(t'')=\boldsymbol{q}''}[D\boldsymbol{q}(t)]\exp\left[\left(-\frac{1}{2\varepsilon}\boldsymbol{S}_{\varepsilon'}(\dot{\boldsymbol{q}},\boldsymbol{q},t;\varepsilon)\right)_{\varepsilon'}\right]. \tag{2.269}$$

Here

$$\boldsymbol{S}_{\varepsilon'}(\dot{\boldsymbol{q}},\boldsymbol{q},t,\varepsilon) = \int_{t'}^{t''} L_{\varepsilon'}(\dot{\boldsymbol{q}},\boldsymbol{q},t;\varepsilon)\,\mathrm{d}t \tag{2.270}$$

is the continuous-time limit of the discrete action (2.209) with

$$L_{\varepsilon'}(\dot{\boldsymbol{q}},\boldsymbol{q},t;\varepsilon) = \|\dot{\boldsymbol{q}}(t) - \boldsymbol{b}_{\varepsilon'}[\boldsymbol{q}(t),t;\varepsilon]\|^2 - \varepsilon\sum_{i=1}^{n} b_{i,i,\varepsilon'}[\boldsymbol{q}(t),t;\varepsilon], \tag{2.271}$$

$$b_{i,i,\varepsilon'}[\boldsymbol{q}(t),t;\varepsilon] = \frac{\partial b_i[\boldsymbol{q}(t),t;\varepsilon]}{\partial q_i};\varepsilon,\varepsilon' \in (0,1], \tag{2.272}$$

as the Lagrangian. From Eq. (2.269) one obtain

$$\left(\mathbf{E}\,[\boldsymbol{x}_{t'',\varepsilon'}^{q',\varepsilon}(\omega)]^2\right)_{\varepsilon'} = \int_{-\infty}^{\infty}\mathrm{d}\boldsymbol{q}''\ \|\boldsymbol{q}''\|^2\left(\boldsymbol{p}_{\varepsilon'}^{\varepsilon}(\boldsymbol{q}',t' \mid \boldsymbol{q}'',t'')\right)_{\varepsilon'} =$$

$$\left(\lim_{p\to\infty}\lim_{\Delta t\to 0}\breve{\boldsymbol{I}}_{N,\varepsilon,\varepsilon',p}(\boldsymbol{q}',t',t'')\right)_{\varepsilon'} \tag{2.273}$$

Here

$$\breve{\boldsymbol{I}}_{N,\varepsilon,\varepsilon',p}(\boldsymbol{q}',t',t'') = \int_{-\infty}^{\infty}\mathrm{d}\boldsymbol{q}''\ \|\boldsymbol{q}''\|^2\boldsymbol{I}_{N,\varepsilon,\varepsilon',p}(\boldsymbol{q}',t' \mid \boldsymbol{q}'',t'') \tag{2.274}$$

Remark 9 (1) Note that for any fixed N in the limit $\varepsilon\to 0$ only one unique minimizing path $\{\breve{\boldsymbol{q}}_0,\breve{\boldsymbol{q}}_1,\cdots,\breve{\boldsymbol{q}}_{N-1},\breve{\boldsymbol{q}}_N\}$ significantly contributes to the multiple integral $\breve{\boldsymbol{I}}_{N,\varepsilon,\varepsilon',p}(\boldsymbol{q}',t' \mid \boldsymbol{q}'',t'')$ given by expression (2.274). The extremality conditions for this minimizing path is

$$\bar{\nabla}\boldsymbol{q}(t_m) = \boldsymbol{b}_{\varepsilon'}(\boldsymbol{q}(t_{m-1}),t_m),m = 1,2,\cdots,N, \tag{2.275}$$

with a boundary condition

$$\boldsymbol{q}(t') = \breve{\boldsymbol{q}}_0 = \boldsymbol{q}'. \tag{2.276}$$

Here $\bar{\nabla}$ is a conjugate of the difference operator $\bar{\nabla}$ defined by formulae[45]:

$$\bar{\nabla}\boldsymbol{q}(t_m) = \frac{\boldsymbol{q}(t_{m+1}) - \boldsymbol{q}(t_m)}{\Delta t},N \geqslant m \geqslant 0, \tag{2.277}$$

$$\bar{\nabla}\boldsymbol{q}(t_m) = \frac{\boldsymbol{q}(t_m) - \boldsymbol{q}(t_{m-1})}{\Delta t},N + 1 \geqslant m \geqslant 1. \tag{2.278}$$

(2) However we note that as was shown in [7] the canonical Laplace approximation[27] is not a valid asymptotic approximation in the limit $\varepsilon\to 0$ for a path-integral (2.273), see also Ref. [42].

From Eq. (2.269) – Eq. (2.270) one obtain

$$\left(\mathbf{E}\left[\boldsymbol{x}_{t'',\varepsilon'}^{q',\varepsilon}(\omega)\right]^2\right)_{\varepsilon'} = \int_{-\infty}^{\infty}\mathrm{d}\boldsymbol{q}''\ \|\boldsymbol{q}''\|^2\left[\left(\boldsymbol{p}_{\varepsilon'}^{\varepsilon}(\boldsymbol{q}',t' \mid \boldsymbol{q}'',t'')\right)_{\varepsilon'}\right] =$$

$$\int_{-\infty}^{\infty}\mathrm{d}\boldsymbol{q}''\ \|\boldsymbol{q}''\|^2\int_{\boldsymbol{q}(t')=\boldsymbol{q}'}^{\boldsymbol{q}(t'')=\boldsymbol{q}''}[D\boldsymbol{q}(t)]\exp\left[\left(-\frac{1}{2\varepsilon}\int_{t'}^{t''}L_{\varepsilon'}(\dot{\boldsymbol{q}},\boldsymbol{q},t;\varepsilon)\,\mathrm{d}t\right)_{\varepsilon'}\right] =$$

$$\int_{\boldsymbol{q}(t')=\boldsymbol{q}'}[D\boldsymbol{q}(t)]\ \|\boldsymbol{q}(t'')\|^2\exp\left[\left(-\frac{1}{2\varepsilon}\int_{t'}^{t''}L_{\varepsilon'}(\dot{\boldsymbol{q}},\boldsymbol{q},t;\varepsilon)\,\mathrm{d}t\right)_{\varepsilon'}\right]. \tag{2.279}$$

Let us consider now the quantity

$$\left(\boldsymbol{p}_{\varepsilon'}^{\varepsilon}(\boldsymbol{q}',t' \mid \boldsymbol{q}'',t'';L,m')\right)_{\varepsilon'} = \left(\lim_{\Delta t\to 0}\boldsymbol{I}_{m',N,\varepsilon,\varepsilon'}^{L}(\boldsymbol{q}',t' \mid \boldsymbol{q}'',t'')\right)_{\varepsilon'}. \tag{2.280}$$

Here

$$\boldsymbol{I}^{L}_{m',N,\varepsilon,\varepsilon'}(\boldsymbol{q}',t' \mid \boldsymbol{q}'',t'') = \breve{\boldsymbol{N}}_N\int_{-L}^{L}\mathrm{d}\boldsymbol{q}_1\cdots\int_{-L}^{L}\mathrm{d}\boldsymbol{q}_{m'}\int_{-\infty}^{\infty}\mathrm{d}\boldsymbol{q}_{m'+1}\cdots\int_{-\infty}^{\infty}\mathrm{d}\boldsymbol{q}_{N-1}\times$$

$$\exp\left[-\frac{1}{2\varepsilon}\left(S_{\varepsilon'}(\boldsymbol{q}_0,\boldsymbol{q}_1,\cdots,\boldsymbol{q}_{m'},\boldsymbol{q}_{m'+1},\cdots,\boldsymbol{q}_{N-1},\boldsymbol{q}_N,\varepsilon)\right)_{\varepsilon'}\right], \quad (2.281)$$

and $m' \ll N, L \gg 1$.

The quantity defined by Eq. (2.280) – Eq. (2.281) is symbolically indicated by the path-integral expression

$$\left(\boldsymbol{p}^{\varepsilon}_{\varepsilon'}(\boldsymbol{q}',t' \mid \boldsymbol{q}'',t'';L,m')\right)_{\varepsilon'} = \int_{\boldsymbol{q}(t')=\boldsymbol{q}'}^{\boldsymbol{q}(t'')=\boldsymbol{q}''}[D\boldsymbol{q}(t;L,m')]\exp\left[\left(-\frac{1}{2\varepsilon}S_{\varepsilon'}(\dot{\boldsymbol{q}},\boldsymbol{q},t;\varepsilon)\right)_{\varepsilon'}\right]. \quad (2.282)$$

Using Eq. (2.280) – Eq. (2.282) we define the quantity

$$\left(\mathbf{E}_{L,m'}\left[\boldsymbol{x}^{q',\varepsilon}_{t'',\varepsilon'}(\omega)\right]^2\right)_{\varepsilon'} = \int_{-\infty}^{\infty}\mathrm{d}\boldsymbol{q}''\,\|\boldsymbol{q}''\|^2\left[\left(p^{\varepsilon}_{\varepsilon'}(\boldsymbol{q}',t' \mid \boldsymbol{q}'',t'';L,m')\right)_{\varepsilon'}\right] =$$

$$\lim_{\Delta t\to 0}\breve{\boldsymbol{I}}_{N,\varepsilon,\varepsilon'}(\boldsymbol{q}',t',t'';L,m'). \quad (2.283)$$

Here

$$\breve{\boldsymbol{I}}_{N,\varepsilon,\varepsilon'}(\boldsymbol{q}',t',t'';L,m') = \int_{-\infty}^{\infty}\mathrm{d}\boldsymbol{q}''\,\|\boldsymbol{q}''\|^2\boldsymbol{I}_{N,\varepsilon,\varepsilon'}(\boldsymbol{q}',t' \mid \boldsymbol{q}'',t'';L,m'). \quad (2.284)$$

The quantity (2.283) is symbolically indicated by the path-integral expression

$$\left(\mathbf{E}_{L,m'}\left[\boldsymbol{x}^{q',\varepsilon}_{t'',\varepsilon'}(\omega)\right]^2\right)_{\varepsilon'} = \int_{-\infty}^{\infty}\mathrm{d}\boldsymbol{q}''\,\|\boldsymbol{q}''\|^2\int_{\boldsymbol{q}(t')=\boldsymbol{q}'}^{\boldsymbol{q}(t'')=\boldsymbol{q}''}[D\boldsymbol{q}(t;L,m')]\times$$

$$\exp\left[\left(-\frac{1}{2\varepsilon}\int_{t'}^{t''}\boldsymbol{L}_{\varepsilon'}(\dot{\boldsymbol{q}},\boldsymbol{q},t)\,\mathrm{d}t\right)_{\varepsilon'}\right] =$$

$$\int_{\boldsymbol{q}(t')=\boldsymbol{q}'}[D\boldsymbol{q}(t;L,m')]\,\|\boldsymbol{q}(t'')\|^2\exp\left[\left(-\frac{1}{2\varepsilon}\int_{t'}^{t''}\boldsymbol{L}_{\varepsilon'}(\dot{\boldsymbol{q}},\boldsymbol{q},t)\,\mathrm{d}t\right)_{\varepsilon'}\right]. \quad (2.285)$$

From Eq. (2.207) and Eq. (2.281) we obtain

$$\boldsymbol{I}_{N,\varepsilon,\varepsilon'}(\boldsymbol{q}',t' \mid \boldsymbol{q}'',t'') = \boldsymbol{I}^{L}_{m',N,\varepsilon,\varepsilon'}(\boldsymbol{q}',t' \mid \boldsymbol{q}'',t'') + \boldsymbol{\Theta}^{L}_{m',N,\varepsilon,\varepsilon'}(\boldsymbol{q}',t' \mid \boldsymbol{q}'',t''). \quad (2.286)$$

Here

$$\boldsymbol{\Theta}^{L}_{m',N,\varepsilon,\varepsilon'}(\boldsymbol{q}',t' \mid \boldsymbol{q}'',t'') = \breve{\boldsymbol{N}}_N\int_{\mathbf{R}^n\setminus\bar{\omega}_L}\mathrm{d}\boldsymbol{q}_1\cdots\int_{\mathbf{R}^n\setminus\bar{\omega}_L}\mathrm{d}\boldsymbol{q}_{m'}\int_{-\infty}^{\infty}\mathrm{d}\boldsymbol{q}_{m'+1}\cdots\int_{-\infty}^{\infty}\mathrm{d}\boldsymbol{q}_{N-1}\times$$

$$\exp\left[-\frac{1}{2\varepsilon}\left(\boldsymbol{S}_{\varepsilon'}(\boldsymbol{q}_0,\boldsymbol{q}_1,\cdots,\boldsymbol{q}_{m'},\boldsymbol{q}_{m'+1},\cdots,\boldsymbol{q}_{N-1},\boldsymbol{q}_N,t,\varepsilon)\right)_{\varepsilon'}\right]. \quad (2.287)$$

Here

$$\bar{\omega}_L = [-L,L]^n. \quad (2.288)$$

From Eq. (2.284) and Eq. (2.286) we obtain

$$\breve{\boldsymbol{I}}_{N,\varepsilon,\varepsilon'}(\boldsymbol{q}',t',t'') = \breve{\boldsymbol{I}}^{L}_{m',N,\varepsilon,\varepsilon'}(\boldsymbol{q}',t',t'') + \breve{\boldsymbol{\Theta}}^{L}_{m',N,\varepsilon,\varepsilon'}(\boldsymbol{q}',t',t''). \quad (2.289)$$

Here

$$\breve{\boldsymbol{\Theta}}^{L}_{m,N,\varepsilon,\varepsilon'}(\boldsymbol{q}',t',t'') = \breve{\boldsymbol{N}}_N\int_{\mathbf{R}^n\setminus\bar{\omega}_L}\mathrm{d}\boldsymbol{q}_1\cdots\int_{\mathbf{R}^n\setminus\bar{\omega}_L}\mathrm{d}\boldsymbol{q}_{m'}\int_{-\infty}^{\infty}\mathrm{d}\boldsymbol{q}_{m'+1}\cdots\int_{-\infty}^{\infty}\mathrm{d}\boldsymbol{q}_{N-1}\int_{-\infty}^{\infty}\mathrm{d}\boldsymbol{q}_N\times$$

$$\| \boldsymbol{q}_N \|^2 \exp\left[-\frac{1}{2\varepsilon} \left(\boldsymbol{S}_{\varepsilon'}(\boldsymbol{q}_0, \boldsymbol{q}_1, \cdots, \boldsymbol{q}_{m'}, \boldsymbol{q}_{m'+1} \cdots, \boldsymbol{q}_{N-1}, \boldsymbol{q}_N, t, \varepsilon) \right)_{\varepsilon'} \right]. \tag{2.290}$$

Remark. We note that : $\forall \varepsilon', \varepsilon' \neq 0$ there exist parameter $L = L(\boldsymbol{\varepsilon}')$ such that $\forall \boldsymbol{q}_m (\| \boldsymbol{q}_m \| \geqslant L)$ the inequality

$$\| \boldsymbol{b}_{\varepsilon'}(\boldsymbol{q}_m, t_m) \|^2 \leqslant \| \boldsymbol{q}_m \|^{-q}, q \geqslant 2 \tag{2.291}$$

is satisfied.

Lemma 1.

$$\breve{\Theta}^L_{m,N,\varepsilon,\varepsilon'}(\boldsymbol{q}', t', t'') \leqslant O[\exp(-L)]. \tag{2.292}$$

Proof. Using inequality (2.291), we will choose parameter L such that the equality

$$S_{\varepsilon'}(\boldsymbol{q}_0, \boldsymbol{q}_1, \cdots, \boldsymbol{q}_m, \cdots, \boldsymbol{q}_{N-1}, \boldsymbol{q}_N, t, \varepsilon) = \Delta t \sum_{m=1}^{N} \left\| \frac{\boldsymbol{q}_m - \boldsymbol{q}_{m-1}}{\Delta t} \right\|^2 +$$

$$O((t'' - t')L^{-q}) \cong \frac{1}{\Delta t} \sum_{m=1}^{N} \| \boldsymbol{q}_m - \boldsymbol{q}_{m-1} \|^2 \tag{2.293}$$

is satisfied. From Eq. (2.290) and Eq. (2.293) we obtain

$$\breve{\Theta}^L_{m',N,\varepsilon,\varepsilon'}(\boldsymbol{q}', t', t'') = \breve{\boldsymbol{N}}_N \int_{\mathbf{R}^n \setminus \bar{\omega}_L} \mathrm{d}\boldsymbol{q}_1 \cdots \int_{\mathbf{R}^n \setminus \bar{\omega}_L} \mathrm{d}\boldsymbol{q}_{m'} \int_{-\infty}^{\infty} \mathrm{d}\boldsymbol{q}_{m'+1} \cdots \int_{-\infty}^{\infty} \mathrm{d}\boldsymbol{q}_{N-1} \int_{-\infty}^{\infty} \mathrm{d}\boldsymbol{q}_N \times$$

$$\| \boldsymbol{q}_N \|^2 \exp\left[-\frac{1}{2\varepsilon} \left(S_{\varepsilon'}(\boldsymbol{q}_0, \boldsymbol{q}_1, \cdots, \boldsymbol{q}_m, \boldsymbol{q}_{m'+1} \cdots, \boldsymbol{q}_{N-1}, \boldsymbol{q}_N, \varepsilon) \right)_{\varepsilon'} \right] \cong$$

$$\breve{N}_N \int_{\mathbf{R}^n \setminus \bar{\omega}_L} \mathrm{d}\boldsymbol{q}_1 \cdots \int_{\mathbf{R}^n \setminus \bar{\omega}_L} \mathrm{d}\boldsymbol{q}_{m'} \int_{-\infty}^{\infty} \mathrm{d}\boldsymbol{q}_{m'+1} \cdots \int_{-\infty}^{\infty} \mathrm{d}\boldsymbol{q}_{N-1} \int_{-\infty}^{\infty} \mathrm{d}\boldsymbol{q}_N \times$$

$$\| \boldsymbol{q}_N \|^2 \exp\left(-\frac{1}{2\varepsilon\Delta t} \sum_{m=1}^{N} \| \boldsymbol{q}_m - \boldsymbol{q}_{m-1} \|^2 \right). \tag{2.294}$$

From Eq. (2.294) we obtain the inequality

$$\breve{\Theta}^L_{m',N,\varepsilon,\varepsilon'}(\boldsymbol{q}', t', t'') \leqslant \breve{N}_N \int_{-\infty}^{\infty} \mathrm{d}\boldsymbol{q}_1 \cdots \int_{-\infty}^{\infty} \mathrm{d}\boldsymbol{q}_{m-1} \int_{\mathbf{R}^n \setminus \bar{\omega}_L} \mathrm{d}\boldsymbol{q}_{m'} \int_{-\infty}^{\infty} \mathrm{d}\boldsymbol{q}_{m'+1} \cdots$$

$$\int_{-\infty}^{\infty} \mathrm{d}\boldsymbol{q}_{N-1} \int_{-\infty}^{\infty} \mathrm{d}\boldsymbol{q}_N \times \| \boldsymbol{q}_N \|^2 \exp\left(-\frac{1}{2\varepsilon\Delta t} \sum_{m=1}^{N} \| \boldsymbol{q}_m - \boldsymbol{q}_{m-1} \|^2 \right). \tag{2.295}$$

Let $\boldsymbol{C}^L_{m',N,\varepsilon,\varepsilon'}(\boldsymbol{q}', t', t'')$ be the multiple integral:

$$\boldsymbol{C}^L_{m',N,\varepsilon,\varepsilon'}(\boldsymbol{q}', t', t'') = \breve{N}_N \int_{-\infty}^{\infty} \mathrm{d}\boldsymbol{q}_1 \cdots \int_{-\infty}^{\infty} \mathrm{d}\boldsymbol{q}_{m'-1} \int_{\mathbf{R}^n \setminus \bar{\omega}_L} \mathrm{d}\boldsymbol{q}_{m'} \int_{-\infty}^{\infty} \mathrm{d}\boldsymbol{q}_{m'+1} \cdots$$

$$\int_{-\infty}^{\infty} \mathrm{d}\boldsymbol{q}_{N-1} \int_{-\infty}^{\infty} \mathrm{d}\boldsymbol{q}_N \| \boldsymbol{q}_N \|^2 \exp\left(-\frac{1}{2\varepsilon\Delta t} \sum_{m=1}^{N} \| \boldsymbol{q}_m - \boldsymbol{q}_{m-1} \|^2 \right) =$$

$$\breve{N}_N \int_{-\infty}^{\infty} \mathrm{d}\boldsymbol{q}_1 \cdots \int_{-\infty}^{\infty} \mathrm{d}\boldsymbol{q}_{m'-1} \exp\left(\frac{1}{2\varepsilon\Delta t} \sum_{m=1}^{m'} \| \boldsymbol{q}_m - \boldsymbol{q}_{m-1} \|^2 \right) \times$$

$$\int_{\mathbf{R}^n \setminus \bar{\omega}_L} \mathrm{d}\boldsymbol{q}_{m'} \int_{-\infty}^{\infty} \mathrm{d}\boldsymbol{q}_{m'+1} \cdots \int_{-\infty}^{\infty} \mathrm{d}\boldsymbol{q}_{N-1} \int_{-\infty}^{\infty} \mathrm{d}\boldsymbol{q}_N \times \| \boldsymbol{q}_N \|^2 \exp\left(\frac{1}{2\varepsilon\Delta t} \sum_{m=m'+1}^{N} \| \boldsymbol{q}_m - \boldsymbol{q}_{m-1} \|^2 \right). \tag{2.296}$$

By simple canonical observation we obtain

$$\int_{-\infty}^{\infty} \mathrm{d}\boldsymbol{q}_1 \cdots \int_{-\infty}^{\infty} \mathrm{d}\boldsymbol{q}_{m'-1} \exp\left(-\frac{1}{2\varepsilon\Delta t} \sum_{m=1}^{m'} \| \boldsymbol{q}_m - \boldsymbol{q}_{m-1} \|^2 \right) =$$

$$(2\pi\varepsilon\Delta t)^{n(m'-1)/2}\exp\left[\frac{1}{2\varepsilon(m'-1)\Delta t}\|\boldsymbol{q}_{m'}-\boldsymbol{q}_0\|^2\right]. \tag{2.297}$$

Substitution Eq. (2.297) into Eq. (2.296) gives

$$\boldsymbol{C}^L_{m',N,\varepsilon,\varepsilon'}(\boldsymbol{q}',t',t'') =$$

$$\breve{N}_{N,m'}\int_{\mathbf{R}^n\setminus\bar{\omega}_L}\mathrm{d}\boldsymbol{q}_{m'}\exp\left[-\frac{1}{2\varepsilon(m'-1)\Delta t}\|\boldsymbol{q}_{m'}-\boldsymbol{q}_0\|^2\right]\times$$

$$\int_{-\infty}^{\infty}\mathrm{d}\boldsymbol{q}_{m'+1}\cdots\int_{-\infty}^{\infty}\mathrm{d}\boldsymbol{q}_{N-1}\int_{-\infty}^{\infty}\mathrm{d}\boldsymbol{q}_N\ \|\boldsymbol{q}_N\|^2\exp\left(-\frac{1}{2\varepsilon\Delta t}\sum_{m=m'+1}^{N}\|\boldsymbol{q}_m-\boldsymbol{q}_{m-1}\|^2\right). \tag{2.298}$$

Here

$$\breve{\boldsymbol{N}}_{N,m'} = (2\pi\varepsilon\Delta t)^{n(m'-1)/2}\breve{\boldsymbol{N}}_N. \tag{2.299}$$

From Eq. (2.298) we obtain

$$\boldsymbol{C}^L_{m',N,\varepsilon,\varepsilon'}(\boldsymbol{q}',t',t'') = \breve{\boldsymbol{N}}_{N,m'}(2\pi\varepsilon\Delta t)^{\frac{n(N-m'-1)}{2}}\int_{\mathbf{R}^n\setminus\bar{\omega}_L}\mathrm{d}\boldsymbol{q}_{m'}\times\exp\left[\frac{1}{2\varepsilon(m'-1)\Delta t}\cdot\right.$$

$$\left.\|\boldsymbol{q}_{m'}-\boldsymbol{q}_0\|^2\right]\int_{-\infty}^{\infty}\mathrm{d}\boldsymbol{q}_N\ \|\boldsymbol{q}_N\|^2\times\exp\left[\frac{1}{2\varepsilon(N-m'+1)\Delta t}\|\boldsymbol{q}_N-\boldsymbol{q}_{m'}\|^2\right] = \breve{\boldsymbol{N}}_{N,m'}\cdot$$

$$(2\pi\varepsilon\Delta t)^{n(N-m'+1)/2}\times\int_{\mathbf{R}^n\setminus\bar{\omega}_L}\mathrm{d}\boldsymbol{q}_{m'}\ \|\boldsymbol{q}_{m'}\|^2\exp\left[\frac{1}{2\varepsilon(m'-1)\Delta t}\|\boldsymbol{q}_{m'}-\boldsymbol{q}_0\|^2\right]. \tag{2.300}$$

Assumption. We assume now that $\boldsymbol{q}_0\notin\mathbf{R}^n\setminus\bar{\omega}_L$.

From Eq. (2.300) using Laplace approximation[43] we obtain

$$\boldsymbol{C}^L_{m',N,\varepsilon,\varepsilon'}(\boldsymbol{q}',t',t'')\leqslant O[\exp(-L)]. \tag{2.301}$$

Theorem 15. (Hölder's inequality) Let $r_1=p, r_2=q\in[1,\infty)$, with $1/p+1/q=1$, let $'\gamma,\boldsymbol{C}_i:(C^1([t',t'']))^n\to\mathbf{R}, i=1,2,3$ be an functional such that $'\gamma='\gamma(\dot{\boldsymbol{q}},\boldsymbol{q},t',t''), \boldsymbol{C}_i=\boldsymbol{C}_i(\dot{\boldsymbol{q}},\boldsymbol{q},t',t''), i=1,2,3, \boldsymbol{C}_3=\boldsymbol{C}_1\boldsymbol{C}_2, '\gamma(\dot{\boldsymbol{q}},\boldsymbol{q},t',t'')>0$. Let $\|\boldsymbol{C}_i(\boldsymbol{q}',t',t'';L,m')\|^{'\gamma}_{r_i}, i=1,2,3$ be the path integral

$$\|\boldsymbol{C}_i(\boldsymbol{q}',t',t'';L,m')\|^{'\gamma}_{r_i} =$$

$$\left(\int_{-\infty}^{\infty}\mathrm{d}\boldsymbol{q}''\int_{\boldsymbol{q}(t')=\boldsymbol{q}'}^{\boldsymbol{q}(t'')=\boldsymbol{q}''}[D\boldsymbol{q}(t;L,m')]'\gamma(\dot{\boldsymbol{q}},\boldsymbol{q},t',t'')\boldsymbol{C}_i^{r_i}\,|(\dot{\boldsymbol{q}},\boldsymbol{q},t',t'')|^{r_i}\right)^{1/r_i}<\infty.$$

Then

$$\|\boldsymbol{C}_3(\boldsymbol{q}',t',t'';L,m')\|^{'\gamma}_1\leqslant$$

$$\|\boldsymbol{C}_1(\boldsymbol{q}',t',t'';L,m')\|^{'\gamma}_p\ \|\boldsymbol{C}_2(\boldsymbol{q}',t',t'';L,m')\|^{'\gamma}_q. \tag{2.302}$$

From Theorem 5 we obtain

Corollary 1. Assume that: ① $\frac{1}{p}+\frac{1}{q}=1$,

$$②\ \boldsymbol{I}_p=\boldsymbol{I}_p(\boldsymbol{q}',t',t'';L,m')=\int_{-\infty}^{\infty}\mathrm{d}\boldsymbol{q}''\ \|\boldsymbol{q}''\|^2\int_{\boldsymbol{q}(t')=\boldsymbol{q}'}^{\boldsymbol{q}(t'')=\boldsymbol{q}''}[D\boldsymbol{q}(t;L,m')]\times$$

$$\exp\left(-\int_{t'}^{t''}[\dot{\boldsymbol{q}}(t)]^2\mathrm{d}t\right)\times\exp\left[p\int_{t'}^{t''}\boldsymbol{G}_1(\dot{\boldsymbol{q}},\boldsymbol{q},t)\,\mathrm{d}t\right]<\infty, \tag{2.303}$$

$$③\ \boldsymbol{I}_q=\boldsymbol{I}_q(t',t'';L,m')=\int_{-\infty}^{\infty}\mathrm{d}\boldsymbol{q}''\ \|\boldsymbol{q}''\|^2\int_{\boldsymbol{q}(t')=\boldsymbol{q}'}^{\boldsymbol{q}(t'')=\boldsymbol{q}''}[D\boldsymbol{q}(t;L,m')]\exp\left\{-\int_{t'}^{t''}[\dot{\boldsymbol{q}}(t)]^2\mathrm{d}t\right\}\times$$

$$\exp\left[q\int_{t'}^{t''}\boldsymbol{G}_2(\dot{\boldsymbol{q}},\boldsymbol{q},t)\,\mathrm{d}t\right] < \infty. \tag{2.304}$$

Then inequality

$$\int_{-\infty}^{\infty}\mathrm{d}\boldsymbol{q}''\ \|\boldsymbol{q}''\|^2\int_{q(t')=q'}^{q(t'')=q''}[D\boldsymbol{q}(t;L,m')]\exp\left\{-\int_{t'}^{t''}[\dot{\boldsymbol{q}}(t)]^2\mathrm{d}t\right\}\times$$
$$\exp\left[\int_{t'}^{t''}\boldsymbol{G}_1(\dot{\boldsymbol{q}},\boldsymbol{q},t)\,\mathrm{d}t\right]\exp\left[\int_{t'}^{t''}\boldsymbol{G}_2(\dot{\boldsymbol{q}},\boldsymbol{q},t)\,\mathrm{d}t\right]\leqslant[\boldsymbol{I}_p]^{\frac{1}{p}}\times[\boldsymbol{I}_q]^{\frac{1}{q}} \tag{2.305}$$

is satisfied.

Theorem 16. (1) Let $\boldsymbol{b}(\boldsymbol{x},t)=(b_1(\boldsymbol{x},t),\cdots,b_n(\boldsymbol{x},t))$ be an vector function, where $b_i(\boldsymbol{x},t), i=1,\cdots,n$ is a polynomial on variable $\boldsymbol{x}$. Let $\hat{\boldsymbol{b}}(\boldsymbol{x},t)$ be the linear part of the vector-function $\boldsymbol{b}(\boldsymbol{x},t)$, i. e.,

$$\hat{b}_i(\boldsymbol{x},t)=\sum_{\alpha,\ |\alpha|\leqslant 1}b_i^{\alpha}(t)\boldsymbol{x}^{\alpha},\ i=1,\cdots,n. \tag{2.306}$$

Let $\hat{\boldsymbol{L}}(\dot{\boldsymbol{q}},\boldsymbol{q},t)$ be the Lagrangian

$$\hat{\boldsymbol{L}}(\dot{\boldsymbol{q}},\boldsymbol{q},t)=\|\dot{\boldsymbol{q}}(t)-\hat{\boldsymbol{b}}(\boldsymbol{q}(t),t)\|^2-\varepsilon\sum_{i=1}^{n}\hat{b}_{i,i}(\boldsymbol{q}(t),t). \tag{2.307}$$

Here

$$\hat{b}_{i,i}(\boldsymbol{q}(t),t)=\frac{\partial\hat{b}_i(\boldsymbol{q}(t),t)}{\partial\boldsymbol{q}_i}. \tag{2.308}$$

and let $\boldsymbol{L}_{\varepsilon'}(\dot{\boldsymbol{q}},\boldsymbol{q},t)$ be

$$\boldsymbol{L}_{\varepsilon'}(\dot{\boldsymbol{q}},\boldsymbol{q},t)=\|\dot{\boldsymbol{q}}(t)-\boldsymbol{b}(\boldsymbol{q}_{\varepsilon'}(t),t)\|^2-\varepsilon\sum_{i=1}^{n}b_{i,i}(\boldsymbol{q}_{\varepsilon'}(t),t). \tag{2.309}$$

Here $\boldsymbol{q}_{\varepsilon'}(t)=(\boldsymbol{q}_{\varepsilon',1}(t),\cdots,\boldsymbol{q}_{\varepsilon',i}(t),\cdots,\boldsymbol{q}_{\varepsilon',n}(t))$ and

$$q_{\varepsilon',i}(t)=\frac{q_i(t)}{1+(\varepsilon')^l\left[q_i^2(t)+\int_{t'}^{t''}q_i^2(t)\,\mathrm{d}t\right]^l},l\geqslant 3. \tag{2.310}$$

(2) Let $\boldsymbol{x}_{t'',\varepsilon'}^{x_0,\varepsilon}(\omega)$ be the solution of the Colombeau-Ito's SDE (2.202) – (2.203).

Then there exist Colombeau constant $\tilde{C}'=(C'_{\varepsilon'})_{\varepsilon'}>0$, such that the inequalities:

$$① \left(\liminf_{\varepsilon\to 0}\mathbf{E}\left[\|\boldsymbol{x}_{t'',\varepsilon'}^{x_0,\varepsilon}(\omega)\|^2\right]\right)_{\varepsilon'}\leqslant(C'_{\varepsilon'})_{\varepsilon'}(\boldsymbol{I}_{\varepsilon'}^{\varepsilon})_{\varepsilon'} \tag{2.311}$$

$$② \left[\left(\liminf_{\varepsilon\to 0}\mathbf{E}\left[\|\boldsymbol{x}_{t'',\varepsilon'}^{x_0,\varepsilon}(\omega)\|^2\right]\right)_{\varepsilon'}\right]\leqslant\tilde{C}'[(\boldsymbol{I}_{\varepsilon'}^{\varepsilon})_{\varepsilon'}],$$

where

$$(\boldsymbol{I}_{\varepsilon'}^{\varepsilon})_{\varepsilon'}=\int_{-\infty}^{\infty}\mathrm{d}\boldsymbol{q}''\ \|\boldsymbol{q}''\|^2\int_{\boldsymbol{q}(t')=\boldsymbol{q}'}^{\boldsymbol{q}(t'')=\boldsymbol{q}''}[D\boldsymbol{q}(t)]\exp\left[\left(-\frac{1}{2\varepsilon}\int_{t'}^{t''}\hat{\boldsymbol{L}}_{\varepsilon'}(\dot{\boldsymbol{q}},\boldsymbol{q},t)\,\mathrm{d}t\right)_{\varepsilon'}\right] \tag{2.312}$$

and

$$③ \liminf_{\varepsilon\to 0}\mathbf{E}\left[\boldsymbol{x}_{t'',\varepsilon'}^{x_0,\varepsilon}(\omega)\right]^2\leqslant\tilde{C}'\ \|\boldsymbol{U}(t'')\|^2 \tag{2.313}$$

is satisfies. Here a vector function $\boldsymbol{U}(t)=(U_1(t),\cdots,U_n(t))$ is the solution of the differential master equation

$$\dot{U}(t) = \mathbf{J}_t U(t) + \boldsymbol{b}(t), U(0) = \boldsymbol{x}_0 = \boldsymbol{q}'. \tag{2.314}$$

Here $\boldsymbol{b}(t) = \boldsymbol{b}(0,t)$ and $\mathbf{J}_t = \mathbf{J}(t)$ is Jacobian, i. e., $\mathbf{J}_t$ is $n \times n$ -matrix:

$$\mathbf{J}(t) = [\partial \boldsymbol{b}_i(\boldsymbol{x},t)/\partial x_j]_{x=0}. \tag{2.315}$$

Proof. For short, we will be considered proof only for the case of the 1-dimensional Colombeau-Ito's SDE, without loss of generality.

Let us consider Feynman's path integral (2.285) corresponding to 1-dimensional Colombeau-Ito's SDE, i. e.,

$$\left(\boldsymbol{I}^{\varepsilon}_{\varepsilon'}(L,m')\right)_{\varepsilon'} = \left(\mathbf{E}_{L,m'}\left[\boldsymbol{x}^{x_0,\varepsilon}_{t'',\varepsilon'}(\omega)\right]\right)_{\varepsilon'} = \int_{-\infty}^{\infty} \mathrm{d}q'' \, \| q'' \|^2 \times$$

$$\int_{q(t')=q'}^{q(t'')=q''} [Dq(t;L,m')] \exp\left[\left(-\frac{1}{2\varepsilon}\int_{t'}^{t''} L_{\varepsilon'}(\dot{q},q,t)\,\mathrm{d}t\right)_{\varepsilon'}\right]. \tag{2.316}$$

Let us rewrite now Lagrangian $L_{\varepsilon'}(\dot{q},q,t)$ in the next equivalent form

$$L_{\varepsilon'}(\dot{q},q,t) = L'_{\varepsilon'}(\dot{q},q,t) - 2\varepsilon b_{1,1}(q_{\varepsilon'}(t),t). \tag{2.317}$$

Here

$$L'_{\varepsilon'}(\dot{q},q,t) = [\dot{q}(t) - b(q_{\varepsilon'}(t),t)]^2 = \left[\dot{q}(t) - \hat{b}(q_{\varepsilon'}(t),t) - b_2(q_{\varepsilon'}(t),t)\right]^2, \tag{2.318}$$

$$\hat{b}(x,t) = b^0(t) + x b^1(t) \tag{2.319}$$

and

$$b_2(x,t) = \sum_{\alpha,2\leqslant|\alpha|\leqslant r} b^{\alpha}(t) x^{\alpha}. \tag{2.320}$$

Let us rewrite now Eq. (2.320) in the next form

$$b_2(x,t) = b_{2,2}(x,t) + b_{2,3}(x,t). \tag{2.321}$$

Here

$$b_{2,2}(x,t) = \sum_{\alpha} b^{\alpha}(t) x^{\alpha}, |\alpha| = 2, \tag{2.322}$$

$$b_{2,3}(x,t) = \sum_{\alpha} b^{\alpha}(t) x^{\alpha}, |\alpha| \geqslant 3. \tag{2.323}$$

From Eq. (2.318) – Eq. (2.320) we obtain

$$\boldsymbol{L}'_{\varepsilon}(\dot{q},q,t) = \left\{[\dot{q}(t) - \hat{b}(q_{\varepsilon'}(t),t)] - b_2(q_{\varepsilon'}(t),t)\right\}^2 =$$

$$\left[\dot{q}(t) - \hat{b}(q_{\varepsilon'}(t),t)\right]^2 - 2b_2(q_{\varepsilon'}(t),t)[\dot{q}(t) - \hat{b}(q(t),t)] +$$

$$b_2^2(q_{\varepsilon'}(t),t) = [\dot{q}(t)]^2 - 2\dot{q}(t)\hat{b}(q_{\varepsilon'}(t),t) + \hat{b}^2(q_{\varepsilon'}(t),t) -$$

$$2\dot{q}(t) b_2(q_{\varepsilon'}(t),t) + 2b_2(q_{\varepsilon'}(t),t)\hat{b}(q(t),t). \tag{2.324}$$

Substituting Eq. (2.322) and Eq. (2.323) into Eq. (2.324) gives

$$\boldsymbol{L}'_{\varepsilon}(\dot{q},q,t) = [\dot{q}(t)]^2 - 2\dot{q}(t) b^0(t) - 2\dot{q}(t) q_{\varepsilon'}(t) b^1(t) +$$

$$[b^0(t)]^2 + 2q_{\varepsilon'}(t) b^0(t) b^1(t) + [q_{\varepsilon'}(t) b^1(t)]^2 -$$

$$2\dot{q}(t) b_2[q_{\varepsilon'}(t),t] + 2b_2[q_{\varepsilon'}(t),t]\hat{b}[q(t),t] =$$

$$[\dot{q}(t)]^2 - 2\dot{q}(t) b^0(t) - 2\dot{q}(t) q_{\varepsilon'}(t) b^1(t) +$$

$$[b^0(t)]^2 + 2q_{\varepsilon'}(t)b^0(t)b_0^1(t) + [q_{\varepsilon'}(t)b^1(t)]^2 - 2\dot{q}(t)b_2[q_{\varepsilon'}(t),t] +$$
$$2b_2[q_{\varepsilon'}(t),t]b^0(t) + 2q_{\varepsilon'}(t)b_2[q_{\varepsilon'}(t),t]b^1(t) =$$
$$[\dot{q}(t)]^2 - 2\dot{q}(t)b^0(t) - 2\dot{q}(t)q_{\varepsilon'}(t)b^1(t) +$$
$$[b^0(t)]^2 + 2q_{\varepsilon'}(t)b^0(t)b_0^1(t) + [q_{\varepsilon'}(t)b^1(t)]^2 -$$
$$2\dot{q}(t)b_{2,2}[q_{\varepsilon'}(t),t] - 2\dot{q}(t)b_{2,3}[q_{\varepsilon'}(t),t] +$$
$$2b_{2,2}[q_{\varepsilon'}(t),t]b^0(t) + 2b_{2,3}[q_{\varepsilon'}(t),t]b^0(t) +$$
$$2q_{\varepsilon'}(t)b_2[q_{\varepsilon'}(t),t]b^1(t). \quad (2.325)$$

Substituting Eq. (2.325) into Eq. (2.316) gives

$$(\boldsymbol{I}_{\varepsilon'}^{\varepsilon}(q',t',t'';L,m'))_{\varepsilon'} =$$
$$\exp\left\{-\frac{1}{2\varepsilon}\int_{t'}^{t}[b^0(t)]^2\mathrm{d}t\right\}\int_{-\infty}^{\infty}\mathrm{d}q''\ \|q''\|^2\int_{q(t')=q'}^{q(t'')=q''}[Dq(t;L,m')]\times$$
$$\exp\left(-\frac{1}{2\varepsilon}\int_{t'}^{t''}\left\{[\dot{q}(t)]^2 - 2\dot{q}(t)b^0(t) - 2\dot{q}(t)q_{\varepsilon'}(t)b^1(t) + 2q_{\varepsilon'}(t)b^0(t)b^1(t) +\right.\right.$$
$$\left[q_{\varepsilon'}(t)b^1(t)\right]^2 - 2\dot{q}(t)b_{2,2}(q_{\varepsilon'}(t),t) - 2\dot{q}(t)b_{2,3}(q_{\varepsilon'}(t),t) + 2b_{2,2}(q_{\varepsilon'}(t),t)b^0(t) +$$
$$\left.\left.2b_{2,3}(q_{\varepsilon'}(t),t)b^0(t) + 2q_{\varepsilon'}(t)b_2(q_{\varepsilon'}(t),t)b^1(t)\right\}\mathrm{d}t\right). \quad (2.326)$$

Using replacement $q(t) = p(t)\sqrt{2\varepsilon}$ into Feynman path integral (2.326), we obtain

$$\left(\boldsymbol{I}_{\varepsilon'}^{3}(q',t',t'';L,m')\right)_{\varepsilon'} =$$
$$\exp\left\{-\frac{1}{2\varepsilon}\int_{t'}^{t''}[b^0(t)]^2\mathrm{d}t\right\}\int_{-\infty}^{\infty}\mathrm{d}q''\ \|q''\|^2\int_{p(t')=\frac{q'}{\sqrt{2\varepsilon}}}^{p(t'')=\frac{q''}{\sqrt{2\varepsilon}}}\left[Dp\left(t;\frac{L}{\sqrt{2\varepsilon}},m'\right)\right]\times$$
$$\exp\left(-\int_{t'}^{t''}\left\{[\dot{p}(t)]^2 - \frac{2}{\sqrt{2\varepsilon}}\dot{p}(t)b^0(t) - 2\dot{p}(t)p_{\varepsilon'}(t)b^1(t) + \frac{2}{\sqrt{2\varepsilon}}p_{\varepsilon'}(t)b^0(t)b^1(t) +\right.\right.$$
$$\left.\left.[p_{\varepsilon'}(t)b^1(t)]^2 - 2\sqrt{2\varepsilon}\dot{p}(t)\breve{b}_2(p_{\varepsilon'}(t),t) + 2\breve{b}_2(p_{\varepsilon'}(t),t)\hat{b}(\sqrt{2\varepsilon}p_{\varepsilon'}(t),t)\right\}\mathrm{d}t\right). \quad (2.327)$$

Here

$$\breve{b}_2(p_{\varepsilon'}(t),t) = \sum_{\alpha}(\sqrt{2\varepsilon})^{|\alpha|-2}b_0^{\alpha}(t)\,[p_{\varepsilon'}(t)]^{\alpha}, 2 \leqslant |\alpha| \leqslant r. \quad (2.328)$$

And

$$p_{\varepsilon'}(t) = \frac{p(t)}{1 + 2(\varepsilon')^l\varepsilon^l\left[p^2(t) + \int_{t'}^{t''}p^2(t)\mathrm{d}t\right]^l}, l \geqslant 1. \quad (2.329)$$

Let us rewrite now Feynman path integral (2.327) in the next equivalent form

$$\left(\boldsymbol{I}_{\varepsilon'}^{\varepsilon}(q',t',t'';L,m')\right)_{\varepsilon'} =$$
$$\exp\left\{-\frac{1}{2\varepsilon}\int_{t'}^{t''}[b^0(t)]^2\mathrm{d}t\right\}\int_{-\infty}^{\infty}\mathrm{d}q''\ \|q''\|^2\int_{p(t')=\frac{q'}{\sqrt{2\varepsilon}}}^{p(t'')=\frac{q''}{\sqrt{2\varepsilon}}}\left[Dp\left(t;\frac{L}{\sqrt{2\varepsilon}},m'\right)\right]\times$$

$$\exp\left\{-\int_{t'}^{t''}[\dot{p}(t)]^2 dt\right\} \times \exp\left\{-\int_{t'}^{t''}\left[-\frac{2}{\sqrt{2\varepsilon}}\dot{p}(t)b^0(t) - 2\dot{p}(t)p_{\varepsilon'}(t)b^1(t) + \frac{2}{\sqrt{2\varepsilon}}p_{\varepsilon'}(t)b^0(t)b^1(t) + [p_{\varepsilon'}(t)b^1(t)]^2\right]dt\right\} \times$$

$$\exp\left\{\int_{t'}^{t''}\left[2\sqrt{2\varepsilon}\dot{p}(t)\breve{b}_2(p_{\varepsilon'}(t),t) - 2\breve{b}_2(p_{\varepsilon'}(t),t)\hat{b}(\sqrt{2\varepsilon}p(t),t)\right]dt\right\}. \quad (2.330)$$

Assume that $1/p + 1/q = 1$ and $q = 1/\varepsilon$. Then

$$p = \frac{1}{1-\varepsilon} = 1 + \varepsilon + o(\varepsilon). \quad (2.331)$$

Using Corollary 1 now, from Eq. (2.330) we obtain

$$\left(\boldsymbol{I}^{\varepsilon}_{\varepsilon'}(q',t',t'';L,m')\right)_{\varepsilon'} \leqslant \exp\left\{-\frac{1}{2\varepsilon}\int_{t'}^{t''}[b^0(t)]^2 dt\right\}\left\{\int_{-\infty}^{\infty} dq'' \, \| q'' \|^2 \int_{p(t')=\frac{q'}{\sqrt{2\varepsilon}}}^{p(t'')=\frac{q''}{\sqrt{2\varepsilon}}} \cdot \left[Dp\left(t;\frac{L}{\sqrt{2\varepsilon}},m'\right)\right] \times \exp\left\{-\int_{t'}^{t''}[\dot{p}(t)]^2 dt\right\} \times\right.$$

$$\exp\left(-(1+\varepsilon)\int_{t'}^{t''}\left(-\frac{2}{\sqrt{2\varepsilon}}\dot{p}(t)b^0(t) - 2\dot{p}(t)p_{\varepsilon'}(t)b^1(t) + \frac{2}{\sqrt{2\varepsilon}}p_{\varepsilon'}(t)b_0^0(t)b_0^1(t) + \left[p_{\varepsilon'}(t)b^1(t)\right]^2\right)dt\right)\Bigg\}^{1-\varepsilon} \times \left\{\int_{-\infty}^{\infty} dq'' \int_{p(t')=\frac{q'}{\sqrt{2\varepsilon}}}^{p(t'')=\frac{q''}{\sqrt{2\varepsilon}}}\left[Dp\left(t;\frac{L}{\sqrt{2\varepsilon}},m'\right)\right] \times\right.$$

$$\exp\left[-\int_{t'}^{t''}[\dot{p}(t)]^2 dt\right] \times \exp\left[\frac{1}{\varepsilon}\int_{t'}^{t''}\left(2\sqrt{2\varepsilon}\dot{p}(t)\breve{b}_2(p_{\varepsilon'}(t),t) - 2\breve{b}_2(p_{\varepsilon'}(t),t)\hat{b}(\sqrt{2\varepsilon}p(t),t)\right)dt\right]\Bigg\}^{\varepsilon}. \quad (2.332)$$

Therefore

$$\left(\boldsymbol{I}^{\varepsilon}_{\varepsilon'}(q',t',t'';L,m')\right)_{\varepsilon'} \leqslant \left([\boldsymbol{I}^{\varepsilon,1}_{\varepsilon'}(q',t',t'';L,m')]^{1-\varepsilon}\right)_{\varepsilon'} \times \left([\boldsymbol{I}^{\varepsilon,2}_{\varepsilon'}(q',t',t'';L,m')]^{\varepsilon}\right)_{\varepsilon'}. \quad (2.333)$$

Here

$$\left(\boldsymbol{I}^{\varepsilon,1}_{\varepsilon'}(q',t',t'';L,m')\right)_{\varepsilon'} = \left(\exp\left[-\frac{1}{2\varepsilon(1+\varepsilon)}\int_{t'}^{t''}[b^0(t)]dt\right]\right) \times \int_{-\infty}^{\infty} dq'' \, \| q'' \|^2 \int_{p(t')=\frac{q'}{\sqrt{2\varepsilon}}}^{p(t'')=\frac{q''}{\sqrt{2\varepsilon}}}\left(Dp\left(t;\frac{L}{\sqrt{2\varepsilon}},m'\right)\right)\exp\left(-\int_{t'}^{t''}[\dot{p}(t)]^2 dt\right) \times$$

$$\exp\left(-(1+\varepsilon)\int_{t'}^{t''}\left\{-\frac{2}{\sqrt{2\varepsilon}}\dot{p}(t)b^0(t) - 2\dot{p}(t)p_{\varepsilon'}(t)b^1(t) + \frac{2}{\sqrt{2\varepsilon}}p_{\varepsilon'}(t)b^0(t)b^1(t) + [p_{\varepsilon'}(t)b^1(t)]^2\right\}dt\right)\Bigg)_{\varepsilon'}. \quad (2.334)$$

And

$$\left(\boldsymbol{I}^{\varepsilon,2}_{\varepsilon'}(q',t',t'';L,m')\right)_{\varepsilon'} = \int_{-\infty}^{\infty} dq'' \int_{p(t')=\frac{q'}{\sqrt{2\varepsilon}}}^{p(t'')=\frac{q''}{\sqrt{2\varepsilon}}}\left[Dp\left(t;\frac{L}{\sqrt{2\varepsilon}},m'\right)\right]\exp\left\{-\int_{t'}^{t''}[\dot{p}(t)]^2 dt\right\} \times$$

$$\exp\left(\frac{1}{\varepsilon}\int_{t'}^{t''}\left(2\sqrt{2\varepsilon}\dot{p}(t)\breve{b}_2(p_{\varepsilon'}(t),t)-2\breve{b}_2(p_{\varepsilon'}(t),t)\hat{b}(\sqrt{2\varepsilon}p(t),t)\right)\mathrm{d}t\right). \tag{2.335}$$

(1) Let us evaluate now path integral $\left(\boldsymbol{I}_{\varepsilon'}^{\varepsilon,1}(q',t',t'';L,m')\right)_{\varepsilon'}$. From Eq. (2.334) using replacement $p(t)=\frac{q(t)}{\sqrt{2\varepsilon}}$ into Feynman path integral in the RHS of the Eq. (2.334), we obtain

$$\left(\boldsymbol{I}_{\varepsilon'}^{\varepsilon,1}(q',t',t'';L,m')\right)_{\varepsilon'}=\exp\left\{-\frac{1}{2\varepsilon}\int_{t'}^{t''}[b^0(t)]^2\mathrm{d}t\right\}\int_{-\infty}^{\infty}\mathrm{d}q''\,\|q''\|^2\int_{q(t')=q'}^{q(t'')=q''}[Dq(t;L,m')]\times\exp\left\{-\frac{1}{2\varepsilon}\int_{t'}^{t''}\left[(\dot{q}(t))^2-2\dot{q}(t)b^0(t)-2\dot{q}(t)q_{\varepsilon'}(t)b^1(t)+2q_{\varepsilon'}(t)b^0(t)b_0^1(t)+[q_{\varepsilon'}(t)b^1(t)]^2+O(\varepsilon)\right]\mathrm{d}t\right\}=\int_{-\infty}^{\infty}\mathrm{d}q''\,\|q''\|^2\int_{q(t')=q'}^{q(t'')=q''}[Dq(t;L,m')]\times$$

$$\exp\left(-\frac{1}{2\varepsilon}\int_{t'}^{t''}\{[\dot{q}(t)-b_0(q(t),t)]^2+O(\varepsilon')\}\mathrm{d}t\right). \tag{2.336}$$

We estimate now path integral in the RHS of the Eq. (2.336), using canonical perturbation expansion of anharmonic systems. Denoting the global minimum of the action

$$\hat{S}=\int_{t'}^{t''}[\dot{q}(t)-b_0(q(t),t)]^2\mathrm{d}t \tag{2.337}$$

by $\check{q}(t)$, it follows that it satisfies the extremality conditions for the minimizing path $\check{q}(t)$

$$\dot{\check{q}}(t)-b_0[\check{q}(t),t]=0.\ \check{q}(t')=\boldsymbol{q}' \tag{2.338}$$

In the limit $\varepsilon\to0,\varepsilon'\to0,\frac{\varepsilon'}{\varepsilon}\to0$ from Eq. (2.336) we obtain

$$(\boldsymbol{I}_{\varepsilon'}^{\varepsilon,1})_{\varepsilon'}=|\check{q}(t)|^2(C_{\varepsilon'})_{\varepsilon'}+(O(\varepsilon'/\varepsilon))_{\varepsilon'}. \tag{2.339}$$

Or in the following form: for any $\varepsilon\approx0,\varepsilon'\approx0,\frac{\varepsilon'}{\varepsilon}\approx0$

$$\left(\lim_{\varepsilon\to0}\boldsymbol{I}_{\varepsilon'}^{\varepsilon,1}\right)_{\varepsilon'}\approx|\check{q}(t)|^2(C_{\varepsilon'})_{\varepsilon'}. \tag{2.340}$$

(2) Let us evaluate now path integral $\left(\boldsymbol{I}_{\varepsilon'}^{\varepsilon,2}(L,m')\right)_{\varepsilon'}$. Let us rewrite Eq. (2.328) in the following form

$$\breve{b}_2[p_{\varepsilon'}(t),t]=\breve{b}_{2,2}[p_{\varepsilon'}(t),t]+\breve{b}_{2,3}[p_{\varepsilon'}(t),t], \tag{2.341}$$

$$\breve{b}_{2,2}[p_{\varepsilon'}(t),t]=b_0^2(t)p_{\varepsilon'}^2(t), \tag{2.342}$$

$$\breve{b}_{2,3}[p_{\varepsilon'}(t),t]=\sum_{\alpha}(\sqrt{2\varepsilon})^{|\alpha|-2}b_0^{\alpha}(t)[p_{\varepsilon'}(t)]^{\alpha},3\leqslant|\alpha|\leqslant r. \tag{2.343}$$

Substituting Eqs. (2.341) - (2.343) into Eq. (2.335) gives

$$\left(\boldsymbol{I}_{\varepsilon'}^{\varepsilon,2}(L,m')\right)_{\varepsilon'}=\int_{-\infty}^{\infty}\mathrm{d}\boldsymbol{q}''\int_{p(t')=\frac{q'}{\sqrt{2\varepsilon}}}^{p(t'')=\frac{q''}{\sqrt{2\varepsilon}}}\left[Dp\left(t;\frac{L}{\sqrt{2\varepsilon}},m'\right)\right]\times$$

$$\exp\left(-\int_{t'}^{t''}\left\{[\dot{p}(t)]^2+\frac{1}{\varepsilon}b^2(t)b^0(t)p_{\varepsilon'}^2(t)+\frac{2\sqrt{2}}{\sqrt{\varepsilon}}b^2(t)\dot{p}(t)p_{\varepsilon'}^2(t)+\right.\right.$$

$$\frac{2\sqrt{2}}{\sqrt{\varepsilon}}b^3(t)b^0(t)p_{\varepsilon'}^3(t)+O[\dot{p}(t)p_{\varepsilon'}^3(t)]+O[p_{\varepsilon'}^4(t)]+o(\sqrt{\varepsilon})\Big\}\,\mathrm{d}t\Big). \tag{2.344}$$

We let now that

$$\sup_{t\in[t',t'']}|b_0^2(t)b_0^0(t)|=\mu. \tag{2.345}$$

From Eq. (2.329) and Eqs. (2.344) – (2.345) one obtain the inequality

$$\left(\boldsymbol{I}_{\varepsilon'}^{\varepsilon,2}(L,m')\right)_{\varepsilon'}\leqslant\Bigg(\int_{-\infty}^{\infty}\mathrm{d}\boldsymbol{q}''\int_{p(t')=\frac{q'}{\sqrt{2\varepsilon}}}^{p(t'')=\frac{q''}{\sqrt{2\varepsilon}}}\Big[Dp\Big(t;\frac{L}{\sqrt{2\varepsilon}},m'\Big)\Big]\times$$

$$\exp\Big(-\int_{t'}^{t''}\Big\{[\dot{p}(t)]^2-\frac{\mu}{\varepsilon}p^2(t)+\frac{2\sqrt{2}}{\sqrt{\varepsilon}}b_0^2(t)\dot{p}(t)p_{\varepsilon'}^2(t)+\frac{2\sqrt{2}}{\sqrt{\varepsilon}}b_0^3(t)b_0^0(t)p_{\varepsilon'}^3(t)+$$

$$O[\dot{p}(t)p_{\varepsilon'}^4(t)]+O[p_{\varepsilon'}^4(t)]+o(\sqrt{\varepsilon})\Big\}\,\mathrm{d}t\Big)\Bigg)_{\varepsilon'}=\left(\breve{\boldsymbol{I}}_{\varepsilon'}^{\varepsilon,2}(L,m')\right)_{\varepsilon'} \tag{2.346}$$

Let us estimate now path integral $\left(\breve{\boldsymbol{I}}_{\varepsilon'}^{\varepsilon,2}(L,m')\right)_{\varepsilon'}$ in the RHS of the inequality (2.346), using saddle point approximation[27]. Denoting the critical path of the action

$$\hat{S}=\int_{t'}^{t''}\Big\{[\dot{p}(t)]^2-\frac{\mu}{\varepsilon}p^2(t)+O[\sqrt{\varepsilon}p_{\varepsilon'}^3(t)]+O[\sqrt{\varepsilon}\dot{p}(t)p_{\varepsilon'}^2(t)]+\cdots\Big\}\mathrm{d}t. \tag{2.347}$$

By $p_{\mathrm{cr},\varepsilon'}(t)$, it follows that it satisfies the Euler equation for the critical path $p_{\mathrm{cr},\varepsilon'}(t)$

$$\omega^{-2}\ddot{p}_{\mathrm{cr},\varepsilon'}(t)+p_{\mathrm{cr},\varepsilon'}(t)+o(\sqrt{\varepsilon}\dot{p}_{\mathrm{cr},\varepsilon'}(t))+O(\sqrt{\varepsilon}p_{\mathrm{cr},\varepsilon'}^2(t))+\cdots=0, \tag{2.348}$$

$$\mathrm{E}\omega^2=\omega^2(\varepsilon)=\Big(\frac{\mu}{\varepsilon}\Big), \tag{2.349}$$

$$p(t')=\frac{q'}{\sqrt{2\varepsilon}}=\tilde{q}',p(t'')=\frac{q''}{\sqrt{2\varepsilon}}=\tilde{q}''. \tag{2.350}$$

Therefore[44]-[45]:

$$p_{\mathrm{cr},\varepsilon'}(t)=\frac{[\tilde{q}''\sin\omega(t-t')+\tilde{q}'\sin\omega(t''-t)]}{\sin\omega(t''-t')}+O(\varepsilon'\varepsilon^{\gamma}),\gamma\geqslant 1.5. \tag{2.351}$$

Let $\hat{S}_2$ be

$$\hat{S}_2=-\int_{t'}^{t''}\{[\dot{p}(t)]^2-\omega^2p^2(t)\}\,\mathrm{d}t. \tag{2.352}$$

Substituting Eq. (2.351) into Eq. (2.352) gives

$$\hat{S}_2=-\frac{\omega}{2\sin(\omega T)}[(\tilde{q}'^2+\tilde{q}''^2)\cos(\omega T)-2\tilde{q}'\tilde{q}''],T=t''-t'. \tag{2.353}$$

Assumption. We assume now that: $\cot(\omega T)>0$.

Remark. Let $\tilde{q}''_s$ be a saddle point of the polynomial $\hat{S}_2(\tilde{q}',\tilde{q}'';T)$ on variable $\tilde{q}''$. Note that a saddle point $\tilde{q}''_s$ of the polynomial $\hat{S}_2$ is:

$$\tilde{q}''_s=\frac{\tilde{q}'}{\cos(\omega T)}. \tag{2.354}$$

Substituting Eq. (2.354) into Eq. (2.353) gives

$$\hat{S}_2(\tilde{q}',\tilde{q}''_s;\omega,T)\Big|_{\tilde{q}''_s=\tilde{q}'/\cos(\omega T)}\triangleq\hat{S}_2^{\#}(\tilde{q}';\omega,T)=$$

$$-\frac{\tilde{q}'^2\omega}{2\sin(\omega T)}[\cos(\omega T)-\cos^{-1}(\omega T)]=\frac{\tilde{q}'^2\omega\tan(\omega T)}{2}. \tag{2.355}$$

Assumption. We assume now that: $\cos(\omega T)\cong 1$, $\sin(\omega T)\cong 0$ such that the condition

$$\tilde{q}'^2\omega\tan(\omega T)=q'\omega^2\tan(\omega T)\cong 0, \tag{2.356}$$

is satisfied, and so

$$\exp[\hat{S}_2^{\#}(\tilde{q}';\omega,T)]=O(1). \tag{2.357}$$

Remark. We note that

$$\int_{t'}^{t''}p^2_{\mathrm{cr},\varepsilon'}(t;\tilde{q}',\tilde{q}'')\mathrm{d}t=O[T/\sin(\omega T)]. \tag{2.358}$$

We are dealing now with the finite Fourier series[39]:

$$\boldsymbol{q}_n=\boldsymbol{q}(t_n)=\boldsymbol{q}_0+\sum_{m=1}^{N}\sqrt{2/(N+1)}\sin[\nu_m(t_n-t')]\boldsymbol{q}(\nu_m). \tag{2.359}$$

Here

$$\nu_m=\frac{\pi m}{T}=\frac{\pi m}{(N+1)\epsilon},\epsilon=\frac{T}{(N+1)}. \tag{2.360}$$

Inserting now the expansion (2.359) into the time-sliced action (2.209), yields

$$\left(\boldsymbol{S}_{\varepsilon'}(\boldsymbol{q}_0,\boldsymbol{q}_1,\cdots,\boldsymbol{q}_{N-1},\boldsymbol{q}_N,\varepsilon)\right)_{\varepsilon'}=$$

$$\Delta t\sum_{n=1}^{N}\left(L_{\varepsilon'}\left(\frac{\boldsymbol{q}_n-\boldsymbol{q}_{n-1}}{\Delta t},\frac{\boldsymbol{q}_n-\boldsymbol{q}_{n-1}}{2},t_n\right)\right)_{\varepsilon'}=$$

$$\epsilon\sum_{n=1}^{N}\left(\frac{\boldsymbol{q}_n(\{\boldsymbol{q}(\nu_m)\}_{m=1}^{N})-\boldsymbol{q}_{n-1}(\{\boldsymbol{q}(\nu_m)\}_{m=1}^{N})}{\epsilon},\right.$$

$$\left.\frac{\boldsymbol{q}_n(\{\boldsymbol{q}(\nu_m)\}_{m=1}^{N})-\boldsymbol{q}_{n-1}(\{\boldsymbol{q}(\nu_m)\}_{m=1}^{N})}{2},t_n,\varepsilon\right)_{\varepsilon'}=$$

$$\epsilon\sum_{n=1}^{N}\left(\left\|\frac{\boldsymbol{q}_n(\{\boldsymbol{q}(\nu_m)\}_{m=1}^{N})-\boldsymbol{q}_{n-1}(\{\boldsymbol{q}(\nu_m)\}_{m=1}^{N})}{\epsilon}+\right.\right.$$

$$\left.\left.\boldsymbol{b}_{\varepsilon'}\left(\frac{\boldsymbol{q}_n(\{\boldsymbol{q}(\nu_m)\}_{m=1}^{N})-\boldsymbol{q}_{n-1}(\{\boldsymbol{q}(\nu_m)\}_{m=1}^{N})}{2},t_m,\varepsilon\right)\right\|^2\right)_{\varepsilon'}=$$

$$(\boldsymbol{S}_{\varepsilon'}(\boldsymbol{q}_0,\boldsymbol{q}(\nu_1),\cdots,\boldsymbol{q}(\nu_{N-1}),\boldsymbol{q}(\nu_N),\boldsymbol{q}_{N+1},\varepsilon))_{\varepsilon'} \tag{2.361}$$

Before performing the integral (2.208), we must transform the measure of integration from the local variables $\boldsymbol{q}_n$, to the Fourier components $\boldsymbol{q}(\nu_m)$. Due to the orthogonality relation[45], the transformation (2.359) has a unit determinant implying that

$$\prod_{n=1}^{N}\mathrm{d}\boldsymbol{q}_n=\prod_{m=1}^{N}\mathrm{d}\boldsymbol{q}(\nu_m)=\prod_{n=1}^{N}\mathrm{d}\boldsymbol{p}_m. \tag{2.362}$$

Substituting Eq. (2.361) and Eq. (2.362) into Eqs. (2.207) – (2.208), yields

$$\left(p^{\varepsilon}_{\varepsilon'}(\boldsymbol{q}',t'\mid\boldsymbol{q}'',t'')\right)_{\varepsilon'}=\lim_{\epsilon\to 0}\boldsymbol{I}_N(\boldsymbol{q}',t'\mid\boldsymbol{q}'',t''). \tag{2.363}$$

Here

$$\boldsymbol{I}_N(\boldsymbol{q}',t'\mid\boldsymbol{q}'',t'')=\check{N}_N\int_{-\infty}^{\infty}\mathrm{d}\boldsymbol{q}_0\int_{-\infty}^{\infty}\mathrm{d}\boldsymbol{q}(\nu_1)\cdots\int_{-\infty}^{\infty}\mathrm{d}\boldsymbol{q}(\nu_m)\cdots\int_{-\infty}^{\infty}\mathrm{d}\boldsymbol{q}(\nu_{N-1})\times\left[p_{\varepsilon'}(\boldsymbol{q}_0-\boldsymbol{q}')\right]_{\varepsilon'}\times$$

$$\exp\left[-\frac{1}{2\varepsilon}\left(\boldsymbol{S}_{\varepsilon'}(\boldsymbol{q}_0,\boldsymbol{q}(\nu_1),\cdots,\boldsymbol{q}(\nu_{N-1}),\boldsymbol{q}(\nu_N),\boldsymbol{q}_{N+1},t_m,\varepsilon)\right)_{\varepsilon'}\right] \tag{2.364}$$

$\boldsymbol{q}_{N+1} = \boldsymbol{q}'', \mathrm{d}\boldsymbol{q}(\nu_m) = \prod_{j=1}^{n} \mathrm{d}q_j(\nu_m), m = 0,\cdots,N, \epsilon = (t''-t')/N+1, t_m = m\Delta t.$

The quantity defined by Eqs. (2.363) – (2.364) is symbolically indicated by the path-integral expression

$$\left(\boldsymbol{p}_{\varepsilon'}^{\varepsilon}(\boldsymbol{q}',t' \mid \boldsymbol{q}'',t'')\right)_{\varepsilon'} =$$

$$\int_{\boldsymbol{q}(t')=\boldsymbol{q}'}^{\boldsymbol{q}(t'')=\boldsymbol{q}''}[D\boldsymbol{q}(\{\nu_m\}_{m=1}^{\infty})]\exp\left[\left(-\frac{1}{2\varepsilon}S_{\varepsilon'}(\dot{\boldsymbol{q}},\boldsymbol{q},t',t'';\varepsilon)\right)_{\varepsilon'}\right]. \tag{2.365}$$

Let us consider the quantity now,

$$\left(\boldsymbol{p}_{\varepsilon'}^{\varepsilon}(\boldsymbol{q}',t' \mid \boldsymbol{q}'',t'';P,m')\right)_{\varepsilon'} = \left(\lim_{\epsilon\to 0}\boldsymbol{I}_{m',N,\varepsilon,\varepsilon'}^{P}(\boldsymbol{q}',t' \mid \boldsymbol{q}'',t'')\right)_{\varepsilon'}. \tag{2.366}$$

Here

$$\boldsymbol{I}_{m',N,\varepsilon,\varepsilon'}^{P}(\boldsymbol{q}',t' \mid \boldsymbol{q}'',t'') = \boldsymbol{I}_{N,\varepsilon,\varepsilon'}(\boldsymbol{q}',t' \mid \boldsymbol{q}'',t'';P,m') =$$

$$\check{N}_N\int_{-P}^{P}\mathrm{d}\boldsymbol{q}(\nu_1)\cdots\int_{-P}^{P}\mathrm{d}\boldsymbol{q}(\nu_m')\int_{-\infty}^{\infty}\mathrm{d}\boldsymbol{q}(\nu_{m'+1})\cdots\int_{-\infty}^{\infty}\mathrm{d}\boldsymbol{q}(\nu_{N-1})\times$$

$$\int_{-\infty}^{\infty}\mathrm{d}\boldsymbol{q}(\nu_N)\exp\left[-\frac{1}{2\varepsilon}(S_{\varepsilon'}(\boldsymbol{q}_0,\boldsymbol{q}(\nu_1),\cdots,\boldsymbol{q}(\nu_{m'}),\right.$$

$$\left.\boldsymbol{q}(\nu_{m'+1}),\cdots,\boldsymbol{q}(\nu_{N-1}),\boldsymbol{q}(\nu_N),\boldsymbol{q}_{N+1},\varepsilon)\right]_{\varepsilon'} \tag{2.367}$$

and $m' \ll N, P \gg 1$.

The quantity defined by Eqs. (2.366) – (2.367) is symbolically indicated by the path-integral expression

$$\left(\boldsymbol{p}_{\varepsilon'}^{\varepsilon}(\boldsymbol{q}',t' \mid \boldsymbol{q}'',t'';P,m')\right)_{\varepsilon'} =$$

$$\int_{\boldsymbol{q}(t')=\boldsymbol{q}'}^{\boldsymbol{q}(t'')=\boldsymbol{q}''}[D\boldsymbol{q}(\{\nu_m\}_{m=1}^{\infty};P,m')]\exp\left[\left(-\frac{1}{2\varepsilon}S_{\varepsilon'}(\dot{\boldsymbol{q}},\boldsymbol{q},t',t'';\varepsilon)\right)_{\varepsilon'}\right]. \tag{2.368}$$

Using Eqs. (2.363) – (2.368) we define the quantity

$$\left(\mathbf{E}_{P,m'}\left[\boldsymbol{x}_{t'',\varepsilon'}^{q',\varepsilon}(\omega)\right]^2\right)_{\varepsilon'} =$$

$$\int_{\infty}^{\infty}\mathrm{d}\boldsymbol{q}'' \,\|\boldsymbol{q}''\|^2\left[\left(\boldsymbol{p}_{\varepsilon'}^{\varepsilon}(\boldsymbol{q}',t' \mid \boldsymbol{q}'',t'';P,m')\right)_{\varepsilon'}\right] = \lim_{\epsilon\to 0}\breve{\boldsymbol{I}}_{N,\varepsilon,\varepsilon'}(\boldsymbol{q}',t',t'';P,m'). \tag{2.369}$$

Here

$$\breve{\boldsymbol{I}}_{N,\varepsilon,\varepsilon'}(\boldsymbol{q}',t',t'';P,m') = \int_{-\infty}^{\infty}\mathrm{d}\boldsymbol{q}''\,\|\boldsymbol{q}''\|^2\boldsymbol{I}_{N,\varepsilon,\varepsilon'}(\boldsymbol{q}',t' \mid \boldsymbol{q}'',t'';P,m'). \tag{2.370}$$

The quantity (2.369) is symbolically indicated by the path-integral expression

$$\left(\boldsymbol{I}_{\varepsilon'}^{\varepsilon}(P,m')\right)_{\varepsilon'} = \left(\mathbf{E}_{P,m'}\left[\boldsymbol{x}_{t'',\varepsilon'}^{q',\varepsilon}(\omega)\right]^2\right)_{\varepsilon'} = \int_{-\infty}^{\infty}\mathrm{d}\boldsymbol{q}''\,\|\boldsymbol{q}''\|^2\int_{\boldsymbol{q}(t')=\boldsymbol{q}'}^{\boldsymbol{q}(t'')=\boldsymbol{q}''}[D\boldsymbol{q}((\nu_m)_{m=1}^{\infty};$$

$$P,m')]\times\exp\left[\left(-\frac{1}{2\varepsilon}\int_{t'}^{t''}L_{\varepsilon'}(\dot{\boldsymbol{q}},\boldsymbol{q},t)\mathrm{d}t\right)_{\varepsilon'}\right]. \tag{2.371}$$

Substituting Eq. (2.325) into Eq. (2.371) gives

$$\left(\boldsymbol{I}_{\varepsilon'}^{\varepsilon}(\boldsymbol{q}',t',t'';P,m')\right)_{\varepsilon'} = \exp\left[-\frac{1}{2\varepsilon}\int_{t'}^{t''}[b^0(t)]^2\mathrm{d}t\right]\int_{-\infty}^{\infty}\mathrm{d}\boldsymbol{q}''\ \|\boldsymbol{q}''\|^2 \times$$

$$\int_{\boldsymbol{q}(t')=\boldsymbol{q}'}^{\boldsymbol{q}(t'')=\boldsymbol{q}''}[Dq(\{\nu_m\}_{m=1}^{\infty};P,m')]\times$$

$$\exp\left(-\frac{1}{2\varepsilon}\int_{t'}^{t''}\left\{[\dot{q}(t)]^2 - 2\dot{q}(t)b^0(t) - 2\dot{q}(t)q_{\varepsilon'}(t)b^1(t) + 2q_{\varepsilon'}(t)b^0(t)b^1(t) +\right.\right.$$

$$[q_{\varepsilon'}(t)b^1(t)]^2 - 2\dot{q}(t)b_{2,2}[q_{\varepsilon'}(t),t] - 2\dot{q}(t)b_{2,3}[q_{\varepsilon'}(t),t] +$$

$$\left.\left.2b_{2,2}[q_{\varepsilon'}(t),t]b^0(t) + 2b_{2,3}[q_{\varepsilon'}(t),t]b^0(t) + 2q_{\varepsilon'}(t)b_2[q_{\varepsilon'}(t),t]b^1(t)\right\}\mathrm{d}t\right). \tag{2.372}$$

Using replacement $q(t) = p(t)\sqrt{2\varepsilon}$ into Feynman path integral (2.372), we obtain

$$\left(\boldsymbol{I}_{\varepsilon'}^{\varepsilon}(q',t',t'';P,m')\right)_{\varepsilon'} = \exp\left\{-\frac{1}{2\varepsilon}\int_{t'}^{t''}[b^0(t)]^2\mathrm{d}t\right\}\int_{-\infty}^{\infty}\mathrm{d}q''\ \|q''\|^2 \times$$

$$\int_{p(t')=\frac{q'}{\sqrt{2\varepsilon}}}^{p(t'')=\frac{q''}{\sqrt{2\varepsilon}}}\left[Dp\left(\{\nu_m\}_{m=1}^{\infty};\frac{P}{\sqrt{2\varepsilon}},m'\right)\right]\times$$

$$\exp\left(-\int_{t'}^{t''}\left\{[\dot{p}(t)]^2 - \frac{2}{\sqrt{2\varepsilon}}\dot{p}(t)b^0(t) - 2\dot{p}(t)p_{\varepsilon'}(t)b^1(t) + \frac{2}{\sqrt{2\varepsilon}}p_{\varepsilon'}(t)b^0(t)b^1(t) +\right.\right.$$

$$\left.\left.[p_{\varepsilon'}(t)b^1(t)]^2 - 2\sqrt{2\varepsilon}\dot{p}(t)\breve{b}_2[p_{\varepsilon'}(t),t] + 2\breve{b}_2[p_{\varepsilon'}(t),t]\hat{b}[\sqrt{2\varepsilon}p_{\varepsilon'}(t),t]\right\}\mathrm{d}t\right). \tag{2.373}$$

Let us rewrite now Feynman path integral (2.373) in the next equivalent form

$$\left(\boldsymbol{I}_{\varepsilon'}^{\varepsilon}(q',t',t'';P,m')\right)_{\varepsilon'} = \exp\left\{-\frac{1}{2\varepsilon}\int_{t'}^{t''}[b^0(t)]^2\mathrm{d}t\right\}\int_{-\infty}^{\infty}\mathrm{d}q''\ \|q''\|^2 \times$$

$$\int_{p(t')=\frac{q'}{\sqrt{2\varepsilon}}}^{p(t'')=\frac{q''}{\sqrt{2\varepsilon}}}\left\{\boldsymbol{D}p\left[(\nu_m)_{m=1}^{\infty};\frac{P}{\sqrt{2\varepsilon}},m'\right]\right\}\times$$

$$\exp\left\{-\int_{t'}^{t''}[\dot{p}(t)]^2\mathrm{d}t\right\}\left[\exp\left(-\int_{t'}^{t''}\left\{-\frac{2}{\sqrt{2\varepsilon}}\dot{p}(t)b^0(t) - 2\dot{p}(t)p_{\varepsilon'}(t)b^1(t) +\right.\right.\right.$$

$$\left.\left.\left.\frac{2}{\sqrt{2\varepsilon}}p_{\varepsilon'}(t)b^0(t)b^1(t) + [p_{\varepsilon'}(t)b^1(t)]^2\right\}\mathrm{d}t\right)\right]\times$$

$$\exp\left(\int_{t'}^{t''}\left\{2\sqrt{2\varepsilon}\dot{p}(t)\breve{b}_2[p_{\varepsilon'}(t),t] - \breve{b}_2[p_{\varepsilon'}(t),t]\hat{b}[\sqrt{2\varepsilon}p(t),t]\right\}\mathrm{d}t\right) \tag{2.374}$$

Assume that $1/p + 1/q = 1$ and $q = 1/\varepsilon$. Then

$$p = \frac{1}{1-\varepsilon} = 1 + \varepsilon + o(\varepsilon). \tag{2.375}$$

Using Corollary 1 now, from Eq. (2.374) we obtain

$$\left(\boldsymbol{I}_{\varepsilon'}^{\varepsilon}(q',t',t'';P,m')\right)_{\varepsilon'} \leqslant \exp\left\{-\frac{1}{2\varepsilon}\int_{t'}^{t''}[b^0(t)]^2\mathrm{d}t\right\}\times$$

$$\left(\int_{-\infty}^{\infty}\mathrm{d}q''\ \|q''\|^2\int_{p(t')=q'}^{p(t'')=\frac{q''}{\sqrt{2\varepsilon}}}\left[Dp\left(\{\nu_m\}_{m=1}^{\infty};\frac{P}{\sqrt{2\varepsilon}},m'\right)\right]\times\exp\left\{-\int_{t'}^{t''}[\dot{p}(t)]^2\mathrm{d}t\right\}\times\right.$$

$$\left\{\exp\left[-(1+\varepsilon)\int_{t'}^{t''}\left(-\frac{2}{\sqrt{2\varepsilon}}\dot{p}(t)b^0(t)-2\dot{p}(t)p_{\varepsilon'}(t)b^1(t)+\frac{2}{\sqrt{2\varepsilon}}p_{\varepsilon'}(t)b_0^0(t)b_0^1(t)+\right.\right.\right.$$

$$\left.\left.\left.[p_{\varepsilon'}(t)b^1(t)]^2\right)\mathrm{d}t\right]\right\}\Bigg)^{1-\varepsilon}\times\left(\int_{-\infty}^{\infty}\mathrm{d}\boldsymbol{q}''\int_{p(t')=q'}^{p(t'')=\frac{q''}{\sqrt{2\varepsilon}}}\left[Dp\left(\{\nu_m\}_{m=1}^{\infty};\frac{P}{\sqrt{2\varepsilon}},m'\right)\right]\times\right.$$

$$\exp\left\{-\int_{t'}^{t''}[\dot{p}(t)]^2\mathrm{d}t\right\}\times\left\{\exp\left[\frac{1}{\varepsilon}\int_{t'}^{t''}[2\sqrt{2\varepsilon}\dot{p}(t)\breve{b}_2(p_{\varepsilon'}(t),t)\breve{b}_2(p_{\varepsilon'}(t),t)-\right.\right.$$

$$\left.\left.\left.\breve{b}_2(p_{\varepsilon'}(t),t)\hat{b}(\sqrt{2\varepsilon}p(t),t)]\mathrm{d}t\right]\right\}\right)^{\varepsilon}. \tag{2.376}$$

Therefore

$$\left\langle \boldsymbol{I}_{\varepsilon'}^{\varepsilon}(q',t',t'';P,m')\right\rangle_{\varepsilon'}\leqslant\left\langle[\boldsymbol{I}_{\varepsilon'}^{\varepsilon,1}(q',t',t'';P,m')]^{1-\varepsilon}\right\rangle_{\varepsilon'}\times$$

$$\left\langle(\boldsymbol{I}_{\varepsilon'}^{\varepsilon,2}(q',t',t'';P,m'))^{\varepsilon}\right\rangle_{\varepsilon'}. \tag{2.377}$$

Here

$$\left\langle \boldsymbol{I}_{\varepsilon'}^{\varepsilon,1}(q',t',t'';P,m')\right\rangle_{\varepsilon'}=\left\langle\exp\left\{-\frac{1}{2\varepsilon(1+\varepsilon)}\int_{t'}^{t''}[b^0(t)]^2\mathrm{d}t\right\}\times\right.$$

$$\int_{-\infty}^{\infty}\mathrm{d}q''\,\|q''\|^2\int_{p(t')=0}^{p(t'')=\frac{q''}{\sqrt{2\varepsilon}}}\left[Dp\left(\{\nu_m\}_{m=1}^{\infty};\frac{P}{\sqrt{2\varepsilon}},m'\right)\right]\times\exp\left\{-\int_{t'}^{t''}[\dot{p}(t)]^2\mathrm{d}t\right\}\times$$

$$\exp\left\{-(1+\varepsilon)\int_{t'}^{t''}\left[-\frac{2}{\sqrt{2\varepsilon}}\dot{p}(t)b^0(t)-2\dot{p}(t)p_{\varepsilon'}(t)b^1(t)+\frac{2}{\sqrt{2\varepsilon}}p_{\varepsilon'}(t)b^0(t)b^1(t)+\right.\right.$$

$$\left.\left.\left.[p_{\varepsilon'}(t)b^1(t)]^2\right]\mathrm{d}t\right\}\right\rangle_{\varepsilon'} \tag{2.378}$$

and

$$\left\langle \boldsymbol{I}_{\varepsilon'}^{\varepsilon,2}(q',t',t'';P,m')\right\rangle_{\varepsilon'}=\int_{-\infty}^{\infty}\mathrm{d}q''\int_{p(t')=0}^{p(t'')=\frac{q''}{\sqrt{2\varepsilon}}}\left[Dp\left(\{\nu_m\}_{m=1}^{\infty};\frac{P}{\sqrt{2\varepsilon}},m'\right)\right]\times$$

$$\exp\left\{-\int_{t'}^{t''}[\dot{p}(t)]^2\mathrm{d}t\right\}\times\left\{\exp\left[\frac{1}{\varepsilon}\int_{t'}^{t''}[2\sqrt{2\varepsilon}\dot{p}(t)\breve{b}_2(p_{\varepsilon'}(t),t)-\right.\right.$$

$$\left.\left.\breve{b}_2(p_{\varepsilon'}(t),t)\hat{b}(\sqrt{2\varepsilon}p(t),t)]\mathrm{d}t\right]\right\}. \tag{2.379}$$

Let us evaluate now path integral $\left\langle \boldsymbol{I}_{\varepsilon'}^{\varepsilon,1}(q',t',t'';L,m')\right\rangle_{\varepsilon'}$. From Eq. (2.378) using replacement $p(t)=\frac{q(t)}{\sqrt{2\varepsilon}}$ into Feynman path integral in the RHS of the Eq. (2.378), we obtain

$$\left\langle \boldsymbol{I}_{\varepsilon'}^{\varepsilon,1}(q',t',t'';P,m')\right\rangle_{\varepsilon'}=\exp\left\{-\frac{1}{2\varepsilon}\int_{t'}^{t''}[b^0(t)]^2\mathrm{d}t\right\}\int_{-\infty}^{\infty}\mathrm{d}q''\,\|q''\|^2\times$$

$$\int_{q(t')=0}^{q(t'')=q''}[Dq(\{\nu_m\}_{m=1}^{\infty};P,m')]\times$$

$$\exp\left(-\frac{1}{2\varepsilon}\int_{t'}^{t''}\left\{[\dot{q}(t)]^2-2\dot{q}(t)b^0(t)-2\dot{q}(t)q_{\varepsilon'}(t)b^1(t)+2q_{\varepsilon'}(t)b^0(t)b_0^1(t)+\right.\right.$$

$$\left.\left.[q_{\varepsilon'}(t)b^1(t)]^2+O(\varepsilon)\right\}\mathrm{d}t\right)=\int_{-\infty}^{\infty}\mathrm{d}q''\,\|q''\|^2\int_{\boldsymbol{q}(t')=0}^{\boldsymbol{q}(t'')=\boldsymbol{q}''}[Dq(\{\nu_m\}_{m=1}^{\infty};P,m')]\times$$

$$\exp\left(-\frac{1}{2\varepsilon}\int_{t'}^{t''}\left\{[\dot{q}(t)-\hat{b}(q_{\varepsilon'}(t),t)]^2+O(\varepsilon)\right\}dt\right). \tag{2.380}$$

In the limit $\varepsilon\to 0,\varepsilon'\to 0,\varepsilon'/\varepsilon\to 0$ by simple calculation, one obtain

$$\lim_{\substack{\varepsilon\to 0\\ \varepsilon'\to 0\\ \varepsilon'/\varepsilon\to 0}}\left(\boldsymbol{I}_{\varepsilon'}^{\varepsilon;1}(q',t',t'';L,m')\right)=[\check{q}(t)]^2 \tag{2.381}$$

where

$$\dot{\check{q}}(t)-b_0(\check{q}(t),t)=0,\ \check{q}(t')=q'. \tag{2.382}$$

Or in the following equivalent form

$$\left(\boldsymbol{I}_{\varepsilon'}^{\varepsilon;1}(q',t',t'';L,m')\right)_{\varepsilon'}=(C)_{\varepsilon'}\left([\check{q}(t)]^2\right)_{\varepsilon'}. \tag{2.383}$$

Here $\varepsilon\approx 0,\varepsilon'\approx 0,\varepsilon'/\varepsilon\approx 0$ and

$$\dot{\check{q}}(t)-\hat{b}(\check{q}(t),t)=0,\ \check{q}(t')=q'. \tag{2.384}$$

Let us evaluate now path integral $\left(\boldsymbol{I}_{\varepsilon'}^{\varepsilon;2}(P,m')\right)_{\varepsilon'}$. Let us rewrite Eq. (2.328) in the following form

$$\breve{b}_2(p_{\varepsilon'}(t),t)=\breve{b}_{2,2}(p_{\varepsilon'}(t),t)+\breve{b}_{2,3}(p_{\varepsilon'}(t),t), \tag{2.385}$$

$$\breve{b}_{2,2}(p_{\varepsilon'}(t),t)=b_0^2(t)p_{\varepsilon'}^2(t), \tag{2.386}$$

$$\breve{b}_{2,3}(p_{\varepsilon'}(t),t)=\sum_{\alpha}(\sqrt{2\varepsilon})^{|\alpha|-2}b_0^{\alpha}(t)[p_{\varepsilon'}(t)]^{\alpha},3\leqslant|\alpha|\leqslant r. \tag{2.387}$$

Substituting Eqs. (2.385) - (2.387) into Eq. (2.378) gives

$$\left(\boldsymbol{I}_{\varepsilon'}^{\varepsilon;2}(P,m')\right)_{\varepsilon'}=\int_{-\infty}^{\infty}d\boldsymbol{q}''\int_{p(t')=\frac{q'}{\sqrt{2\varepsilon}}}^{p(t'')=\frac{q''}{\sqrt{2\varepsilon}}}\left[Dp\left(\{\nu_m\}_{m=1}^{\infty};\frac{P}{\sqrt{2\varepsilon}},m'\right)\right]\times$$

$$\exp\left(-\int_{t'}^{t''}\left\{[\dot{p}(t)]^2+\frac{1}{\varepsilon}b^2(t)b^0(t)p_{\varepsilon'}^2(t)+\frac{2\sqrt{2}}{\sqrt{\varepsilon}}b^2(t)\dot{p}(t)p_{\varepsilon'}^2(t)\right.\right.$$

$$\left.\left.+\frac{2\sqrt{2}}{\sqrt{\varepsilon}}b^3(t)b^0(t)p_{\varepsilon'}^3(t)+O(\dot{p}(t)p_{\varepsilon'}^3(t))+O(p_{\varepsilon'}^4(t))+o(\sqrt{\varepsilon})\right\}dt\right). \tag{2.388}$$

We let now that

$$\sup_{t\in[t',t'']}|b_0^2(t)b_0^0(t)|=\mu. \tag{2.389}$$

From Eq. (2.329) and Eqs. (2.388) - (2.389) one obtain the inequality

$$\left(\boldsymbol{I}_{\varepsilon'}^{\varepsilon;2}(P,m')\right)_{\varepsilon'}\leqslant\left(\int_{-\infty}^{\infty}d\boldsymbol{q}''\int_{p(t')=\frac{q'}{\sqrt{2\varepsilon}}}^{p(t'')=\frac{q''}{\sqrt{2\varepsilon}}}\left[Dp\left(\{\nu_m\}_{m=1}^{\infty};\frac{P}{\sqrt{2\varepsilon}},m'\right)\right]\times\right.$$

$$\exp\left(-\int_{t'}^{t''}\left\{[\dot{p}(t)]^2-\frac{\mu}{\varepsilon}p^2(t)+\frac{2\sqrt{2}}{\sqrt{\varepsilon}}b_0^2(t)\dot{p}(t)p_{\varepsilon'}^2(t)+\frac{2\sqrt{2}}{\sqrt{\varepsilon}}b_0^3(t)b_0^0(t)p_{\varepsilon'}^3(t)+\right.\right.$$

$$\left.\left.\left.O(\dot{p}(t)p_{\varepsilon'}^3(t))+O(p_{\varepsilon'}^4(t))+o(\sqrt{\varepsilon})\right\}dt\right)\right)_{\varepsilon'}=\left(\breve{\boldsymbol{I}}_{\varepsilon'}^{\varepsilon;2}\left(\frac{P}{\sqrt{2\varepsilon}},m'\right)\right)_{\varepsilon'}. \tag{2.390}$$

Let us estimate now path integral $\left(\breve{\boldsymbol{I}}_{\varepsilon'}^{\varepsilon;2}\left(\frac{P}{\sqrt{2\varepsilon}},m'\right)\right)_{\varepsilon'}$ in the RHS of the inequality

(2.390). Denoting the critical path of the action

$$\hat{S} = \int_{t'}^{t''}\left\{[\dot{p}(t)]^2 - \frac{\mu}{\varepsilon}p^2(t) + O(\sqrt{\varepsilon}p_{\varepsilon'}^3(t)) + O(\sqrt{\varepsilon}\dot{p}(t)p_{\varepsilon'}^2(t)) + \cdots\right\}\mathrm{d}t. \tag{2.391}$$

By $p_{\mathrm{cr}}(t)$, it follows that it satisfies the Euler equation for the critical path $p_{\mathrm{cr},\varepsilon'}(t)$

$$\omega^{-2}\ddot{p}_{\mathrm{cr},\varepsilon'}(t) + p_{\mathrm{cr},\varepsilon'}(t) + o(\sqrt{\varepsilon}\dot{p}_{\mathrm{cr},\varepsilon'}(t)) + O(\sqrt{\varepsilon}p_{\mathrm{cr},\varepsilon'}^2(t)) + \cdots = 0, \tag{2.392}$$

where

$$\omega^2 = \omega^2(\varepsilon) = \frac{\mu}{\varepsilon}, \tag{2.393}$$

$$p(t') = \frac{q'}{\sqrt{2\varepsilon}} = \tilde{q}', p(t'') = \frac{q''}{\sqrt{2\varepsilon}} = \tilde{q}''. \tag{2.394}$$

We estimate now path integral $\left(\breve{\boldsymbol{I}}_{\varepsilon'}^{\varepsilon,2}\left(\frac{P}{\sqrt{2\varepsilon}}, m'\right)\right)_{\varepsilon'}$ using canonical perturbation expansion of anharmonic systems (see [45] Chap. 3, sect. 15). Let us rewrite action (2.391) in the following form

$$\hat{S} = \int_{t'}^{t''}\left\{[\dot{p}(t)]^2 - \frac{\mu}{\varepsilon}p^2(t) + O(\sqrt{\varepsilon}p_{\varepsilon'}^3(t)) + O(\sqrt{\varepsilon}\dot{p}(t)p_{\varepsilon'}^2(t)) + \cdots\right\}dt =$$

$$-\int_{t'}^{t''}\left\{[\dot{p}(t)]^2 - \omega^2p^2(t)\right\}\mathrm{d}t + \int_{t'}^{t''}\left[O(\sqrt{\varepsilon}p_{\varepsilon'}^3(t)) + O(\sqrt{\varepsilon}\dot{p}(t)p_{\varepsilon'}^2(t)) + \cdots\right]\mathrm{d}t =$$

$$\hat{\boldsymbol{S}}_2 + \int_{t'}^{t''}V_{\varepsilon'}[p(t), \dot{p}(t), t]\mathrm{d}t'. \tag{2.395}$$

Thus corresponding perturbation expansion of the path integral $\left(\breve{\boldsymbol{I}}_{\varepsilon'}^{\varepsilon,2}\left(\frac{P}{\sqrt{2\varepsilon}}, m'\right)\right)_{\varepsilon'}$ is

$$\left(\breve{\boldsymbol{I}}_{\varepsilon'}^{\varepsilon,2}\left(\frac{P}{\sqrt{2\varepsilon}}, m'\right)\right)_{\varepsilon'} = \left(\int_{-\infty}^{\infty}\mathrm{d}q''\int_{p(t')=\frac{q'}{\sqrt{2\varepsilon}}}^{p(t'')=\frac{q''}{\sqrt{2\varepsilon}}}\left[Dp\left(\{\nu_m\}_{m=1}^{\infty}; \frac{P}{\sqrt{2\varepsilon}}, m'\right)\right] \times \exp(\hat{\boldsymbol{S}}_2)\right)_{\varepsilon'} +$$

$$\left(\int_{-\infty}^{\infty}\mathrm{d}q''\int_{p(t')=\frac{q'}{\sqrt{2\varepsilon}}}^{p(t'')=\frac{q''}{\sqrt{2\varepsilon}}}\left[Dp\left(\{\nu_m\}_{m=1}^{\infty}; \frac{P}{\sqrt{2\varepsilon}}, m'\right)\right] \times\right.$$

$$\left.\int_{t'}^{t''}V_{\varepsilon'}[p(t), \dot{p}(t), t]\mathrm{d}t'\exp[\hat{\boldsymbol{S}}_2]\right)_{\varepsilon'} + \cdots \tag{2.396}$$

Let us consider now Gaussian path integral

$$\boldsymbol{I}_{\omega}^2(L, m') = \int_{-\infty}^{\infty}\mathrm{d}\boldsymbol{q}''\int_{p(t')=\frac{q'}{\sqrt{2\varepsilon}}}^{p(t'')=\frac{q''}{\sqrt{2\varepsilon}}}\left[Dp\left(t; \frac{P}{\sqrt{2\varepsilon}}, m'\right)\right]\exp(\hat{S}_{2,N}) =$$

$$\lim_{N\to\infty}\boldsymbol{I}_N^{\omega}(P, m'), \tag{2.397}$$

$$\boldsymbol{I}_N^{\omega}(P, m') = \breve{N}_N\int_{-P/\sqrt{2\varepsilon}}^{P/\sqrt{2\varepsilon}}\mathrm{d}p_1\cdots\int_{-P/\sqrt{2\varepsilon}}^{P/\sqrt{2\varepsilon}}\mathrm{d}p_{m'}\int_{-\infty}^{\infty}\mathrm{d}p_{m'+1}\times\cdots\int_{-\infty}^{\infty}\mathrm{d}p_{N-1}\times\exp(\hat{S}_{2,N}). \tag{2.398}$$

Here

$$\hat{S}_{2,N} = -\frac{\epsilon}{2}\sum_{m=1}^{N+1}\left[(\bar{\nabla}p_m)^2 - \omega^2 p_m\right]. \tag{2.399}$$

Note that: $\sum_{m=1}^{N+1}(\bar{\nabla}p_m)^2 = -\sum_{m=1}^{N+1}p_m\bar{\nabla}\nabla p_m$ [45].

We now turn to the fluctuation factor[45] of the path integral $\boldsymbol{I}_N^{\omega}(P,m')$. With the matrix notation for the lattice fluctuation operator: $\nabla\bar{\nabla}+\omega^2$, we have to solve the multiple integral

$$\mathcal{F}_N^{\omega}(T) = \frac{1}{\sqrt{2\pi\epsilon}}\prod_{m=1}^{m'}\left(\int_{-P/\sqrt{2\varepsilon}}^{P/\sqrt{2\varepsilon}}\frac{dp_m}{\sqrt{2\pi\epsilon}}\right)\prod_{m=m'+1}^{N}\left(\int_{-\infty}^{\infty}\frac{dp_m}{\sqrt{2\pi\epsilon}}\right)\times$$

$$\exp\left[\epsilon\sum_{m=1}^{N}\delta p_m\left(\nabla\bar{\nabla}+\omega^2\right)_{mm''}\right]\delta p_{m''}. \tag{2.400}$$

The eigenvalues of the fluctuation operator $\nabla\bar{\nabla}+\omega^2$ can be refered to [45]

$$\sigma_m(\omega) = \omega^2 - \Omega_m\bar{\Omega}_m = \omega^2 - \frac{1}{\epsilon^2}[2-2\cos(\epsilon\nu_m)], \tag{2.401}$$

$$\nu_m = \frac{\pi m}{T} = \frac{\pi m}{(N+1)\epsilon}, \epsilon = \frac{T}{(N+1)}. \tag{2.402}$$

Thus

$$\sigma_m(\omega) = \omega^2 - \frac{1}{\epsilon^2}\left[2-2\cos\left(\frac{\pi m}{(N+1)}\right)\right]. \tag{2.403}$$

We set now

$$m' = \sup_{m\geq 1}\left[m \mid \omega^2 - 2\left(\frac{\pi m}{T}\right)^2 \geq 0 \text{ and } \pi m/(N+1) < 1\right]. \tag{2.404}$$

Therefore for all $m \leqslant m'$ one obtain

$$\sigma_m(\omega) = \omega^2 - \frac{1}{\epsilon^2}\left[2-2\left(1-\left(\frac{\pi m}{N+1}\right)^2+O(\epsilon^4)\right)\right] =$$

$$\omega^2 - 2\left(\frac{\pi m}{T}\right)^2 + O(\epsilon^2). \tag{2.405}$$

Thus for all $m \leqslant m'$ the inequality

$$\sigma_m(\omega) = \omega^2 - 2\left(\frac{\pi m}{T}\right)^2 + O(\epsilon^2) \geq 0 \tag{2.406}$$

is satisfied and consequently ① all the eigenvalues $\sigma_m(\omega)$ with $m \leqslant m'$ are positive and ② all the eigenvalues $\sigma_m(\omega)$ with $m \gg m'+1$ are negative. We have chose now number $r = m'$ in Eq. (2.369) such that the inequalities

$$\sigma_r(\omega) = \omega^2 - 2\left(\frac{\pi r}{T}\right)^2 + O(\epsilon^2) \geq 0 \tag{2.407}$$

And

$$\sigma_{r+1}(\omega) = \omega^2 - 2\left(\frac{\pi(r+1)}{T}\right)^2 + O(\epsilon^2) < 0 \tag{2.408}$$

is satisfied and therefore $m' = O(\omega)$. We have choose now the number $\epsilon \in \mathbf{R}_+$ such that the equalities

$$\rho(m) = (P/\sqrt{2\varepsilon})\sqrt{\epsilon\sigma_m(\omega)} = 0(1), m \leqslant m' \tag{2.409}$$

is satisfied. From Eq. (2.400), inequalities (2.407) – (2.408) and Eq. (2.409) one obtain

$$\mathcal{F}_N^{\omega}(T) = \frac{1}{\sqrt{2\pi\epsilon}}\prod_{m=1}^{m'}\left(\int_{-P/\sqrt{2\varepsilon}}^{P/\sqrt{2\varepsilon}}\frac{dp_m}{\sqrt{2\pi\epsilon}}\right)\prod_{m=m'+1}^{N}\left(\int_{-\infty}^{\infty}\frac{dp_m}{\sqrt{2\pi\epsilon}}\right)\times\exp\left[-\epsilon\sum_{m=1}^{N}(\delta p_m)^2\sigma_m(\omega)\right] =$$

$$\frac{1}{\sqrt{2\pi\epsilon}}\prod_{m=1}^{m'}\left(\int_{-\rho(m)}^{\rho(m)}\frac{dp_m}{\sqrt{2\pi\,|\sigma_m(\omega)|\,\epsilon^2}}\right)\exp\left[\sum_{m=1}^{m'}(\delta p_m)^2\right]\times$$

$$\prod_{m=m'+1}^{N}\left[\int_{-\infty}^{\infty}\frac{dp_m}{\sqrt{2\pi\sigma_m(\omega)\epsilon^2}}\right]\exp\left[\sum_{m=m'+1}^{N}-(\delta p_m)^2\right] =$$

$$\frac{1}{\sqrt{2\pi\epsilon}}\prod_{m=1}^{m'}\left[\frac{1}{\sqrt{2\pi\,|\sigma_m(\omega)|\,\epsilon^2}}\right]\prod_{m=m'+1}^{N}\left[\frac{1}{\sqrt{2\pi\epsilon^2\sigma_m(\omega)}}\right]\times\exp(O(m')). \quad (2.410)$$

Note that $\exp\,(O(m'))\simeq\exp(O(\omega^2))$. From Eq. (2.410) we obtain

$$\mathcal{F}_N^{\omega}(T) = \frac{1}{\sqrt{2\pi\epsilon}}\prod_{m=1}^{m'}\left[\frac{1}{\sqrt{2\pi\epsilon^2\,|\sigma_m(\omega)|}}\right]\prod_{m=m'+1}^{N}\left[\frac{1}{\sqrt{2\pi\epsilon^2\sigma_m(\omega)}}\right]\times\exp(O(\omega^2)) =$$

$$\frac{\exp(O(\omega^2))}{\sqrt{2\pi\epsilon}}\prod_{m=1}^{N}\frac{1}{\sqrt{2\pi\epsilon^2\,|\,\Omega_m\bar{\Omega}_m-\omega^2\,|}}. \quad (2.411)$$

The product of these eigenvalues, as well known [45] is found by introducing an auxiliary frequency $\bar{\omega}$ satisfying

$$\sin\frac{\epsilon\bar{\omega}}{2} = \frac{\epsilon\omega}{2}. \quad (2.412)$$

Now we decompose the product as [46]:

$$\prod_{m=1}^{N}(\epsilon^2\,|\,\Omega_m\bar{\Omega}_m-\omega^2\,|\,) = \prod_{m=1}^{N}(\epsilon^2\,|\Omega_m\bar{\Omega}_m|)\prod_{m=1}^{N}\left(\frac{\epsilon^2\,|\,\Omega_m\bar{\Omega}_m-\omega^2\,|}{\epsilon^2\,|\,\Omega_m\bar{\Omega}_m\,|}\right)\prod_{m=1}^{N}(\epsilon^2\,|\Omega_m\bar{\Omega}_m|)$$

$$\prod_{m=1}^{N}\left[\,\left|1-\frac{\sin^2\left(\frac{\epsilon\bar{\omega}}{2}\right)}{\sin^2\left(\frac{m\pi}{2(N+1)}\right)}\right|\,\right] = \frac{\sin(\bar{\omega}T)}{\sin(\epsilon\bar{\omega})}. \quad (2.413)$$

From Eq. (2.382) and Eq. (2.383) we obtain[46]:

$$\mathcal{F}_N^{\omega}(T) = \frac{1}{\sqrt{2\pi\epsilon}}\sqrt{\frac{\sin(\epsilon\bar{\omega})}{|\sin(\bar{\omega}T)|}}\exp(O(\omega^2)). \quad (2.414)$$

In the limit $\epsilon\to 0$ finally we obtain[45]:

$$\mathcal{F}^{\omega}(T)\lim_{N\to\infty}F_N^{\omega}(T) = \frac{1}{\sqrt{2\pi}}\sqrt{\frac{\omega}{|\sin(\omega T)|}}\exp(O(\omega^2)). \quad (2.415)$$

Remark. For a given ω we can choose some real T, such that the inequality

$$\boldsymbol{C}_T = \frac{(\varepsilon')\varepsilon^{2l}}{\sin(\omega T)} \gg 1, \quad (2.416)$$

where $\omega = \omega(\varepsilon)$, is satisfied.

From Eq. (2.396) and inequality (2.416) we obtain:

$$\left(\breve{\boldsymbol{I}}_{\varepsilon'}^{\varepsilon,2}\left(\frac{P}{\sqrt{2\varepsilon}},m'\right)\right)_{\varepsilon'} = \boldsymbol{I}_N^{\omega}(P,m') + O(1/\boldsymbol{C}_T). \quad (2.417)$$

From inequality (2.390) and Eq. (2.417) we obtain:

$$\left(\boldsymbol{I}_{\varepsilon'}^{\varepsilon,2}(P,m') \right)_{\varepsilon'} \leqslant \boldsymbol{I}_N^{\omega}(P,m') + O(1/\boldsymbol{C}_T). \tag{2.418}$$

From Eq. (2.415) and Eqs. (2.353) – (2.357) we obtain:

$$\boldsymbol{I}_N^{\omega}(P,m') = \frac{1}{\sqrt{2\pi}}\sqrt{\frac{\omega}{|\sin(\omega T)|}}\exp(O(\omega^2)) \times \int_{-\infty}^{\infty} d\tilde{q}''\exp[\hat{S}_2(\tilde{q}',\tilde{q}'')] =$$

$$\frac{1}{\sqrt{2\pi}}\exp[\tilde{q}'^2\omega\tan(\omega T)/2]\exp(O(\omega^2))\sqrt{\frac{\omega}{|\sin(\omega T)|}}\Bigg/\sqrt{\frac{\partial^2\hat{S}_2(\tilde{q}',\tilde{q}'')}{\partial\tilde{q}''^2}} =$$

$$\exp(O(\omega^2)). \tag{2.419}$$

Substituting equality (2.419) into inequality (2.418) gives

$$\left(\boldsymbol{I}_{\varepsilon'}^{\varepsilon,2}(L,m') \right)_{\varepsilon'} \leqslant \exp(O(\omega^2)). \tag{2.420}$$

Substituting equality (2.383) and inequality (2.420) into inequality (2.377) gives

$$\left(\boldsymbol{I}_{\varepsilon'}^{\varepsilon}(q',t',t'';P,m') \right)_{\varepsilon'} \leqslant \left([\boldsymbol{I}_{\varepsilon'}^{\varepsilon,1}(q',t',t'';P,m')]^{1-\varepsilon} \right)_{\varepsilon'} \times$$

$$\left([\boldsymbol{I}_{\varepsilon'}^{\varepsilon,2}(q',t',t'';P,m')]^{\varepsilon} \right)_{\varepsilon'} \leqslant (C)_{\varepsilon'} \left([\check{q}(t)]^{2(1-\varepsilon)} \right)_{\varepsilon'} \exp(\varepsilon \times O(\omega^2)). \tag{2.421}$$

Here $\varepsilon \approx 0, \varepsilon' \approx 0, \varepsilon'/\varepsilon \approx 0$ and

$$\dot{\check{q}}(t) - b_0(\check{q}(t),t) = 0, \check{q}(t') = q'.$$

Inequality (2.421) completed the proof.

2.7 Conclusions

(1) We pointed out that the canonical Laplace approximation[27] is not a valid asymptotic approximation in the limit $\varepsilon \to 0$ for a path-integral (2.273), see also Ref. [42].

(2) Supporting technical analysis. Let us consider optimal control problem from Example. 1. Corresponding Bellman equation is:

$$\min_{\alpha_1 \in [-\rho,\rho]} \left\{ \max_{\alpha_1 \in [-\rho,\rho]} \left[\frac{\partial V}{\partial t} + x_2 \frac{\partial V}{\partial x_1} + (-x_2^3 + \alpha_1 + \alpha_2)\frac{\partial V}{\partial x_2} \right] \right\} = 0$$

$$V(T,x_1,x_2) = x_1^2 + x_2^2, t \in [0,T]. \tag{2.422}$$

Complete constructing the exact analytical solution for PDE is a complicated unresolved classical problem, because PDE is not amenable to analytical treatments. Even the theorem of existence classical solution for boundary problems such PDE is not proved. Thus, even for simple cases a problem of construction feedback optimal control by Bellman Eq. (2.422) complicated numerical technology or principal simplification is needed[46]. However as one can see complete constructing feedback optimal control from Theorem 6 is simple. In this paper, the generic imperfect dynamics model of air-to-surface missiles are given in addition to the related simple guidance law.

Chapter 3 Exact Quasi-Classical Asymptotic Beyond Maslov Canonical Operator and Quantum Jumps Nature

In this chapter exact quasi-classical asymptotic beyond WKB-theory and beyond Maslov canonical operator to the Colombeau solutions of the n-dimensional Schrodinger equation is presented. Quantum jumps nature is considered successfully. We pointed out that an explanation of quantum jumps can be found to result from Colombeau solutions of the Schrödinger equation alone without additional postulates.

3.1 Introduction

A number of experiments on trapped single ions or atoms have been performed in recent years[1-5]. Monitoring the intensity of scattered laser light off of such systems has shown abrupt changes that have been cited as evidence of "quantum jumps" between states of the scattered ion or atom. The existence of such jumps was required by Bohr in his theory of the atom. Bohr's quantum jumps between atomic states[6] were the first form of quantum dynamics to be postulated. He assumed that an atom remained in an atomic eigenstate until it made an instantaneous jump to another state with the emission or absorption of a photon. Since these jumps do not appear to occur in solutions of the Schrodinger equation, something similar to Bohr's idea has been added as an extra postulate in modern quantum mechanics.

Stochastic quantum jump equations[7-9] were introduced as a tool for simulating the dynamics of a dissipative system with a large Hilbert space and their links with quantum measurement were also noted[10-14]. This measurement interpretation is generally known as quantum trajectory theory[15]. By adding filter cavities as part of the apparatus, even the quantum jumps in the dressed state model can be interpreted as approximations to measurement-induced jumps[16].

The question arises whether an explanation of these jumps can be found to result from an Colombeau solution[17]-[19] $(\Psi_\varepsilon(x,t;\hbar))_\varepsilon$ of the Schrödinger equation alone without additional postulates. We found exact quasi-classical asymptotic of the quantum averages with position variable with localized initial data.

$$(\langle i,t,x_0;\hbar,\varepsilon\rangle)_\varepsilon = \left(\int x_i \mid \Psi_\varepsilon(x,t,x_0;\hbar) \mid^2 dx\right)_\varepsilon, \varepsilon \in (0,1], x, x_0 \in \mathbf{R}^d, i = 1,\cdots,d, \tag{3.1}$$

i. e, we found the limiting Colombeau quantum averages (limiting Colombeau quantum trajectories) such that[19]:

$$(\langle i,t,x_0;\varepsilon\rangle)_\varepsilon = (\lim_{\hbar\to 0}(i,t,x_0;\hbar,\varepsilon))_\varepsilon = \left(\lim_{\hbar\to 0}\int x_i\,|\,\Psi_\varepsilon(x,t,x_0;\hbar)\,|^2\mathrm{d}x\right)_\varepsilon,$$

$$\varepsilon \in (0,1],x \in \mathbf{R}^d,i = 1,\cdots,d,t \in [0,T] \tag{3.2}$$

and limiting quantum trajectories $\boldsymbol{q}(t,x_0) = \{q_1(t,x_0),\cdots,q_d(t,x_0)\} \in \mathbf{R}^d,t \in [0,T]$ such that

$$q_i(t,x_0) = \langle i,t,x_0\rangle = \lim_{\varepsilon\to 0}\lim_{\hbar\to 0}\langle i,t,x_0;\hbar,\varepsilon\rangle = \lim_{\varepsilon\to 0}\lim_{\hbar\to 0}\int x_i\,|\,\Psi_\varepsilon(x,t,x_0;\hbar)\,|^2\mathrm{d}x,$$

$$x \in \mathbf{R}^d,i = 1,\cdots,d \tag{3.3}$$

if limit in LHS of Eq. (3.3) exists.

The physical interpretation of these asymptotic given below, shows that the answer is "yes" for the limiting quantum trajectories with localized initial data.

Note that an axiom of quantum measurement is: if the particle is in some state $|\,\Psi_{t,\varepsilon}\rangle$ that the probability $P(x,\delta x)$ of getting a result $x \in \mathbf{R}^d$ at instant t with an accuracy of $\|\delta x\| \ll 1$ will be given by

$$P_t(x,\delta x,x_0;\hbar,\varepsilon) = \int_{\|x-x'\|\leqslant\|\delta x\|}|\langle x'\,|\,\Psi_{t,\varepsilon}\rangle|^2\mathrm{d}x' =$$

$$\int_{\|x-x'\|\leqslant\|\delta x\|}|\,\Psi_\varepsilon(x',t,x_0;\hbar)\,|^2\mathrm{d}x'. \tag{3.4}$$

We rewrite now Eq. (3.4) of the form

$$P_t(x,\delta x,x_0;\hbar,\varepsilon) = \frac{1}{(2\pi)^{d/2}\,\|\delta x\|^d}\int_{\mathbf{R}^d}|\,\Psi_\varepsilon(x',t,x_0;\hbar)\,|^2\exp\left(-\frac{\|x-x'\|^2}{\|\delta x\|^2}\right)\mathrm{d}x'. \tag{3.5}$$

We define well localized limiting quantum trajectories $\boldsymbol{q}(t) = \boldsymbol{q}(t,x_0) = \{q_1(t,x_0),\cdots,q_d(t,x_0)\} \in \mathbf{R}^d,\boldsymbol{q}(0,x_0) = x_0,t \in [0,T]$ such that:

$$\left(\lim_{\|\delta\boldsymbol{q}(t)\|\to 0}\lim_{\hbar\to 0}P_t(\boldsymbol{q}(t),\delta\boldsymbol{q}(t),x_0;\hbar,\varepsilon)\right)_\varepsilon = 1 \tag{3.6}$$

and well localized limiting quantum trajectories

$$\boldsymbol{q}(t) = \boldsymbol{q}(t,x_0) = \{q_1(t,x_0),\cdots,q_d(t,x_0)\} \in \mathbf{R}^d,$$

$$\boldsymbol{q}(0,x_0) = x_0,t \in [0,T]$$

such that:

$$\lim_{\varepsilon\to 0}\lim_{\|\delta\boldsymbol{q}(t)\|\to 0}\lim_{\hbar\to 0}P_t(\boldsymbol{q}(t),\delta\boldsymbol{q}(t),x_0;\hbar,\varepsilon) = 1 \tag{3.7}$$

if limit in LHS of Eq. (3.7) exists.

3.2 Colombeau solutions of the Schrödinger equation and corresponding path integral representation

Let **H** be a complex infinite dimensional separable Hilbert space, with inner product $\langle\cdot,\cdot\rangle$

and norm $|\cdot|$.

Let us consider Schrödinger equation:

$$-i\hbar\left(\frac{\partial\Psi(t)}{\partial t}\right)+\hat{H}(t)\Psi(t)=0,\Psi(0)=\Psi_0(x), \tag{3.8}$$

$$H(t)=-\left(\frac{\hbar^2}{2m}\right)\Delta+V(x,t). \tag{3.9}$$

Here operator $H(t):\mathbf{R}\times\mathbf{H}\to\mathbf{H}$ is essentially self-adjoint, $\hat{H}(t)$ is the closure of $H(t)$.

Theorem 3.2.1.[21],[22]. Assume that: (1) $\Psi_0(x)\in L_2(\mathbf{R}^d)$, (2) $V(x,t)$ is continuous and $\sup_{x\in\mathbf{R}^d,t\in[0,T]}|V(x,t)|<+\infty$. Then corresponding solution of the Schrödinger Eqs. (3.8) – (3.9) exist and can be represented via formula

$$\Psi(t,x)=\lim_{n\to\infty}\left(\frac{nm}{4\pi it\hbar}\right)^{d(n+1)/2}\int_{\mathbf{R}^d}\cdots\int_{\mathbf{R}^d}\mathrm{d}x_0\mathrm{d}x_1\cdots\mathrm{d}x_n\Psi_0(x_0)\times$$

$$\exp\left[\frac{i}{\hbar}S(x_0,x_1,\cdots,x_n,x_{n+1};t)\right], \tag{3.10}$$

where we have set $x_{n+1}=x$ and

$$S(x_0,x_1,\cdots,x_n,x_{n+1};t)=\sum_{i=1}^{n}\left[\frac{m}{4}\frac{|x_{i+1}-x_i|^2}{(t/n)^2}-V(x_{i+1},t_i)\right], \tag{3.11}$$

where $t_i=\frac{it}{n}$. Let $q_n(t)$ be a trajectory; that is, a function from $[0,t]$ to $\mathbf{R}^d$ with $q_n(0)=x_0$ and set $q_n(t_i)=x_i, i=1,\cdots,n+1$. We rewrite Eq. (3.6) for a future application symbolically for short of the following form

$$\Psi(t,x)=\lim_{n\to\infty}\int_{q_n(t)=x}D[q_n(t)]\Psi_0[q_n(0)]\exp\left[\frac{i}{\hbar}S(q_n(t),x;t)\right], \tag{3.12}$$

where we have set (i) $S(q_n(t),x;t)=S(x_0,x_1,\cdots,x_n,x_{n+1};t)$ and (ii) $D[q_n(t)]$ that is, a

$$D[q_n(t)]=\left(\frac{nm}{4\pi it\hbar}\right)^{d(n+1)/2}\prod_{j=0}^{n}\mathrm{d}x_j. \tag{3.13}$$

Trotter and Kato well known classical results give a precise meaning to the Feynman integral when the potential $V(x,t)$ is sufficiently regular[19-21]. However if potential $V(x,t)$ is a non-regular this is well known problem to represent solution of the Schrödinger Eqs. (3.8) – (3.9) via formula (3.3), see [21].

We avoided this difficulty using contemporary Colombeau framework[17-19]. Using replacement $x_i\to\frac{x_i}{1+\varepsilon^{2k}|x|^{2k}},\varepsilon\in(0,1],k\geqslant 1$, we obtain from potential $V(x,t)$ regularized potential $V_\varepsilon(x,t),\varepsilon\in(0,1]$, such that $V_{\varepsilon=0}(x,t)=V(x,t)$ and

(1) $(V_\varepsilon(x,t))_\varepsilon\in G(\mathbf{R}^d)$,

(2) $\sup_{x\in\mathbf{R}^d,t\in[0,T]}|V_\varepsilon(x,t)|<+\infty,\varepsilon\in(0,1]$. $\quad$ (3.14)

Here $G(\mathbf{R}^d)$ is Colombeau algebra of Colombeau generalized functions[17-19].

Finally we obtain regularized Schrödinger equation of Colombeau form[17-19]:

$$-i\hbar\left(\frac{\partial\Psi_\varepsilon(t)}{\partial t}\right)_\varepsilon+\left(\hat{H}_\varepsilon(t)\Psi_\varepsilon(t)\right)_\varepsilon=0,(\Psi_\varepsilon(0))_\varepsilon=\Psi_0(x),\tag{3.15}$$

$$H_\varepsilon(t)=-\left(\frac{\hbar^2}{2m}\right)\Delta+V_\varepsilon(x,t).\tag{3.16}$$

Using the inequality (3.14) Theorem 3.2.1 asserts again that corresponding solution of the Schrödinger Eqs. (3.15) – (3.16) exist and can be represented via formulae[19]:

$$\left(\Psi_\varepsilon(t,x)\right)_\varepsilon=$$
$$\left(\lim_{n\to\infty}\left(\frac{nm}{4\pi it\hbar}\right)^{d(n+1)/2}\int_{\mathbf{R}^d}\cdots\int_{\mathbf{R}^d}dx_0dx_1\cdots dx_n\Psi_0(x_0)\exp\left[\frac{i}{\hbar}S_\varepsilon(x_0,x_1,\cdots,x_n,x_{n+1};t)\right]\right)_\varepsilon,\tag{3.17}$$

where we have set $x_{n+1}=x$ and

$$S_\varepsilon(x_0,x_1,\cdots,x_n,x_{n+1};t)=\sum_{i=1}^{n}\left[\frac{m}{4}\frac{|x_{i+1}-x_i|^2}{(t/n)^2}-V_\varepsilon(x_{i+1},t_i)\right],\tag{3.18}$$

where we have set $t_i=\frac{it}{n}$.

We rewrite Eq. (3.7) for a future application symbolically of the following form

$$(\Psi_\varepsilon(t,x))_\varepsilon=\left(\lim_{n\to\infty}\int_{q_n(t)=x}D[q_n(t)]\Psi_0[q_n(0)]\exp\left\{\frac{i}{\hbar}S_\varepsilon[q_n(t);t]\right\}\right)_\varepsilon,\tag{3.19}$$

or of the following form

$$(\Psi_\varepsilon(t,x))_\varepsilon=(\lim_{n\to\infty}\Psi_{\varepsilon,n}(t,x))_\varepsilon=\left(\lim_{n\to\infty}\int_{q(t)=x}D_n[q(t)]\Psi_0[q_n(0)]\times\right.$$
$$\left.\exp\left[\frac{i}{\hbar}S_\varepsilon(\dot{q}(t),q(t);t)\right]\right)_\varepsilon.\tag{3.20}$$

For the limit in RHS of Eqs. (3.19) and (3.20) we will be used canonical path integral notation

$$(\Psi_\varepsilon(t,x))_\varepsilon=\left(\int_{q(t)=x}D[q(t)]\Psi_0[q(0)]\exp\left[\frac{i}{\hbar}S_\varepsilon(\dot{q}(t),q(t))\right]\right)_\varepsilon,\tag{3.21}$$

where

$$S_\varepsilon(\dot{q}(t),q(t))=\int_0^t\left[\frac{m}{4}\dot{q}^2(s)-V_\varepsilon(q(s),s)\right]ds.$$

Substitution $n=8k+7$ into RHS of the Eq. (3.7) gives

$$(\Psi_\varepsilon(t,x))_\varepsilon=$$
$$\left(\lim_{k\to\infty}\left(\frac{(8k+7)m}{4\pi t\hbar}\right)^{d(4k+4)}\int_{\mathbf{R}^d}\cdots\int_{\mathbf{R}^d}dx_0dx_1\cdots dx_{8k+7}\Psi_0(x_0)\times\right.$$
$$\left.\exp\left[\frac{i}{\hbar}S_\varepsilon(x_0,x_1,\cdots,x_{8k+7},x_{8k+8};t)\right]\right)_\varepsilon.\tag{3.22}$$

We rewrite Eq. (3.22) for a future application symbolically of the following form

$$(\Psi_\varepsilon(t,x))_\varepsilon=\left(\lim_{n\to\infty}\int_{q_n(t)=x}D^+[q_n(t)]\Psi_0[q_n(0)]\exp\left[\frac{i}{\hbar}S_\varepsilon(q_n(t);t)\right]\right)_\varepsilon,\tag{3.23}$$

or of the following Eq:

$$(\Psi_\varepsilon(t,x))_\varepsilon = \left(\lim_{n\to\infty}\int_{q(t)=x} D_n^+[q(t)]_n \Psi_0[q(0)]\exp\left[\frac{i}{\hbar}S_\varepsilon(\dot q(t),q(t);t)\right]\right)_\varepsilon. \quad (3.24)$$

For the limit in RHS of Eqs. (2.23) and (2.24) we will use following path integral notation

$$(\Psi_\varepsilon(t,x))_\varepsilon = \left(\int_{q(t)=x} D^+[q(t)]\Psi_0[q(0)]\exp\left[\frac{i}{\hbar}S_\varepsilon(\dot q(t),q(t))\right]\right)_\varepsilon. \quad (3.25)$$

Let us consider now regularized oscillatory integral

$$(\mathbf{J}_{\varepsilon,n}(t;\hbar))_\varepsilon =$$

$$\left(\int_{\mathbf{R}^d}\cdots\int_{\mathbf{R}^d}\mathrm{d}x_0\mathrm{d}x_1\cdots\mathrm{d}x_n f(x_0,x_1,\cdots,x_n)\Psi_0(x_0)\exp\left[\frac{i}{\hbar}S_\varepsilon(x_0,x_1,\cdots,x_n;t)\right]\right)_\varepsilon. \quad (3.26)$$

Lemma 2.1 (Localization Principle[27-28]) Let Ω be a domain in $\mathbf{R}^{d\times n}$ and $f\in C_0^\infty(\Omega)$ be a smooth function of compact support, $S_\varepsilon\in C^\infty(\Omega)$, $\varepsilon\in(0,1]$ be a real valued smooth function without stationary points in supp(f), i. e. $\partial_x S_\varepsilon(x)\neq 0$ for $x\in\Omega$. Let L be a differential operator

$$L(f) = -\sum_{i=1}^{n}\frac{\partial}{\partial x_i}\left(|S'_{\varepsilon,x}(x)|^{-2}\frac{\partial S_\varepsilon}{\partial x_i}f\right). \quad (3.27)$$

Then

$$\left(|\mathbf{J}_{\varepsilon,n}(t;\hbar)|\right)_\varepsilon \leqslant \hbar\int_\Omega L(f(x))\,\mathrm{d}x.$$

$$\forall n,m\in\mathbf{N},\forall\hbar\leqslant 1.$$

there exist c_m such that

$$\left(\mathbf{J}_{\varepsilon,n}(t;\hbar)\right)_\varepsilon \leqslant c_m\hbar^m\|f\|,\quad \|f\| = \sup_{x\in\Omega}\sum_{|\alpha|\leqslant m}|D^\alpha f|. \quad (3.28)$$

Lemma 2.2 (Generalized Localization Principle) Let Ω_n be a domain in $\mathbf{R}^{d\times n}$ and $f_n\in C_0^\infty(\Omega_n)$ be a real valued smooth function without stationary points in supp(f), i. e. $\partial_x S_\varepsilon(x)\neq 0$ for $x\in\Omega_n$ and let $(£_\varepsilon(t;\hbar))_\varepsilon$ be infinite sequence $n\in\mathbf{N}$:

$$\left(£_{\varepsilon,n}(t;\hbar)\right)_\varepsilon = \left(\left(\frac{nm}{4\pi it\hbar}\right)^{d(n+1)/2}\int_{\Omega_n}\mathrm{d}x_0\mathrm{d}x_1\cdots\mathrm{d}x_n f_n(x_0,x_1,\cdots,x_n)\Psi_0(x_0)\times\right.$$

$$\left.\exp\left[\frac{i}{\hbar}S_\varepsilon(x_0,x_1,\cdots,x_n;t)\right]\right)_\varepsilon. \quad (3.29)$$

Then there exist infinite sequence $\{\hbar_k\}_{k\in\mathbf{N}}$, $\lim_{k\to\infty}\hbar_k = 0$ such that

$$\left(\lim_{\substack{n\to\infty\\k\to\infty}}£_{\varepsilon,n}(t;\hbar_k)\right)_\varepsilon = 0. \quad (3.30)$$

Proof. Eq. (2.30) immediately follows from (3.28).

Remark 2.1 From Lemma 2.2 follows that stationary phase approximation is not a valid asymptotic approximation in the limit $\hbar\to 0$ for a path-integral of Eqs. (3.21) and (3.22).

3.3 Exact quasi-classical asymptotic beyond Maslov canonical operator

Theorem 3.1 Let us consider Cauchy problem (3.15) with initial data $\Psi_0(x)$ is given via formula

$$\Psi_0(x) = \frac{\eta^{d/4}}{(2\pi)^{d/4}\hbar^{d/4}}\exp\left[-\frac{\eta(x-x_0)^2}{2\hbar}\right], \tag{3.31}$$

where $0 < \hbar \ll \eta \ll 1$ and $x^2 = \langle x, x\rangle$.

(1) We assume now that: (i) $(V_\varepsilon(x,t))_\varepsilon \in G(\mathbf{R}^d)$, (ii) $V_{\varepsilon=0}(x,t) = V(x,t): \mathbf{R}_+ \times \mathbf{R}^d \to \mathbf{R}$ and (iii) $\forall t \in \mathbf{R}_+$ function $V(x,t)$ is a polynomial on variable $x = (x_1, \cdots, x_d)$, i.e.

$$V(x,t) = \sum_{\|\alpha\| \leqslant m} g_\alpha(t)x^\alpha, \alpha = (i_1, \cdots, i_d), x^\alpha = x_1^{i_1} \times \cdots \times x_d^{i_d}, \|\alpha\| = \sum_{r=1}^{d} i_r. \tag{3.32}$$

(2) Let $u(\tau,t,\lambda,x,y) = (u_1(\tau,t,\lambda,x,y), \cdots, u_d(\tau,t,\lambda,x,y))$ be the solution of the boundary problem

$$\frac{\partial^2 u^T(\tau,t,\lambda,x,y)}{\partial\tau^2} = \text{Hess}[V(\lambda,\tau)]u^T(\tau,t,\lambda,x,y) + [V'(\lambda,\tau)]^T, \tag{3.33}$$

$$u(0,t,\lambda,x,y) = y, u(\tau,t,\lambda,x,y) = x. \tag{3.34}$$

Here

$$\lambda = (\lambda_1, \cdots, \lambda_d) \in \mathbf{R}^d, u^T(\tau,t,\lambda,x,y) = (u_1(\tau,t,\lambda,x,y), \cdots, u_d(\tau,t,\lambda,x,y))^T,$$

$$V'(\lambda,\tau) = \left(\left[\frac{\partial V(x,t)}{\partial x_1}\right]_{x=\lambda}, \cdots, \left[\frac{\partial V(x,t)}{\partial x_d}\right]_{x=\lambda}\right) \text{ and Hess}[V(\lambda,\tau)] = \left[\frac{\partial^2 V(x,t)}{\partial x_i \partial x_j}\right]_{x=\lambda}. \tag{3.35}$$

(3) Let $S(t,\lambda,x,y)$ be the master action given via formula

$$S(t,\lambda,x,y) = \int_0^t L(\dot{u}(\tau,t,\lambda,x,y), u(\tau,t,\lambda,x,y), \tau)\mathrm{d}\tau, \tag{3.36}$$

where master Lagrangian $L(\dot{u},u,\tau)$ are

$$L(\dot{u},u,\tau) = \frac{m}{2}\dot{u}^2(\tau,t,\lambda,x,y) - \hat{V}(u(\tau,t,\lambda,x,y),\tau), \dot{u} = \left(\frac{\partial u_1}{\partial\tau}, \cdots, \frac{\partial u_d}{\partial\tau}\right), \dot{u}^2 = \langle \dot{u}, \dot{u}\rangle, \tag{3.37}$$

$$\hat{V}((\tau,t,\lambda,x,y),\tau) = u(\tau,t,\lambda,x,y)\text{Hess}[V(\lambda,\tau)]u^T(\tau,t,\lambda,x,y) + V'(\lambda,\tau)u^T(\tau, t,\lambda,x,y). \tag{3.38}$$

Let $y_{\text{cr}} = y_{\text{cr}}(t,\lambda,x) \in \mathbf{R}^d$ be solution of the linear system of the algebraic equations

$$\left[\frac{\partial S(t,\lambda,x,y)}{\partial y_i}\right]_{y=y_{\text{cr}}} = 0 (i = 1, \cdots, d). \tag{3.39}$$

(4) Let $\hat{x} = \hat{x}(t,\lambda,x_0) \in \mathbf{R}^d$ be solution of the linear system of the algebraic equations

$$y_{\text{cr}}(t,\lambda,\hat{x}) + \lambda - x_0 = 0. \tag{3.40}$$

Assume that: for a given values of the parameters t,λ,x_0, the point $\hat{x} = \hat{x}(t,\lambda,x_0)$ is not a focal point on a corresponding trajectory is given by corresponding solution of the boundary problem (3.33). Then for the limiting quantum average given via formulae of Eq. (3.1) the inequalities is satisfied:

$$\lim_{\substack{\hbar\to 0\\ \varepsilon\to 0}} |\langle i,t,x_0;\hbar\rangle - \lambda_i| \leqslant$$

$$2\left[\,|\det S_{y_{cr}y_{cr}}(t,\lambda,\hat{x}(t,\lambda,x_0),y_{cr}(t,\lambda,\hat{x}(t,\lambda,x_0)))|\,\right]^{-1}|\hat{x}_i(t,\lambda,x_0)|, i = 1,\cdots,d. \tag{3.41}$$

Thus, one can calculate the limiting quantum trajectory corresponding to potential $V(x,t)$ by using transcendental master equation

$$\hat{x}_i(t,\lambda,x_0) = 0 (i = 1,\cdots,d). \tag{3.42}$$

Proof. From inequality (B.15) of Appendix B and Theorem B1, using inequalities (B.53.a) and (B.53.b) of Appendix B we obtain

$$\lim_{\substack{\varepsilon\to 0\\ \sigma\to 0}}\lim_{\hbar\to 0} |\langle \hat{x}_i, T;\sigma,l,\lambda,\varepsilon\rangle - \lambda_i| \leqslant \lim_{\hbar\to 0}[R_1(T,\lambda) + R_2(T,\lambda)](i = 1,\cdots,d), \tag{3.43}$$

where

$$R_1(T,\lambda) = \int dx \left\{\int_{q(T)=x} D^+[q(t)]\Psi[q(0)][|q_i(T)|]^{\frac{1}{2}}\cos\left[\frac{1}{\hbar}S_1(\dot{q},q,\lambda,T)\right]\right\}^2, \tag{3.44}$$

$$R_2(T,\lambda) = \int dx \left\{\int_{q(T)=x} D^+[q(t)]\Psi[q(0)][|q_i(T)|]^{\frac{1}{2}}\sin\left[\frac{1}{\hbar}S_1(\dot{q},q,\lambda,T)\right]\right\}^2. \tag{3.45}$$

We note that

$$R_1(T,\lambda) = \int dx\,[\breve{R}_1(x,T,\lambda)]^2,\ R_2(T,\lambda) = \int dx\,[\breve{R}_2(x,T,\lambda)]^2, \tag{3.46}$$

where

$$\breve{R}_1(x,T,\lambda) = \int_{q(T)=x} D^+[q(t)]\Psi[q(0)][|q_i(T)|]^{\frac{1}{2}}\cos\left[\frac{1}{\hbar}S_1(\dot{q},q,\lambda,T)\right] =$$

$$\int dy\int_{q(0)=y}^{q(T)=x} D^+[q(t)]\Psi[q(0)][|q_i(T)|]^{\frac{1}{2}}\cos\left[\frac{1}{\hbar}S_1(\dot{q},q,\lambda,T)\right] = \int dy\breve{R}_1(x,y,T,\lambda), \tag{3.47}$$

$$\breve{R}_1(x,y,T,\lambda) = \int_{q(0)=y}^{q(T)=x} D^+[q(t)]\Psi[q(0)][|q_i(T)|]^{\frac{1}{2}}\cos\left[\frac{1}{\hbar}S_1(\dot{q},q,\lambda,T)\right] \tag{3.48}$$

and

$$\breve{R}_2(x,T,\lambda) = \int_{q(T)=x} D^+[q(t)]\Psi[q(0)][|q_i(T)|]^{\frac{1}{2}}\sin\left[\frac{1}{\hbar}S_1(\dot{q},q,\lambda,T)\right] =$$

$$\int dy\int_{q(0)=y}^{q(T)=x} D^+[q(t)]\Psi[q(0)][|q_i(T)|]^{\frac{1}{2}}\sin\left[\frac{1}{\hbar}S_1(\dot{q},q,\lambda,T)\right] = \int dy\breve{R}_2(x,y,T,\lambda), \tag{3.49}$$

$$\breve{R}_2(x,y,T,\lambda) = \int_{q(0)=y}^{q(T)=x} D^+[q(t)]\Psi[q(0)][|q_i(T)|]^{\frac{1}{2}}\sin\left[\frac{1}{\hbar}\boldsymbol{S}_1(\dot{q},q,\lambda,T)\right]. \tag{3.50}$$

From Eq. (3.46) one obtain

$$\breve{R}_1(x,y,T,\lambda) = \frac{1}{2}[\breve{R}_{1,1}(x,y,T,\lambda) + \breve{R}_{1,2}(x,y,T,\lambda)], \tag{3.51}$$

where

$$\breve{R}_{1,1}(x,y,T,\lambda) = \int_{q(0)=y}^{q(T)=x} D^+[q(t)]\Psi[q(0)][|q_i(T)|]^{\frac{1}{2}}\exp\left[\frac{i}{\hbar}\boldsymbol{S}_1(\dot{q},q,\lambda,T)\right], \tag{3.52}$$

$$\breve{R}_{1,2}(x,y,T,\lambda) = \int_{q(0)=y}^{q(T)=x} D^+[q(t)]\Psi[q(0)][|q_i(T)|]^{\frac{1}{2}}\exp\left[-\frac{i}{\hbar}\boldsymbol{S}_1(\dot{q},q,\lambda,T)\right]. \tag{3.53}$$

Let us calculate now path integral $\breve{R}_{1,1}(x,y,T,\lambda)$ and path integral $\breve{R}_{1,2}(x,y,T,\lambda)$, using stationary phase approximation. From Eq. (B.23) of Appendix B follows directly that action $\boldsymbol{S}_1(\dot{q},q,\lambda,T)$ coincide with master action $S(t,\lambda,x,y)$ is given via formulae Eqs. (3.6) – (3.8) and therefore from of Eq. (3.52) and Eq. (3.53) one obtain

$$\breve{R}_{1,1}(x,y,T,\lambda) = \int_{q(0)=y}^{q(T)=x} D^+[q(t)]\Psi[q(0)][|q_i(T)|]^{\frac{1}{2}}\exp\left[\frac{i}{\hbar}\boldsymbol{S}_1(\dot{q},q,\lambda,T)\right] =$$

$$\breve{R}_{1,1}(x,y,T,\lambda) = [|x_i|]^{\frac{1}{2}}\Psi(y)\exp\left[\frac{i}{\hbar}S(t,\lambda,x,y)\right] \tag{3.54}$$

and

$$\breve{R}_{1,2}(x,y,T,\lambda) = \int_{q(0)=y}^{q(T)=x} D^+[q(t)]\Psi[q(0)][|q_i(T)|]^{\frac{1}{2}}\exp\left[\frac{i}{\hbar}\boldsymbol{S}_1(\dot{q},q,\lambda,T)\right] =$$

$$\breve{R}_{1,2}(x,y,T,\lambda) = [|x_i|]^{\frac{1}{2}}\Psi(y)\exp\left[-\frac{i}{\hbar}\boldsymbol{S}(t,\lambda,x,y)\right]. \tag{3.55}$$

From Eq. (3.47) and Eq. (3.54) we obtain

$$\breve{R}_1(x,T,\lambda) = \int dy\breve{R}_1(x,y,T,\lambda). \tag{3.56}$$

Substitution Eq. (3.55) into Eq. (3.57) gives

$$\breve{R}_{1,1}(x,T,\lambda) = [|x_i|]^{\frac{1}{2}}\int dy\Psi(y)\exp\left[\frac{i}{\hbar}S(t,\lambda,x,y)\right]. \tag{3.57}$$

Similarly one obtain

$$\breve{R}_{1,2}(x,T,\lambda) = [|x_i|]^{\frac{1}{2}}\int dy\Psi(y)\exp\left[-\frac{i}{\hbar}S(t,\lambda,x,y)\right]. \tag{3.58}$$

Let us calculate now integral $\breve{R}_{1,1}(x,T,\lambda)$ and integral $\breve{R}_{1,2}(x,T,\lambda)$ using stationary phase approximation. Let $y_{cr} = y_{cr}(t,\lambda,x) \in \mathbf{R}^d$ be the stationary point of master action $S(t,\lambda,x,y)$ and therefore Eq. (3.39) is satisfied. Having applied stationary phase approximation one obtain

$$\breve{R}_{1,1}(x,y_{cr}(t,\lambda,x),T,\lambda) = [|\det S_{y_{cr}y_{cr}}(t,\lambda,x,y_{cr}(t,\lambda,x))|]^{-\frac{1}{2}} \times$$

$$[|x_i|]^{\frac{1}{2}}\Psi(y_{cr}(t,\lambda,x))\exp\left[\frac{i}{\hbar}S(t,\lambda,x,y_{cr}(t,\lambda,x))\right], \tag{3.59}$$

$$\breve{R}_{1,2}(x,y_{cr}(t,\lambda,x),T,\lambda) = \left[\left|\det S_{y_{cr}y_{cr}}(t,\lambda,x,y_{cr}(t,\lambda,x))\right|\right]^{-\frac{1}{2}} \times$$
$$[|x_i|]^{\frac{1}{2}}\Psi(y_{cr}(t,\lambda,x))\exp\left[-\frac{i}{\hbar}S(t,\lambda,x,y_{cr}(t,\lambda,x))\right]. \tag{3.60}$$

Substitution Eq. (3.59) – Eq. (3.60) into Eq. (3.51) gives

$$\breve{R}_1(x,y_{cr}(t,\lambda,x),T,\lambda) = \frac{1}{2}\left[\breve{R}_{1,1}(x,y_{cr}(t,\lambda,x),T,\lambda) + \breve{R}_{1,2}(x,y_{cr}(t,\lambda,x),T,\lambda)\right]$$

$$= \left[\left|\det S_{y_{cr}y_{cr}}(t,\lambda,x,y_{cr}(t,\lambda,x))\right|\right]^{-\frac{1}{2}}[|x_i|]^{\frac{1}{2}}\Psi(y_{cr}(t,\lambda,x)) \times$$

$$\left\{\exp\left[\frac{i}{\hbar}S(t,\lambda,x,y_{cr}(t,\lambda,x))\right] + \exp\left[-\frac{i}{\hbar}S(t,\lambda,x,y_{cr}(t,\lambda,x))\right]\right\} =$$

$$\left[\left|\det S_{y_{cr}y_{cr}}(t,\lambda,x,y_{cr}(t,\lambda,x))\right|\right]^{-\frac{1}{2}}[|x_i|]^{\frac{1}{2}}\Psi(y_{cr}(t,\lambda,x)) \times$$

$$\cos\left[\frac{1}{\hbar}S(t,\lambda,x,y_{cr}(t,\lambda,x))\right]. \tag{3.61}$$

Substitution Eq. (3.61) into Eq. (3.46) gives

$$R_1(T,\lambda) = \int dx\int dx\,[\breve{R}_1(x,T,\lambda)]^2 =$$

$$\int dx\left[\left|\det S_{y_{cr}y_{cr}}(t,\lambda,x,y_{cr}(t,\lambda,x))\right|\right]^{-1}|x_i|\Psi^2(y_{cr}(t,\lambda,x)) \times$$

$$\cos^2\left[\frac{1}{\hbar}S(t,\lambda,x,y_{cr}(t,\lambda,x))\right]. \tag{3.62}$$

Similarly one obtain

$$R_2(T,\lambda) = \int dx\int dx\,[\breve{R}_2(x,T,\lambda)]^2 =$$

$$\int dx\left[\left|\det S_{y_{cr}y_{cr}}(t,\lambda,x,y_{cr}(t,\lambda,x))\right|\right]^{-1}|x_i|\Psi^2(y_{cr}(t,\lambda,x)) \times$$

$$\sin^2\left[\frac{1}{\hbar}S(t,\lambda,x,y_{cr}(t,\lambda,x))\right]. \tag{3.63}$$

Therefore

$$R(T,\lambda) = R_1(T,\lambda) + R_2(T,\lambda) =$$

$$2\int dx\left[\left|\det S_{y_{cr}y_{cr}}(t,\lambda,x,y_{cr}(t,\lambda,x))\right|\right]^{-1}|x_i|\Psi^2(y_{cr}(t,\lambda,x)). \tag{3.64}$$

Substitution Eq. (3.1) into Eq. (3.64) gives

$$R_2(T,\lambda) = 2\frac{\eta^{d/2}}{(2\pi)^{d/2}\hbar^{d/2}}\int dx\left[\left|\det S_{y_{cr}y_{cr}}(t,\lambda,x,y_{cr}(t,\lambda,x))\right|\right]^{-1} \times$$

$$|x_i|\exp\left[-\frac{\eta(y_{cr}(t,\lambda,x)-x_0)^2}{\hbar}\right]. \tag{3.65}$$

Let us calculate now integral (3.65) using Laplace's approximation. It is easy to see that corresponding stationary point $\hat{x} = \hat{x}(t,\lambda,x_0) \in \mathbf{R}^d$ is the solution of the linear system of the algebraic equations (3.40). Therefore finally we obtain

$$R(T,\lambda) = 2|\hat{x}_i(t,\lambda,x_0)|\left[\left|\det S_{y_{cr}y_{cr}}\left(t,\lambda,\hat{x}(t,\lambda,x_0),y_{cr}(t,\lambda,\hat{x}(t,\lambda,x_0))\right)\right|\right]^{-1} + O(\hbar^d)\ (i=1,\cdots,d), \tag{3.66}$$

Substitution Eq. (3.64) into inequality (3.43) gives the inequality (3.41). The inequality (3.41) completed the proof.

3.4 Quantum anharmonic oscillator with a cubic potential supplemented by additive sinusoidal driving

In this subsection we calculate exact quasi-classical asymptotic for quantum anharmonic oscillator with a cubic potential supplemented by additive sinusoidal driving. Using Theorem 3.1 we obtain corresponding limiting quantum trajectories given via Eq. (1.3).

Let us consider quantum anharmonic oscillator with a cubic potential

$$V(x) = \frac{m\omega^2}{2}x^2 - ax^3 + bx, x \in \mathbf{R}; a,b > 0 \tag{3.67}$$

supplemented by an additive sinusoidal driving. Thus

$$V(x,t) = \frac{m\omega^2}{2}x^2 - ax^3 + bx - [A\sin(\Omega t)]x. \tag{3.68}$$

The corresponding master Lagrangian given by Eq. (3.37), are

$$L(\dot{u},u,\tau) = \left(\frac{m}{2}\right)\dot{u}^2 - m\left[\left(\frac{\omega^2}{2}\right)+\left(\frac{3a\lambda}{m}\right)\right]u^2 - [m\omega^2\lambda + 3a\lambda^2 - b - A\sin(\Omega t)]u. \tag{3.69}$$

We assume now that: $\frac{\omega^2}{2} + \frac{3a\lambda}{m} \geqslant 0$ and rewrite Eq. (3.69) of the form

$$L(\dot{u},u,\tau) = (m/2)\dot{u}^2 - (m\bar{\omega}^2\lambda/2)u^2 + g(\lambda,t)u, \tag{3.70}$$

where $\bar{\omega}(\lambda) = \sqrt{2\left|\frac{\omega^2}{2} + \frac{3a\lambda}{m}\right|}$ and $g(\lambda,t) = -[m\omega^2\lambda + 3a\lambda^2 - b - A\sin(\Omega t)]$.

The corresponding master action $S(t,\lambda,x,y)$ given by Eq. (3.36), are

$$S(t,\lambda,x,y) = \frac{m\omega}{2\sin\omega t}\left\{(\cos\omega t)(y^2+x^2) - 2xy + \frac{2x}{m\omega}\int_0^t g(\lambda,\tau)\sin(\omega\tau)\mathrm{d}\tau + \frac{2y}{m\omega}\int_0^t g(\lambda,\tau)\sin[\omega(t-\tau)]\mathrm{d}\tau - \frac{2}{m^2\omega^2}\int_0^t\int_0^\tau g(\lambda,\tau)g(\lambda,s)\sin\omega(t-\tau)\sin(\omega s)\mathrm{d}s\mathrm{d}\tau\right\}. \tag{3.71}$$

The linear system of the algebraic Eq. (3.39) is

$$\frac{\partial S(t,\lambda,x,y)}{\partial y} = 2y\cos\omega t - 2x + \frac{2}{m\omega}\int_0^t g(\lambda,t)\sin[\omega(t-\tau)]\mathrm{d}\tau = 0. \tag{3.72}$$

Therefore

$$y_{cr}(t,\lambda,x) = \frac{x}{\cos\omega t} - \frac{1}{m\omega\cos\omega t}\int_0^t g(\lambda,t)\sin[\omega(t-\tau)]\mathrm{d}\tau \tag{3.73}$$

The linear system of the algebraic Eq. (3.40) is

$$\frac{x}{\cos\omega t} - \frac{1}{m\omega\cos\omega t}\int_0^t g(\lambda,t)\sin[\omega(t-\tau)]\,d\tau + \lambda - x_0 = 0. \tag{3.74}$$

Therefore the solution of the linear system of the algebraic Eq. (3.40) is

$$\hat{x}(t,\lambda,x_0) = \frac{1}{m\omega}\int_0^t g(\lambda,t)\sin[\omega(t-\tau)]\,d\tau + [\lambda(t) - x_0]\cos\omega t. \tag{3.75}$$

Transcendental master Eq. (3.41) is

$$\frac{1}{m\omega}\int_0^t g(\lambda(t),t)\sin[\omega(t-\tau)]\,d\tau + [\lambda(t) - x_0]\cos\omega t = 0 \tag{3.76}$$

Finally from Eq. (3.76) one obtain

$$d(\lambda(t))\left[\frac{\cos(\omega t)}{\omega} - \frac{1}{\omega}\right] + \frac{A[\omega\sin(\Omega t) - \Omega\sin(\omega t)]}{\omega^2 - \Omega^2} - [\lambda(t) - x_0]m\omega\cos(\omega t) = 0, \tag{3.77}$$

where $d(\lambda) = m\omega^2\lambda + 3a\lambda^2 - b$.

Numerical Examples.

Example 1. $x = 0, m = 1, \Omega = 0, \omega = 9, a = 3, b = 10, A = 0.$

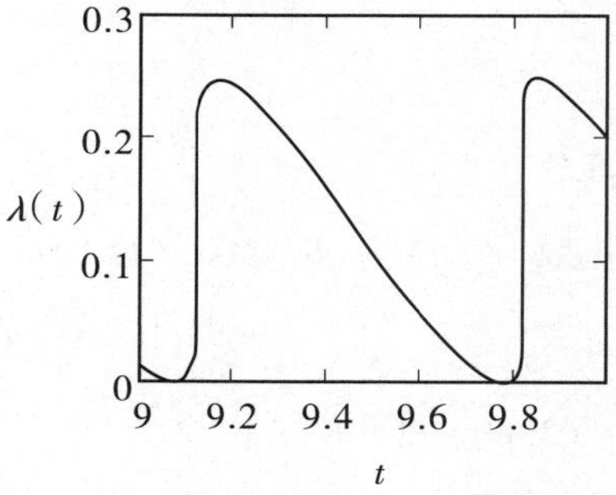

Fig. 3.1 Limiting quantum trajectory $\lambda(t)$ with a jump

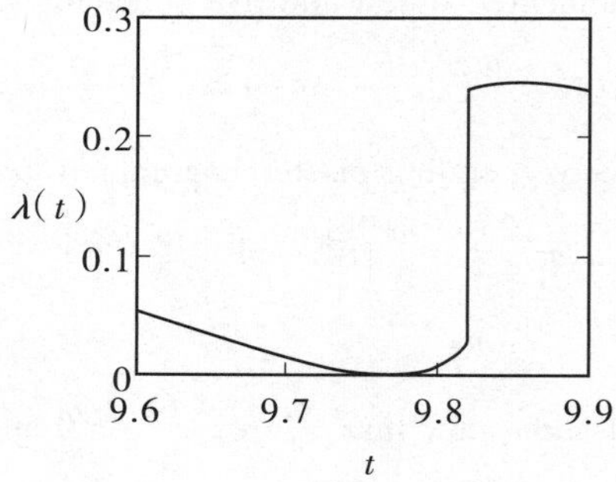

Fig. 3.2 Limiting quantum trajectory $\lambda(t)$ with a jump

3.5 Comparison exact quasi-classical asymptotic with stationary-point approximation

We set now $d = 1$. Let us consider now path integral (3.21) with $S_\varepsilon(\dot{q}(t), q(t); t)$ given via formula

$$\boldsymbol{S}_\varepsilon(\dot{q}(t), q(t), t) = \frac{1}{4}\int_0^t [\dot{q}(\tau) - F_\varepsilon(q(\tau),\tau)]^2\,d\tau. \tag{3.78}$$

Note that for corresponding propagator $K_\varepsilon(x,t \mid y,0)$ the time discretized path-integral representation $K_{\varepsilon,N}(x,t \mid y,0)$ is:

$$K_{\varepsilon,N}(x,t \mid y,0) = \int \frac{dx_1 dx_2 \cdots dx_{N-1}}{(4\pi\hbar\Delta t)^{N/2}}\exp\left[\frac{i}{\hbar}\boldsymbol{S}_{\varepsilon,N}(x_0,\cdots,x_N)\right], \tag{3.79}$$

where $S_{\varepsilon,N}(x_0,\cdots,x_N)$ are:

$$\boldsymbol{S}_{\varepsilon,N}(x_0,\cdots,x_N) = \frac{\Delta t}{4}\sum_{n=0}^{N-1}\left[\frac{x_{n+1}-x_n}{\Delta t} - F_\varepsilon(x_n,t_n)\right]^2. \tag{3.80}$$

Here the initial-x_0 and end-points x_N are fixed by the prescribed x_0 and by the additional constraint $x_N = y$.

Let us calculate now integral (3.79) using stationary-point approximation. Denoting an critical points of the discrete-time action (3.80) by $\boldsymbol{x}_{\varepsilon,k} = (x_{1,\varepsilon,k},\cdots,x_{N-1,\varepsilon,k})$, it follows that $\boldsymbol{x}_{\varepsilon,k}$ satisfies the critical point conditions

$$\frac{\partial S_{\varepsilon,N}(x_{0,\varepsilon,k},\cdots,x_{N-1,\varepsilon,k},x_{N,\varepsilon,k})}{\partial x_{n,\varepsilon,k}} = 0, n = 1,\cdots,N-1, \tag{3.81}$$

supplemented by the prescribed boundary conditions for $n = 0, n = N$: $x_{0,\varepsilon,k} = x_0$, $x_{N,\varepsilon,k} = x$.

From Eq. (3.79) in the limit $\hbar \to 0$ using formally stationary-point approximation one obtain

$$K_{\varepsilon,N}(x,t \mid x_0,0) \cong \boldsymbol{Z}_{\varepsilon,N-1}(\boldsymbol{x}_{\varepsilon,k},x_{0,\varepsilon,k},x_{N,\varepsilon,k})\exp\left[\frac{i}{\hbar}\boldsymbol{S}_{\varepsilon,N}(x_{\varepsilon,k},x_{0,\varepsilon,k},x_{N,\varepsilon,k})\right] + O(\hbar). \tag{3.82}$$

Here the pre-factor $\boldsymbol{Z}_{\varepsilon,N}(\boldsymbol{x}_{\varepsilon,k})$ is given via N-dimensional Gaussian integral of the canonical form as

$$\boldsymbol{Z}_{\varepsilon,N-1}(\boldsymbol{x}_{\varepsilon,k},x_{0,\varepsilon,k},x_{N,\varepsilon,k}) = \int\frac{\mathrm{d}y_1\mathrm{d}y_2\cdots\mathrm{d}y_{N-1}}{(4\pi\hbar\Delta t)^{N/2}}\exp\left[\frac{i}{2\hbar}\sum_{n,m=1}^{N-1} y_n\frac{\partial^2 S_{\varepsilon,N}(x_{\varepsilon,k},x_{0,\varepsilon,k},x_{N,\varepsilon,k})}{\partial x_{n,\varepsilon,k}\partial x_{m,\varepsilon,k}}y_m\right]. \tag{3.83}$$

The Gaussian integral in (3.83) is given via canonical formula

$$\boldsymbol{Z}_{\varepsilon,N-1}(\boldsymbol{x}_{\varepsilon,k},x_{0,\varepsilon,k},x_{N,\varepsilon,k}) = \left[4\pi\hbar\Delta t\det\left(2\Delta t\frac{\partial^2 S_{\varepsilon,N}(x_{\varepsilon,k},x_{0,\varepsilon,k},x_{N,\varepsilon,k})}{\partial x_{n,\varepsilon,k}\partial x_{m,\varepsilon,k}}\right)\right]^{-1/2}; n,m = 1,\cdots,N-1. \tag{3.84}$$

Here $\det(A_{n,m})$ denote the determinant of an $N-1 \times N-1$ matrix with elements $A_{n,m}$.

Let us consider now Cauchy problem (3.15) with initial data $\Psi_0(x)$ is given via formula

$$\Psi_0(x) = \frac{\eta^{1/4}}{(2\pi)^{1/4}\hbar^{1/4}}\exp\left[-\frac{\eta(x-z_0)^2}{2\hbar}\right]. \tag{3.85}$$

Note that for corresponding Colombeau solution $\Psi_\varepsilon(t,x)$ given via path-integral (3.21), the time discretized path-integral representation $\Psi_{\varepsilon,N}(t,x)$ are

$$\Psi_{\varepsilon,N}(t,x) = \int\Psi_0(x_0)K_{\varepsilon,N}(x,t \mid x_0,0)\mathrm{d}x_0 =$$

$$\sum_k\int\mathrm{d}x_0\boldsymbol{Z}_{\varepsilon,N-1}(\boldsymbol{x}_{\varepsilon,k},x_0,x_{N,\varepsilon,k})\Psi_0(x_0)\exp\left[\frac{i}{\hbar}\boldsymbol{S}_{\varepsilon,N}(\boldsymbol{x}_{\varepsilon,k},x_0,x_{N,\varepsilon,k})\right][1+O(\hbar)]. \tag{3.86}$$

Let us calculate now integrals in RHS of Eq. (3.86) using stationary-point approximation. Corresponding critical point conditions are

$$\frac{\partial S_{\varepsilon,N}(x_{0,\varepsilon,k},\cdots,x_{N-1,\varepsilon,k},x_{N,\varepsilon,k})}{\partial x_{0,\varepsilon,k}} = 0. \tag{3.87}$$

From Eq. (3.86) we obtain

$$\Psi_{\varepsilon,N}(t,x) = \sum_k \boldsymbol{Z}_{\varepsilon,N}(\boldsymbol{x}_{\varepsilon,k}, x_{0,\varepsilon,k}, x)\Psi_0(x_{0,\varepsilon,k})\exp\left[\frac{i}{\hbar}\boldsymbol{S}_{\varepsilon,N}(\boldsymbol{x}_{\varepsilon,k}, x_{0,\varepsilon,k}, x)\right][1 + O(\hbar)]. \tag{3.88}$$

$$\boldsymbol{Z}_{\varepsilon,N}(\boldsymbol{x}_{\varepsilon,k}, x_{0,\varepsilon,k}, x) = \left[\Delta t \det\left(2\Delta t \frac{\partial^2 S_{\varepsilon,N}(x_{\varepsilon,k}, x_{0,\varepsilon,k}, x)}{\partial x_{n,\varepsilon,k}\partial x_{m,\varepsilon,k}}\right)\right]^{-1/2} \quad (n,m = 0,1,\cdots,N-1). \tag{3.89}$$

Let us denote

$$\boldsymbol{x}_{\varepsilon,0} = (x_{0,\varepsilon,0}, x_{1,\varepsilon,0}, \cdots, x_{N-1,\varepsilon,0}) = (x_{0,\varepsilon,0}(x), x_{1,\varepsilon,0}(x), \cdots, x_{N-1,\varepsilon,0}(x)),$$

the critical point for which the critical point conditions (3.82) are

$$\frac{x_{n+1,\varepsilon,0} - x_{n,\varepsilon,0}}{\Delta t} - F_\varepsilon(x_{n,\varepsilon,0}, t_n) = 0 \, (n = 0,1,\cdots,N-1). \tag{3.90}$$

Therefore the time discretized path-integral representation of the Colombeau quantum averages given by Eq. (3.1) are

$$(\langle 1,t,z_0;\hbar,\varepsilon\rangle)_\varepsilon = \left(\int x\,|\Psi_{\varepsilon,N}(x,t;\hbar)|^2 \mathrm{d}x\right)_\varepsilon =$$

$$\frac{\eta^{1/2}}{(2\pi)^{1/2}\hbar^{1/2}}\left(\int \mathrm{d}x x \boldsymbol{Z}^2_{\varepsilon,N}(\boldsymbol{x}_{\varepsilon,k}, x_{0,\varepsilon,0}, x)\exp\left[-\frac{\eta(x_{0,\varepsilon,0}(x) - z_0)^2}{\hbar}\right]\right)_\varepsilon [1 + O(\hbar^2)] +$$

$$\frac{\eta^{\frac{1}{2}}}{(2\pi)^{\frac{1}{2}}\hbar^{\frac{1}{2}}}\left(\sum_{k\geqslant 1}\int \mathrm{d}x x \boldsymbol{Z}^2_{\varepsilon,N}(\boldsymbol{x}_{\varepsilon,k}, x_{0,\varepsilon,0}, x)\exp\left[-\frac{\eta(x_{0,\varepsilon,0}(x) - z_0)^2}{\hbar}\right]\right)_\varepsilon [1 + O(\hbar^2)] +$$

$$O[\exp(-c\hbar^{-1})], \tag{3.91}$$

where $c > 0, \varepsilon \in (0,1], x \in \mathbf{R}$. Let us calculate now integrals in RHS of Eq. (3.91) using stationary-point approximation. Corresponding critical point conditions are

$$x_{0,\varepsilon,0}(x_{N,\varepsilon,0}) - z_0 = 0, \tag{3.92}$$

$$x_{0,\varepsilon,k}(x_{N,\varepsilon,k}) - z_0 = 0, k \geqslant 1. \tag{3.93}$$

Here $x_{N,\varepsilon,0}$ can be calculated using linear recursion (3.90) with initial data $x_{0,\varepsilon,0} = z_0$. From Eq. (3.91) – Eq. (3.92) one obtain

$$(\langle 1,t,z_0;\hbar,\varepsilon\rangle)_\varepsilon \cong (x_{N,\varepsilon,0}\boldsymbol{Z}^2_{\varepsilon,N}(\boldsymbol{x}_{\varepsilon,k}, x_{0,\varepsilon,0}, x_{N,\varepsilon,0}))_\varepsilon \tag{3.94}$$

and

$$\boldsymbol{Z}_{\varepsilon,N}(\boldsymbol{x}_{\varepsilon,k}, x_{0,\varepsilon,k}, x_{N,\varepsilon,0}) = \left[\Delta t \det\left(2\Delta t \frac{\partial^2 S_{\varepsilon,N}(\boldsymbol{x}_{\varepsilon,k}, x_{0,\varepsilon,k}, x_{N,\varepsilon,0})}{\partial x_{n,\varepsilon,k}\partial x_{m,\varepsilon,k}}\right)\right]^{-1/2} \quad (n,m = 0,1,\cdots,N). \tag{3.95}$$

As demonstrated in Ref. [26] the determinant appearing in Eq. (3.89) can be calculated using second order linear recursion:

$$\frac{Q_{n+1,\varepsilon,k} - 2Q_{n,\varepsilon,k} - Q_{n-1,\varepsilon,k}}{(\Delta t)^2} = 2\frac{Q_{n,\varepsilon,k}F'_x(x_{n,\varepsilon,k}, t_n) - Q_{n-1,\varepsilon,k}F'_x(x_{n-1,\varepsilon,k}, t_{n-1})}{\Delta t} -$$

$$Q_{n,\varepsilon,k}\left[\frac{x_{n+1,\varepsilon,k} - x_{n,\varepsilon,k}}{\Delta t} - F_\varepsilon(x_{n,\varepsilon,k}, t_n)\right]F''_{\varepsilon,x^2}(x_{n,\varepsilon,k}, t_n) +$$

$$Q_{n,\varepsilon,k}\left[F'_{\varepsilon,x}(x_{n,\varepsilon,k},t_n)\right]^2 - Q_{n-1,\varepsilon,k}\left[F'_{\varepsilon,x}(x_{n-1,\varepsilon,k},t_{n-1})\right]^2,$$

$$F'_{\varepsilon,x}(x,t) = \frac{\partial F_\varepsilon(x,t)}{\partial x}, F''_{\varepsilon,x}(x,t) = \frac{\partial^2 F_\varepsilon(x,t)}{\partial x^2}, \tag{3.94}$$

with initial data

$$Q_{1,\varepsilon,k} = \Delta t, Q_{2,\varepsilon,k} = Q_{1,\varepsilon,k} + \Delta t + O((\Delta t)^2) \tag{3.97}$$

from which the pre-factor $\mathbf{Z}_{\varepsilon,N}(\boldsymbol{x}_{\varepsilon,k}, x_{0,\varepsilon,k}, x_{N,\varepsilon,0})$ in Eq. (3.94) follows as

$$\mathbf{Z}_{\varepsilon,N}(\boldsymbol{x}_{\varepsilon,k}, x_{0,\varepsilon,k}, x_{N,\varepsilon,0}) = \sqrt{Q_{N,\varepsilon,k}}. \tag{3.98}$$

In the limit $\Delta t \to 0$ from critical point conditions (3.90) and (3.92) one obtain

$$\dot{x}(t) - F_\varepsilon(x(t),t) = 0, x(0) = z_0. \tag{3.99}$$

In the limit $\Delta t \to 0$ from a second order linear recursion Eq. (3.96) one obtain the second order linear differential equation

$$\ddot{Q}_{\varepsilon,0}(t) = 2\frac{\mathrm{d}}{\mathrm{d}t}[Q_{\varepsilon,0}(t)F'_{\varepsilon,x}(x(t),t)] \tag{3.100}$$

with initial data

$$Q_{\varepsilon,0}(0) = 0, \dot{Q}_{\varepsilon,0}(0) = 1. \tag{3.101}$$

By integration Eq. (3.100) one obtain the first order linear differential equation

$$\dot{Q}_{\varepsilon,0}(t) = 2Q_{\varepsilon,0}(t)F'_{\varepsilon,x}(x(t),t) + 1, Q_{\varepsilon,0}(0) = 0. \tag{3.102}$$

In the limit $\Delta t \to 0$ from Eq. (3.94), Eq. (3.98) – Eq. (3.99) and Eq. (3.102) one obtain

$$E(t) = (\langle 1, t, z_0; \hbar, \varepsilon\rangle)_\varepsilon \cong x(t)Q_{\varepsilon,0}^{-2}(t). \tag{3.103}$$

We set now in Eq. (3.78)

$$F_\varepsilon(q,\tau) = -aq^3 + bx + A\sin(\Omega\tau). \tag{3.104}$$

Corresponding differential master equation are

$$\dot{q} = -(3a\lambda^2 - b)q - (a\lambda^3 - b\lambda) + A\sin(\Omega\tau), q(0) = z_0 - \lambda. \tag{3.105}$$

From Eq. (3.105) one obtain that corresponding transcendental master equation are

$$[z_0 - \lambda(t)] - \frac{\Delta[\lambda(t)]}{\Theta[\lambda(t)]}(\exp\{t\Delta[\lambda(t)]\} - 1) + A\frac{\exp(t\Delta[\lambda(t)])}{\Omega^2 + \Theta^2[\lambda(t)]}\{\Theta[\lambda(t)]\sin(\Omega t)$$

$$-\Omega\cos(\Omega t)\} + A\frac{\Omega}{\Omega^2 + \Theta^2[\lambda(t)]} = 0. \tag{3.106}$$

Numerical Examples

Comparison of the: (1) classical dynamics calculated by using Eq. (3.78), (2) limiting quantum trajectory $\lambda(t)$ calculated by using master equation Eq. (3.106) and (3) limiting quantum trajectory calculated by using stationary-point approximation given by Eq. (3.103) (Fig 3.3 and Fig 3.4).

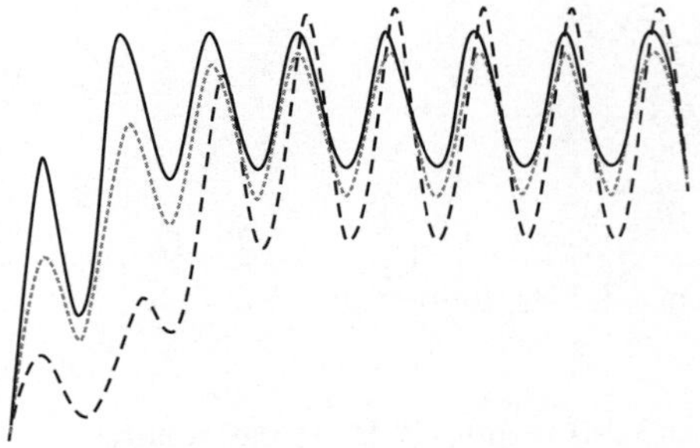

Fig. 3. 3 Limiting quantum trajectory $\lambda(t)$ without jumps: $a = 0.3, b = 1, A = 2$

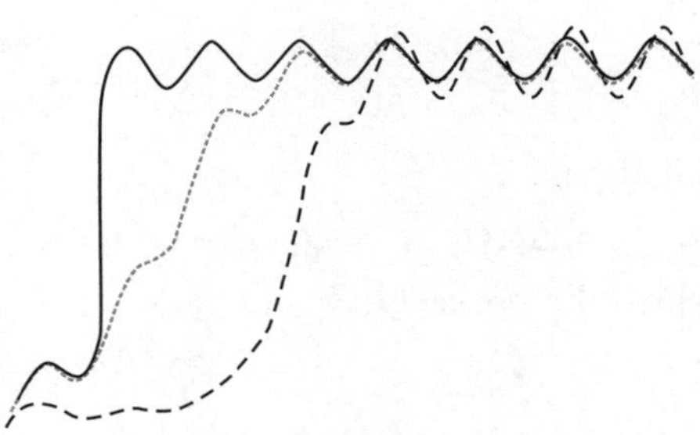

Fig. 3. 4 Limiting quantum trajectory $\lambda(t)$ with a jump: $a = 1, b = 1, A = 0.3$

3. 6 Conclusions

We pointed out that there exist limiting quantum trajectories given via Eq. (3. 3) with a jumps. Such jumps do not depend on any single measurement of particle position $q(t)$ at instant t and obtained without any reference to a phenomenological master-equation of Lindblad form.

An axiom of quantum mechanics is that we cannot predict the result of any single measurement of an observable of a quantum mechanical system in a superposition of eigenstates. However we can predict the result of any single measurement of particle position $q(t)$ at instant t with a probability $P(t,q,\delta q) \cong 1$ if valid the condition: $\lim_{\varepsilon \to 0} \lim_{\|\delta q(t)\| \to 0} \lim_{\hbar \to 0} P_t(q(t), \delta q(t), x_0; \hbar, \varepsilon) = 1$, where P_t given via Eq. (3. 5).

Chapter 4 Creation and Development of Fractal-Scaling or Scale-Invariant Radiolocation for Detection and Recognition of Low-Contrast Targets Against the Background of High Intensity Earth and Sea Disturbances

4.1 Introduction

A richness of content and the role of classical statistical radio physics, radio engineering and radiolocation permits simultaneously enormous possibilities to develop alternative techniques. This chapter presents alternative solutions of actual problems of modern radiolocation which are based on the ideas and methods of new scientific fundamental direction Fractal radio physics and fractal radio electronics: Designing of fractal radio systems. This direction was initiated by Prof. A. Potapov since 1980 at the IRE RAS and currently it is widely developed in his works and acknowledged in the scientific world.

Intensive development of modern radar technology establishes new demands to the radiolocation theory. Some of these demands do not touch the theory basis and reduce to the precision increase, improvement and development of new calculation methods. Others are fundamental and related to the basis of the radiolocation theory. The last demands are the most important both in theory and in practice.

The purpose of this chapter is both determination and offering a new radiolocation method based on the fractal-scaling or scale-invariant principles, and to attract attention to classical problems of general statistic theory from the viewpoint of the modern fractal analysis, fractional calculus and global fractal-scaling method created and developed by the author.

As compared with usual detection methods, the fractal-scaling or scale-invariant methods proposed by Prof. A. Potapov, can effectively improve the signal/interference relation and considerably increase the probability of target detection. Methods under consideration are suitable both for usual radars and for synthetic aperture radar (SAR), and also for MIMO (multiple-inputmultiple-output) systems for multipositioning radiolocation.

4.2 Necessity of new solution methods of modern radio physical and radar problems

It is well known that the soviet "non-linear" physical, mathematical and mechanical school in XX century in the field of the oscillation theory and radio physics is high on the list of the world science. The ideas and solutions of the author (Prof. A. Potapov) about the necessity of the transition to his "fractal thinking" underlie in the fundamental of this work. Paraphrasing S. M. Ritov's words about the "non-linear thinking" form his interview on 19th of February, 1991[1]246. This means to find mathematical tools, learn to think with conceptions of this mathematical theory, adopt these conceptions and get used to them. Then a new intuition will appear.

The mathematical apparatus already exists. This is the fractional calculus, the fractional measure and the fractal theory[2-6]. The fractional mathematical analysis has a long history and extraordinarily rich content.

All modern radio engineering is based on the classical theory of the integer-valued measure and the integer-valued calculus. Thus, a large area of mathematical analysis called the fractional calculus dealing with the derivatives and integrals of the random (real or complex) order and the fractal theory were counted out (!).

Currently in radio physics, radio electronics and multidimentional signal processing integer-valued measures (integrals and derivatives of entire order), Gaussian statistics and Markovian processes are used[7,8]. I note that the Markovian process theory in applications reached satiation and investigations are carried out at the level of the sharp complication of synthesized algorithms. Improvement of classical radar signal detectors and their software reach saturation. As the experience has shown in barely noticeable and low-contrast target detection (small q_0^2 signal/background relation) at the terrestrial cover background, use of traditional classical filtration algorithms is not always possible, because they demand a long-term accumulation of radar echo. To solve these problems successfully under present-day conditions, the search and development on principle of different approaches to traditional radar problems and radio electronics are needed. This enforces to search out on principal of new methods of problem solution.

4.3 Textural measures in radio physics and radiolocation

At present an urgent necessity to consider modern radar systems in the aggregate with the high-frequency propagation channel and probing objects from the viewpoint of the theory of complex non-equilibrium systems opened to the power, information and entropy flows (i. e., from synergetic attitudes) exist. That determined the development of new information technologies in radio physics and radiolocation at IRE RAS using the textural and fractal measures on the basis of synergetic principles of nonlinear dynamics[9,10]. As a result of these investigations world the concepts of textural and fractal signatures were first introduced by Prof. A. Potapov[3,4,9,10].

As opposed to the tone and color which are related to separate image patches, the texture is connected with more than one fragment. The author formulated a definition[9,10], that the texture is the matrix or the fragment of spatial properties of the image subsections with homogeneous statistical performances. These textural features are based on the statistical characteristics of intensity levels of image elements and they are related to the probabilistic indication. Random values of which are dispersed over all classes of nature objects. One can make a decision about the texture belonging to one or another class only on the basis of specific indication values of the texture under consideration. In that case one says in the accepted phrase about the texture signature. Classical radiolocation signatures include the time, spectral and polarization features of returned signal. From the author's standpoint, the texture signature is the universe distribution of measurements for the texture under consideration in such type of scenes like a given one.

When texture decomposition is applicable, two main factors are revealed. The first one correlates the texture with non-derivative elements from which the image is consisted of, and the second one describes the spatial dependence in between. Tone non-derivative elements represent the image areas characterized with defined brightness values proportional to the returned signal intensity. In turn it depends on values of the specific radar cross section (RCS) σ_* of earth's surface. Since the conception of specific RCS is significant only for a spatially uniform object therefore the image texture of a real terrestrial cover is determined by σ_* space variations.

All stated above permits to estimate interconnections between the conceptions of specific RCS of the underlying cover and its texture. When the little subsection of the image is characterized by small changing of type non-derivative elements then the value of specific RCS is the predominant property of this subsection. When brightness of these elements changes noticeably, this predominant characteristic is inserted in the texture. In other words when the number of distinguishable type non-derivative elements in the image reduces, the role of energy characters in particular σ_* increases. Actually for one resolution cell the only energy characters exist. If the number of distinguishable type non-derivative elements increases, then textural features will prevail.

In statistical texture analysis one uses first-order or second-order statistics. When second-order statistics are used directly, textural features are taken by means of matrix of space dependence probability distribution of tonal gradation **P** called gradient distribution matrix. This method is proposed in Paper [11]. It is shown experimentally[9,10,12-14], that features based on correlation functions do not well estimate the image texture as ones determined in accordance with the gradient distribution matrix **P**.

The first calculation of full ensemble consisting of 28 textural features and their detailed synchronic analysis for real [optical and radar at millimetric wave (MMW) bands at wave 8.6 mm] and synthesized textures on the basis of autoregressive models was carried out by Prof. A. Potapov at IRE RAS at the beginning of 1980s[9,10]. Long-term natural experiments were carried out by Prof. A. Potapov and Central Design Department "Almaz" and other USSR lead industrial agencies. When extraction of millimetric waves (MMW) signal scattered by different terrains, Prof. A. Potapov carried out first experiments for selection of lots of frequency and temporal scaling in 1985. Determined fractal properties of the accepted selection suppose the presence of these lots. The problem of calculation of textural features taking into account the signature drift when seasons are changed was stated and solved. Estimations of window extent influence on determination accuracy of textural features to image different types of terrains were optimized. Donkey's years these works at the field of research of terrains radar images at MMW using textural information were the only and in accordance with the specialist opinions these ones are urgent until now[12,14].

Analysis of obtained data permitted to prove very important peculiarity[9,10]: one-dimensional regions of radar image textural features existence at MMW band (assembly R) are almost completely inlayed into corresponding regions of aerial photography features (assembly A): $R \in A$. Thus, regions of radar image features as if shrink as compared with regions of aerial photography features. This is happening because of the smoothing in radar image the fine structure of terrains under investigation characteristic for aerial photography. Consequently with considerable credibility value one can forecast the assembly R on assembly A.

Compactness of existence domains of textural features for radar image textures enables to propose that in accordance with radar images the terrain classification and target detection sometimes is carried out more precise. Complexification of optical and radio engineering systems and detectors[9,10] complement their basic qualities, increases the probability of low-contrast object correct detection and increases general system self-descriptiveness. Invariance with respect to the scale and turn is reached by means of the selection of specific sampling increment when texture digitization (usually it is of the order of autocorrelation interval) and averaging operation of feature values on four scanning directions when electronic data processing.

Round the detected target background reflection plots are always present integrated by general texture conception. It enables to propose new approaches to detect extended low-contrast target against the terrain background in obtained radar images or multidimentional signals. On the basis of results obtained at the IRE RAS Prof. A. Potapov first proposed and implemented following original and effective signal detection methods at small q_o^2 signal-background relations of the order of unity or smaller than unity: dispersion method, detection method by means of linear simulated standards and the method with direct application of texture feature ensemble[9,10,13-16]. Moreover the important quality of textural processing methods is the possibility of the speckle neutralization on the coherent Earth surface-mapping obtained by means of synthetic aperture radar (SAR). Complete description of textural measures at MMW and in optics are presented in Work [17, 18].

4.4 Non-Gaussian statistics, fractal-scaling methods, invariants and fractional measurers in radio physics and radiolocation

It is well known that at the early and more latter stages of experimental works on dispersion of meter, decimeter, centimeter and millimeter waves investigators faced with the Gaussian model applicability problems. Soon numerous artificial attempts of creation of dispersion models to increase probabilistic distribution tail area of reflected signals began.

Long and fruitful participation of Prof. A. Potapov in multiscale experimental works (1979 – 1990) jointly with lead USSR agencies resulted in the necessity of principle refusal from Gaussian statistics in case of sufficiently high radar resolution. All investigations were carried out at the wave length $\lambda = 2.2$ mm and 8.6 mm (active radiation) and $\lambda = 3.5$ mm (passive radiation)[9,10]. It is in physical nature experiments the author showed in 1979 – 1980 that the Gaussian statistics are inapplicable to almost over all angle of sight range θ of electromagnetic wave[9,10,12,14,17-19].

Experimental investigations in 1979 – 1990 permitted both the solution of traditional scattering problems and to determine general law of modulated signal fine structure patterning over the MMW range and to propose of principle new feature class based on the fine structure of modulated signals scattered by statistically rough surface (see Fig. 1.18, Chapter 1)[9,10,20].

Fluctuations inside pulse, their statistics, correlated and spectrum and average pulse (– width) broadening determined by the value reciprocal coherence band Δf_k [9,10,21] were related to the fine structure characteristics of radar echo. These features in the aggregate with dependences of terrain scattering cross sections at MMW present determined type of radiolocation signatures.

Analysis of experimentally obtained extensive database (more than 30 terrain categories[9,10]) in the aggregate with visual study of complexity factor of scattered radiation isolines results in ideas of ensemble introduction of principal new features based on the scaling indices and fractional dimension characteristics, i. e., introduction of fractal signatures in practice.

Fractal (Hausdorff) dimension D or its signature at different surface-mapping parts are simultaneously and texture measure[3,4,9,10,17], i. e., properties of spatial correlation of radio scattering by corresponding surface patches.

The important role of fractal stable distributions and their value for practical radiolocation problems was shown. Probabilistic distribution tail area thickening of instantaneous fractal dimensions of analyzable signal occurs because of power law influence. Experimental data determine the characteristic species of fractal distribution (paretian) and role of the heavy (or reinforced) tails which text all useful information at ultra-low signal/disturbance relationships (Fig. 4.1). Using this information permits Low-Contrast Target Detection at the background of terrestrial and sea disturbances.

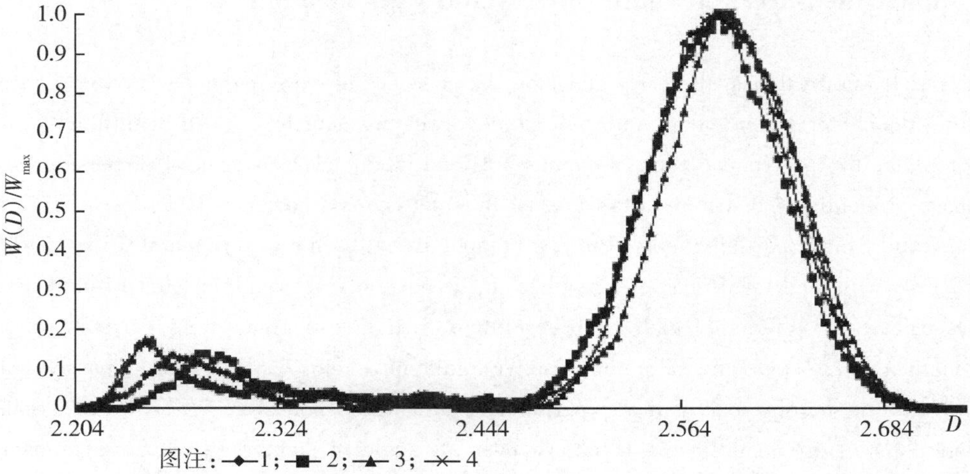

Fig. 4. 1 Empirical fractal distributions with heavy tails for images observed in the presence of an intense Gaussian noise: (1 and 3) scene A, (2 and 4) scene B, (1 and 2) q_0^2 = − 10 dB, and (3 and 4) q_0^2 = − 20 dB

Textural and fractal methods, developed by Prof. A. Potapov (Fig. 4. 2) allow partially overcome prior uncertainty in radiolocation problems by means of the geometry and sampling topology—one-dimensional or multidimentional[3,4,22,23]. Topological sampling singularities but not averaged implementations have exclusive value. One can realize target detection on the basis of features such as full fractal signature, its part or slope.

In fractal approach one focuses our attention on description and radio physical signal and fields processing exclusively at the space of fractional measure using the hypothesis of physical scaling and distributions with heavy tails or stable distributions[3,4,14,22-31]. Fractal methods can function at all signal levels: amplitude, frequency, phase and polarized. There is no such thing in literature before the author's investigations and works.

Results of numerous experiments show that the fractal-scaling methods proposed by Prof. A. Potapov are effective for the object filtration at images in optical and radio frequency band at negative signal/background relationship $q_0^2 \approx -10, \cdots, -15$ dB. Gaussian fluctuations of model right-angled object areas with roof-mean-square deviation (standard deviation) of the order of 30% practically do not influence on their detection quality.

Analysis of quantity influence of objects with different topology on image characteristics was carried out by means of two methods: classical (running window) and locally dispersive. The fractal dimension invariance of complex images relatively object orientation was proved. The functional relationship between fractal dimension value of complex image and quantity and size of artificial objects was revealed.

The fractal signature characterizes the spatial fractal cepstrum of image. It was proved that

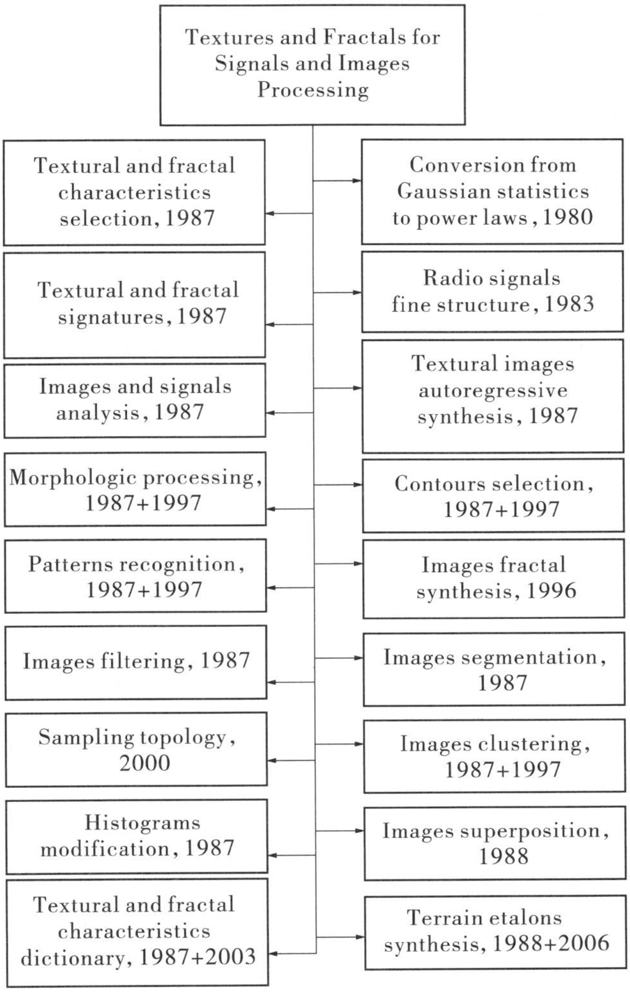

Fig. 4. 2 Textural and fractal processing methods of low-contrast images and poop signals in non-Gaussian interferences

the fractal cepstrum is on the one hand a convenient topological invariant (preliminary orientation/scaling is not needed) and on the other hand it is insensitive to an image contrast. Characteristic point locations at fractal cepstrum permit to determine the target class (in accordance with some rule), its sizes and target quantity. Relative change of characteristic point location enables to solve the problem of deterministic target detection even at very low contrast.

4. 5 First fractal detectors of weak radar signals

Creation of first standard dictionary of target classes fractal features and improvement of knoware are milestones in development and breadboarding of the first fractal non-parametric detector of radar signals (FNDRS)[9,18,21-50]. We point out that both in Russian and abroad books of fractal detectors using feature space fractional metric have never been considered except Prof. A. Potapov's publications. Fractal pattern recognition algorithms are based on the use of the paradigm "target topology is its fractal dimension".

Fractal signatures including spectra of fractal dimensions and fractal cepstrum represent the attribute vector uniquely determining wide class of targets and objects than the use of fractal dimension values. Thus, we can specify the propose structure of the fractal radiolocation detector of target classes consisting of edge detector and fractal signature calculator. Obtained signatures are compared with the signature database and the decision concerning the presence or absence of the object is made in accordance with some criterion. Some original variants of generalized structures of radar fractal detectors are presented in Fig. 4. 3. One can synthesize all kinds of fractal detectors from these schemes.

The structure aggregated scheme of the fractal detector of radar signals is presented in Fig. 4. 3a. It includes the contour filter and the fractal cepstrum calculator. After comparison with the database of standard fractal cepstrum one makes a decision at the compare facility. Further concretization of the FNDRS structure scheme is presented in Fig. 4. 3b. Input signal (radar image, 1-D sampling) comes into the input transducer. It is intended for preliminary preparation of analyzed sampling. This preparation includes either compulsory noise (in the case when radar low-resolution analog-digital converter is used) or, for example, contrast compression (in the case of sampling with high dynamic range).

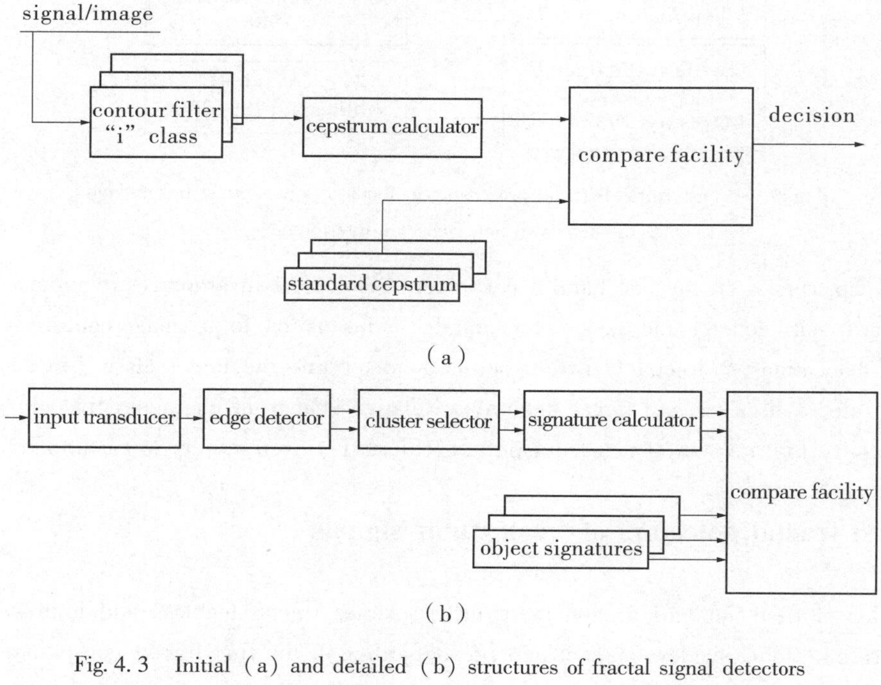

Fig. 4. 3 Initial (a) and detailed (b) structures of fractal signal detectors

Prepared input sampling comes into the edge detector. Operation of this facility is based on the measurement of the local fractal dimension over all sampling elements. It is important that one can execute the preliminary detection in accordance with the empiric distribution of local fractal dimensions obtained at the edge detector output over all sampling elements (Fig. 4.1)[9,18,22-24,32-34].

After the edge detector the input sampling actually represents a binary array. "Units" in this array mean the belonging of corresponding sampling element to the contour of some object. If several objects are present in the sampling, the question about their division arises. One can solve the problem both on the basis of the fractal theory—repeatedly dividing the sampling and by known algorithms. The case under consideration is when these contours are not connected. In our facility the cluster selector realizes this task.

Obtained subset of initial sampling containing the contours of one object comes to the input of the signature calculator. This facility creates several "smoothed" samplings in accordance with the expected observation scales and calculates the "length" and "area". At the facility output we have a function $\lg(S) = f(\lg\delta)$, which is the necessary fractal signature of analyzed sampling. The detection process is realized by means of the comparison of the fractal sampling signature with the database fractal signatures (or data bank). At positive comparison result (threshold crossing) the decision about the object presence in the sampling is made. Permanent improvement of the knoware of digital fractal processing methods is one of the basic factors of development of different scheme variants of fractal detectors.

On the basis of investigations carried out the generalized structure scheme of the first operating radar fractal detector was developed (Fig. 4.4). We used the following notations: RFA is the radio frequency amplifier, M is the mixer, G is the heterodyne, BPF is the band-pass filter, IFA is the intermediate frequency amplifier, QD is the quadrature detector (spectrum transfer into zero frequency), Re and Im are the real and imaginary quadrature, ADC is the analog-digital converter, CP is the central processor, RAM is the random-access memory, I/O is the input/output device (monitor, keyboard, mouse, printer, network adapter), HDD is the hard-disk drive, OS is the operating system, SW is the software.

In FNDRS HDD the following software (SW) is kept: (i) "SW-1" is the calculation of fractal dimension instantaneous values D; (ii) "SW-2" is the calculation of full signature $\Lambda(\delta)$ of treated data array; (iii) "SW-3" is the calculation in real time of observed data in different time scales, decimation, interpolation (crowding); (iv) "SW-4" is the data selection in accordance with their fractional measure; (v) "SW-5" is the wanted signal extraction from the input mixture by means of multiresolution fractal analysis. One can expand FNDRS SW in accordance with current tasks. Detailed FNDRS operation description, results of fractal real radio signal processing and detection performances are considered in more detail in Work [3, 4, 9, 18, 22-24, 32-34].

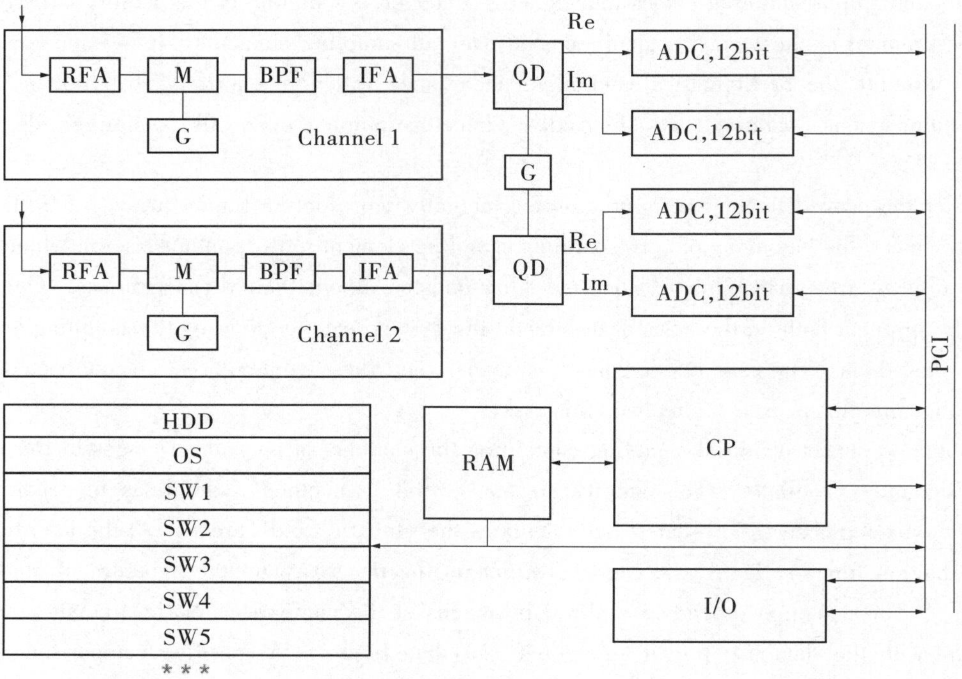

Fig. 4. 4 Block diagram of FNDRS model

Extraordinary motivating factor is that in spite of very small prior information content applied in construction of non-parametric fractal detectors they possess a high efficiency, notably: (i) loss of efficiency when transition from parametric to non-parametric procedures often make up several percent; (ii) efficiency of non-parametric procedure as compared with fixed parametric one race up in deflection of true distributions from calculating distributions; (iii) in many cases non-parametric procedures are asymptotically optimal.

The main source of these properties is that at non-parametric approach only reliable accessible a priori information is used. Inadequate information, for example, the assumption about the presence of Gaussian process, in a manner is equivalent to the introduction of accessory chance disturbing factors.

Thus, application of multiscale fractal analysis permits to synthesize sufficiently simple but effective algorithms of coherent and incoherent detection of radar signals. These algorithms successfully compete with the classical detection algorithms. Realized software permits to carry out the calculation of various fractal indicators in real time. One can easily thread these created algorithms to raise the speed capability.

4. 6 Creation of breakthrough fractal scaling technologies and fractal radio systems

A fundamental difference of the fractal-scaling methods proposed by Prof. A. Potapov from classical ones in connected with the principle of different (fractional) approach to the main constituents of physical signal. This permitted to move on new level of data of real non-Markovian signals and fields. Thus, this is the brand new radio engineering and radiolocation.

In the case of the fractal approach one needs to find, realize and use the rules that the fractional (complex) topology of considered images is obeyed. Algorithms of the fractal image recognition are based on the use of the paradigm "the target topology is its fractal dimension"[3,4,23].

Over period of thirty five years of scientific investigations the global fractal-scaling technique created by Prof. A. Potapov fully proved its value. It found the numerous appendices (Fig. 4.5). This is a kind of time call. That is briefly and expressively designated as the fractal paradigm[3,4,10,14,17,21 - 75].

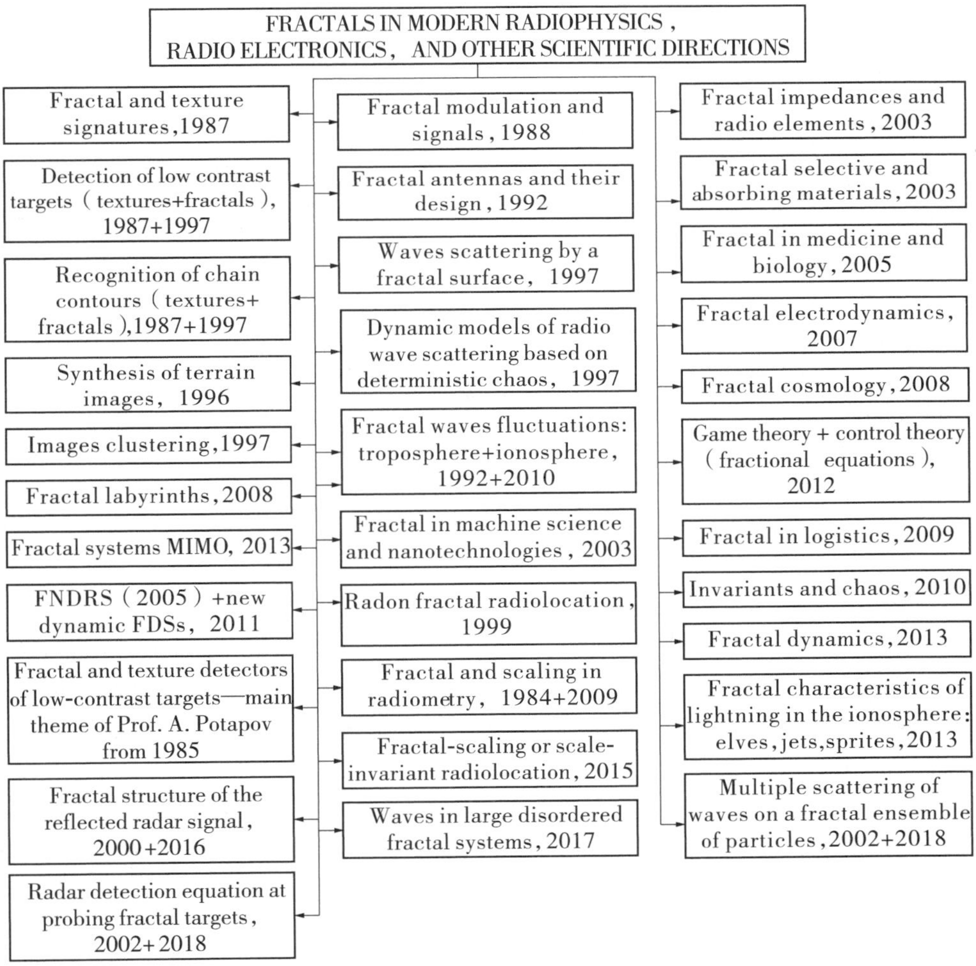

Fig. 4.5 Sketch of Prof. A. Potapov's development of new information technologies based on fractals, fractional operators and scaling effects

The fractal geometry is the colossal Benois Mandelbrot of genius desert. But its radio physical/radio engineering and practical application is the desert of the Russian (now International) scientific school of fractal techniques and fractional operators under the leadership of Prof. A. Potapov (see his site www. potapov-fractal. com). A few exaggerating one can say that the fractals made up the thin amalgam on the strong science frame of the XX century ending. In the modern situation it attempts to humiliate their significance and rely only on the classical knowledge suffered an intellectual fiasco.

In the fractal investigations the author always bases on three global theses:

(1) Processing of information distorted by non-Gaussian noise in a fractional measure space with using of scaling and stable non-Gaussian probabilistic distributions (1981) (Fig. 4. 2 and Fig. 4. 5).

(2) Application of continuous nondifferentiable functions.

(3) Fractal radio systems (2005) (Fig. 4. 5 and Fig. 4. 6).

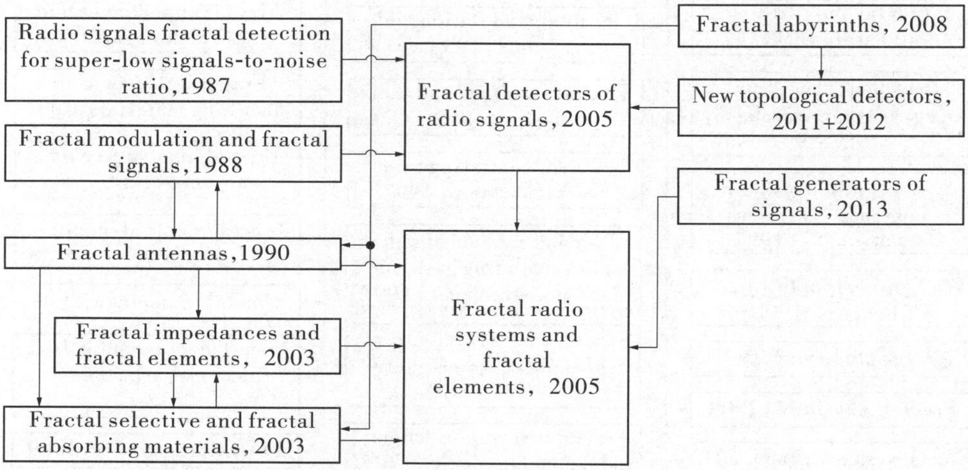

Fig. 4. 6 Prof. A. Potapov's conception of fractal radio systems, devices and radioelements

Logical integration of problem triad above mentioned into the general "fractal analysis and synthesis" creates the basis of the fractal-scaling technique (2006) and single global idea of the fractal natural science and the fractal paradigm (2011) proposed and intensively developed at present by Prof. A. Potapov.

Let us describe the conception of the fractal radar set on the basis of the material mentioned above and the application problems of its scale-invariant principles in others modern radio monitoring systems.

4. 7 About fractal-scaling or scale-invariant radiolocation and its modern applications

At present fractal radar works are widely carried out at IRE RAS. In accordance with perspective radar set demands we will consider the generalized functional diagram of classical system (Fig. 4.7). On the one hand it is fairly simple and on the other hand it contains all of principle necessary elements. Also one can speak about both single-channel radar station (RS) and about multichannel one.

Synchronizer block provides the operation coordination of all elements of RS scheme. Generation and radiation of electromagnetic energy is realized by means of transmitter consisting of modulator, radio frequency (RF) generator and transmitting antenna. Reflected signals come to the receiving antenna.

Receiving device executes all necessary input signal transforms connected with their division, amplification and noise separation. Output device is intended to execute final operations of signal processing and their transformation into the required form. Output device depending on measurement results can effect on the receiver (B line), antenna (C line) and transmitter (D and E lines), that permits automatic variation of radiated oscillation mode, reception and signal processing approaching them to the best. Connection of output device with receiving antenna provides the possibility of automatic measurement of angular coordinates and antenna control when target location. In turn, the data about the antenna angular position (F line) are entered into the output device from the antenna assembly.

From Fig. 4.7 one can directly pass to the fractal radar. Practically all applications of hypothetic or currently projectible fractal algorithms, elements, units and processes that one can introduce into the diagram in Fig. 4.7 are presented in Fig. 4.8. The ideology of the fractal radar is based on the fractal radio system concept (Fig. 4.6).

The ideology of the transition to the fractal-scaling detectors is also based on the fractal radio system concept (Fig. 4.6). The main types of selected diagram clusters of new dynamic fractal signal detector (FSD) proposed by Prof. A. Potapov in 2011 – 2016 are presented in Fig. 4.9.

In the author's investigations for the first time the functional postulate "topology maximum at energy minimum" permitting more effectively application of advantages of the fractal-scaling information processing is introduced. Application of the fractal principle results in the soul-searching in the detection field of movable and immovable objects at the intensive disturbance and noise background[3,4,9,10,14,17,21 – 75].

In particular, for the versions of fractal MIMO-systems earlier proposed by the author the multifrequency operation is used, as the fractal antennas permit simultaneously radiation a few wave lengths. For the pilotless vehicle (unmanned aerial vehicle, UAV) the creation of diminutive fractal radar set (radio system) with fractal elements and/or parametrons is also possible (Fig. 4.6 and Fig. 4.8). Simultaneously fractal processing of transmitted information from UAV on the board or the control center will permit considerable improvement and automation of detection processes, clustering and target and object recognition. In addition the fractal coating at UAV reduces the probability of its inflight detection.

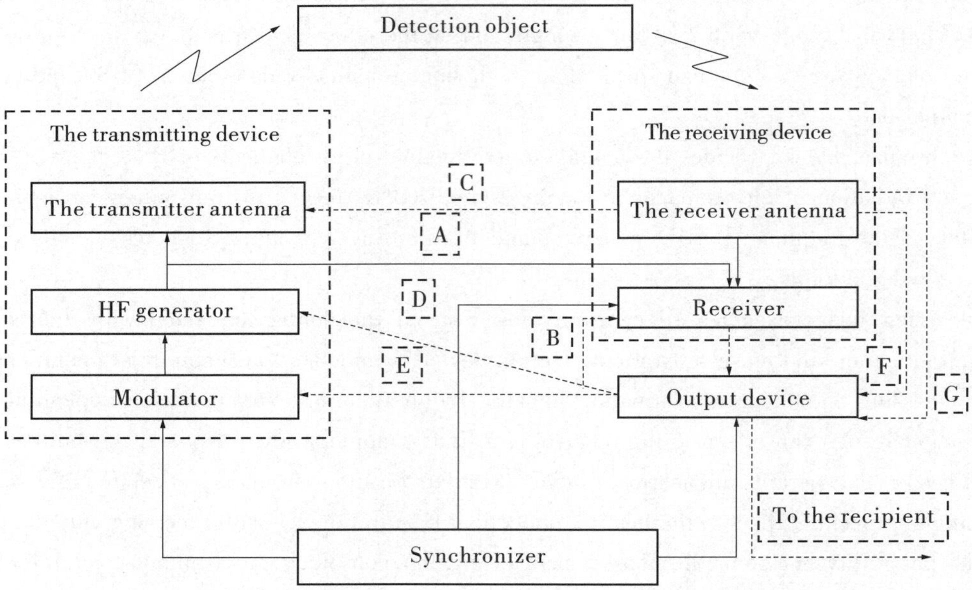

Fig. 4. 7 Generalized functional diagram of classical radar set

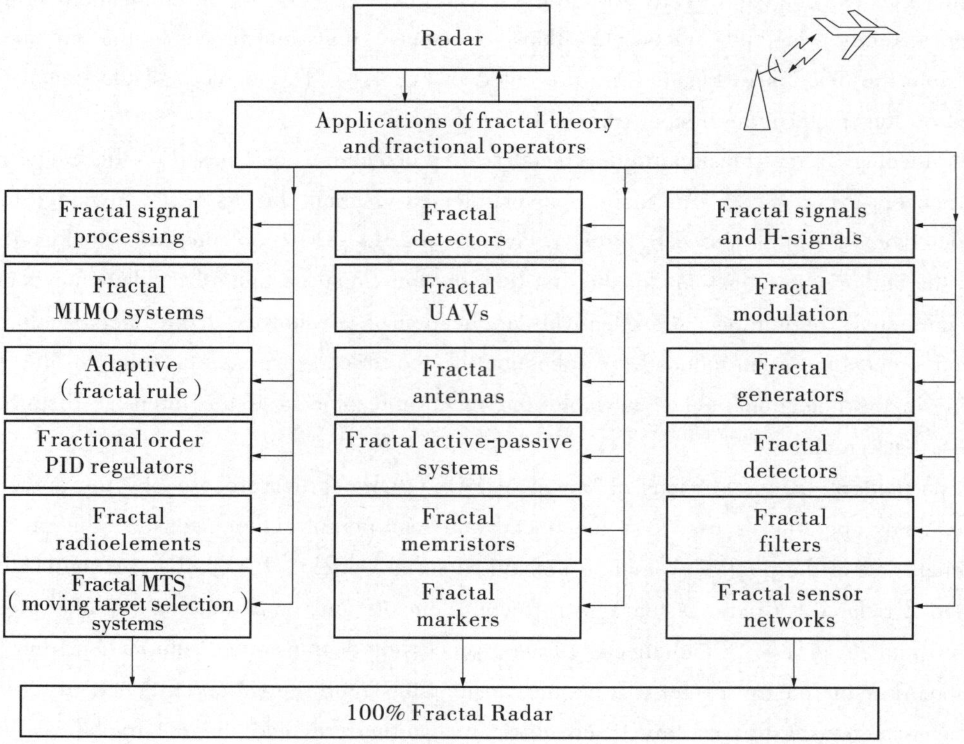

Fig. 4. 8 Application points of fractals, scaling and fractional operators for proceeding to the fractal radar

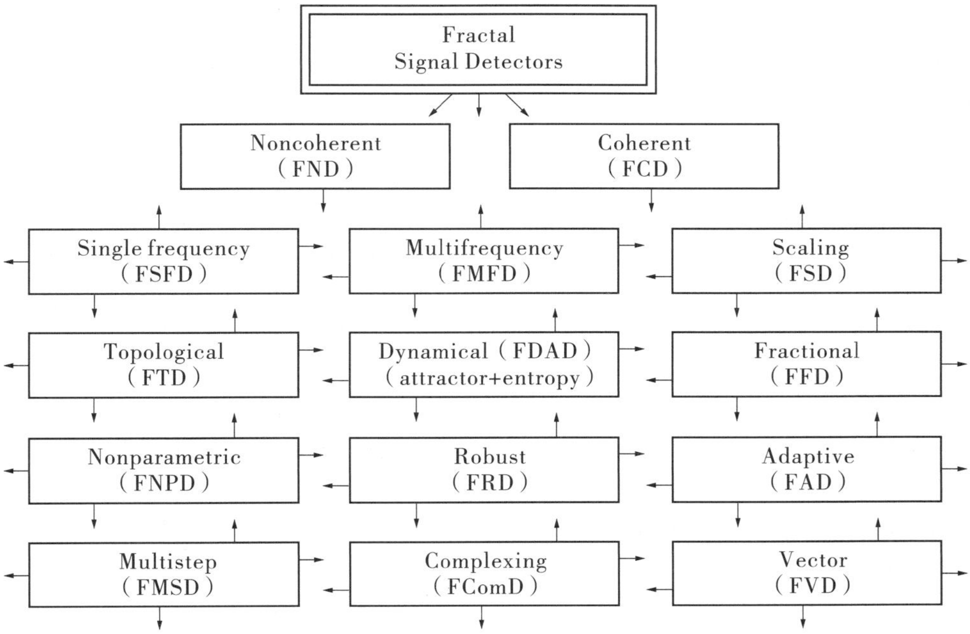

Fig. 4. 9 The main kinds of new dynamical fractal signal detectors (FSD) proposed by the author

4. 8 Conception of fractal radio elements (fractal capacitor), and fractal antennas

The third stage of the work on creation and development of breakthrough informational technologies for solving modern problems of radio physics and radio electronics, which was begun in IRE RAS in 2005, is characterized by transformation to design of fractal element base of fractal radio systems on the whole. Creation of the first reference dictionary of fractal signs of targets classes and permanent improvement of algorithmic supply were the main points during the development and prototyping of the FNDRS in the form of a back-end processor (Fig. 4. 4). Basing on the obtained results, we can talk about design of not only fractal blocks (devices) but also about design of a fractal radio system itself. Such fractal radio systems (Fig. 4. 6) which structurally include (beginning with the input) fractal antennas and digital fractal detectors are based on the fractal methods of information processing and they can use fractal methods of modulation and demodulation of radio signals in the long view[3,4].

Fractal antennas are extremely effective during development of two-frequency or multi-frequency radio location and telecommunication systems[3,4,17,24,51,52,63]. The structures form of such antennas is invariant to certain scale transformations that is an electrodynamics similarity is observed. As it is known, spiral and log periodic antennas are the most obvious examples of frequency-independent antennas. Fractal antennas were the next step in building of new ultra broadband and multiband antennas. The scaling of fractal structures gives them multiband properties in an electromagnetic sense. Multifrequency radio measurements along with fractal processing of the obtained information are a serious alternative to existing methods of enhancing the signal-to-noise ratio. Since every target has its own typical scales then one can directly determine a new signs class (except for the pointed above) in the form of fractal-and-frequency signatures by selecting the search frequency grid.

Unlike the classical methods when smooth antenna diagrams (AD) are synthesized an idea of realization of radiation characteristics with a repetitive structure at arbitrary scales initially underlies the fractal synthesis theory. It gives a possibility to design new regimes in the fractal radio dynamics, to obtain fundamentally new properties and fractal radio elements as well (for example, a fractal capacitor)[71].

Application of a recursive process theoretically allows to create a self-similar hierarchical structure up to separate conductive tracks in a microchip and in nanostructures[3,4]. In practice the sum of random values converges not to Gaussian distributions but to Levy stable distributions with heavy tails (i. e. fractal distributions-paretians) quite frequently. Simulation of Levy distributed random values can lead to processes of anomalous diffusion which is described with fractional derivatives on space and/or time variables. In substance, equations with fractional derivatives describe non-Markov processes with memory.

Physical simulation of fractional integral and differential operators allows to create radio elements with passive elements, simulating nonlinear impedances $Z(\omega)$ with frequency scaling

$$Z(\omega) = A\,(j\omega)^{-\eta}, \tag{4.1}$$

where $0 \leqslant \eta \leqslant 1$, A —const, ω —angular frequency basing on the modern nanotechnologies.

For that purpose the model of impedance $Z(\omega)$ was created in the form of an unlimited chain (continuous) fraction. In case of a finite stage of building the equivalent circuit for *RC* chains with using the n-th matching fraction for the given continuous fraction, one can adjust frequency ranges which the necessary power law of impedance of the form $\omega^{-\eta}$ will be observed in.

Thus, independently of the Work [2], our model of the impedance $Z(\omega)$ in the form of an endless chain (continuous) fraction was created. In the case of the final stage of construction of the equivalent electrical circuit for *RC* chains when the corresponding n-th fraction of the considered continued fraction is used, we can adjust the frequency bands in which there will be a power-law dependence of the impedance $\omega^{-\eta}$ (Fig. 4. 10). In this case, we first implement in practice nonlinear "fractal capacitor" or the fractal impedance[71].

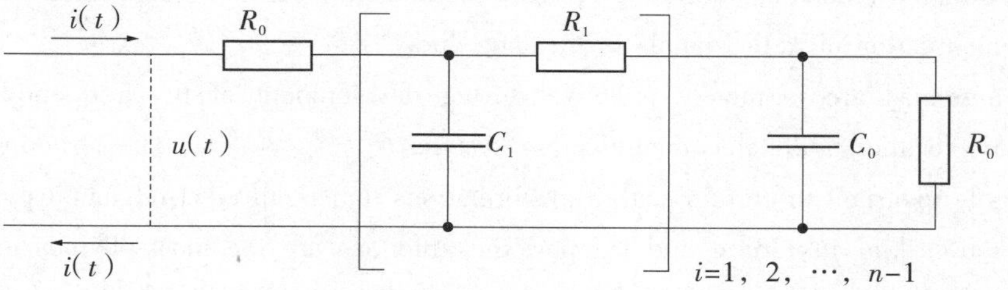

Fig. 4. 10 An example of the implementation of the fractional operator $d^{-1/2}/dt^{-1/2}$ or fractal capacitor

Basing on nanophase materials one can also create planar and volume nanostructures which simulate the considered above "fractal" radio elements and radio devices of microelectronics i. e. , the question is about building an element base of new generation. In particular, an elementary generalization of Cantor set at physical level allows to proceed to so called Cantor blocks in the planar technology of molecular nanostructures.

Application of fractal structures also allows to create media which show complex reflecting and transmitting properties in a wide frequency range and able to simulate three-dimensional photon and magnon crystals which are the new media of information transfer (for more details see Work [4, 17, 71]). One can select a configuration and sizes of fractal structures and check such unusual properties for a frequency range on the scheme in Fig. 4. 11. The pickup antenna (is not shown) was placed closely to the fractal plates.

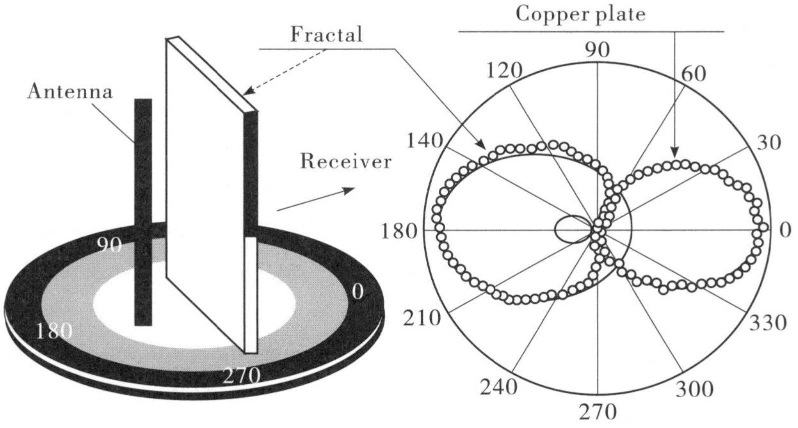

Fig. 4. 11 Scheme of "fractal" experiments

On the right of Fig. 4. 11, pictures of a secondary electromagnetic field for fractal and copper plates are shown. One can see that the "superwave" fractal structure slows down the directional radiation while a metallic plate does not reveal this function. Such "superwave" properties mean that a fractal plate can act as a compact reflector.

Thus fractal structures always have a self-similar series of resonances which lead to logarithmic periodicity of working zones. The related topologic fractal structure allows to modulate the electromagnetic waves transmission coefficient. The lowest frequency of weakening corresponds to wave lengths which can significantly enhance the outer sizes of the fractal plate and makes such fractal structures be the superwave reflectors. The obtained results allow to extend the applied above calculation method on the basis of algorithms of a numerical solution of hyper-singular integral equations to a wide class of electrodynamic problems which appear during researches of fractal magnon crystals, fractal resonators, fractal screens, fractal radar barriers and also other fractal frequency-selective surfaces and volumes which are required for realizing the fractal radio systems.

Promising elements of fractal radio electronics include also functional elements fractal impedances which are implemented based on the fractal geometry of the conductors on the surface (fractal nano-structures) and in space (the fractal antenna), the fractal geometry of micro-relief surface of substrates or fractal structure of polymer composites, etc.

The fractal radio systems proposed by the author reveal new opportunities in the modern radio electronics and can have the widest outlooks of practical application[3,4,17,71].

4.9 Processing of images obtained from unmanned aerial vehicles in the regime of flight over inhomogeneous terrain with fractal-scaling and integral methods

At present unmanned aerial vehicles (UAV) are actively developed in the world. They are used to solve a broad spectrum of scientific and practical problems, and UAV can become the main element of formation of integrated information field[3,4,7,8]. By means of optical, radar and IR-sensors located at UAV one can obtain digital images with high resolution and high efficiency at any time. The following modes are possible: scanning of predetermined ground; radar search in accordance with linear itinerary; route mapping; detailed image acquisition of interest small grounds and locating objects. Wide sight field of sensors enables to scan a ground regardless of UAV manoeuvres.

UAV flight pass depends on the search problem. If an object information is unknown, a flight pass can be presented in the form of usual scanning. Such flight can be programmed and it realized automatically without operator intervention. If one needs to detect an object on the basis of the preliminary information the flight pass can have an intricate profile: horizontal turns, ups and downs. An operator can intervene in control and change a flight mode and profile. UAV with high altitude and flight time can compete with satellites in the field of creation of telecommunications networks and navigation systems. As rule light UAV are heavily influenced by atmospheric disturbances. From presented examples one can see the most important role of data processing and correction of the UAV information at all flight stages.

The following are the results of processing multiple real images from the UAV in flight mode over heterogeneous terrains with different objects. The processing was carried out by our own and long ago established fractal-scaling and integral methods[3,4,34,48,50,66,74,75]. Methods of successive improvement of image quality (when background distortions, brightness weakening and distortions of image form itself are present) were developed by the author. These methods are based on the successive application of the iterative Fourier procedures with the phase and amplitude correction by fractional degree filters[34].

Processing results are presented in Fig. 4. 12 – Fig. 4. 13. The photo processing was carried out in accordance with the local estimations of fractal dimension D and using the fractal signature method on two observation scales.

In Fig. 4. 12a the tunnel image in mountain terrain obtained with UAV is shown. Filtration results in accordance with estimations of fractal dimension D at two scales 3 – 5 (Fig. 4. 12b) and 8 – 11 (Fig. 4. 12c) are presented next at the first line. Data of the integral method are presented in Fig. 4. 12d.

In Fig. 4. 13a the urban image with moving vehicles along the highway is presented. Filtration results in accordance with estimations of fractal dimensions D at scale 4 – 6 (Fig. 4. 13b) are presented at the first line. Windows are chosen to obtain the best results of vehicle selection. In Fig. 4. 13c the estimation map of fractal dimension D without a filtration is presented. Data of the integral method are presented in Fig. 4. 13d.

In Fig. 4. 14a the UAV image of some access point is shown. Results of fractal filtration in accordance with fractal dimension estimations at two scales 4 – 6 are presented in Fig. 4. 14b. Measuring and scaling windows were selected to obtain the best results of the fractal filtration. In Fig. 4. 14c the integrated map of fractal dimension estimations D without a filtration is presented. Results of integral method application are presented in Fig. 4. 14d.

Some chosen results of UAV image processing on the basis of the integer-valued Lebesgue measure are presented in Fig. 4. 15. Selection of moving and immovable vehicles is presented in Fig. 4. 15b and Fig. 4. 15d.

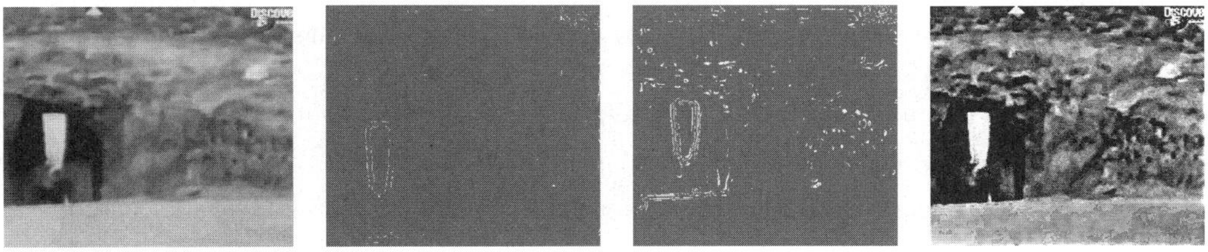

(a) (b) (c) (d)

Fig. 4. 12 Fragment of the mountain terrain with the tunnel (a), filtration results in accordance with estimations of fractal dimension at two scales 3 – 5 (b), and at scales 8 – 11 (c), and results of integral method application (d)

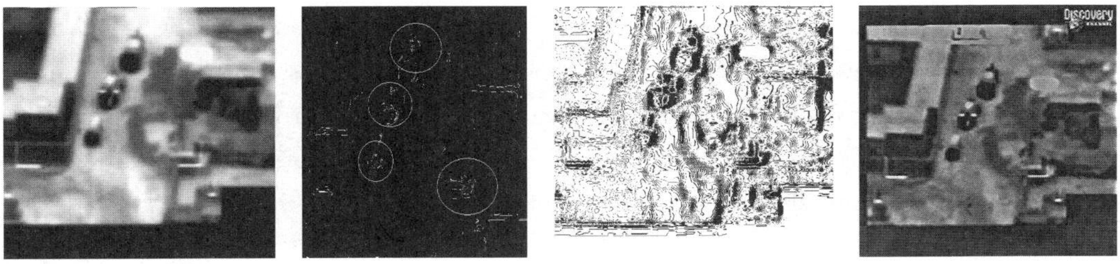

(a) (b) (c) (d)

Fig. 4. 13 Urban image with moving vehicles along the highway (a), results of fractal filtration in accordance with estimations of fractal dimension D at scale 4 – 6 (b), complete fractal estimation map D over the field without a filtration (c), and results of integral method application (d)

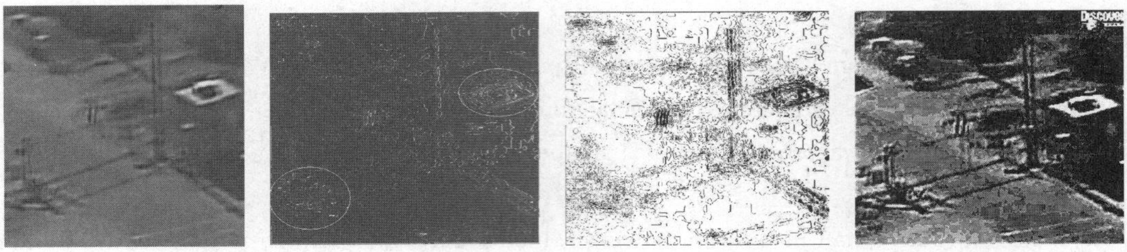

(a) (b) (c) (d)

Fig. 4. 14 The access point image of some building (a), results of fractal filtration in accordance with fractal dimension estimations at two scales 4 – 6 (b), complete fractal map of estimations D over the field without a filtration (c), and results of integral method application (d)

(a) (b) (c) (d)

Fig. 4. 15 Image of vehicle accumulation (a) and processing results on the basis of the integral method (b); Image of a vehicle caravan (c) and processing results on the basis of the integral method (d)

Cited data show the functionality of processing methods of images obtained at the passive observation. If in practical applications very fast processing is needed (in accordance with the frame input), then one can restrict only by our modified method of brightness and contrast increase and also by the fractal contouring.

This investigation continues Prof. A. Potapov's cycle of urgent works on the substantiation of the application of the theory of fractals, physical scaling and fractional functionals in the problems of radar and radio physics. These works were begun by Prof. A. Potapov first in USSR at the IRE of Academy of Sciences in the late 1970s.

Results of processing of different UAV image types presented in this work show that developed modern processing methods have high performance and increase quality and detailing of treated images in passive and active illumination mode approximately a few times (in coherent mode the diffraction-limited resolution is reached). These methods can be successfully applied to the information processing from space and other aircraft complexes.

4.10 Strategic directions in synthesis of new topological radar detectors of low-contrast targets against the background of high-intensity noise from the ground, sea and the fall of rain

Intensive development of modern radar technology establishes new demands to the radiolocation theory. Some of these demands do not touch the theory basis and reduce to the precision increase, improvement and development of new calculation methods. Other ones are fundamental and related to the basis of the radiolocation theory. The last demands are the most important both in the theory and in practice.

Radar detection of unobtrusive and small objects near the ground and sea surface and also in meteorological precipitations is an extremely hard problem[3,4,7,9,10,13,43,44,46,49-52,59,61,68,72,74,75]. One should take into account that the noise from the sea surface and vegetation has nonstationary and multi-scale behavior especially at high incidence angles of the sensing wave.

Often, variety of subjacent coverings, conditions of radar observation and maintenance of the objects mentioned above leads to the fact that almost always signal-to-noise ratio q_0^2 for these tasks fills in the area of negative (in decibels) values, that is $q_0^2 < 0$ dB. It makes the classical radar methods and algorithms of detection non-applicable in most cases, that is use of energy detectors (when likelihood ratio is exclusively defined by the energy of an input signal) is impossible.

Detection of low-contrast objects against the background of natural high-intensity noise mentioned above inevitably requires us to be able to propose and then to calculate some fundamentally new property which differs from the functionals related to the noise and signal energy.

We think that the initial information comes from different radio systems in the form of a one-dimensional signal and a radar image—Fig. 4.16[68,74]. The system of initial radio systems and consideration of a radar image and a one-dimensional signal in the MMW range was already presented in Work[3,4,9,10,13,17]. Now a fractal radar, a MIMO-radar and UAVs are included into the pattern of Fig. 4.16.

The fractal radar conception is presented in Work [49, 61] in detail, the MIMO-radar conception is considered in Work [50, 72, 75]. The main idea of fractal MIMO-radars is the use of fractal antennas and fractal detectors[3,4]. An ability of fractal antennas of simultaneous operating at several frequencies or radiating a wideband sensing signal drastically increases the number of degrees of freedom. It determines many important advantages of such a kind of radio location and sufficiently broadens opportunities of adaptation.

All the currently existing methods and signs of detection of unobtrusive objects against the background of high-intensity reflections from the sea, ground and meteorological formations are presented on Fig. 4.17[3,4,10,68]. Correlations between various signs and methods are also shown. The work on the classification of such methods, algorithms and signs was begun by

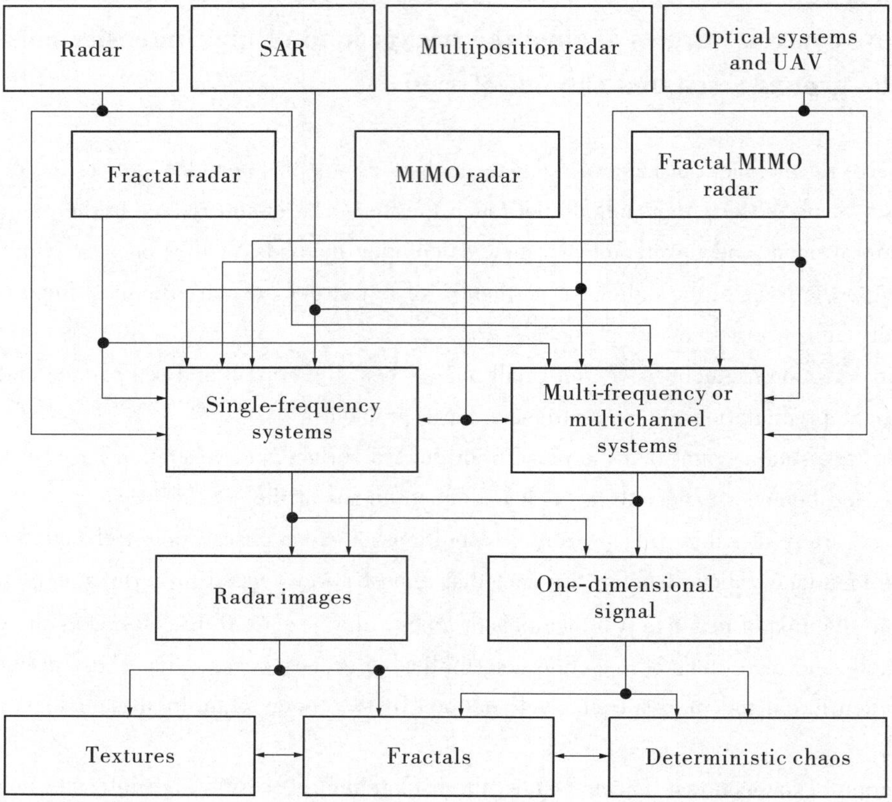

Fig. 4. 16 Radio systems of the initial information

Prof. A. Potapov in China in May 2015 during the presentation of project "Leading Talents of Guangdong Province" and was completed in China in March 2016.

As compared with usual detection methods, the fractal-scaling or scale-invariant methods proposed by Prof. A. Potapov can effectively improve the signal/interference relation and considerably increase the probability of target detection. Methods under consideration are suitable both for usual radars and for SAR, and also for MIMO systems for multipositioning radiolocation.

Also in terms of Weierstrass function for one-dimensional fractal scattering surface we obtained scattering field absolute value dependences on incident angle and surface fractal dimension D. In subsequent computer calculations, we used the above expression for the coherence function (CF) Ψ_k (Fig. 4. 17):

$$\Psi_k = \langle E_s(k_1) E_s(k_2) \rangle \tag{4.2}$$

of the fields $E_s(k)$ scattered by the fractal surface[3,4,21,50,72,75].

We can show that the tail intensity of signals reflected by a fractal surface is described by power functions:

$$I(t) \sim 1/(t')^{3-D}. \tag{4.3}$$

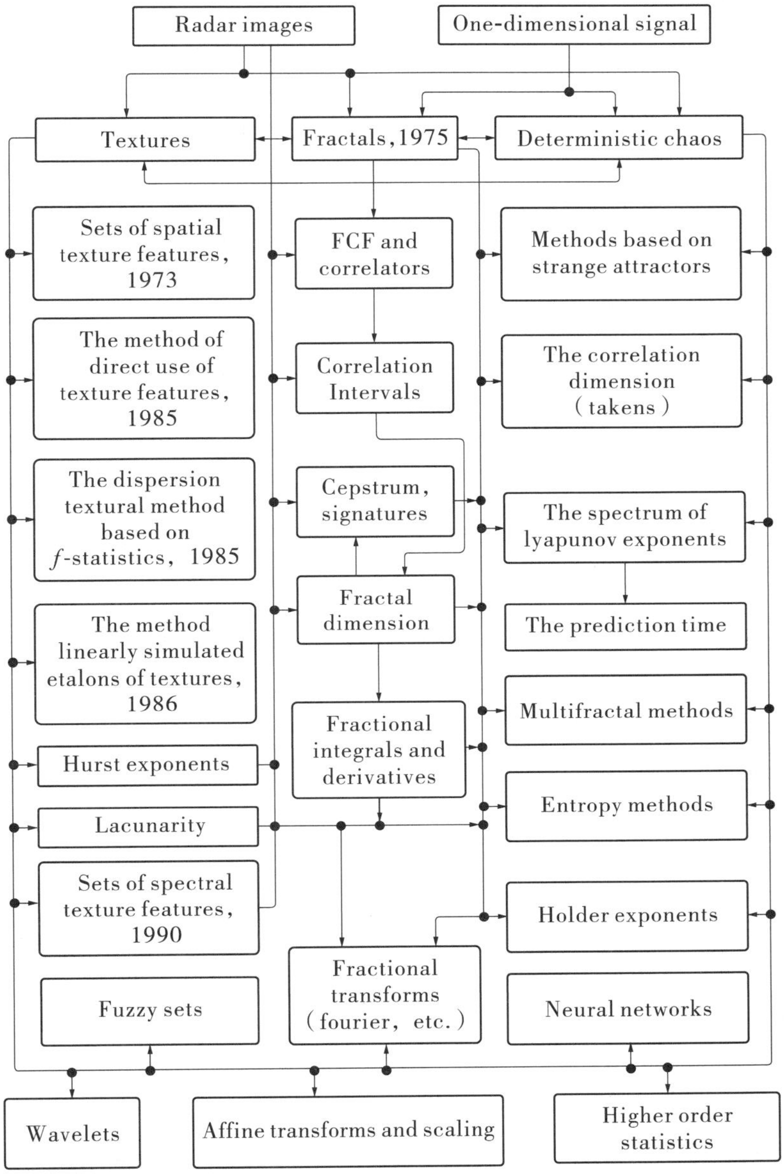

Fig. 4. 17 Signs and methods of detection of low-contrast objects against the background of high-intensity ground noise

Result (4. 3) is very important because for standard cases the intensity of a reflected quasi-monochromatic signal decreases exponentially. Thus, the shape of a signal scattered by a fractal statistically rough surface substantially differs from the shape of a scattered signal obtained with allowance for classical effects of diffraction by smoothed surfaces[3,4,21,50,68,72,75].

Fractal (Hausdorff) dimension D or its signature at different surface-mapping parts are simultaneously and texture measure[3,4,68,75], i. e. , properties of spatial correlation of radio scattering by corresponding surface patches.

Fractal signatures including spectra of fractal dimensions and fractal cepstrum represent the attribute vector uniquely determining wide class of targets and objects than the use of fractal dimension values. Thus, we can specify the propose structure of the fractal radiolocation detector of target classes consisting of edge detector and fractal signature calculator. Obtained signatures are compared with the signature database and the decision concerning the presence or absence of the object is made in accordance with some criterion.

The general conception of the textural or fractal detector is presented in Fig. 4. 18. The set of textural or fractal signs is determined basing on the received radio signal or image. Then a decision of signal presence H_1 or its absence H_0 is done in the threshold device at threshold value Π and certain level of probability of a false alarm F.

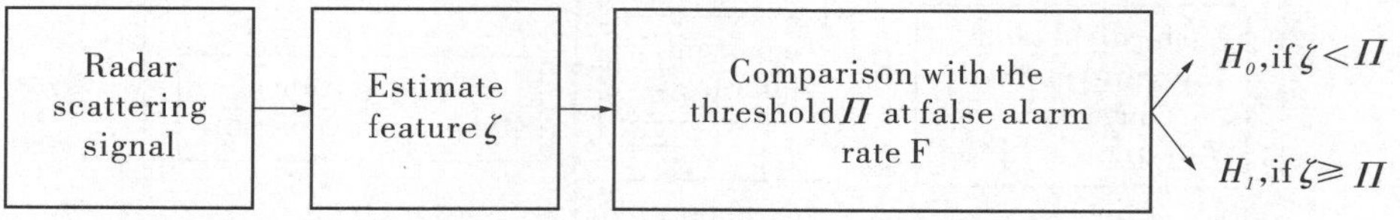

Fig. 4. 18 Conception of textural or fractal detector

Values of fractal dimension D, Hurst exponents H for multi-scale surfaces, Holder exponents, values of lacunarity and so on can be also used as signs. Hurst exponent

$$H = 3 - D \tag{4.4}$$

for a radar image and

$$H = 2 - D \tag{4.5}$$

for a one-dimensional signal.

High sensitivity of estimation of functionals of non-integral dimension to the presence of a continuous contour in images suggests a large potential of fractal filtering of the contours of objects in strong interference (Fig. 4. 19). The observation was made using ground-based telescope, and the distance between it and objects was about 800 km. These data are presented in [34]. None of the modern methods of digital processing can provide comparable objects resolution!

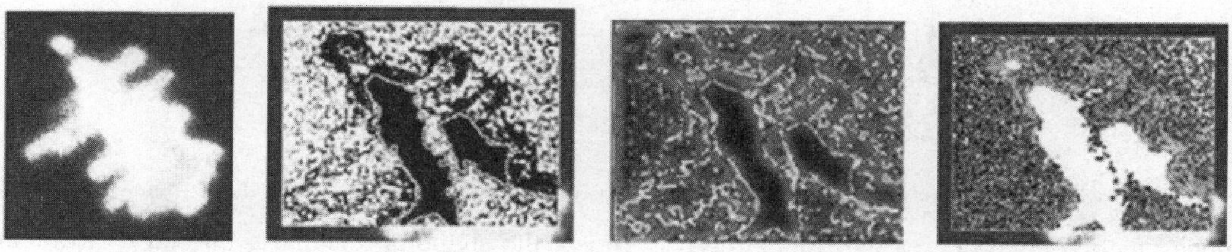

(a) (b) (c) (d)

Fig. 4. 19 The initial image of space complex at the time of joining "Shuttle" – "Peace" (a) and the results (b – d) of the fractal processing (targets clustering) for various values of the threshold D of topological fractal nonparametric detector

Figure 4. 20 shows selected results of fractal nonparametric filtering of low-contrast objects. Aircraft images (Fig. 4. 20a) were masked by an additive Gaussian noise. In this case, the SNR ratio q_0^2 was −3 dB (Fig. 4. 20b). It is seen in the figures that all desired information is hidden q_0^2 in the noise. The optimum mode of filtering of necessary contours or objects is chosen by the operator using the spatial distribution of fractal dimensions D of a scene (Fig. 4. 20c). This distribution is determined automatically and is shown in the right panel of the computer display[4,13,17].

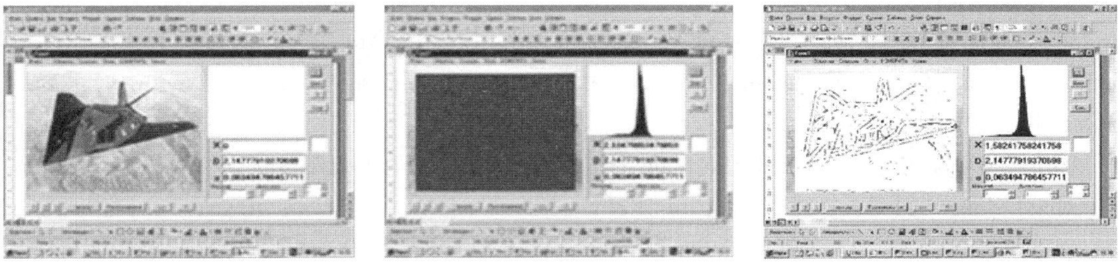

(a) (b) (c)

Fig. 4. 20 Source image (a), source image and noise SNR ratio was −3 dB (b), results of fractal filtration (c)

4. 11 Postulates of fractal radar and cognitive radar in fractal-scaling design

Introduction of conceptions of "texture", "determined or dynamical chaos" and "fractal" into scientific use of radio physics allowed us for the first time in the world to propose and then to apply fundamentally new topological (and not energetic) properties (the invariants) which were combined by the author under a general conception— "Sampling Topology"[3,4,10,23,50,75]. We think that the initial information comes from different radio systems in the form of a one-dimensional signal and a radar image (Fig. 4. 16).

Conceptions introduced in Work [76] were completely in tune with author's ideas.

Cognitive radar defined in Work [76] is based on 3 main postulates:

(1) intelligent signal processing, which builds on learning through interactions of the radar with the surrounding environment;

(2) feedback from the receiver to the transmitter, which is a facilitator of intelligence;

(3) preservation of the information content of radar returns, which is realized by the Bayesian approach to target detection through tracking.

We focus on future possibilities of radars with the special emphasis on the fractal and chaos.

Fractal radar defined in Work [3, 4, 10, 13, 14, 17, 21 − 75] is based on 4 main postulates:

(1) intelligent signal processing based on the theory of fractional measure, scaling effects and fractional operator's theory;

(2) Hausdorff dimension or fractal dimension D of a signal or a radar image (RI) is di-

rectly connected with the topological dimension;

(3) robust non-Gaussian probability distributions of the fractal dimension of the processed signal;

(4) "Maximum topology with a minimum of energy" for the received signal. It allows to take advantages of fractal scaling information processing more effectively.

The key point of fractal approach is to focus on describing and processing of radar signal (fields) exclusively in the space of fractional measure with the use of the scaling hypothesis and distributions with heavy-tailed or stable distributions (non-Gaussian). Fractal-scaling processing methods of signals, wave fields and images are in a broad sense based on the pieces of information, which isn't usually taken into account and irretrievably lost if classical methods of processing is applied.

This work is concerned with the main radio physical area-radiolocation and it aims to ascertain what's done and things to do in this field on the basis of the fractal theory. Investigations carried out showed the correctness of the path chosen by Prof. A. Potapov (since 1980) to improve the radiolocation technique.

It's necessary to think about the processing of the input signals with a low threshold at high levels of false alarm and then a transition to a low level of false alarms. Moreover, the false alarm probability is never measured in real time. In principle [3, 4, 10, 50, 68, 72, 75, 76], we need a new metric, and the new parameters of radar detection.

4.12 Officially accepted results of fractal investigations

Careful bibliographic search shows our full and absolutely priority in all "fractal" directions (Fig. 4.2 and Fig. 4.5) in USSR and Russia, no less than in world science. There is serious acknowledgement:

• In the book *Progress Report of Presidium of the Russian Academy of Sciences. Scientific Achievements of the Russian Academy of Sciences in 2007* (Moscow: Nauka, 2008, p. 41) in the subsection "Radar systems" of section "Information Technology and Computing Systems" the following text is presented. "The reference dictionary of fractal attributes of optical and RF images, necessary for realization of essentially new fractal methods of radar information processing and creation of highly informative devices for detection and recognition of weak signals in intensive non-Gaussian noise, is created. It was established, that for effective solving of radiolocation problems and multi-dimensional signal fractal detectors design, fractional dimension, fractal signatures and cepstrums and, also, texture signatures of area backgrounds has significant importance (IRE RAS)".

• In the book *Progress Report of Presidium of the Russian Academy of Sciences. Scientific Achievements of the Russian Academy of Sciences in 2009* (Moscow: Nauka, 2010, p. 24) in subsection "Radar systems. Geoinformational technologies and systems" of section "Nanotechnology and Information Technology" the following text is presented: "For the first time in world practice constructing principles of fractal adaptive radio systems and fractal radioelements for modern radio-engineering and radiolocation problems are proposed and experimentally proven. Operation principle of these systems and elements is based on the introduction of radiating and received signals fractional transform in non-integer dimension space, considering their scaling effects and non-Gaussian statistics. This achievements allow a new level of real non-Markov's signals and fields informational structure (IRE RAS)".

• In the book *Progress Report of Presidium of the Russian Academy of Sciences. Scientific Achievements of the Russian Academy of Sciences in 2011* (Moscow: Nauka, 2012, p. 199) and book *Report to Government of Russian Federation. About* 2011 *Year Results of Realization of the Fundamental Scientific Researches of State Scientific Academies in 2008 – 2012*. (Moscow: Nauka, 2012, p. 242) In three volumes in subsection "Radar systems. Geoinformational technologies and systems" of section "Informatics and Information Technology" the following text is presented: "Basing on fractal analysis systematic research of fractal antennas electrodynamics properties is performed. Wideband and multi-range properties of fractal antennas and resonances quantity dependence on fractal iteration number are confirmed. It is shown, that basing on miniature fractal antennas, the effective realization of frequency-selective environments and shields, deforming radiolocation target image is possible. Fractal frequency-selective 3D-environments or fractal 'sandwiches' (engineered radio-electronic micro- and nanoconstructions) at IRE RAS are investigated".

• In the book *Progress report of Presidium of the Russian Academy of Sciences. Scientific achievements of Russian Academy of Sciences in 2012* (Moscow: Nauka, 2013, p. 195) in subsection "Component base of microelectronics, nanoelectronics and quantum computers. Materials for micro- and nano-electronics. Nano-and micro-system technics. Solid-state electronics" of section "Nanotechnology and information technology" the following text is presented: "It is established that in physical base of memristor operation there is integral quantum Hall effect. The voltage to current relations for any memristor type are obtained. The results are directed to practical realization of memristors as new elements of electronic schematics" (SRI AMA KBSC of RAS, IRE RAS).

4.13 Personal meetings with Benois Mandelbrot

As mentioned above the author uses the "fractal" term almost everywhere. In conclusion I'd like to share my impressions about the meeting with B. Mandelbrot with readers. The meeting occurred in his house near NewYork in December 2005. At that time I was responsible for the international project. I had to visit America frequently. My personal meeting with the founder of the fractal geometry B. Mandelbrot occurred on Friday, December 16th, 2005. Before this the intensive correspondence was going on both when I was still in Moscow and when I flew across

USA from south to north with my lectures on the results of 5 years international project. Mandelbrot himself was in an extensive trip but his secretary phoned and told that the maitre would come back home ad hoc. He was extremely interested in meeting at home and talking with the Russian physicist who dealt with various "fractal" experiments and applications of the fractal theory in radio physics and radio electronics.

My interpreter and I got to NewYork central railroad terminal by taxi and then we got on an electric train leaving at 9: 30 local time. After a while we got off in a small station and went to B. Mandelbrot's house by taxi. As approaching to the house we saw a silhouette of a high strong grayish man with glasses on a lace appeared behind the door. He dressed in home clothing. While we were getting off the car he'd already opened the front glassed-in door. Mandelbrot was looking at me, smiling, holding out a hand first. Then he suggested us to undress and all of us went to his room. He asked me to sit in front of him explaining that this way was better for him to talk and there was a comfortable arm chair for me. There was his world-famous fundamental book *The Fractal Geometry of Nature* on the table. I took out my monograph on fractals and presented to Mandelbrot. Then I told him about scientific work and my results on fractal applications in radio physics and answered his questions. Mandelbrot listened with a keen interest and very attentively. It was a surprise for him that there was such a success on fractals' applications in Russia and there was already the direct approach to the fractal technologies. He became very interested in my proposed conception of fractal radio systems and in designing an essentially new fractal elemental base.

He was well familiar with fractal antennas. Suddenly he said that sometime the matter would be in producing a fractal capacitor! I replied with enthusiasm that I was already "caught" with this idea for a lot of years and a big paper about physical modeling of fractal impedances, fractional operators and production of fractal capacitors was ready for the press. This was incredibly: We were thinking about realizing the same idea on the opposite sides of the world! Mathematical questions were less interesting for Mandelbrot. He was getting more and more interested in disciplines created at IRE RAS: the fractal radio physics and the fractal radio electronics with its development.

Fig. 4. 21 Personal meeting with Benois Mandelbrot

Human simplicity, openness, interest in the surrounding world and wisdom—these particular properties were peculiar to B. Mandelbrot. Sometime they told about B. Mandelbrot's arrogance. I could assert only the reverse. He did not make me feel the difference between our statuses during all the conversation. He first inquired about all fractal developments.

40 minutes later Mandelbrot stood up, after apologizing, he went out to other room. Then he came back with a pile of his books. He asked me if I already had some of these books. Mandelbrot said that he liked my works. He inquired when and how my book was written. I replied that I prepared the first version as early as in the beginning of 1990s. Then the search for publishers began and at the same time improving and significant rework of the monograph text was going on. Mandelbrot said that he had two books on the go: one was in Italy and the other was in America. With a smile he admitted that he wrote slowly, thoroughly using all his old works. Our conversation lasted for almost 2 hours. Tempus fugit. At a certain moment he was called by the phone. He suggested giving us a lift to the railway station on his car. Mandelbrot drove the car on his own. We were at the railway terminal at noon. I said goodbye to B. Mandelbrot and we waved to each other. This was an unforgettable meeting. These were all the minutest details of the meeting with the great scientist in my memory.

4.14 Conclusion

This chapter is concerned with the main radio physical area—radiolocation and it aims to ascertain what's done and things to do in this field on the basis of the fractal theory. Investigations carried out showed the correctness of the path chosen by Prof. A. Potapov (since 1980) to improve the radiolocation technique. In particular over period of thirty five years this resulted in invention, creation and development of the new kind and method of radiolocation, namely, fractal-scaling or scale-invariant radiolocation. This implies radical changes in the structure of theoretical radiolocation itself and in its mathematical apparatus also.

Earlier fractals made up the thin amalgam on the strong science frame of the XX century ending. In the modern situation it attempts to humiliate their significance and rely only on the classical knowledge suffered an intellectual fiasco.

The detailed analysis of all works published by Prof. A. Potapov is not an aim of this chapter. Nevertheless the acquaintance with Prof. A. Potapov investigations in this area should substantially help to large group of experts and more accurately determine the practical application ways of the fractal theory to solve the radio physical and radiolocation problems. Prof. A. Potapov considers that the "sampling topology" problem[3,4,17] is one of the most important in radio electronics, and he also convinced that without fractals and scaling all signal-detection theory loses its causal meaning for the signal and noise conceptions.

The functional principle "topology maximum at energy minimum" for receivable signal permitting effective application of advantages of the fractal-scaling information processing was introduced by the author. This refers to the adaptive target signal processing. Application of the fractal principle results in the soul-searching in the detection field of movable and immovable objects at the intensive disturbance and noise background.

In this chapter, the author touched upon only the most important problems connected with the application of the fractals and scaling effects in statistical radio physics and radiolocation. In the development of fractal directions many important periods have already passed including the establishment period of this field of science. However we will have to solve many problems. It is the solution technique (approach) which is the most valuable but not results and implementations. This method was created by Prof. A. Potapov[3,4,9,10,12-75].

As a result, in the scientific community the new sense space with unusual for classical physics properties and problems was created. The field of fractal sciences was opened for creative search. However prior to create a clear understanding of the ideas underlaid in the basis of the fractal-scaling method is needed. Prof. A. Potapov defined problems mentioned above in 1980 and for thirty five years he is successful in working on their solution and development. Scientific results obtained in recent years are the initial material for further development and substantiation of practical application of fractal methods in modern fields of radio physics, radio engineering, radiolocation, electronics and information controlling systems.

Acknowledgments This chapter was written by Prof. A. Potapov (President of cooperative Chinese-Russian laboratory of informational technologies and signals fractal processing of Jinan University) in March 2016 when working on the project "Leading Talents of Guangdong Province" The program number is 00201502, 2016 - 2020.

Chapter 5 Multiple Scattering of Waves in Fractal Discrete Randomly-Inhomogeneous Media from the Point of View of Radiolocation of the Multiple Targets

5.1 Introduction

The propagation and scattering of waves in disordered systems is thought to be one of the most difficult sections of theoretical physics. Natural and artificial environments are countless examples of diverse wave scattering by discrete particles or targets. It is well known that, if a wave propagates in a randomly inhomogeneous medium, multiple scattering must be taken into account. Phenomenological and statistical approaches are used to consider multiple scattering. The phenomenological approach is based on the theory of radiation transfer in a scattering medium. The transfer equation expresses the radiation energy conservation law or the condition for the brightness balance of light beams with allowance for their polarization. Statistical analysis of multiple wave scattering uses a stochastic wave equation or a system of such equations for which the problem of wave diffraction by a statistical ensemble of particles is studied.

The theory of multiple waves scattering in media containing random scatterers has been developed by many authors[1-6]. We should note that, since the fundamental publication[7], fractals are suitable models for various physical phenomena[8-21]. The propagation and scattering of waves in fractal randomly inhomogeneous media is of great physical interest for radio physics, radar, remote sensing, optics, acoustics, communication technology, biology, medicine, and nanotechnology. One of original theoretical and experimental studies of scattering and diffraction on fractals was carried out in paper [8] (see also Ref. [14, 15]).

The theoretical basis for description of single and multiple scattering of waves by particles is based on classical electrodynamics. In this paper, we consider in detail issues of the general theory of multiple scattering of electromagnetic waves in fractal randomly inhomogeneous media based on modifications of the Foldy-Tversky theory[1-3]. We introduce main concepts of a fractal medium and formulate mathematics of multiple scattering of electromagnetic waves in a fractal medium simultaneously with physics of the scattering process. Modifications of the Foldy-Tversky integral equation for the coherent field and the Tversky integral equation for the second moment of the field are presented.

We have studied the dependence of the received power on the fractal signature of the probed object or target for the case of radar. It is also shown that a fractal signature can be used to study the dependence of volume scattering on distance. Theoretical studies are consistent with the results obtained by foreign authors.

The obtained results are applicable to solution of various problems of condensed matter physics, radio physics, and radar, in particular, to description of the dynamics of a growing fractal surface in nanotechnology and remote sensing (for radar) as well as description of the snow cover profile[21-23].

5.2 A classical solution

Two basic approaches to the problem of wave propagation in a random cloud of scatterers are known: rigorous (analytical) theory and transfer theory[4-6]. The rigorous theory or the multiple scattering theory is based on fundamental differential equations for fields involving statistical approximations. The results of preceding studies (see Ref. [1] and the bibliography herein) were generalized by Tversky, who obtained a closed system of integral equations[2,3]. In the case of moving scatterers, the field becomes a function of time; therefore, field correlations are observed in both space and time. We give here in some detail the main results of the Tversky classical theory based mainly on studies[2,3,5,6].

Let us consider a cloud of N particles with coordinates $\vec{r}_1, \vec{r}_2, \cdots, \vec{r}_N$ randomly distributed in volume V. The particles can differ in both shape and size. Scalar field ψ^a at point of space $\vec{r}_a$ which is not occupied by particles satisfies the wave equation:

$$(\nabla^2 + k^2)\psi = 0, \tag{5.1}$$

where $k = 2\pi/\lambda$ is the wave number in the surrounding medium, and λ is the wavelength.

Quantity ψ in Eq. (5.1) can describe one of components of the electric or the magnetic field. We denote the incident wave at point $\vec{r}_a$ in the absence of particles by φ_i^a. For fields of the φ_i^a type, the superscript means the point at which the field is considered, and the subscript indicates the origin of this field. Then field ψ^a at point $\vec{r}_a$ (Fig. 5.1) represents the sum of incident wave φ_i^a and U_s^a contributions from each of the N particles located at points $\vec{r}_s$, $s = 1, 2, \cdots, N$:

$$\psi^a = \varphi_i^a + \sum_{s=1}^{N} U_s^a, \tag{5.2}$$

$$U_s^a = u_s^a \Phi^s. \tag{5.3}$$

Here, U_s^a is the wave at point $\vec{r}_a$ scattered by the scatter located at point $\vec{r}_s$.

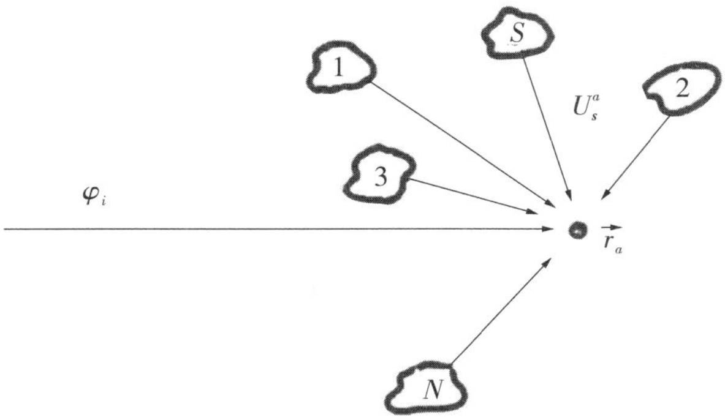

Fig. 5. 1 Field at point $\vec{r}_a$

In accordance with Eq. (5. 3), the value of U_s^a is determined by the action of scattering operator u_s^a on incident wave Φ^s for the particle at point $\vec{r}_s$ and observation point $\vec{r}_a$ (Fig. 5. 2) The expression $u_s^a \Phi^s$ is the operator writing of the field at point $\vec{r}_a$ formed due to the incidence of wave Φ^s on the scatterer located at point $\vec{r}_s$.

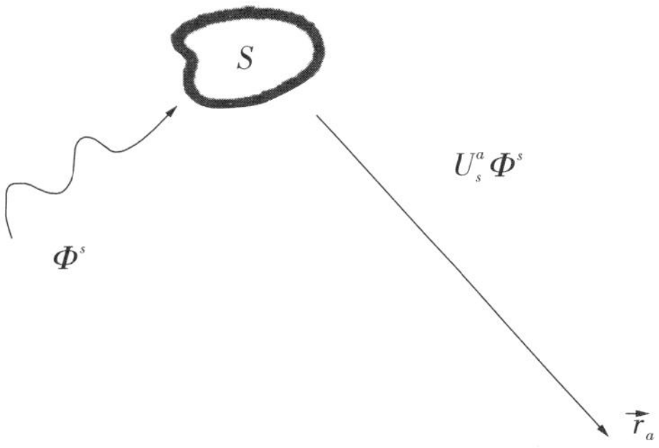

Fig. 5. 2 Contribution of the s-th particle to the total field

In the case of plane wave Φ^s propagating in the direction of ort $\hat{i}$ and in the far-zone approximation (the distance between points $\vec{r}_s$ and $\vec{r}_a$ is large),

$$\Phi^s = \exp(i\vec{k}\vec{r}), \vec{k} = k\hat{i}, \quad (5.4)$$

the following relation can be written for operator u_s^a:

$$u_s^a = f(\hat{0}, \hat{i}) \exp(ikr)/r, \quad (5.5)$$

where $\hat{0}$ is the unit vector in the direction $\vec{r}_a - \vec{r}_s$, $r = |\vec{r}_a - \vec{r}_s|$, and $f(\hat{0}, \hat{i})$ is the scattering amplitude.

Effective field Φ^s consists of incident wave φ_i^s and the field scattered from all (Fig. 5. 3, particles 1, 2, 3, …), except for the scatterer at point $\vec{r}_s$. Then we obtain:

$$\Phi^s = \varphi_i^s + \sum_{t=1,t\neq s}^{N} U_t^s. \tag{5.6}$$

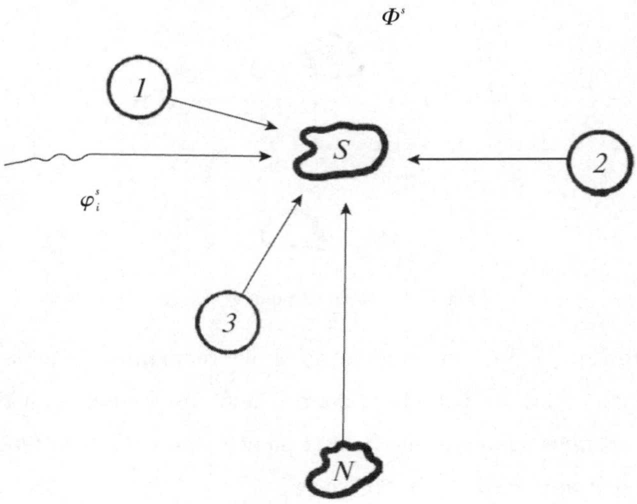

Fig. 5. 3 Effective field Φ^s for the s-th particle

Eqs. (5. 2) and (5. 6) form the fundamental pair of equations:

$$\psi^a = \varphi_i^a + \sum_{s=1}^{N} u_s^a \Phi^s, \tag{5.7a}$$

$$\Phi^s = \varphi_i^s + \sum_{t=1,t\neq s}^{N} u_t^s \Phi^t. \tag{5.7b}$$

Substituting Eq. (5. 7b) in Eq. (5. 7a) and repeating this process, we exclude Φ from these equations. In this case, we obtain:

$$\psi^a = \varphi_i^a + \sum_{s=1}^{N} u_s^a(\varphi_i^s + \sum_{t=1,t\neq s}^{N} u_t^s \Phi^t) = \varphi_i^a + \sum_{s=1}^{N} u_s^a \varphi_i^s + \sum_{s=1}^{N}\sum_{t=1,t\neq s}^{N} u_s^a u_t^s \varphi_i^t +$$
$$+ \sum_{s=1}^{N}\sum_{t=1,t\neq s}^{N}\sum_{m=1,m\neq t}^{N} u_s^a u_t^s u_m^t \varphi_i^m + \cdots. \tag{5.8}$$

Let us now consider each term of Eq. (5. 8). The first term is the incident wave φ_i^a. The next term in this series

$$\sum_{s=1}^{N} u_s^a \varphi_i^s \tag{5.8a}$$

takes into account all single-scattering events (Fig. 5. 4a). The next double sum

$$\sum_{s=1}^{N}\sum_{t=1,t\neq s}^{N} u_s^a u_t^s \varphi_i^t \tag{5.8b}$$

describes all processes of double scattering (Fig. 5. 4b).

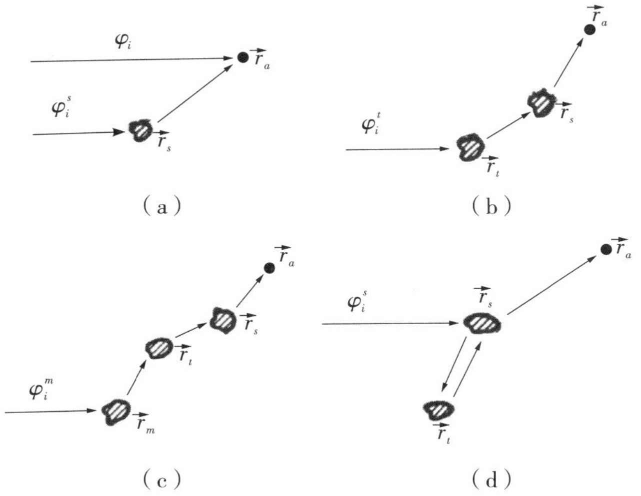

Fig. 5.4 Schemes for (a) single scattering, (b) double scattering, (c) triple scattering by various particles, and (d) triple scattering in the case when the wave passes the same particle more than once

The third sum is triple. It does not contain terms with $t = s$ and $m = t$ in the sum while the term with $s = m$ is present. This sum can be expanded so as to separate the terms with different indices:

$$\sum_{s=1}^{N}\sum_{t=1,t\neq s}^{N}\sum_{m=1,m\neq t}^{N} u_s^a u_t^s u_m^t \varphi_i^m = \sum_{s=1}^{N}\sum_{t=1,t\neq s}^{N}\sum_{m=1,m\neq t,m\neq s}^{N} u_s^a u_t^s u_m^t \varphi_i^m + \sum_{s=1}^{N}\sum_{t=1,t\neq s}^{N} u_s^a u_t^s u_s^t \varphi_i^s. \quad (5.8c)$$

The first triple sum in Eq. (5.8c) is shown in Fig. 5.4c. There are scatterers only at points $\vec{r}_s$ and $\vec{r}_t$ in the second triple sum in Eq. (5.8c); its plot is shown in Fig. 5.4d.

In the general case, total field ψ^a at point $\vec{r}_a$, which is a superposition of the incident wave and all multiply scattered waves, can be divided into two parts.

1. The first part, which is described by the first sum in Eq. (5.8c), contains all multiply scattered waves with successive scattering by different scatterers, as illustrated by the diagram in Fig. 5.5a. Note that s is the current index for all scatterers. Therefore, there are N terms with different s; the index t is the current index of all scatterers except for s, that is, there are $(N - 1)$ terms with different t. Similarly, there are $(N - 2)$ terms with different m.

2. The second part described by the second sum in Eq. (5.8c) corresponds to all wave trajectories that pass through a particle more than once. This situation is shown in Fig. 5.5b.

In the Tversky theory, all members belonging to the first group are taken into account (Fig. 5.5a) and members belonging to the second group are neglected (Fig. 5.5b). Obviously, the first group describes practically all multiply scattered waves, and the Tversky theory should give excellent results if backscattering is small in comparison with scattering in other directions.

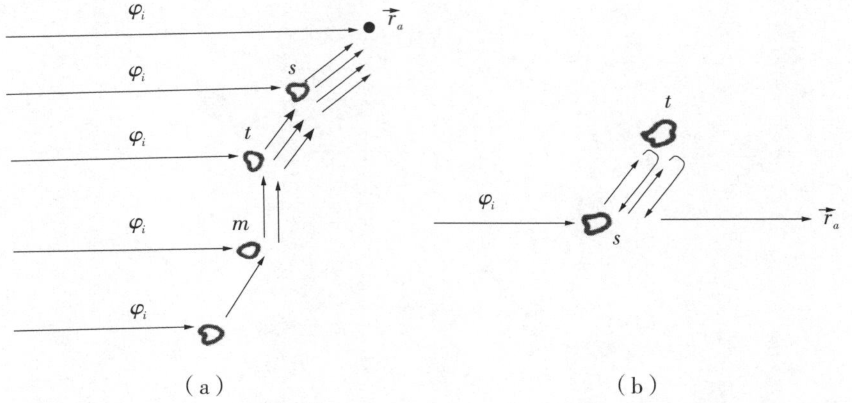

Fig. 5.5 (a) Scheme of the paths of scattered waves passing through different scatterers, and (b) the scheme of the paths of scattered waves passing through the same scatterer more than once

From the mathematical point of view, the Tversky theory with consideration for the above comments is based on the representation of the field in the following form:

$$\psi^a = \varphi_i^a + \sum_{s=1}^{N} u_s^a \varphi_i^s + \sum_{s=1}^{N} \sum_{t=1,t\neq s}^{N} u_s^a u_t^s \varphi_i^t + \sum_{s=1}^{N} \sum_{t=1,t\neq s}^{N} \sum_{m=1,m\neq t,m\neq s}^{N} u_s^a u_t^s u_m^t \varphi_i^m + \cdots \tag{5.9}$$

Table 5.1 presents the results of comparison (based on the data from book [5]) for the numbers of terms taken into account in the exact description of multiple scattering Eq. (5.8) and in the Tversky description (5.9). As can be seen from the table, the difference between the exact description and the Tversky description is negligible at large N.

Table 5.1 Errors in description of multiple scattering for the exact and Tversky solutions

Scattering process	Exact solution[(5.8)] $-E$	Tversky equation[(5.9)] $-T$	$(E-T)/E$
φ_i^a incident field	1	1	0
Single scattering	N	N	0
Double scattering	$N(N-1)$	$N(N-1)$	0
Triple scattering	$N(N-1)^2$	$N(N-1)(N-2)$	$1/(N-1)$
Fourfold scattering	$N(N-1)^3$	$N(N-1)(N-2)(N-3)$	$(3N-5)/(N-1)^2$

Eq. (5.9), which is called the Tverskoi expansion, is useful in understanding the physics of scattering processes, but it is inconvenient in calculation of the desired values. For this case, Foldy and Tversky obtained closed integral equations.

5.3 Statistical averaging for the case of discrete scatterers

Following book [5, 6], let us consider random function f (for example, field ψ^a or a product of fields) that depends on parameters of all scatterers. Then, we consider ensemble averaging of this function. Applying probability density function $W(\underline{1}, \underline{2}, \underline{3}, \cdots, \underline{N})$, we can write the mean value of function f in the following form:

$$\langle f\rangle = \iint\cdots\int f\, W(\underline{1},\underline{2},\underline{3},\cdots,\underline{s},\cdots,\underline{N})\,\mathrm{d}\underline{1}\,\mathrm{d}\underline{2}\,\mathrm{d}\underline{3}\cdots\mathrm{d}\underline{s}\cdots\mathrm{d}\underline{N}, \tag{5.10}$$

$$\mathrm{d}\underline{s} = \mathrm{d}\vec{r}_s\mathrm{d}\zeta_s = \mathrm{d}x_s\mathrm{d}y_s\mathrm{d}z_s\mathrm{d}\zeta_s\ , \tag{5.11}$$

where $\mathrm{d}\vec{r}_s = \mathrm{d}x_s\mathrm{d}y_s\mathrm{d}z_s$ is the elementary volume and the value of $\mathrm{d}\zeta_s$ takes into account all remaining characteristics of the scatterer: $\mathrm{d}\zeta_s = \mathrm{d}$ (the shape of the s-th scatterer) $\times\,\mathrm{d}$ (its orientation in space) $\times$ d (its size).

At small particle concentrations and small dimensions in comparison with the distance between them, all particles can be considered as point particles and the influence of their size affects only the scattering characteristics. Under this assumption, we have

$$W(\underline{1},\underline{2},\underline{3},\cdots,\underline{s},\cdots,\underline{N}) = w(\underline{1})\,w(\underline{2})\,w(\underline{3})\cdots w(\underline{s})\cdots w(\underline{N}). \tag{5.12}$$

In the case of identical statistical characteristics of particles,

$$w(\underline{s}) = w(\vec{r}_s,\zeta_s) \tag{5.13}$$

and, integrating over all ζ_s, we obtain:

$$\langle f\rangle = \iint\cdots\int [f]_\zeta w(\vec{r}_1)w(\vec{r}_2)\cdots w(\vec{r}_s)\cdots w(\vec{r}_N)\,\mathrm{d}\vec{r}_1\mathrm{d}\vec{r}_2\cdots\mathrm{d}\vec{r}_N \tag{5.14}$$

where $[f]_\zeta$ is the mean value of f corresponding to the average scatterer characteristics (in shape and orientation).

Probability density function $w(\vec{r}_s)$ can be interpreted as follows:

$w(\vec{r}_s)\mathrm{d}\vec{r}_s$ = the probability of occurrence of the s-th scatterer in the elementary volume

$$= \mathrm{d}\vec{r}_s = \frac{\text{the number of scatterers inside } \mathrm{d}\vec{r}_s = \mathrm{d}x_s\mathrm{d}y_s\mathrm{d}z_s}{\text{the total number of scatterers in } V} = \frac{\rho(\vec{r}_s)\mathrm{d}\vec{r}_s}{N}, \tag{5.15}$$

where $\rho(\vec{r}_s)$ is the local concentration of particles, i. e., the number of scatterers in a unit volume. Therefore,

$$w(\vec{r}_s) = \frac{\rho(\vec{r}_s)}{N}. \tag{5.16}$$

For a constant concentration of particles in volume V, we have $\rho = N/V$ and $w(\vec{r}_s) = 1/V$. In this case, mean value (5.14) is given by

$$\langle f\rangle = \iint\cdots\int [f]_\zeta \frac{\rho(\vec{r}_1)\rho(\vec{r}_2)\cdots\rho(\vec{r}_N)}{N^N}\mathrm{d}\vec{r}_1\mathrm{d}\vec{r}_2\cdots\mathrm{d}\vec{r}_N. \tag{5.17}$$

If the mean value $[f]_\zeta$ depends only on the position of the s-th scatterer and is independent on the position of the other scatterers, then $[f]_\zeta = f(\vec{r}_s)$ and it is possible to integrate (5.17) over all $\vec{r}_1, \cdots, \vec{r}_N$ except for $\vec{r}_s$. Since

$$\int w(\vec{r}_1)\mathrm{d}\vec{r}_1 = \int\frac{\rho(\vec{r}_1)\mathrm{d}\vec{r}_1}{N} = 1,$$

we obtain:

$$\langle f(\vec{r}_s)\rangle = \int f(\vec{r}_s)\frac{\rho(\vec{r}_s)}{N}\mathrm{d}\vec{r}_s. \tag{5.18}$$

Spatial integrations are performed over entire volume V. If mean value $[f]_\zeta$ depends on the position of two different scatterers (the s-th and the t-th), then, upon writing $[f] = \mathrm{f}(\vec{r}_s, \vec{r}_t)$, we obtain:

$$\langle f(\vec{r}_s, \vec{r}_t) \rangle = \iint f(\vec{r}_s, \vec{r}_t) \frac{\rho(\vec{r}_s)\rho(\vec{r}_t)}{N^2} \mathrm{d}\vec{r}_s \mathrm{d}\vec{r}_t. \tag{5.19}$$

Relations (5.18) and (5.19) can be easily generalized to any number of scatterers. At large scatterer concentrations, a two-point probability distribution function must be introduced[2,6,15].

5.4 The Foldy-Tversky basic integral equation for a coherent field

Let us suppose now that the particles filling volume V move randomly and consider the field at the internal point $\vec{r} \subset V$. In the general case, field ψ^a varies in time because of random time variations of the particle coordinates, although this variation is much slower than that due to time factor $\exp(-\mathrm{i}\omega t)$. A typical measurement of the amplitude takes a large amount of time during which the electromagnetic signal is averaged over a representative set of positions and states of the particles. Consequently, it is often convenient to decompose field ψ^a into the mean (or coherent) field $\langle \psi^a \rangle$ and a fluctuating (or noncoherent) field ψ_f^a.

Statistical averaging is performed along those coordinates and states of all particles that are physically realizable during the measurement time. It is very important to recognize that coherent field $\langle \psi^a \rangle$ defined in this way is not a real physical field and it is a purely mathematical construction. Indeed, if we restore the time harmonic factor $\exp(-\mathrm{i}\omega t)$, which we omitted for brevity so far, we should conclude that the mean value of the actual electric field is zero

$$\frac{1}{T}\int_t^{t+T} \mathrm{d}t' \exp(-\mathrm{i}\omega t')_{T \gg 2\pi/\omega} = 0.$$

On the contrary, the coherent field does not disappear since it is defined as the time-averaged part of the electric field, which does not contain the $\exp(-\mathrm{i}\omega t)$ factor. The only reason for introducing the coherent field is that it appears eventually in formulae for quantities describing the multiply scattered radiation and can be actually measured using a suitable device. These quantities are determined in such a way that factor $\exp(-\mathrm{i}\omega t)$ naturally disappears when it is multiplied by its complex conjugate analogue.

The square of the mean field amplitude is the coherent intensity $|\langle \psi^a \rangle|^2$. The mean square of the fluctuating field amplitude is the incoherent intensity $\langle |\psi_f^a|^2 \rangle$. The total intensity is the mean square of the total field amplitude $\langle |\psi^a|^2 \rangle$; it is equal to the sum of the coherent and incoherent intensities:

$$\langle |\psi^a|^2 \rangle = \langle |\langle \psi^a \rangle + \psi_f^a|^2 \rangle = |\langle \psi^a \rangle|^2 + \langle |\psi_f^a|^2 \rangle. \tag{5.20}$$

Following references [2, 5, 6], let us consider briefly and at a qualitative level the normal incidence of a plane wave onto a semi – infinite region with random scatterers. The coherent intensity decreases due to scattering and absorption according to the following law:

$$C = \text{the coherent intensity} = \exp(-\rho\sigma_t z), \tag{5.21}$$

where σ_t is the sum of scattering and absorption cross sections.

The dissipated power is an incoherent power that contributes to the total intensity. As a result, total intensity T depends essentially on only the absorption:

$$T = \text{the total intensity} \approx \exp(-\rho\sigma_a z). \tag{5.22}$$

Therefore, incoherent intensity I can be approximated by the following expression:

$$I \approx \exp(-\rho\sigma_a z) - \exp(-\rho\sigma_t z). \tag{5.23}$$

In the transfer theory, coherent and incoherent intensities correspond to the attenuated incident and the diffuse intensity, respectively. Let us consider a coherent field using the Tversky theory and Eq. (5.9):

$$\langle\psi^a\rangle = \varphi_i^a + \sum_{s=1}^{N}\langle u_s^a\varphi_i^s\rangle + \sum_{s=1}^{N}\sum_{t=1,t\neq s}^{N}\langle u_s^a u_t^s\varphi_i^t\rangle + \sum_{s=1}^{N}\sum_{t=1,t\neq s}^{N}\sum_{m=1,m\neq t,m\neq s}^{N}\langle u_s^a u_t^s u_m^t\varphi_i^m\rangle + \cdots \tag{5.24}$$

or using Eqs. (5.17) – (5.19):

$$\langle\psi^a\rangle = \varphi_i^a + \sum_{s=1}^{N}\int u_s^a\varphi_i^s w(\vec{r}_s)\,\mathrm{d}\vec{r}_s + \sum_{s=1}^{N}\sum_{t=1,t\neq s}^{N}\iint u_s^a u_t^s\varphi_i^t w(\vec{r}_s)w(\vec{r}_t)\,\mathrm{d}\vec{r}_s\mathrm{d}\vec{r}_t +$$

$$\sum_{s=1}^{N}\sum_{t=1,t\neq s}^{N}\sum_{m=1,m\neq t,m\neq s}^{N}\iiint u_s^a u_t^s u_m^t\varphi_i^m w(\vec{r}_s)w(\vec{r}_t)w(\vec{r}_m)\,\mathrm{d}\vec{r}_s\mathrm{d}\vec{r}_t\mathrm{d}\vec{r}_m + \cdots \tag{5.25}$$

Taking into account Eq. (5.16), we obtain:

$$\langle\psi^a\rangle = \varphi_i^a + \int u_s^a\varphi_i^s\rho(\vec{r}_s)\,\mathrm{d}\vec{r}_s + \frac{N(N-1)}{N^2}\iint u_s^a u_t^s\varphi_i^t\rho(\vec{r}_s)\rho(\vec{r}_t)\,\mathrm{d}\vec{r}_s\mathrm{d}\vec{r}_t +$$

$$\frac{N(N-1)(N-2)}{N^3}\iiint u_s^a u_t^s u_m^t\varphi_i^m\rho(\vec{r}_s)\rho(\vec{r}_t)\rho(\vec{r}_m)\,\mathrm{d}\vec{r}_s\mathrm{d}\vec{r}_t\mathrm{d}\vec{r}_m + \cdots \tag{5.26}$$

In the limit $N\to\infty$, we have:

$$\langle\psi^a\rangle = \varphi_i^a + \int u_s^a\varphi_i^s\rho(\vec{r}_s)\,\mathrm{d}\vec{r}_s + \iint u_s^a u_t^s\varphi_i^t\rho(\vec{r}_s)\rho(\vec{r}_t)\,\mathrm{d}\vec{r}_s\mathrm{d}\vec{r}_t +$$

$$\iiint u_s^a u_t^s u_m^t\varphi_i^m\rho(\vec{r}_s)\rho(\vec{r}_t)\rho(\vec{r}_m)\,\mathrm{d}\vec{r}_s\mathrm{d}\vec{r}_t\mathrm{d}\vec{r}_m + \cdots \tag{5.27}$$

In the course of derivation of formula (5.27), the following relations were used:

$$\sum_{s=1}^{N}\langle u_s^a\varphi_i^s\rangle = \sum_{s=1}^{N}\int(u_s^a\varphi_i^s)\frac{\rho(\vec{r}_s)}{N}\mathrm{d}\vec{r}_s = \int u_s^a\varphi_i^s\rho(\vec{r}_s)\,\mathrm{d}\vec{r}_s,$$

$$\sum_{s=1}^{N}\sum_{t=1,t\neq s}^{N}\langle u_s^a u_t^s\varphi_i^t\rangle = \sum_{s=1}^{N}\sum_{t=1,t\neq s}^{N}\iint u_s^a u_t^s\varphi_i^s\frac{\rho(\vec{r}_s)\rho(\vec{r}_t)}{N^2}\mathrm{d}\vec{r}_s\mathrm{d}\vec{r}_t =$$

$$\frac{(N-1)}{N}\iint u_s^a u_t^s\varphi_i^s\rho(\vec{r}_s)\rho(\vec{r}_t)\,\mathrm{d}\vec{r}_s\mathrm{d}\vec{r}_t, \tag{5.28}$$

which, in the limit $N\to\infty$, produces

$$\iint u_s^a u_t^s\varphi_i^s\rho(\vec{r}_s)\rho(\vec{r}_t)\,\mathrm{d}\vec{r}_s\mathrm{d}\vec{r}_t.$$

We should note that indices s, t, $\cdots$ are now not summation indices, and they are used to denote only different integration variables. Eq. (5.27) is the complete vector version of the solution obtained by Tversky for scalar waves in 1964.

Eq. (5.27) is equivalent to the Foldy-Tversky integral equation:

$$\langle \psi^a \rangle = \varphi_i^a + \int u_s^a \langle \psi^s \rangle \rho(\vec{r}_s) \mathrm{d}\vec{r}_s, \tag{5.29}$$

since integration of Eq. (5.29) leads to Eq. (5.27).

Integral Eq. (5.29) is the basic equation for the coherent field in the Tversky theory. Foldy derived it as a certain approximation, and Tversky determined its physical meaning. Quantity $\langle \psi^a \rangle$, determined by integral Eq. (5.29) coincides essentially with the mean value of field which ψ^a, is shown in Fig. 5.5a.

5.5 Tversky integral equation for the correlation function

Let us consider the physical meaning of the Tversky integral equation for the intensity, which agrees with Foldy-Tversky integral Eq. (5.29) for the coherent field. The Tversky integral equation can be written in the following form:

$$\langle \psi^a \psi^{b*} \rangle = \langle \psi^a \rangle \langle \psi^{b*} \rangle + \int v_s^a v_s^{b*} \langle |\psi^s|^2 \rangle \rho(\vec{r}_s) \mathrm{d}\vec{r}_s, \tag{5.30}$$

where contribution v_s^a satisfies the integral equation:

$$v_s^a = u_s^a + \int u_t^a v_s^t \rho(\vec{r}_t) \mathrm{d}\vec{r}_t, \tag{5.31}$$

and the asterisk designates transition to the complex conjugate.

The field moment $\langle \psi^a \psi^{b*} \rangle$ is determined by the pair of integral Eqs. (5.30) and (5.31). To explain the physical meaning of these equations, it is necessary to take their iterations. For Eq. (5.31), we have

$$v_s^a = u_s^a + \int u_t^a u_s^t \rho(\vec{r}_t) \mathrm{d}\vec{r}_t + \int u_t^a u_m^t u_s^m \rho(\vec{r}_t) \rho(\vec{r}_m) \mathrm{d}\vec{r}_t \mathrm{d}\vec{r}_m + \cdots \tag{5.32}$$

Here, first term u_s^a describes scattering by scatterer s at point $\vec{r}_a$ (Fig. 5.6). In the limit $N \to \infty$, the second term has the following form:

$$\int u_t^a u_s^t \rho(\vec{r}_t) \mathrm{d}\vec{r}_t = \sum_{t=1, t \neq s}^{N} \langle u_t^a u_s^t \rangle_s, \tag{5.33}$$

where the angle brackets $\langle \rangle_s$ denote averaging over the characteristics of scatterer t under the assumption that the parameters of particle s are fixed.

Expression (5.33) describes the wave scattered first by particle s, then by particle t, and reaching point $\vec{r}_a$. The third term in Eq. (5.32) describes the wave propagating from particle s to particle m, then to particle t, and finally to point $\vec{r}_a$. Thus, contribution v_s^a describes all processes of multiple scattering from particle s to point a with participation of different scatterers, as shown in Fig. 5.6.

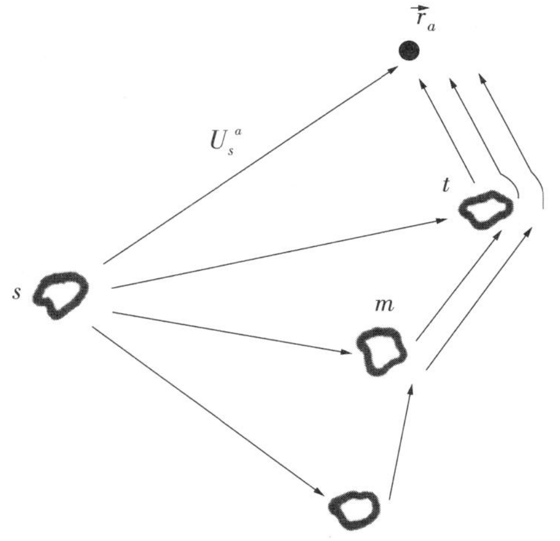

Fig. 5. 6 Scattering for the contribution of v_s^a

Similarly, we integrate Eq. (5. 30):

$$\langle \psi^a \psi^{b*} \rangle = \langle \psi^a \rangle \langle \psi^{b*} \rangle + \int v_s^a v_s^{b*} \, |\langle \psi^s \rangle|^2 \rho(\vec{r}_s) \mathrm{d}\vec{r}_s + \int v_s^a v_s^{b*} v_t^s v_t^{s*} \, |\langle \psi^t \rangle|^2 \rho(\vec{r}_s) \cdot$$

$$\rho(\vec{r}_t) \mathrm{d}\vec{r}_s \mathrm{d}\vec{r}_t + \int v_s^a v_s^{b*} v_t^s v_t^{s*} v_m^t v_m^{t*} \, |\langle \psi^m \rangle|^2 \rho(\vec{r}_s) \rho(\vec{r}_t) \rho(\vec{r}_m) \mathrm{d}\vec{r}_s \mathrm{d}\vec{r}_t \mathrm{d}\vec{r}_m + \cdots \quad (5.34)$$

The first term in (5. 34) is the product of the coherent field at point a and the complex conjugate coherent field at point b. Since quantity $\langle \psi^a \rangle$ is the mean value of field, which corresponds to the sum of all multiple scatterings shown in Fig. 5. 5a, this term can be represented as shown in Fig. 5. 7a.

The next term

$$\int v_s^a v_s^{b*} \, |\langle \psi^s \rangle|^2 \rho(\vec{r}_s) \mathrm{d}\vec{r}_s$$

represents the wave at point a generated by the scattering (Fig. 5. 6) of the coherent field at point s and the wave at point b caused by scattering of the complex conjugate field v_s^{b*}. This term is shown in Fig. 5. 7b. The diagram of the third term is shown in Fig. 5. 7c.

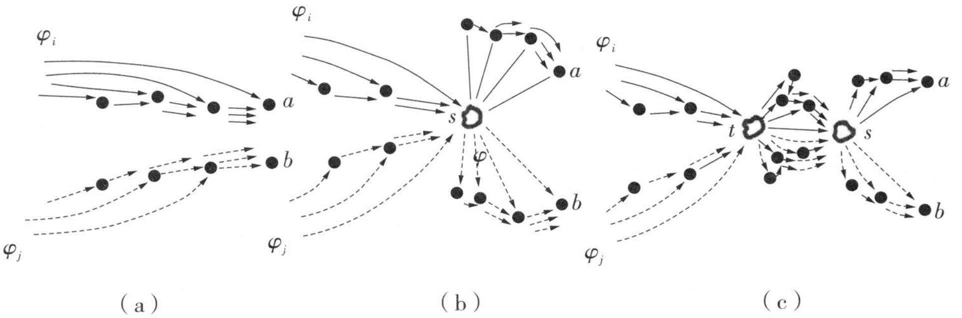

Fig. 5. 7 Scattering corresponding to the (a) first, (b) second, and (c) third terms of Eq. (5. 34)

Continuing this process, we conclude that the Tversky integral equation can be obtained by averaging the product of fields ψ^a and $\langle \psi^{b*} \rangle$, which the main scattering processes (5.9) describe illustrated in Fig. 5.5a. Thus, both the Foldy-Tversky integral equation for a coherent field and the Tversky integral equation for the intensity take into account the same scattering processes described by the expression (5.9) and, therefore, these equations are consistent with each other.

We should note that these equations correspond to the first smoothed approximation in more rigorous Dyson and Bethe-Salpeter equations, which can be derived by using diagram methods[4-6].

5.6 Coherent field

Let us consider the case of the normal incidence of a plane wave onto a layer with thickness d containing a large number of scatterers[5,6]. The incident wave propagating along the z axis is given by

$$\varphi_i(r) = \exp(ikz). \tag{5.35}$$

It is necessary to determine coherent field $\langle \psi \rangle$ inside the layer, which satisfies Foldy-Tversky integral Eq. (5.29). Coherent field $\langle \psi \rangle$ and the layer geometry are independent of coordinates x and y; therefore, the value of $\langle \psi \rangle$ should correspond to a plane wave propagating in the $+z$ direction. In the case of point $\vec{r}_a$ located in the far zone with respect to the scatterer, at point $\vec{r}_a$, we approximately have

$$u_s^a \langle \psi^s \rangle = f(\hat{0},\hat{i}) \frac{\exp(ik|\vec{r}_a - \vec{r}_s|)}{|\vec{r}_a - \vec{r}_s|} \langle \psi^s \rangle, \tag{5.36}$$

where $\hat{i}$ is the unit vector in the direction of propagation of wave $\langle \psi^s \rangle$ and $\hat{0}$ is the unit vector in the direction $\vec{r}_a - \vec{r}_s$.

Using Eqs. (5.35) and (5.36), we bring the integral Eq. (5.29) to the following form:

$$\langle \psi(z) \rangle = \exp(ikz) + \int_0^d dz_s \int_{-\infty}^{\infty} dx_s \int_{-\infty}^{\infty} dy_s f(\hat{0},\hat{i}) \frac{\exp(ik|\vec{r} - \vec{r}_s|)}{|\vec{r} - \vec{r}_s|} \rho(z_s) \langle \psi(z_s) \rangle . \tag{5.37}$$

When finding the coherent field inside the layer of $0 < z < d$, integration in Eq. (5.37) with respect to variables x_s and y_s is performed by using the stationary phase method. Then, neglecting the integral with factor $f(-\hat{i},\hat{i})$ (the backscattering amplitude) in the final expression, we write integral Eq. (5.37) in the following form:

$$\langle \psi(z) \rangle = \exp(ikz) + \int_0^z dz_s \frac{2\pi i}{k} \exp[ik(z - z_s)] f(\hat{i},\hat{i}) \rho(z_s) \langle \psi(z_s) \rangle. \tag{5.38}$$

Further, assuming that particle density $\rho(z_s)$ is constant, we obtain:

$$\langle \psi(z) \rangle = \exp(ikz)\left[1 + \frac{2\pi i}{k} f(\hat{i},\hat{i}) \rho \int_0^z \exp(-ikz_s) \langle \psi(z_s) \rangle dz_s \right]. \tag{5.39}$$

Integral Eq. (5.39) is solved exactly using the substitution

$$\langle \psi(z) \rangle = A\exp(iKz). \tag{5.40}$$

As a result, we obtain

$$A = 1, K = k + \frac{2\pi f(\hat{i},\hat{i})\rho}{k}. \tag{5.41}$$

The solution (5.40) and (5.41) means that, when a plane wave is incident onto a layer, the mean field propagates in the layer with propagation constant K.

In the general case of an arbitrary wave incident on the layer, mean field $\langle \psi \rangle$ can be described assuming that it satisfies the following wave equation:

$$(\nabla^2 + K^2)\langle \psi(\vec{r}) \rangle = 0, \tag{5.42}$$

where the complex wave number $K = k + [2\pi f(\hat{i},\hat{i})/k]\rho$.

We should note in Ref. [5] that scattering amplitude $f(\hat{i},\hat{i})$ is a complex quantity even in the case of non-absorbing scatterers; therefore, coherent field $\langle \psi(\vec{r}) \rangle$ is attenuated during propagation. This type of attenuation is due to scattering and is related to the scattering cross section. For clarification, let us consider the coherent intensity of a plane incident wave. In this case, we have:

$$|\langle \psi(z) \rangle|^2 = \exp\left\{ -\left[\frac{4\pi\rho}{k}\mathrm{Im}f(\hat{i},\hat{i}) \right] z \right\}. \tag{5.43}$$

According to the optical theorem, which follows from the energy conservation law during scattering of waves, the energy withdrawn from the incident wave is spent for scattering and absorption:

$$\frac{4\pi}{k}\mathrm{Im}f(\hat{i},\hat{i}) = \sigma_s + \sigma_a \tag{5.44}$$

where σ_s is the scattering cross section, and σ_a is the absorption cross section.

Then formula (5.43) takes the following form:

$$|\langle \psi(z) \rangle|^2 = \exp[-\rho(\sigma_s + \sigma_a)z], \quad 0 < z < d. \tag{5.45}$$

For the region outside the layer at $z > d$, it is necessary to substitute (5.40) in (5.39) and replace the upper limit of the integral by d. As a result, we obtain:

$$\langle \psi(z) \rangle = \exp[iKd + ik(z - d)], \tag{5.46}$$

$$|\langle \psi(z) \rangle|^2 = \exp[-\rho(\sigma_s + \sigma_a)d], \quad z > d. \tag{5.47}$$

Thus, the coherent intensity attenuates exponentially, and the attenuation constant is proportional to the scatterer density and the total cross section of $(\sigma_s + \sigma_a)$.

The analysis carried out is related to the case of the incidence of a plane wave onto a layer, but the generalization of this approach using Eq. (5.42) is a good approximation for many practical situations.

5.7 Fractal discrete randomly inhomogeneous media

At first, we present a brief review of existing theoretical methods for determining the electrodynamic characteristics of a single particle. These quantities are also necessary for description of wave scattering by an individual random particle as well as a small group of random particles or targets.

Scattering of light by one particle depends on three key factors: the size of the scatterer (in comparison with the wavelength) and its shape and refractive index. An unlimited variability of particles in natural and anthropogenic environments is an insurmountable problem in the theoretical description of propagation and scattering of waves by such particles or scatterer clusters. The case of spherically symmetric particles is an exception since the classical Lorentz – Mie theory (very efficient and numerically exact) or one of its extensions is used to solve it. Most existing exact theoretical approaches belong to one of two broad categories.

Methods of differential equations make it possible to calculate the scattered field by solving the Maxwell equations or the vector wave equation in the frequency or time domain while methods of integral equations are based on volume or surface integral analogs of the Maxwell equations.

Approximate theories are still the main source of physical understanding of scattering and absorption by non-spherical particles. In addition, the geometric optics approximation will likely never become obsolete since its accuracy increases as the particle size parameter grows, whereas all exact theoretical methods for non-spherical particles cease to be practically usable when the size parameter exceeds a certain threshold. However, this method is approximate by definition, and the range of its applications under the conditions of the parameter of the smallest size should be verified by comparing the obtained results with exact numerical solutions of the Maxwell equations.

A review of theoretical and experimental methods for determining single-particle characteristics is presented in Ref. [21] (see also the bibliography herein). In this paper, we consider the problem of multiple scattering only for radar problems, excluding the highresolution optics and lidar systems; therefore, fractality must first manifest itself in the spatial distribution of scattering particles, i. e., in large space-time fractal clusters[14,15]. A snow layer can be an example of such a fractal scattering medium.

Regarding the fractal shape of a single particle (ice crystals in clouds and snow cover), it can be noted that modern theoretical and experimental studies in optics and the IR range[6,14,15,21] have shown that a rather uniform scattering amplitude $f(\hat{0},\hat{i})$ prevails in comparison with other shapes of particles in the whole range of scattering angles from 0° to 180°.

5.8 Modification of the classical Foldy-Tversky theory for fractal discrete randomly inhomogeneous media

As is well known, the idea of fractality is based on the absence of a characteristic length, that is, on selfsimilarity[7,14-16]. Fractal dimension D shows how densely the configuration of the medium or the object in which they are located fills the metric space. For scattering and diffraction, application of fractal ideas is widely represented in Ref. [14, 15].

Let us consider some details of modification of the classical Foldy-Tversky theory for fractal discrete randomly inhomogeneous media.

Let us suppose that there is a fractal object of size l_0. Based on the definition of the D fractal dimension, we find that, when the length or scale varies from l_0 to l ($l < l_0$), N objects that contain some part of the fractal are obtained as follows:

$$N = \left(\frac{l_0}{l}\right)^D. \tag{5.48}$$

If all fractals are similar, then we have a homogeneous fractal. We denote the Euclidean dimension by E and rewrite Eq. (5.48) in the form

$$Nl^E = \left(\frac{l_0}{l}\right)^D l^E. \tag{5.49}$$

Probability density function w (5.15) for a fractal medium can be estimated as follows. The probability w (l) of occupying by a fractal of a part of space is its volume Nl^E, divided by the total volume l_0^E. Then,

$$w(l) = \frac{Nl^E}{l_0^E} = \left(\frac{l}{l_0}\right)^{E-D}. \tag{5.50}$$

When the Foldy-Tversky theory is modified for a fractal scattering medium, expression (5.50) should be taken into account in the first approximation in integral Eqs. (5.26) and (5.27) for the coherent field. Then, we obtain

$$\langle \psi^a \rangle = \varphi_i^a + \int u_s^a \varphi_i^s w(\vec{r}_s)\,\mathrm{d}\vec{r}_s + \iint u_s^a u_t^s \varphi_i^t w(\vec{r}_s) w(\vec{r}_t)\,\mathrm{d}\vec{r}_s \mathrm{d}\vec{r}_t +$$

$$\iiint u_s^a u_t^s u_m^t \varphi_i^m w(\vec{r}_s) w(\vec{r}_t) w(\vec{r}_m)\,\mathrm{d}\vec{r}_s \mathrm{d}\vec{r}_t \mathrm{d}\vec{r}_m + \cdots \tag{5.51}$$

Similarly, the same mathematical operations can be performed for a cascade of fractals nested into each other, or for a chain of fractals. These types of complex fractal and multifractal clusters are widely distributed in nanotechnologies and nature[14-20].

Eqs. (5.30) and (5.34) of second field moment $\langle \psi^a \psi^{b*} \rangle$ for a fractal scattering medium take the following form:

$$\langle \psi^a \psi^{b*} \rangle = \langle \psi^a \rangle \langle \psi^{b*} \rangle + \int v_s^a v_s^{b*} \langle |\psi^s|^2 \rangle w(\vec{r}_s)\,\mathrm{d}\vec{r}_s,$$

$$\langle \psi^a \psi^{b*} \rangle = \langle \psi^a \rangle \langle \psi^{b*} \rangle + \int v_s^a v_s^{b*} |\langle \psi^s \rangle|^2 w(\vec{r}_s)\,\mathrm{d}\vec{r}_s + \int v_s^a v_s^{b*} v_t^s v_t^{s*} |\langle \psi^t \rangle|^2 w(\vec{r}_s) \cdot$$

$$w(\vec{r}_t)\mathrm{d}\vec{r}_s\mathrm{d}\vec{r}_t + \int v_s^a v_s^{b*} v_t^s v_t^{s*} v_m^t v_m^{t*} |\langle\psi^m\rangle|^2 w(\vec{r}_s)w(\vec{r}_t)w(\vec{r}_m)\mathrm{d}\vec{r}_s\mathrm{d}\vec{r}_t\mathrm{d}\vec{r}_m + \cdots, \tag{5.52}$$

where

$$v_s^a = u_s^a + \int u_t^a v_s^t w(\vec{r}_t)\mathrm{d}\vec{r}_t. \tag{5.53}$$

We calculate coherent field $\langle\psi\rangle$ for the normal incidence of a plane wave onto a fractal layer with thickness d containing a large number of scatterers. The incident field has the form of plane wave (5.35) propagating along the z axis. In view of the foregoing, we have:

$$\langle\psi(z)\rangle = \mathrm{e}^{ikz} + \int_0^d \mathrm{d}z_s \int_{-\infty}^{\infty} \mathrm{d}x_s \int_{-\infty}^{\infty} \mathrm{d}y_s w(z_s) f(\hat{0},\hat{i}) \frac{\exp(ik|\vec{r}-\vec{r}_s|)}{|\vec{r}-\vec{r}_s|}\langle\psi(z_s)\rangle. \tag{5.54}$$

Using the above procedure (see Section 5.5), it can be seen mean field $\langle\psi\rangle$ in a fractal scattering medium satisfies the following wave equation:

$$(\nabla^2 + K^2)\langle\psi(\vec{r})\rangle = 0, \tag{5.55}$$

where the complex wave number

$$K = k + \frac{2\pi f(\hat{i},\hat{i})w(\vec{r})}{k}. \tag{5.56}$$

In accordance with Eq. (5.37), integral Eq. (5.29) for the fractal layer has the following form:

$$\langle\psi(z)\rangle = \exp(ikz) + \int_0^d \mathrm{d}z_s \int_{-\infty}^{\infty} \mathrm{d}x_s \int_{-\infty}^{\infty} \mathrm{d}y_s f(\hat{0},\hat{i}) \frac{\exp(ik|\vec{r}-\vec{r}_s|)}{|\vec{r}-\vec{r}_s|} w(z_s)\langle\psi(z_s)\rangle. \tag{5.57}$$

Based on the solution of wave Eq. (5.55), we find that the coherent intensity of a plane incident wave

$$|\langle\psi(z)\rangle|^2 = \exp\left\{-\left[\frac{4\pi w(z)}{k}\mathrm{Im}f(\hat{i},\hat{i})\right]z\right\}. \tag{5.58}$$

For a fractal layer with thickness d, the coherent intensity of a plane incident wave taken with allowance for optical theorem (5.44) and Eqs. (5.45) and (5.47) has the following form:

$$|\langle\psi(z)\rangle|^2 = \exp[-w(r)(\sigma_s + \sigma_a)d] \text{ for } z > d, \tag{5.59}$$

$$|\langle\psi(z)\rangle|^2 = \exp[-w(r)(r)(\sigma_s + \sigma_a)z] \text{ for } 0 < z < d. \tag{5.60}$$

Based on the modification of the classical Foldy-Tversky theory, we obtained a developed general theory for multiple wave scattering in fractal randomly inhomogeneous media.

5.9 Radar equation in two perfect cases of probing

In the radar probing of a target in the atmosphere, the power dependence of the received signal on range r is well known in at least two perfect cases. The power of the signal from a point target is inversely proportional to the fourth degree of the observation range and the square of the range for a homogeneous medium. Inhomogeneous natural media, such as the average atmosphere, have intermittent or heterogeneous structures that do not correspond to any of these cases. Thin, vertically stratified undulating turbulence layers are often observed in the entire average atmosphere; therefore, the validity of the exact dependence of the signal power on r^{-4} or r^{-2} in radar experiments of the middle atmosphere remains questionable[14,15,24].

Properties of natural objects that have structural elements in the hierarchy of space-time scales cannot be described by smooth functions. Their structure can be represented by fractals[14,15]. Therefore, intermediate cases of inhomogeneous or randomly inhomogeneous media are of interest for fractal modeling. Such real fractal models that contain stratification effects can also be useful in radio physical and radar studies.

Variation in power P_s of the scattered signal in the two perfect cases mentioned above is determined by the radar equation. Let us consider a monostatic radar experiment at wavelength λ_0 (frequency f_0), size d of the antenna aperture, and the antenna gain $G = 4\pi A/\lambda^2$, where A is the effective antenna area. The width of the antenna pattern $\theta_a = \alpha\lambda_0/d$, where $\alpha \approx 1$. We assume that the target is in the far zone.

First, we consider a point target with effective cross section σ situated at range r. The received signal power is determined by the radar equation:

$$P_s = \frac{P_t G}{4\pi r^2}\sigma \frac{A}{4\pi r^2}L = \frac{P_t A^2 L\sigma}{4\pi\lambda_0^2 r^4}, \tag{5.61}$$

where P_t is the transmitter power and L is the coefficient that takes into account all losses. The P_s received power is seen to fall with the r distance as r^{-4}.

Next, let us consider a homogeneous ensemble of a set of point targets randomly distributed in space. If we assume that the point targets are statistically independent, then received signal power P_s is obtained by summing the energy contributions from all point targets in the ensemble. The efficiency of scattering by an ensemble of electromagnetic waves is determined by cross section per unit volume σ_s. The region of the medium that contributes to received power P_s is located at range r and is determined by beamwidth θ_a and radial resolution Δr. Thus, the effective volume contributing to P_s is $V = r^2\delta\Omega\Delta r$, where $\delta\Omega = \theta_a^2/4$ is the solid angle of the ray. Then, the radar equation becomes as follows:

$$P_s = \frac{P_t G}{4\pi r^2}V\sigma_v \frac{A}{4\pi r^2}L = \frac{P_t A\alpha^2 L}{64 r^2}\Delta r\sigma_v. \tag{5.62}$$

Resulting functional dependence (5.62) is due to the fact that volume V increases as r^2 and the energy contribution of each point target in volume V decreases as r^{-4}. Note that received power P_s is independent of radar wavelength λ_0, except for via σ_v.

Additionally, we should note that, when probing the Earth's surface at low grazing angles θ, the area of the distributed ground target illuminated by the antenna beam increases linearly with range r. Assuming the static independence of the elementary scatterers of the distributed target, we find that the power P_s decreases with increasing distance as r^{-3}.

The statistical independence of point targets in the flat area or volume should be usually assumed with caution. Thus, there is the statement: "… the problems associated with the influence of local objects are quite complex and have not yet been fully resolved, and only recently meteorological radar specialists have begun to pay attention to this field of research." [24]192. The exact power-law dependence of P_s on range r is a function of the measure of filling of the scattering region by inhomogeneities or irregularities. This dependence varies from r^{-4} for a point target to r^{-2} for a homogeneous medium completely filling entire scattering region V.

Intermediate cases are very important when probing real randomly inhomogeneous media, but they are difficult to analyze. The only computer modeling study on wave scattering by media with partial filling of the probing space based on fractal approximation is known to our best is Ref. [25] (see also Ref. [14, 15]). In Section 5. 9, we present the main results of this work for further research.

5. 10 Wave scattering in a fractal medium: First numerical simulation

Numerical simulation of planar fractal targets was carried out in Ref. [25]. In this study, the space is successively divided into smaller cells, which are reduced eventually to isotropic point targets. The process starts with a unit square (Euclidean dimension $E = 2$), then the square is divided into n^2 equal subcells with side n^{-1}, and this process is repeated. Dimension of the Euclidean space $E = 2$ is preserved since $E = 2 = \log(n^2)/\log(n)$.

Let us suppose now that only $p \geqslant n^{-2}$ subcells are filled at the division stage and the process recursively continues for these subcells. After a cycle of iterations, the unit square is filled with a random pattern of points with finite dimensions. The fractal dimension D of this pattern is at most E, but it may be close to zero. At each iteration, the number of subcells encompassed by the pattern is pn^2 and their side is reduced by factor n. The fractal dimension of D can now be defined as $\{2 + \log(p)/\log(n)\}$. The additional factor is practically negative or zero because $p \leqslant 1$. Fraction p can be also considered as a probability. The resulting pattern is really random and statistically self-similar as the iteration order increases. The degree at which a random point pattern fills a unit square can be controlled by choosing the value of p.

At $p = n^{-2}$, the value of D becomes zero. The extension of the obtained results to $E = 3$ is obvious; in this case, $D = \log(pn^3)/\log(n)$. Examples of realizations of random points on a plane are shown in Fig. 5. 8.

At each stage of division, the cell is divided into $4 \times 4 = 16$ subcells that are then selected with probability $p > 4^{-2}$. The steps are repeated four times because of the final size of the pixels. The p and D parameters increase from left to right. Two different implementations are shown for each case. Such patterns model specific realizations of point targets in a distributed random medium. Fractal dimension D controlled by probability p determines the degree to which

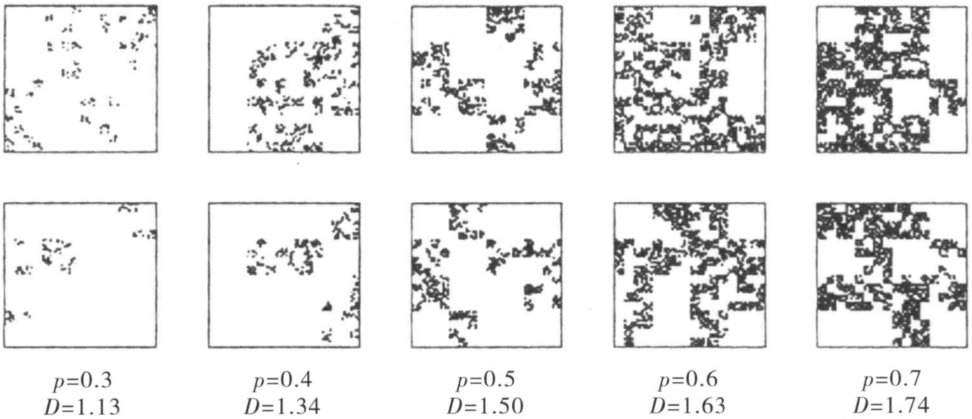

Fig. 5. 8 Random point patterns in the 256 × 256 grid; two different implementations are shown for each case[25]

the medium fills the plane[25]. These points can be considered as random point radar targets. Such examples serve as the basis for the numerical experiments described below.

The value of the effective scattering cross section is assigned to each point target. Point target s_i located at point r_i is illuminated by a beam uniform over θ. The scattering field E_i at the receiving antenna varies with the distance r_i to the point target under the assumption of the far zone as $(ri)^{-2}$, and its phase varies as $-4\pi r_i/\lambda_0$. Complex signal v_i at the receiver output is linearly connected with E_i. Power P_s of the signal from an ensemble of point targets is determined from the accumulated complex voltage $v=\Sigma_i v_i$, as $P_s = vv^*$. It should be noted that this method of determining P_s gives the "true" signal power for any arbitrary target structure. It does not assume the statistical independence of point targets, and all phase factors are directly included in the calculations. In real modeling, only planar targets that are $2D$ ensembles of point targets (Fig. 5. 8) were considered to remove computational constraints; therefore, we consider $D \leqslant 2$ inside the Euclidean space with dimension $E=3$.

The orientation of the planar target relative to the radar beam is shown in Fig. 5. 9. The radar beam is directed at a θ angle from the zenith. The flat target is oriented vertically in the plane determined by the zenith and the axis of the radar beam. Only the part of the planar target (indicated by hatching in Fig. 5. 9) that intersects with the radar volume at range r forms the signal received at this range. A flat target is formed by a fractal implementation of points, such as those shown in Fig. 5. 8. Fractal implementations of planar targets with $D \leqslant 2$ are generated by first division of the area of 4096 × 4096 points into 16 cells with a size of 4 × 4. Each subcell is then similarly subdivided iteratively. The total number of iterations in this division is six ($4096 = 4^6$). The subcell is included with probability p in each iteration. It should be noted that the subcells have various horizontal and vertical boundaries as an artifact of a simple fractal model used here. Full modeling of the volume target with 4096^3 or ~ 69 billion points is clearly impractical for computational reasons.

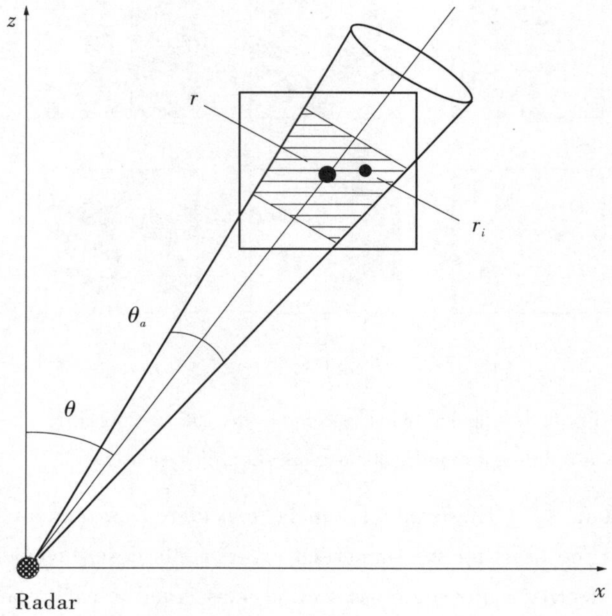

Fig. 5. 9 Geometry of probing of a fractal target

From the physical point of view, a linear dimension of 0. 1 m is associated with each point target. The linear dimension of the scattering region is ~ 0. 4 km. To simplify calculations, the beamwidth and the radial resolution are kept constant, namely, $\theta_a = 0.9°$ and $\Delta r = 0.32$ km. Five values are used for each of the following parameters: nominal range r changes from 5 to 20 km, zenith angle θ changes from 0° to 20°, and wavelength λ_0 changes from 3 to 1 m. Probability parameter p, which controls fractal dimension D is varied from 0. 3 to 0. 7 with a step of 0. 1. For each of the 625 individual cases, the value of P_s was averaged over 20 different implementations. Scaling for the dependence of P_s on any parameter, for example, the range, is then obtained by means of regression analysis. Below, only the dependence of signal power P_s on range r is discussed.

The parameter values are characteristic of typical experimental radars of the middle atmosphere[24]. At some radars with large antennas, near-field effects are significant. They are simple to include and are of interest for future work since the radar equations given above are valid only for the far zone.

For the dependence of the scattered signal power on distance, we determine this relation in the form of a power law[14,15,24,25]:

$$P_s = r^{-\beta}. \tag{5.63}$$

where exponent β is 4. 0 for a point target and 2. 0 for a target that fills spatially the beam, i. e. , $D = 3$.

Exponent β was obtained in Ref. [25] with the help of linear regression of log (P_s) by log (r) in the numerical experiment described above for five different values of D in the range from 1.1 to 1.74 at wavelength $\lambda_0 = 2.5$ m and probing angle $\theta = 10°$. Fig. 5.10 shows the values of parameter β for the five values of fractal dimension D obtained by means of computer simulation.

Points are mean values for more than 20 realizations in the numerical experiment. Parameter β first falls linearly with increasing D and slows down near $D \sim 2$ in the case of volume scattering.

The extreme cases of a point target ($D = 0$) and volume scattering by a homogeneous random medium ($D = 3$) are also identified and shown by circles in Fig. 5.10. The linear regressions (dotted line) by the method of least squares corresponding to five calculated points and the cubic spline (solid line) corresponding to all seven points are shown. For planar targets with $D \sim 1.9$, $\beta \sim 2.22$ was obtained in Ref. [25]. Consequently, the bulk case ($\beta = 2$) is rather common for targets that only intend to fill a plane.

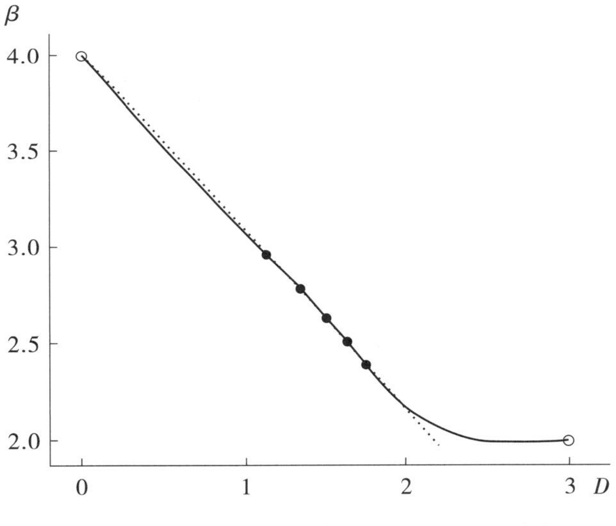

Fig. 5.10 Variation in parameter β as a function of fractal dimension D of the scattering medium: the dashed line is the least squares method, and the solid line is the cubic spline over all seven points

The spline fit shows that the limit of volume scattering at $\beta = 2.0$ is determined to within 5% at $D \sim 2.2$, and this limit is reached if $D > 2.4$.

It is of interest to find out how the results shown in Fig. 5.10 correspond to the r^{-3} range dependence of the terrestrial noise power P_s for radars, which was briefly discussed in Section 5.8. This situation corresponds to an almost horizontal antenna beam ($\theta \sim 90°$) and a horizontal flat target that contains statistically independent elements. The orientation of this planar target, however, is orthogonal to the position shown in Fig. 5.9. This situation can be considered as a set of many statistically independent linear targets, all of which are parallel to the ray. Each of these linear targets corresponds to fractal dimension $D=1$ in Fig. 5.10. Because of the assumed statistical independence of these linear targets, the exponent of the power law for P_s should be the same as for $D=1$. The actual value of this exponent β in Fig. 5.10 is ~ 3.07. The difference by 0.07 in the exponent β may be due to some two-point correlations within the subcell despite the assumed statistical independence of point targets and possible angular (θ) dependence in numerical experiments[25].

5.11 Wave scattering in a fractal medium and the radar equation

Fundamentals of the theory of multiple scattering of electromagnetic waves in fractal discrete randomly inhomogeneous media and disordered artificial dynamic fractal systems are presented in Ref. [26] – [29]. Here, we demonstrate on the basis of the results of study [30] how it is possible to use the results obtained by the analysis of radar signals when the scattering volume depends on fractal dimension D or fractal signature $D(\vec{r}, t)$. In accordance with the theory of wave scattering, it is possible to determine scattering cross section Σ_s and absorption cross section Σ_a (which can be measured experimentally) in the following form:

$$\Sigma_s = w_s\sigma_s \text{ и } \qquad \Sigma_a = w_s\sigma_a, \tag{5.64}$$

where w_s is relative volume of the fractal (5.50) introduced in Ref. [26] – [30].

Let us calculate the value of the backscattered signal from the fractal medium (Fig. 5.9) using the classical radar equation. The power of the received signal is determined from the radar equation[30]

$$P_s = \iint \frac{P_t}{4\pi r_1^2}\sigma_s \frac{1}{4\pi r_2^2}\exp[-2ik(r_1^2 - r_2^2)]\mathrm{d}r_1\mathrm{d}r_2. \tag{5.65}$$

Taking into account formulae (5.64), we obtain:

$$P_s = \frac{P_t\Sigma_s}{4\pi^2}\iint \frac{1}{4\pi r_1^2 r_2^2 w_s}\exp[-2ik(r_1^2 - r_2^2)]\mathrm{d}r_1\mathrm{d}r_2. \tag{5.66}$$

Using formula (5.50) for the relative volume of the fractal and the results from Ref. [30], we have:

$$P_s = \frac{AP_t\Sigma_s}{4\pi^2}\iint \frac{1}{r_1^2 r_2^2}\left(\frac{l}{l_0}\right)^{D-E}\exp[-2ik(r_1^2 - r_2^2)]\mathrm{d}r_1\mathrm{d}r_2 =$$

$$\frac{AP_t\Sigma_s}{4\pi^2}\int r_1\mathrm{d}r_1\mathrm{d}\theta_1\int r_2\mathrm{d}r_2\mathrm{d}\theta_2\frac{1}{r_1^2 r_2^2}\left(\frac{l}{l_0}\right)^{D-E}\exp[-2ik(r_1^2 - r_2^2)], \tag{5.67}$$

where l_0 is the size of the fractal target, l is the interval comparable to the wavelength, and $l < l_0$.

For the far zone $r_1 \approx r_2 \approx r$, and $E=2$ for a flat target. Then,

$$P_s = \frac{AP_t\Sigma_s}{4\pi^2}\iint \frac{1}{r_1^2 r_2^2}\left(\frac{|R|}{l_0}\right)^{D-2}\exp[-2ik(r_1^2 - r_2^2)]\mathrm{d}r_1\mathrm{d}r_2, \tag{5.68}$$

where

$$|R| = [r_1^2 + R_2^2 - 2r_1r_2\cos(\theta_1 - \theta_2)]^{1/2} \approx \sqrt{2}r\sqrt{1 - \cos(\theta_1 - \theta_2)}. \tag{5.69}$$

Consequently, the power of the scattered signal taken with consideration for the results from Ref. [30] is

$$P_s = \frac{AP_t\Sigma_s}{4\pi^2}\int_{\theta_{01}}^{\theta'_{01}}\int_{\theta_{02}}^{\theta'_{02}}[1 - \cos(\theta_1 - \theta_2)]^{(D-2)/2}\mathrm{d}\theta_1\mathrm{d}\theta_2 \times$$

$$\int_{r-(\Delta r/2)}^{r+(\Delta r/2)}\mathrm{d}r_1\int_{r-(\Delta r/2)}^{r+(\Delta r/2)}\frac{2^{(D-2)/2}r^{D-2}}{r^2 l_0^{D-2}}\exp[-2ik(r_1^2 - r_2^2)]\mathrm{d}r_2 = \frac{2^{(D-2)/2}AP_t\Sigma_s}{4\pi^2 l_0^{D-2}r^{4-D}}I_r I_\theta. \tag{5.70}$$

In expression (5.70), integrals I_r and I_θ have the following form:

$$I_r = \int_{r-(\Delta r/2)}^{r+(\Delta r/2)}\mathrm{d}r_1\int_{r-(\Delta r/2)}^{r+(\Delta r/2)}\exp[-2ik(r_1^2 - r_2^2)]\mathrm{d}r_2 = \frac{\sin^2(k\Delta r)}{16k^2}, \tag{5.71}$$

$$I_\theta = \int_{\theta_{01}}^{\theta'_{01}}\mathrm{d}\theta_1\int_{\theta_{02}}^{\theta'_{02}}[1 - \cos(\theta_1 - \theta_2)]^{(D-2)/2}\mathrm{d}\theta_2. \tag{5.72}$$

Hence, we can conclude on the basis of Eq. (5.70) that the following relation is fulfilled for the fractal medium:

$$P_s \propto \frac{1}{r^{4-D}}. \tag{5.73}$$

The obtained result (5.73) coincides with the experimental data in Fig. 5.10 (the linear part of the plot)[25].

For the 3D medium, it is easy to obtain ($E = 3$) the following relation[30]:

$$P_s \propto \frac{1}{r^{5-D}}. \tag{5.74}$$

Result (5.74) agrees with experimental data in Fig. 5.10 (the curvilinear part of the plot)[25].

The results show that it is possible to estimate fractal dimension D or fractal signature $D(\vec{r}, t)$ of the probed fractal medium or the fractal target [such as a dynamic snow layer (see also [14, 15, 22, 26-31])] based on the reflected radar signal.

5.12 "Thermodynamics" of clusters of unmanned aerial vehicles

Unmanned aerial vehicles (UAVs) are currently used to solve scientific and practical problems. They can be the main element of formation of an integrated information field. High-resolution digital images can be promptly obtained at any time of the day using optical, radar, and IR sensors located at UAV. The following modes are possible: scanning of a predetermined area of the terrain, search for targets along a linear route; route mapping; obtaining detailed images of small areas of interest and objects on them. Solution of complex problems is possible only as a result of grouped application of UAVs. A group is a collection of UAVs of similar or different types focused on a common target. In practice, it is possible to implement a significant number of types of UAV group flights. Their classification in the process of accomplishing the assigned tasks is shown in Fig. 5. 11[27,28,32]. In the figure, we first introduced fractal groupings of UAVs and fractal trajectories of UAV flights.

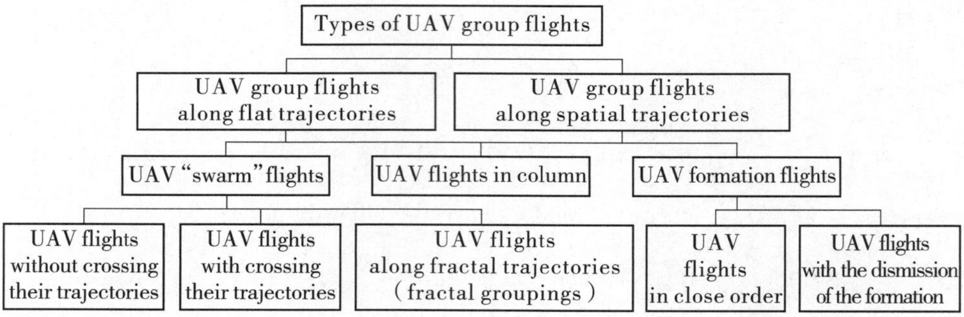

Fig. 5. 11 Classification of group flights of UAVs

Under real conditions, UAVs operate in a nondeterministic non-predictable environment and under the conditions of countermeasures. Each of the UAVs performs a number of actions aimed at solving a common problem. In this case, the problem of dispensing control arises in the field of solving problems. This is possible to ensure the network operation without the traffic congestion with minimal central control. It is necessary to draw a parallel with biological research, in which the researchers attempt to answer the question of the emergence of cooperative behavior in the process of evolution or the so-called evolutionary strategy[26-29,31-35]. Intellectual or population algorithms (methods) belong to the class of stochastic search optimization algorithms. Population algorithms are included in the class of heuristic algorithms for which convergence to the global solution is not proved, but it has been experimentally established that they give a fairly good solution. Properties of the members or agents of the population (in our case, UAVs): autonomy, stochasticity, limited representation, decentralization, communication skills. These properties of agents, even with their simplest behavioral strategies, contribute to formation of the population's swarm intelligence, i. e., "the synergetic paradigm with laws of complex systems" is activated. For poorly formalized problems with frequent multimodality and high dimensionality, population algorithms provide high probability of localization of the global extremum with a suboptimal solution. In formulation and improvement of solutions to the problems of collective interaction of UAVs, it is possible to apply new concepts on the theory of games with incomplete information presented (see Chapter 2).

The topical modern problem is to teach UAVs to fly like a flock of birds or a swarm of bees. Then, we can control one device, and the remaining devices will be controlled according to this technology. And if the leader dies for some reason, then the function of the pack leader is automatically transferred to the next. And it will be so until the last device. For a complex network of many micro- (nano-) UAVs, global monitoring of the territory and objects located on it is carried out. The problem can be considered within the framework of the concept of a distributed measuring medium, where each point of a certain dynamic medium is capable of performing sensory, measuring, and information functions. The fractal-graph approach allows one to study the growth of complex networks and manipulate such networks at a global level without detailed description. In addition, an excessive number of sensors (UAVs) does not guarantee the optimal distribution in/over the studied non-deterministic medium. Introduction of the fractal topology of such networks with consideration for the configuration of the investigated territory will allow more accurate monitoring of this territory and object detection with smaller means (number of UAVs).

5.13 Conclusions

The modified Foldy-Tversky method for multiple wave scattering in fractal discrete randomly inhomogeneous media has a very general character and allows consideration in a unified way of a large number of wave phenomena that can be explained by the fractal theory. The developed modification of the multiple scattering theory allowed one to include in consideration of values of fractal dimension D and fractal signature $D(\vec{r}, t)$ of the disordered large system.

Based on the modified Foldy-Tversky method for multiple wave scattering in fractal discrete randomly inhomogeneous media[27,28], the radar equation for a purely fractal medium is analytically considered. It is shown that the fractal signature $D(\vec{r}, t)$ can be used to study the dependence of volume scattering on range. Theoretical studies are consistent with early results obtained by foreign authors[25,30].

Similarly, it is possible to justify the solution for anisotropic disordered large fractal systems: cascades of fractals nested into each other, graphs formed from chains of fractals, percolation systems, nanosystems, cosmic debris, crowds of drones or small spacecrafts (SSCs) including mini and micro classes, dynamic synthesized space antenna groupings (cluster apertures), low-altitude and high-altitude pseudo satellites (HAPS) and their groupings, and spatially-distributed space systems (clusters) of small SSCs solving emergency monitoring problems.

Chapter 6 Examples of Fractal Devices and Their Theory

6.1 Introduction

The increasing complexity of the modern radio electronic hardware and functions it performs makes it essential to consider opportunities of the new physical principles of creation of new elementary base and the new devices and radio engineering systems. For this, the fractals theory and the deterministic chaos theory are of the great importance. Proposed in IREE RAS fractal devices, fractal radio systems and fractal radioelements bring many improved capabilities in modern radio electronics and can be widely used in practice in future.

6.2 Theoretical foundations of the created fractal-scaling methods

In the fractal-scaling approach proposed and having been developed in the V. A. Kotelnikov IREE of the RAS for 40 years, description and processing of signals and fields is carried out exclusively in fractional measure space using scaling hypotheses, heavy-tailed non-Gaussian stable distributions[1] and, as far as possible, using the apparatus of fractional integral derivatives[2-11]. Note that if an equation includes a time fractional derivative, it is interpreted so as there is memory or, in the case of a stochastic process, non-Markovism.

The main property of fractals is the non-integral value of their dimension D. Development of the dimension theory started from the works of Poincare, Lebesgue, Brauer, Uryson and Menger. In various areas of mathematics, there occur sets that are negligible in one sense or another and are indistinguishable in terms of Lebesgue measure. To distinguish between such sets with a hugely complicated topological structure, it is necessary to involve nontraditional characteristics of smallness, for example, capacity, potential, measures, and Hausdorff dimension, etc. The use of the Hausdorff fractional dimension, closely related to the concepts of entropy, fractals and strange attractors in the theory of dynamical systems, turned to be the most optimal[3,5,7,12].

The concept of a measure and Hausdorff dimension is defined by a p-dimensional measure with an arbitrary real positive number p introduced by Hausdorff in 1919. The concepts introduced by Hausdorff are based on the Carathéodory construct (1914). The Hausdorff dimension

$\dim_H A$ is defined in terms of the Hausdorff α-measure of the set $\text{mes}_{H,\alpha}$ as

$$\text{mes}_{H,\alpha} = \lim_{\varepsilon\to 0}\inf_{\Gamma(A)} \sum [d(U)]^{\alpha}, \quad (6.1)$$

where the lower bound inf is taken with respect to finite or counting coverings Γ of the set A by balls U, the diameters of which are $d(U) < \varepsilon$.

The dimension $\dim_H A$ is defined as such α_0 number that measure (6.1) is equal to zero for $\alpha > \alpha_0$, and for $\alpha < \alpha_0$ it is equal to infinity. In the general case, the concept of measure is not connected with either the metric or the topology. However, the Hausdorff measure can be developed in an arbitrary metric space based on its metric, and the Hausdorff dimension itself is connected with the topological dimension.

The basics of the modern theory of probability are the limit theorems on the convergence of distributions of sums of independent random variables to the so-called stable distributions: Gaussian or non-Gaussian. The former ones base on the central limit theorem, and the latter (non-Gaussian) ones base on the limit theorem proved by B. V. Gnedenko (1939) and V. Döblin (1940)[1]. In this case, the limit theorem imposes restrictions on the form of non-Gaussian distributions. In order for the distribution law $F(x)$ to belong to the domain of attraction of a stable law with a characteristic exponent $\alpha(0 < \alpha < 2)$, different from the Gaussian one, it is necessary and sufficient that

$$(1)\ \frac{F(-x)}{1-F(x)} \to \frac{c_1}{c_2} \quad \text{for } x \to \infty, \quad (6.2)$$

(2) for each constant $k > 0$

$$\frac{1-F(x)+F(-x)}{1-F(kx)+F(-kx)} \to k^{\alpha} \quad \text{for } x \to \infty, \quad (6.3)$$

where the coefficients $c_1 \geqslant 0$, $c_2 \geqslant 0$, $c_1 + c_2 > 0, 0 < \alpha < 2$.

To prove Eq. (6.2) and Eq. (6.3), it is necessary and sufficient that with a certain selection of constants, the following conditions were met[1]:

$$nF(B_n x) \to \frac{c_1}{|x|^{\alpha}}\ (x < 0),$$

$$n[1 - F(B_n x)] \to \frac{c_2}{x^{\alpha}}\ (x > 0), \quad (6.4)$$

$$\lim_{\varepsilon\to 0}\overline{\lim_{n\to\infty}}\, n\left\{\int_{|x|<\varepsilon} x^2 \mathrm{d}F(B_n x) - \left[\int_{|x|<\varepsilon} x\mathrm{d}F(B_n x)\right]^2\right\} = 0.$$

The smaller the α value, the longer the distribution tail and the more it differs from the Gaussian distribution. For $1 < \alpha < 2$, stable laws have a mathematical expectation; for $0 < \alpha \leqslant 1$, stable laws have neither dispersions nor mathematical expectations. Conditions (6.2) – (6.4) determine the so-called non-Gaussian statistics.

In ordinary statistics, fluctuations tend to zero when the sample size or the number of N terms increases. This guarantees the asymptotically exact repeatability of averages and is the source of the traditional successes of classical statistical methods in radiolocation. For Levy statistics, the situation may differ radically. With an increase in the sample size, the accuracy of statistical estimations does not improve! The standard form of the central limit theorem predicts disappearing fluctuations for large N, and from the generalized central limit theorem (for $\alpha < 1$) it follows that the fluctuations are significant for arbitrarily large N. At the same time, for $\alpha < 1$, a case of global nonergodicity of processes is observed.

Note one more fact. Non-integral values of the α index in the range of $1 < \alpha \leqslant 2$ correspond to the generalized Brownian motion with long-term correlations and statistical self-similarity, i. e., fractal process. Self-similarity is mathematically expressed by power laws. The fractal dimension of the probability space of the time series is equal to α index:

$$\alpha = 1/H, \tag{6.5}$$

where H is the Hurst exponent.

It is necessary to distinguish the "ordinary" fractal dimension D of the signal or image under study and the fractal dimension determined by the α index. If D characterizes the "curvedness" of objects, then α characterizes the tail thickness of probability distributions[3,5,7].

In V. A. Kotelnikov IREE of the RAS, various original methods for measuring the fractal dimension D have been developed; including the dispersion method, the method taking into account singularities, functionals, triad, based on the Hausdorff metric, sample subtraction, based on the operation "Exclusive OR", etc.[5,7,13]. The local dispersion method for measuring the fractal dimension D is based on measuring the dispersion of the intensity / brightness σ_i^2 of optical or radar image fragments by two spatial scales δ_i^2:

$$D \approx \frac{\ln\sigma_2^2 - \ln\sigma_1^2}{\ln\delta_2 - \ln\delta_1}, \quad i = 1 \text{ or } 2. \tag{6.6}$$

In the Gaussian case, the dispersive dimension of a random sequence converges to the Hausdorff dimension of the corresponding stochastic process. The principal difficulty is that any numerical method involves discretization (or discrete approximation) of the process or object being analyzed; and discretization destroys fractal properties. To resolve this conflict, it is necessary to develop a special theory based on the methods of fractal interpolation and approximation. The fractal dimension D or its signature $D(t, f, \boldsymbol{r})$ in different parts of the surface image is a texture measure. Fractal methods can function at all signal levels: amplitude, frequency, phase and polarization.

Fractional mathematical analysis has a long history and extremely rich content[4,11]. Ideas about fractional integro-differentiation interested many prominent scientists: Leibniz, Euler, Liouville, and others. Interest in fractional mathematical analysis arose almost simultaneously with the origin of classical analysis (as early as in 1695 G. Leibniz mentioned this fact in letters to G. Lopital when considering differentials and derivatives of 1/2 order). Note the set of papers by the associate member of the Petersburg Academy of Sciences (1884), A. V. Letnikov, who, during his 20 years of scientific work, developed a complete theory of differentiation with an arbitrary index[5,7,11]. At present, the expression for the fractional derivative D_{at}^{α} in the form proposed by Riemann and Liouville (${}_{RL}D_{at}^{\alpha}$) is most frequently used.

The operator of integro-differentiation in the sense of Riemann-Liouville of the fractional order $\alpha \in R$ originated at the point a is defined as follows[3-8,10,11]:

$$ {}_{RL}D_{at}^{\alpha}f(t) = \frac{\operatorname{sign}(t-a)}{\Gamma(-\alpha)}\int_a^t \frac{f(\tau)}{|t-\tau|^{\alpha+1}}\mathrm{d}\tau, \alpha < 0, \tag{6.7} $$

$$ {}_{RL}D_{at}^{\alpha} = f(t), \alpha = 0, \tag{6.8} $$

$$ {}_{RL}D_{at}^{\alpha} = \operatorname{sign}^n(t-a)\frac{\mathrm{d}^n}{\mathrm{d}t^n}D_{at}^{\alpha-n}f(t) = \frac{1}{\Gamma(n-\alpha)}\frac{\mathrm{d}^n}{\mathrm{d}t^n}\int_a^t (t-\tau)^{n-\alpha-1}f(\tau)\mathrm{d}\tau, \tag{6.9} $$

where $n-1 < \alpha \leqslant n$, $n \in \mathbf{N}$; $\operatorname{sign}(z)$ is determined by the equalities $\operatorname{sign} 0 = 0$, $\operatorname{sign} z = z/|z|$, $(z \neq 0)$; $\Gamma(\alpha)$ is a gamma function.

For functions differentiable on the interval $[a, b]$, the definitions of fractional derivatives according to Riemann-Liouville and Letnikov are equivalent. Currently, the Caputo formula[6,8,11] is widely used:

$$ {}_{C}D_{at}^{\alpha}f(t) = \operatorname{sign}^n(t-a)\,{}_{RL}D_{at}^{\alpha-n}f^{(n)}(t), \quad n-1 < \alpha \leqslant n, \quad n \in \mathbf{N}. \tag{6.10} $$

The Riemann-Liouville and Caputo derivatives are associated by the formula

$$ {}_{C}D_{at}^{\alpha}f(t) = {}_{RL}D_{at}^{\alpha}f(t) - \sum_{k=0}^{n-1}\frac{f^{(k)}(\tau)}{\Gamma(k-\alpha+1)}|\tau - t|^{k-\alpha}, \quad n-1 < \alpha \leqslant n, n \in \mathbf{N}. \tag{6.11} $$

In the case $\alpha = n$, we get

$$ {}_{RL}D_{at}^{n}f(t) = {}_{C}D_{at}^{n}f(t) = \operatorname{sign}^n(t-a)\frac{\mathrm{d}^n}{\mathrm{d}t^n}f(t), \quad n \in \mathbf{N}. \tag{6.12} $$

The Caputo derivative has the same physical interpretation as the Riemann-Liouville derivative. In particular, for $f(0) = 0$ and $0 < \alpha < 1$, there is the exact equality

$$ {}_{C}D_{0t}^{\alpha}f(t) = {}_{RL}D_{0t}^{\alpha}f(t). \tag{6.13} $$

When comparing these derivatives, pay attention to the fact that in order to compute the Riemann-Liouville derivative, it is necessary to know the function values. And as for the Caputo derivative, one should know the derivative values, which is much more complicated. Some advantage of the Caputo derivative is that it is zero for a constant function, which is more usual for a researcher.

6.3 Fractal labyrinths as fractal broadband antennas development base

Recent several years Fractal labyrinth topology has became fast-growing interest object of scientists. This structures (systems) description can not been presented within traditional derivative equations of integer order. More exactly this processes and objects are quantitatively describing by integration-differentiation operators of fractional order $D_{at}^{\alpha}[f(t)]$, where $-1 < \alpha < 1$. Presence the fractional derivative in equations is decided to interpret as reflection of specific peculiarity of process or system which is memory or non-Markovian attribute (hereditarity).

Author's definition: fractal labyrinth ≡ labyrinth fractal is topologically connected structure with fractal dimension $D > 1$ and having scaling character of conducting lines.

Model constructions of ref. [14] are usually employed as examples for mentioned structures investigation. Wide simulation abilities of stochastic fractal labyrinth require tools development which is to automate needed geometry synthesis operation significantly. The transition of synthesis process from manual to detailed parametric allows to take a step to more complex tasks using the application. These tasks area as noted can be very wide. In particular there is planned to apply fractal labyrinths as miniature ultra-high frequencies (UHF) radiator geometry which able to become a new class of fractal antennas and fractal antenna arrays.

During our software base scheme development, we tried to take into account first of all the primary needs of possible software user, meanwhile trying to complete it by different additional features making development process easier and make results more accurate. The software was named "Fractalizer"[15,16].

The software (window is shown on Fig. 6. 1) contains first of all graphical plot area to draw a form of fractal curve generator with last action cancel function and choose angle step bar. On the right of generator panel there are tools to parameters specify such as main branch iteration number, branches amount, branches iteration number, width and height of lines, minimum clearance between non-consequent elements (elementary lines) of structure.

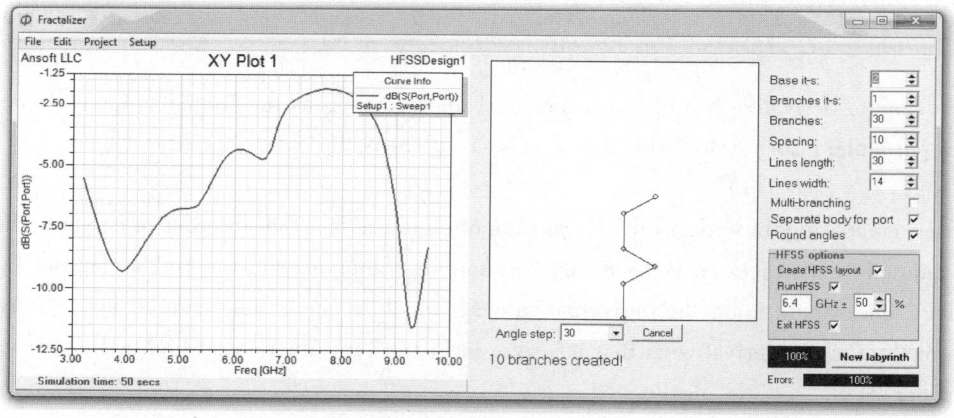

Fig. 6. 1　Window for the software

The software is able to save fractal structures in universal and well-known drawing format Autodesk DXF. DXF file can be imported as geometry in most modern computer designing and simulating software such as ANSYS, Solid Works, etc. Moreover, "Fractalizer" software has settings panel for launch Ansoft HFSS as simulation software for generated fractal structure.

It is shown on Fig. 6. 2 that "Fractalizer" software creates the curves of finite width (case b) instead of zero-width one (case a). Moreover by using angles rounding structure keeps constant width everywhere.

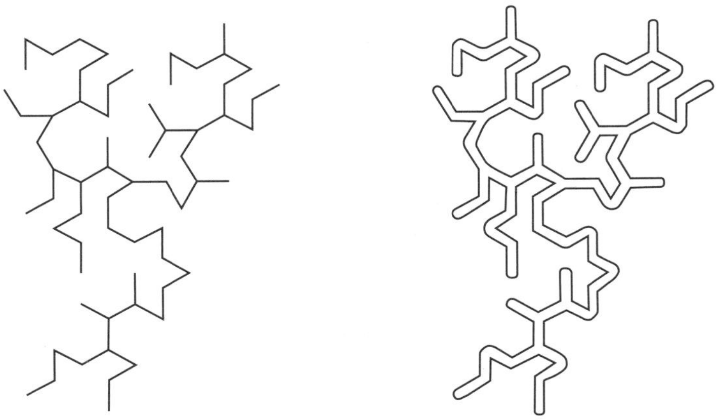

(a) (b)

Fig. 6. 2 An example of the generated curves

Today the software is able to create up to several hundreds of branches with satisfactory performance. Every branch here can contain up to 40 elementary lines.

However a kind of fractal antennas which takes the specific form of stochastic (or nondeterministic) fractals still has little presented. In most cases this kind has branching form and presents a fractal, which form is trending to self-similarity only statistically but generally one stochastic branched fractal can have many random realizations. Therefore the assumption is seems to be rightly that stochastic labyrinth fractal antennas would become to new generation of fractal antennas.

Operation principle and features of "Fractalizer" are improving now. In particular the process is automated for electrodynamical simulation of fractal antennas. An operator labor is not necessary in this stage by default. The exchange scheme is shown on Fig. 6. 3[15,16].

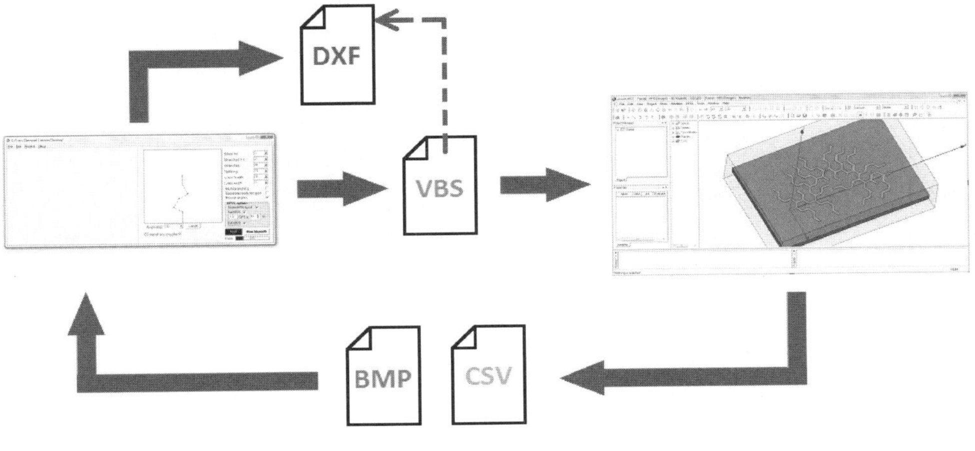

Fig. 6. 3 The exchange scheme

"Fractalizer" operation principle contains several following steps. First the program builds main curve basing on user given generator which is complete user specified iteration of fractal. Next the program calculates amount of curve break points (number of angles) and remembers they as points of possible branch base. Then by using random number generator the program in cycle selects the branch base point and builds this branch before user specified iteration is complete or before an obstacle (another curve) is reached. The number of cycles must be specified by user. If a point is already busy (as other branch base) then random number generator will be re-launched. The branch direction is between two lines forming an angle and from external side of the angle. After structure is ready the software reports quantity of successfully created branches and save the structure as an DXF file. Examples of these resulting structures are shown on Fig. 6. 4.

Results obtained by using the software was automatically imported by it to Ansoft HFSS 12 simulation environment based on finite element method (FEM).

Simulation results of device from Fig. 6. 4b namely reflection coefficient graph in frequency domain and 3D radiation pattern are shown on Fig. 6. 5a and Fig. 6. 5b, correspondingly.

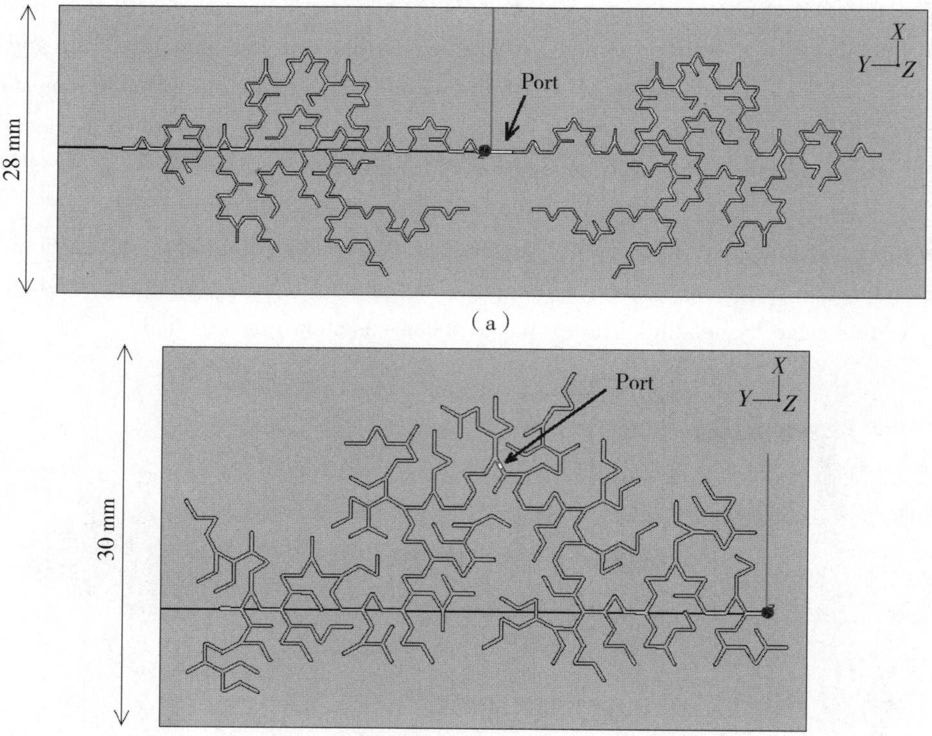

Fig. 6. 4 Results structures for fractal antennas

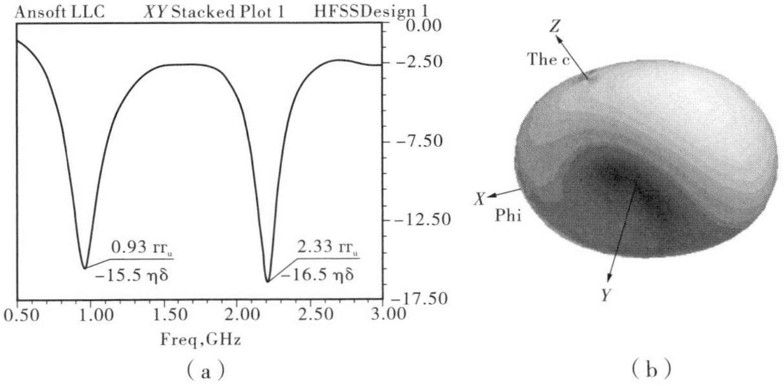

Fig. 6. 5 Reflection coefficient in frequency domain (a) and 3D radiation pattern (b)

As it can be seen, the antenna has two resonances. The first one corresponds to frequency lower than 1 GHz. It means that the antenna is able to receive the wave of 0. 32 meter length.

The structure which shown on Fig. 6. 4b was synthesized stochastically and therefore it is reasonable to assume better results if this one will be optimized by some algorithm. Labyrinth fractals able to be genetically optimized can become universal tool for achievement of different required parameters such as gain, multi-range and wide range as well as radiation pattern form. But the main requirement is to decrease antenna size while its operating frequency is constant. Fig. 6. 6 below represents several UHF fractal antennas and their S11 graph in frequency domain. In a base of layouts there are geometry which synthesized by "Fractalizer"[15,16].

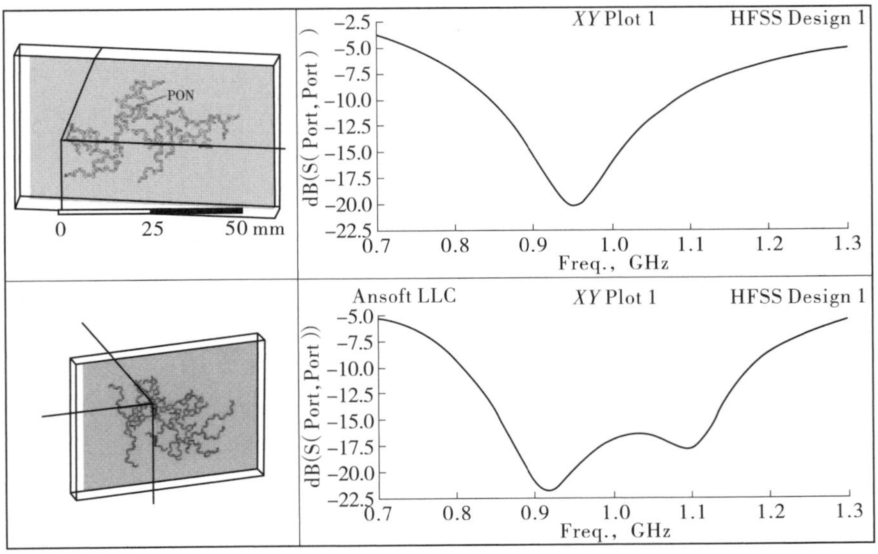

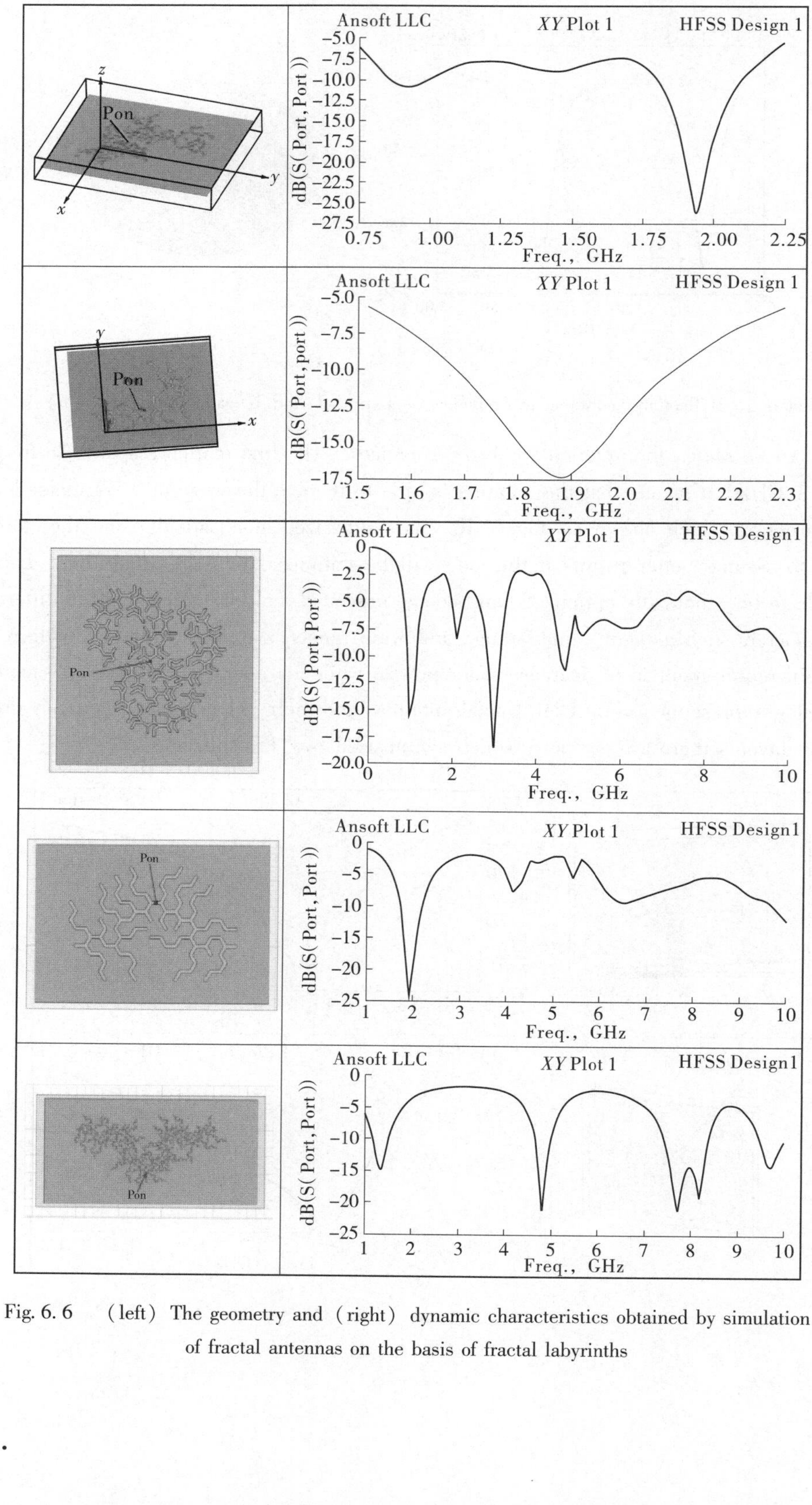

Fig. 6. 6 (left) The geometry and (right) dynamic characteristics obtained by simulation of fractal antennas on the basis of fractal labyrinths

Some designsamples were manufactured (see Fig. 6.7).

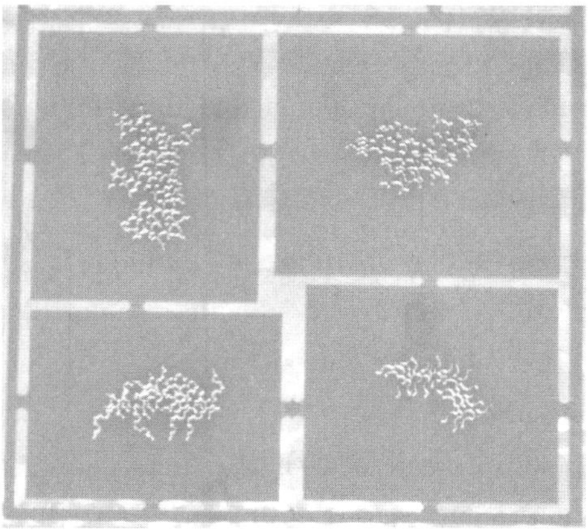

Fig. 6.7 Samples of fractal antennas on the basis of fractal labyrinths

Antenna arrays is one of main components of modern radiosystems. The theory of fractals and fractional calculus application allows to unite accomplishments of amplitude and stochastic arrays.

The first type of antenna arrays has relatively small side lobes but it is sensitive to elements placement errors and excitation currents values. The second type of arrays is robust to elements placement errors and they's failures, but the type characterized by relatively high side lobes level. The scaling principle application for antenna arrays allows more flexible control of radiation pattern in the lobes area.

We suggest to synthesize the stochastic robust antenna arrays by dint of fractal labyrinths properties use. It will enable to control of radiation pattern side lobes energy. The second step is union of several fractal labyrinths clusters with different fractal dimension D in antenna array synthesis space. Therefore by natural way we comes to adaptive fractal antennas. Here it is necessary to use of genetic algorithms for optimize the spatio-temporal big antenna apertures corresponding to specified criteria. The same principles can be used for synthesis of nano-antennas.

We think that proposed software with some further improvements would able to solve basic problems of genetic synthesis and optimization of multi-scale robust antenna arrays with big amount of elements based on fractal labyrinths (and surely other modern nano-constructions and micro-constructions).

Recall that the fractal antennas applications already can be founded in medicine, military technology, physical experiments and cellular communication systems on base stations and mobile terminals.

6.4 Nanostructures and fractals

It is getting possible nowadays to create planar and three-dimensional nanostructure modeling "fractal" radioelements and radio devices of microelectronics based on nanophase materials[5,7,17-19]. In other words, we are talking about the construction of the element base of new generation based on fractal effects and properties. In particular, elementary extension of the Cantor set to the physical level allows us to go to so-called Cantor blocks in planar technology of molecular nanostructures. The percolation synthesis suggested by the author in 2007 for nanostructured composites is also possible.

Recursive process also allows to create self-similar hierarchical structure, down to separate conductive tracks on the chip and nanostructures. It is necessary to take into account and learn how to calculate the mutual and collective influence of electromagnetic fields with all components of the chip: the conductive tracks, semiconductor, insulator, etc.

Considerable attention of the most advanced experts is currently paid to modeling of fractal objects with the complex dynamics by various dissipative systems. The most natural way of modeling is to use the Feigenbaum scenario of transition to chaos through period doubling. In the context of our work, Julia, Fatou and Mandelbrot sets are interesting objects of development of new types and forms of fractal antennas and others fractal nanostructures and Metamaterials based on them[5,7,17-19].

6.5 Fractal photon and magnon crystals

Conventional materials with photon band gaps uses Bragg scattering to create band gaps[7,20]. As a result of Bragg scattering mechanism, width and transverse sizes of photon crystals should be several wavelengths. Systems based on photon band gaps and frequency-selective surfaces usually works in a single frequency band with a suitable wavelength in the volume damped periodically arranged basic functional blocks. Fractal photon and magnon crystals have a number of advantages over their classical counterparts, and are essentially new media for transmitting information[7,21,22].

An N-order fractal should theoretically have N inherent resonances. Each resonance is defined by the excitation current in wire circuits of a certain order of iterations and this current flows towards structures of a higher order. The fractal that interacts with an electromagnetic wave at normal incidence is pretty exactly simulated by a thin homogeneous plate of the same thickness with the effective permittivity $\varepsilon_{\text{eff}}(f)$:

$$\varepsilon_{\text{eff}}(f) = \varepsilon_0 + \sum_l \frac{\beta_l}{f_l^2 - f^2}, \tag{6.14}$$

where f is the frequency, index l defines the resonances number, ε_0 , f_l and β_l are parameters obtained from the calculated spectrum. It follows from (6.14) that when changing from one resonance frequency f_l to another f_{l+1} there is always a point which it is $\varepsilon_{\text{eff}}(f) = 1$ in and consequently certain transmission bands exist. It is always desirable that the reflection/transmission coefficient of a fractal structure was controllable using an external "control knob". Each line segment in a fractal is connected with each other. The external electric current which is supplied to the center of the first level line with a certain phase can be a secondary source. Modulation of the transmittance is determined by the phase shift (or the time lag τ of the sensed signal with respect to the main incident beam) between the incident wave and the feeding current.

In this case we can talk about modelling intellectual surfaces with focused control of its scattering characteristics or a field of the transmitted wave in a wide frequency band. When imposing two identical fractal samples if one is turned on 90° with respect to another one can obtain a structure which is invariant under rotation. Thus such an "active" fractal structure can simulate the total reflection which does not depend on the incidence angle and polarization state and what is usually the characteristic of 3D photonic crystals.

It should be noted that sizes of a traditional 3D photonic crystal must form at least several wavelengths before it will be able fully show its photonic bandgap properties. Thus, for wave 1 GHz, the structure thickness must be about 1 meter. On the other hand, plain fractal structures are such as their transmission band $\Delta f/f_0$ determined by the similarity law ($\Delta f/f_0 \rightarrow$bandgap / middle of the bandgap and $\Delta f / f_0 \sim 5\%$ for one fractal plate) can be significantly enhanced using imposition of identical fractals on each other. Increase of the fractal plates thickness leads to the growth of steepness of transmission bands bounds. Attenuation bands can also be extended using wider metallic conductors of the fractal plates.

The resonance wavelengths may be much larger than the sample sizes. It happens because the low frequency resonance is determined by the longest metallic line in a fractal and such a line is just much longer than linear dimensions of the fractal itself. It gives a fractal its "Superwave" properties that is a fractal plate can effectively reflect electromagnetic waves with lengths much larger than the lateral dimensions. "Superwave" properties imply that the fractal plate can act as a compact reflector. For such technologies authors have developed algorithms and programs which allow calculate different configurations of fractal structures of the crystals under consideration. As an example on Fig. 6.8 there are samples of some drafts on the basis of Sierpinski curve (a) and Cayley tree (b) respectively. Thus the fractal structures always have a self-similar series of resonances leading to the logarithmic periodicity of working zones.

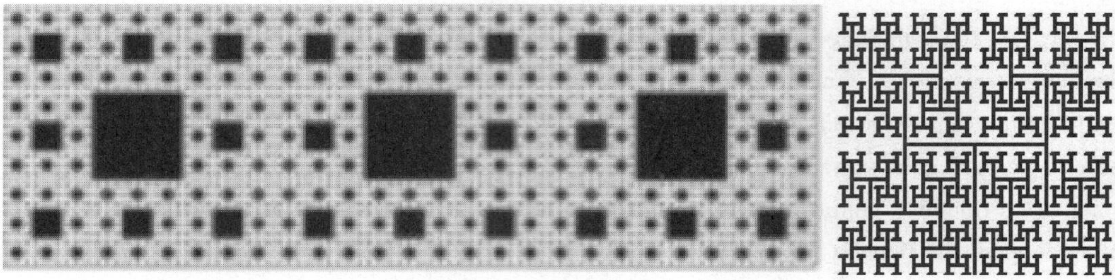

(a) (b)

Fig. 6. 8 Samples of the first (a) and second (b) fractal drafts

The linked topological fractal structure makes it possible to modulate the transmittance coefficient of electromagnetic waves. The lowest attenuation frequency corresponds to wavelengths, which may significantly exceed external dimensions of a fractal plate making such fractal structures to be superwave reflectors. For controllable intellectual surfaces one may also use the principle of reconfigurable fractal arrays with electronic switching of sub arrays which was described in detail in Ref. [5, 7, 17].

6. 6 Fractal signatures in problems of estimation of microrelief of processed surfaces

Basing on the conducted experiments we were the first who proposed estimation methods using different fractal characteristics of the quality of articles surface and properties of microrelief of modern structural materials[7,17,23-27]. Due to intensive development of methods of processing of concentrated energy streams-CES (laser, plasmic, electro erosive), as well as nanotechnologies (chemical assembly, sol-gel processes, metals vapor-phase deposition, atomic layered epitaxy) significant difficulties in description and estimation of the roughness with a profile method arise. In these cases the roughness elements form, elements distribution over the processing square strongly differs from its conventional conception which was formed in the framework of the processing by cutting as a periodic alternation of "juts" and "cavities" which are described in the framework of the Euclidean geometry - Fig. 6. 9[23].

Consequently, now the problems of forming the surface quality including such an important quality characteristics like roughness become particularly vital due to creation of the new technologies of metalworking. These problems come into sharp focus in the field of nanotechnologies which the roughness topology is considered for not as a secondary property being a "response" of the surface layer structure to the influence of a certain physical process (for example, like in the cutting work) but as a property of the structure itself all the more dimensions of such layers are comparable with the electrons mean free path. In Ref. [23], at microrelief level of such processed surfaces, we demonstrated existence of fractal clusters with irregularities distributed by the power laws with heavy tails (Fig. 6. 10).

Name	Relief elements	Name	Relief elements
Mushroom-shaped		Dimples	
T-shaped		Globules	
Spades		Whispers (globulous-whisperous)	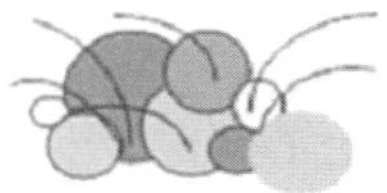
Splats	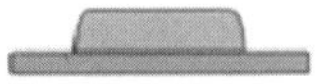	Ridges	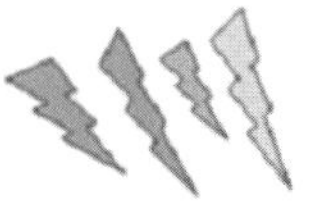
Botiroidal	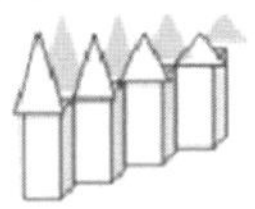	Moire	

Fig. 6. 9 The types of micro surfaces relief elements

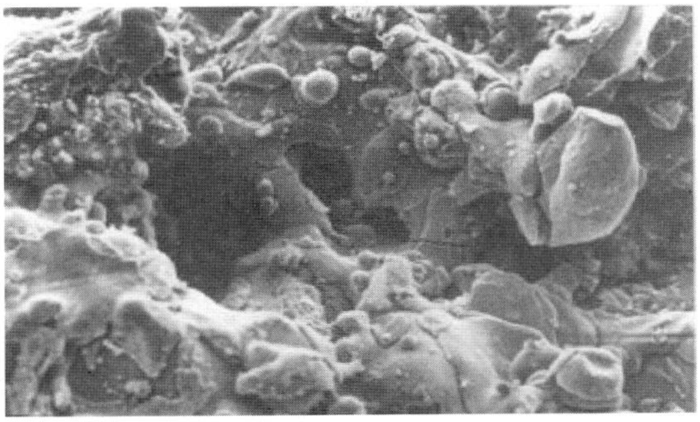

(a)

(b)

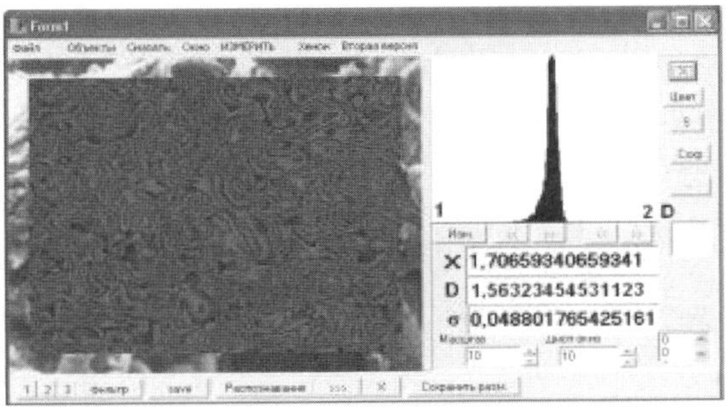

(c)

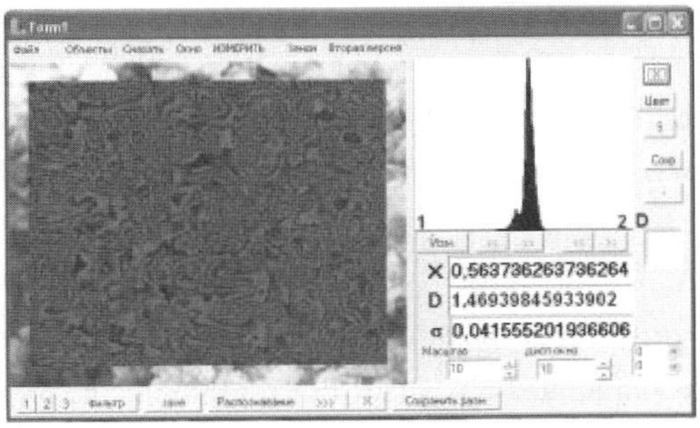

(d)

Fig. 6. 10 Fractal analysis of the samples surfaces with a plasma-sprayed covering of hydroxyapatite: (a, b) 2D-image of samples surface; (c, d) the field and histogram of local fractal dimensions (note the distributions' heavy tails)

Presence of fractality in such different media can be controlled in particular on a change of the skin effect and impedance. Exactly the spatial/time evolution of the current allows the electromagnetic field "feel" fractal properties (fractal signatures) of the physical medium under investigation. Scaling models of the rough layer of solid's surface can be represented as an electric circuits analogy which has the form of the Cantor dust for example and so on Ref. [7, 24 - 27].

6.7 Fractal memristor

In 1971, L. O. Chua[28] proposed a new passive element: a memristor. Until recently, the memristor was considered to be a theoretical element, which can be implemented only using fairly cumbersome transistor or operational amplifier electronic circuits; therefore, memristors found little application. In recent years, however, this viewpoint has been changed[29]; therefore, of interest is thorough investigation of the properties of the memristor.

Let us consider the concept of the memristor. Fig. 6.11 shows a square whose corners correspond to such general quantities of the circuit theory as charge Q, magnetic flux Φ, electric current I, and electric voltage U.

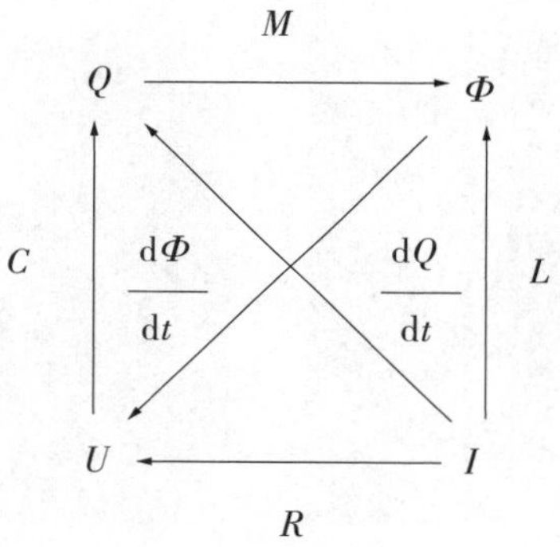

Fig. 6.11 Relations between four basic characteristics of the electric circuits

The square corners are connected by the lines with arrows that point out the relations between two individual characteristics. As follows from the symmetry considerations, there should be six such relations[28,29] specified by the equations

$$Q = CU, \quad \Phi = LI, \quad U = RI, \quad \Phi = MQ, \quad I = \frac{dQ}{dt}, \quad U = \frac{d\Phi}{dt}, \tag{6.15}$$

where C is the capacitance, L is the inductance, R is the resistance, and M is the memristivity (the parameter absent in the traditional theory).

As seen from Eq. (6.15), memristivity relates the charge and the magnetic flux and has the dimensionality of resistance. Based on Eq. (6.1), we can write two expressions for memristivity:

$$M = \frac{L}{RC}, \quad M = \frac{\Phi(0) + \int_0^t U(\tau)\mathrm{d}\tau}{Q(0) + \int_0^t I(\tau)\mathrm{d}\tau}. \tag{6.16}$$

The first expression describes a series or a parallel *RLC* circuit. Here, memristivity characterizes the circuit Q factor, which is equal to $\sqrt{M/R}$ or $\sqrt{R/M}$. The second expression allows us to conclude that the simplest memristor is a conductor with current and memristivity is the resistance of this conductor. In the general case, however, memristivity is a nonlocal function of time, i. e., it may depend on the signal duration.

In this case, the element itself should no longer be just a simple connection of resistors, capacitors, and inductors. The equation describing the relation between voltage and current that follows from expressions (6.16) is

$$U(t) = M(t)\left\{ I(t) + \left[Q(0) + \int_0^t I(\tau)\mathrm{d}\tau \right] \frac{\mathrm{d}\ln M(t)}{\mathrm{d}t} \right\}. \tag{6.17}$$

Similarly to the classical circuit elements, memristors, being actually implemented elements, can exhibit nonlinear and inertial properties (memory, hysteresis of the I-V characteristic, etc). In view of this fact, the attempts were made to develop the memristor on the basis of analog circuits by applying the mathematical tool of fractional integro-differentiation, which is widely used in synergetics and fractal physics[7].

In our opinion, it is reasonable to introduce operators that contain fractional time derivatives of the electric charge and magnetic flux directly in definition (6.15). We write the fractional operators in the form[30]

$$z(x) = D_{0x}^{m-1}\frac{\mathrm{d}y(x)}{\mathrm{d}x} = \partial_{0x}^{m} y(x), 0 \leqslant m \leqslant 1, \tag{6.18}$$

where

$$D_{sx}^{m} y(x) = \begin{cases} \dfrac{\mathrm{sign}(x-s)}{\Gamma(-m)} \displaystyle\int_s^x \frac{y(x')\mathrm{d}x'}{|x-x'|^{m+1}}, & m < 0, \\ y(x), & m = 0, \\ \mathrm{sign}^n(x-s)\dfrac{\mathrm{d}^n}{\mathrm{d}x^n} D_{sx}^{m-n} y(x), & n-1 < m \leqslant n,\ n \in \mathbf{N}, \end{cases}$$

$$\partial_{sx}^{m} y(x) = \mathrm{sign}^n(x-s) D_{sx}^{m-n}\frac{\mathrm{d}^n y(x)}{\mathrm{d}x^n}, \quad n-1 < m \leqslant n,\ n \in \mathbf{N},$$

$\Gamma(-m)$ is the Euler gamma function, and D_{sx}^{m} and ∂_{sx}^{m} are the fractional integro-differential Riemann-Liouville and Caputo operators of order m.

Finally we get[30]

$$z(x) \approx \Gamma(1+m)\frac{\Delta y}{(\Delta x)^m}. \tag{6.19}$$

The ratio $\Delta y/(\Delta x)^m$ in Eq. (6.19) determines the so-called Holder derivative introduced for fractal functions instead of an ordinary derivative and expresses the nonlinear increment law.

This derivative and, consequently, expression (6.18) can be used quite reasonably in definition (6.15) to take into account possible fractional dynamics of the charge and the magnetic flux during operation of the memristor.

Thus, replacing the first derivatives in Eq. (6.15) by operators (6.18) and performing some transformations, we find

$$U(x) = D_{0x}^{\alpha}\left\{M(x)\left[\frac{Q(0)}{t_0} + D_{0x}^{-\beta}I(x)\right]\right\} - \frac{\Phi(0)}{t_0}\frac{1}{\Gamma(1-\alpha)x^{\alpha}}, \tag{6.20}$$

where exponents α and β characterize randomness and fractality of the time changes in the magnetic flux and electric charge, $x = t/t_0$ is the dimensionless time, and t_0 is some characteristic time. In the special case of M = const, Eq. (6.22) is simplified to

$$U(x) = M\left[\frac{Q(0)}{t_0}\frac{1}{\Gamma(1-\beta)x^{\beta}} + D_{0x}^{\alpha-\beta}I(x)\right] - \frac{\Phi(0)}{t_0}\frac{1}{\Gamma(1-\alpha)x^{\alpha}}. \tag{6.21}$$

Obtained Eqs. (6.16) and (6.20) and (6.21) describe the processes with perfect and partial memory, respectively. The presence of memory means that the voltage across the memristor will depend on the features of current flow from switching on to time point t. At $\alpha = \beta = 1$, Eq. (6.20) transforms to Eq. (6.17) and Eq. (6.21) transforms to the Ohm's law. It is shown that the integral quantum Hall effect can be the physical basis of memristor's operational principle[30].

6.8 Fractal oscillator with fractional differential positive feedback

Further investigated fractal oscillator with slow degree of non-linearity, namely generalized sinusoidal signal auto generator[31].

Model operation (Fig. 6.12a) after several assumptions, concerning some concrete schematic solution, in classical theory given by movement equation:

$$u'' + \lambda(a)u' + \omega^2 u = 0, \tag{6.22}$$

where ω is system oscillation frequency, a is it's magnitude and $\lambda(a)$ is equivalent attenuation coefficient[32], described as

$$\lambda(a) = \frac{\omega}{Q}[1 - k(a)R] = \frac{1}{C}\left[\frac{1}{R} - k(a)\right], \tag{6.23}$$

where Q is reactance factor, R is load resistance and C is loop capacitance.

Admit, that generator's positive feedback represents the fractional-derivative circuit based on long RC-line[33]. It have an order of fractional impedance equal to 0.5, corresponding to 45-degree phase shift. It is value between zero-shift for direct feedback (Fig. 6.12a) and 90 degrees for conventional RC integrating circuit. It may be interpreted as some distribution of the positive feedback characteristics, for example as shown at Fig. 6.12b.

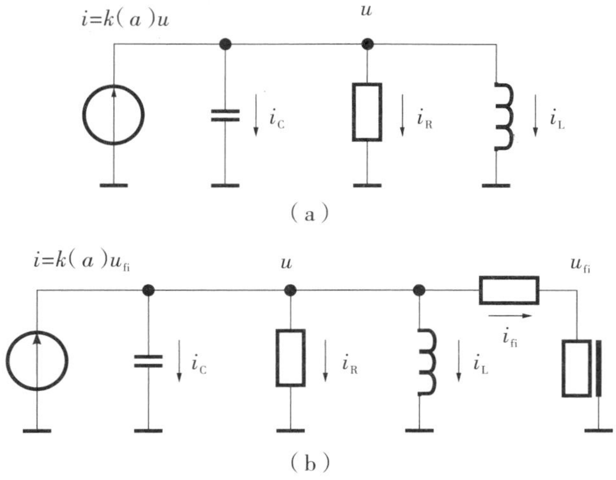

Fig. 6. 12 Classical oscillator (a), oscillator with fractional positive feedback (b)

In this case, assuming fractional integrator current i_{fi} many times greater than resistor current i_R, movement equation becomes this form:

$$u'' + \frac{1}{RC}u' + \omega^2 u = \frac{1}{C}k(a)_L D^{\alpha}(u), \tag{6.24}$$

where $_L D^{\alpha}$ is left-hand side α-order Liouville derivative.

Considered that $u \approx a\cos(\omega t)$, we have:

$$\lambda_{\partial}(a) = \frac{1}{C}\left(\frac{1}{R} - k(a)k_{\partial u}\sin(\alpha\pi/2)\right). \tag{6.25}$$

In Eq. (6.25), $\alpha = 0, \cdots, 1$ is order of fractional impedance[5,7,17,33,34] of positive feedback circuit. Received equation (6.25) shows a large attenuation of fractional system oscillations in compare with conventional at equal L, R, C and $k_{fi} = 1$.

Circuit with fractional integrator (Fig. 6. 12b) can serve as universal fractional oscillator model of any nature. For example, in mechanics, Eq. (6.24) will according to system of shaken swing: most effective swing reached when peak of swinging effort occurs synchronously with swing transition thru balance state, where potential energy equals zero and kinetic energy reaches its maximum. But in reality, peak of swinging effort occurs before or after mentioned moment and system must be described by equation like Eq. (6.25).

Thus, movement magnitude increase, in particular, increase of signal amplitude, as slow time function in according with method of equivalent linearization[31] defined as:

$$a' = \frac{\lambda_{\partial}(a)}{2}a = \frac{a}{2C}\left[\frac{1}{R} - gS(a)\sin(\alpha\pi/2)\right], \tag{6.26}$$

where g is gain constant, $S(a) \leqslant 1$ is non-linearity value and $gS(a) = k(a)$.

Magnitude increase with different values of g shown at Fig. 6. 13. From first to third diagram it equals 0. 166, 0. 175 and 0. 184.

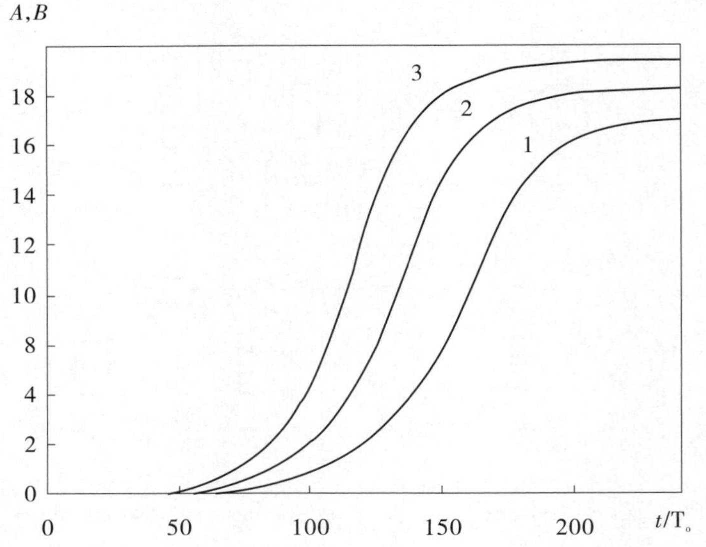

Fig. 6. 13 The dependence of the amplitude of the oscillations on time at different g

So, fractional of fractal positive feedback equals delayed positive feedback. Note that delayed feedback may be only particularly case of fractal dynamic system. With the help of Eqs. (6. 24) – (6. 26) describing of wide range phenomena becomes possible.

6. 9 Nonstationary regimes in electric circuits with ferroelectric negative capacitance

Experimental observation of thermodynamically stable negative capacitance effect in a nanoscale bilayer of ferroelectric lead zirconate titanate $Pb(Zr_{0.2}Ti_{0.8})O_3$ and nonlinear dielectric strontium titanate $SrTiO_3$ at room temperature was first described in Ref. [35]. Later the same effect was found in another system namely in bilayer of ferroelectric barium titanate $BaTiO_3$ and paraelectric $SrTiO_3$ [36]. We will call such bilayer ferroelectric-dielectric systems with negative capacitance by "ferroelectric negative capacitors" (FNC) or "negative capacitors" (NC). The initial push for investigation of NC was done by Ref. [37] where it was proposed that if one be able to replace the ordinary gate oxide by another material with an effective negative capacitance then it will open the way to a radical reduction in the power consumed by field effect transistors and the devices containing them. But research of application of capacitors with negative capacitance in various radio engineering devices is very perspective[38] too because in microelectronics volume of output product is closely related with lowering of it's manufacturing cost.

Let us consider series combination of NC and a resistor R under the action of an arbitrary input voltage $U(t)$. This electric circuit is shown in Fig. 6. 14.

Dependence of voltage on NC V from it's charge Q can be calculated in the framework of

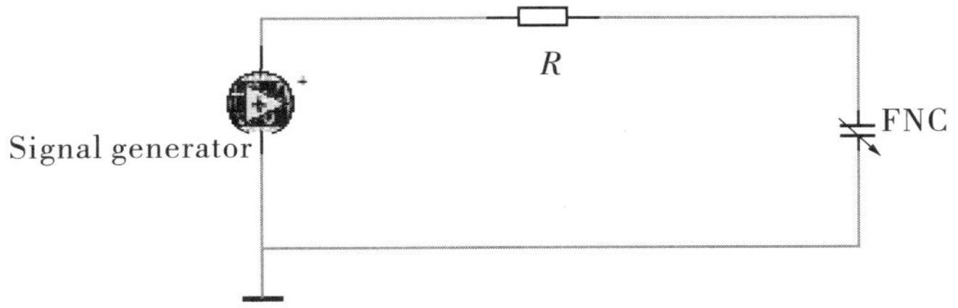

Fig. 6. 14 Electric circuit

Landau's mean field theory for monodomain sample[35,36]:

$$V = -\alpha Q + \beta Q^3,\ \alpha > 0,\ \beta > 0, \tag{6.27}$$

where coefficients α and β depend on properties of materials of ferroelectric-dielectric bilayer and on geometry of FNC. In particular α linearly depends on the Curie temperatures of both materials. For the sample Pb $(ZrO_2TiO_8)O_3$ with a thickness 28 nm located on $SrTiO_3$ with a thickness 48 nm and area 30 × 30 microns parameters α and β, included in the formula (6. 27), can be estimated as $\alpha \sim 10^{10}\,V\cdot C^{-1}$ and $\beta \sim 0.5\cdot 10^{29}\,V\cdot C^{-3}$[35].

To satisfy the Kirchhoff's voltage law in the system presented on Fig. 6. 14 we obtain the following ordinary differential equation:

$$R\cdot\frac{dQ}{dt} - \alpha Q + \beta Q^3 = U(t). \tag{6.28}$$

Introducing the next dimensionless variables:

$$\tau = \frac{\alpha t}{R},\ x(\tau) = \sqrt{\frac{\beta}{\alpha}}\cdot Q(t),\ u(\tau) = \frac{\sqrt{\beta}}{\alpha^{3/2}}\cdot U(t), \tag{6.29}$$

we find that Eq. (6. 29) is reduced to the Abelian equation of the first kind:

$$\frac{dx}{d\tau} = x - x^3 + u(\tau) \tag{6.30}$$

with initial condition $x(0) = x_0$ on dimensionless charge.

Going over to the new variables:

$$\xi = \frac{1}{2}[1 - \exp(2\tau)],\ x(\tau) = \exp(\tau)\cdot\eta(\xi), \tag{6.31}$$

one can rewrite Eq. (6. 30) in the form:

$$\frac{d\eta}{d\xi} = \eta^3 - \exp(-3\tau)u(\tau). \tag{6.32}$$

From Eq. (6. 32) one can see that if $\exp(-3\tau)u(\tau) = A_0^3 \equiv$ const that is $u(\tau) = -A_0^3\exp(3\tau)$ then Eq. (6. 32) is reduced to the following equation with separating variables:

$$\frac{d\eta}{d\xi} = \eta^3 + A_0^3. \tag{6.33}$$

Exact solution of Eq. (6. 33) with initial condition $\eta|_{\xi=0} = x_0$ is equal to:

$$\xi = \frac{F(\eta, A_0) - F(x_0, A_0)}{3\cdot A_0^2}, \tag{6.34}$$

where function $F(\eta, A_0) = \ln \dfrac{|\eta + A_0|}{\sqrt{\eta^2 - A_0 \cdot \eta + A_0^2}} + \sqrt{3} \cdot \arctan \dfrac{2\eta - A_0}{\sqrt{3} \cdot A_0}$.

It means that in this case exact solution of input Eq. (6.30) can be represented in implicit form:

$$\tau(\eta) = \ln \sqrt{1 - 2\xi(\eta)}, x(\eta) = \eta \cdot \sqrt{1 - 2\xi(\eta)}. \tag{6.35}$$

Let us now consider series combination of NC and a resistor R (see Fig. 6.15). In this situation dimensionless charge of NC obeys to the following equation:

$$\frac{dx}{d\tau} = x - x^3. \tag{6.36}$$

It is easy to see that Eq. (6.36) possesses by three stationary states: $x = 0, \pm 1$. In particular $x = 0$ is unstable state and $x = \pm 1$ are stable states.

The exact solution of Eq. (6.36) with initial condition $x(0) = x_0$ is equal to:

$$x(\tau) = \frac{x_0}{\sqrt{x_0^2 + (1 - x_0^2) \cdot \exp(-2\tau)}}. \tag{6.37}$$

Combining solutions (6.35) and (6.37) one can research behaviour of our system in Fig. 6.14 under the action of the following periodical (with period θ) input voltage:

$$U(t) = \begin{cases} U_0 \cdot \exp\left(\dfrac{3\alpha t}{R}\right), & 0 < t < T_1 \\ 0, & T_1 < t < T \end{cases}, U(t + \theta) = U(t). \tag{6.38}$$

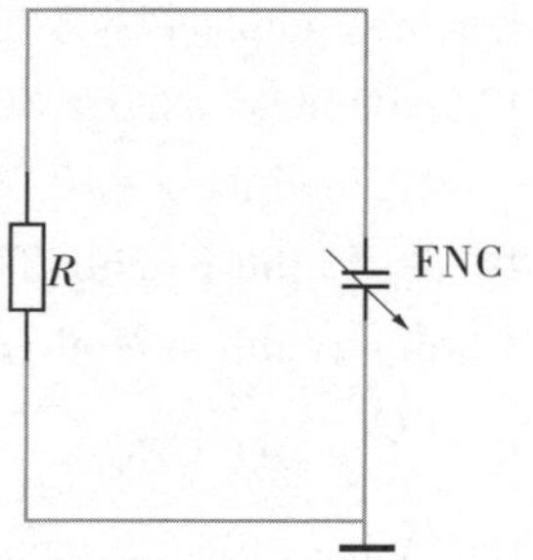

Fig. 6.15 Combination of FNC and a resistor R

In Fig. 6.16 dimensionless input voltage u, dimensionless charge of NC x, dimensionless voltage on NC v and dimensionless current through the resistor i depending on dimensionless time τ are presented in the simplest situation when $x_0 = 1$ and $2\alpha(T - T_1)/R \gg 1$.

Let us consider the semi-infinite transmission line with a filling by NC. It is well known that the behaviour of current $I(x,t)$ and voltage $U(x,t)$ in the line can be described by following equations:

$$\frac{\partial I}{\partial x} = -\frac{\partial Q}{\partial t}, \quad \frac{\partial U}{\partial x} = -L\frac{\partial I}{\partial t}. \tag{6.39}$$

where L is linear inductance and $Q(x,t)$ is linear charge in the line. If $Q(x,t)$ is quite small then the next condition will be fulfilled in the environment of the FNC:

$$U = -Q/C, \tag{6.40}$$

where $C > 0$.

Relation (6.40) immediately leads to changing the type of equation for the current $I(x,t)$ in the line namely the equation of the hyperbolic type becomes the equation of the elliptical type:

$$\frac{\partial^2 I}{\partial x^2} + \frac{1}{v^2} \cdot \frac{\partial^2 I}{\partial t^2} = 0, \tag{6.41}$$

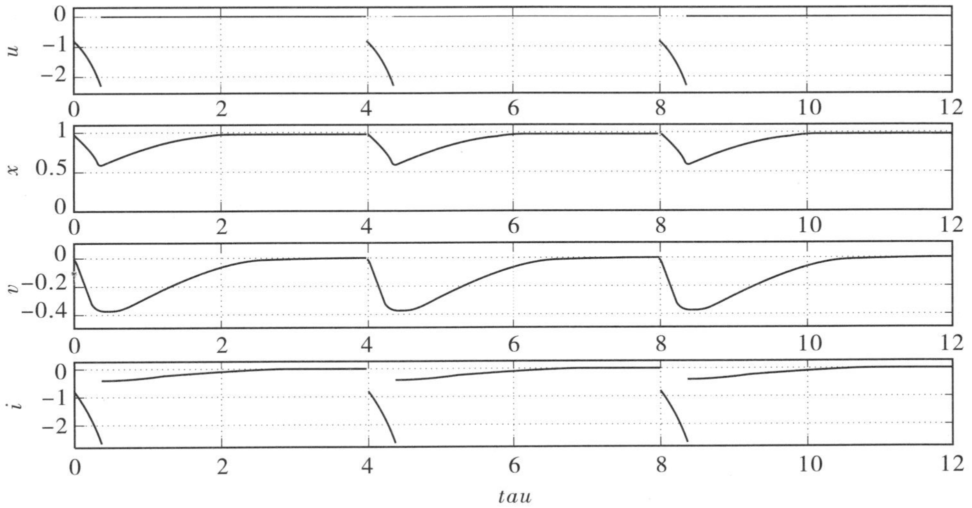

Fig. 6. 16 Functional dependencies on dimensionless time, $A_0 = 0.92$

where $v = \dfrac{1}{\sqrt{L \cdot C}}$ is the so-called phase velocity.

Further let the semi-infinite line at coordinate $x = 0$ is open, i. e., the current is equals to zero:

$$I(0,t) = 0 \tag{6.42}$$

and the voltage at coordinate $x = 0$ of the semi-infinite line is equal to $U(0,t) = -U_0\cos(\omega \cdot t)$ with magnitude $U_0 = \dfrac{Z_0 I_0}{(\omega \cdot T)^2}$, where $Z_0 = \sqrt{\dfrac{L}{C}}$ is wave impedance of the line and time $T > 0$, then

$$\left.\frac{\partial I}{\partial x}\right|_{x=0} = \frac{I_0}{v \cdot T}\frac{\sin(\omega \cdot t)}{\omega \cdot T}. \tag{6.43}$$

Exact solution of the Cauchy problem on variable $x > 0$ along the line for Eq. (6.41) with initial conditions (6.42) and (6.43) is equal to:

$$I(x,t) = I_0 \cdot \frac{\mathrm{sh}(\omega \cdot x/v)}{(\omega \cdot T)^2}\sin(\omega \cdot t). \tag{6.44}$$

On the other hand under increasing frequency ω formula (6.18) is the well-known Hadamard example which demonstrates the incorrect formulation of the above-described Cauchy problem (6.42) – (6.43) because its solution (6.44) don't tend to zero under frequency $\omega \to +\infty$ despite the fact that derivative (6.43) tends to zero uniformly under $\omega \to +\infty$.

Therefore, we have found that in electric circuit with NC there are fundamentally new non-stationary regimes[38]. This untrivial behaviour of observables can be applied in new automatic control systems using NC. Taking into account the last example we underline that researchers ought to be neat and careful under application of NC.

6.10 Harmonic distorsions in an oscillatory circuit with a ferroelectric capacitor with negative capacitance

Consider oscillating circuit including as a capacitor a two-layer ferroelectric structure exhibiting negative capacitance is considered[39,40]. The electric network of the oscillating circuit with a NC capacitor is shown in Fig. 6.17. Applying the second Kirchhoff law to it and using the formula (6.27), we obtain an ordinary differential equation for charge of the NC capacitor:

$$L\frac{d^2q}{dt^2} - \alpha q + \beta q^3 = 0, \tag{6.45}$$

where L is the inductance value of the circuit.

Eq. (6.45) is the Duffing equation with a homoclinic eight[6]. We make it dimensionless by introducing new variables:

$$\tau = \sqrt{\frac{\alpha}{L}}t, \ x = \sqrt{\frac{\beta}{\alpha}}q \tag{6.46}$$

as well as dimensionless energy in the circuit:

$$h = \frac{H}{\alpha^2/\beta}, \tag{6.47}$$

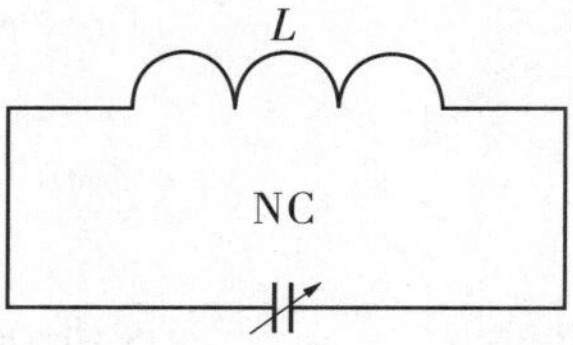

Fig. 6.17 An oscillating circuit with a NC capacitor

where $H = \frac{LI_0^2}{2} - \frac{\alpha q_0^2}{2} + \frac{\beta q_0^4}{4}$ is the total energy in the circuit, which is preserved in consequence of equation (6.45), q_0 is the charge at the NC capacitor and I_0 is the current in terms of the inductance at the initial instant of time.

The typical scale of energy in the circuit is $\alpha^2/\beta \sim 1$. In variables (6.46), the equation (6.45) is written as follows:

$$\ddot{x} - x + x^3 = 0, \tag{6.48}$$

where the point above the dimensionless charge x means its differentiation with respect to the dimensionless time τ.

The solutions of the Duffing equation are expressed in terms of Jacobi elliptic functions $dn(\cdots)$ and $cn(\cdots)$[41]. Namely, in case that $-1/4 < h < 0$, the solution of equation (6.48) is[6]:

$$x(\tau) = Adn\left(\frac{2K(k_1)\tau}{T_1}, k_1\right), \tag{6.49}$$

where $A(h) = \sqrt{1+\sqrt{1+4h}}$, $k_1(h) = \sqrt{2[1-1/A^2(h)]}$, $T_1(h) = 2\sqrt{2}K[k_1(h)]/A(h)$ is the dimensionless period of oscillations of a dimensionless charge at negative energy, and in the case that $h > 0$ the solution of equation (6.48) is[41]:

$$x(\tau) = Acn\left(\frac{4K(k_2)\tau}{T_2}, k_2\right), \tag{6.50}$$

where $k_2(h) = 1/k_1(h)$, $T_2(h) = 4K[k_2(h)]/\sqrt{A^2(h) - 1}$ is the dimensionless period of oscillations of the dimensionless charge at positive energy. In both formulae (6.49) and (6.50) the starting point is chosen so that at that moment the current in the circuit is zero; and $K(k)$ means a complete elliptic integral of the first kind:

$$K(k) = \int_0^1 \frac{\mathrm{d}w}{\sqrt{(1 - w^2)(1 - k^2 \cdot w^2)}}. \tag{6.51}$$

Further, the dimensionless voltage at the NC capacitor is:

$$u(\tau) = -x(\tau) + x^3(\tau). \tag{6.52}$$

The graphs of dependence on the dimensionless voltage time (6.52) at the NC capacitor under negative and positive energy, constructed using formulae (6.49) and (6.50) are shown in Fig. 6.18. These graphs show that the voltage at the NC-capacitor is essentially anharmonic.

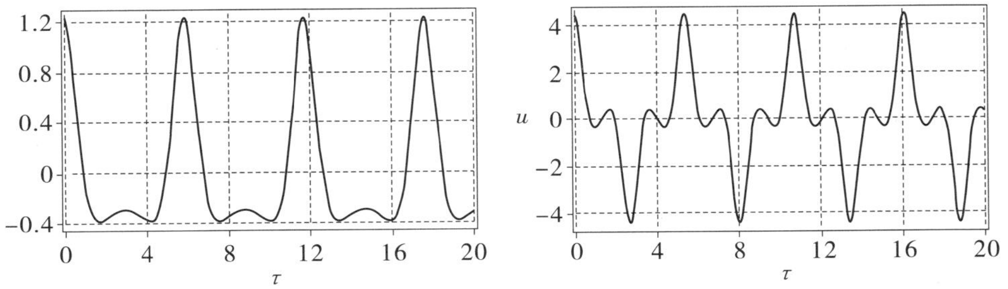

Fig. 6.18 Time dependence of the dimensionless voltage at the NC capacitor: on the left for $h = -0.05$, on the right for $h = 1.2$

In order that the circuit in Fig. 6.17 could be used in radio engineering, we should define the anharmonicity of the voltage quantitatively. The generally accepted parameter for estimating the anharmonicity of oscillations is the coefficient of nonlinear distortions (CND) with respect to the first harmonic:

$$K_u(h) = \frac{1}{u_1(h)} \cdot \sqrt{\sum_{n=2}^{\infty} u_n^2(h)}, \tag{6.53}$$

where $u_n(h)$ ($n \in \mathbf{N}$) are the amplitudes of Fourier harmonics of the dimensionless voltage (6.52).

In consequence of equation (6.48) $u(\tau) = -\ddot{x}(\tau)$, therefore, to determine the quantities $u_n(h)$, first we expand the solutions of (6.49) and (6.50) in Fourier series. The application of the theory of Jacobi elliptic functions gives us for $-1/4 < h < 0$:

$$x(\tau,h) = x_0(h) + \sum_{n=1}^{\infty} x_n(h) \cdot \cos \frac{2\pi n\tau}{T_1(h)}, \tag{6.54}$$

where

$$x_0(h) = \frac{\pi A(h)}{2K[k_1(h)]},\ x_n(h) = \frac{2x_0(h)}{\mathrm{ch}\{n \cdot \gamma[k_1(h)]\}}, \tag{6.55}$$

and for $h > 0$:

$$x(\tau,h) = \sum_{n=1}^{\infty} x_n(h) \cdot \cos \frac{2\pi \cdot (2n-1)\tau}{T_2(h)}, \tag{6.56}$$

where

$$x_n(h) = \frac{\pi A(h)}{k_2(h) K[k_2(h)]} \cdot \frac{\exp\{\gamma[k_2(h)]/2\}}{\operatorname{ch}\{n\gamma[k_2(h)]\}}. \tag{6.57}$$

In the expressions (6.55) and (6.57) the γ parameter depends on the k modulus of elliptic functions as follows: $\gamma(k) = \pi K(\sqrt{1-k^2})/K(k)$.

The Fourier series for the stress (6.52) is obtained by twice differentiating with respect to expansion time (6.54) and (6.56), in particular for $-1/4 < h < 0$:

$$u(\tau,h) = \sum_{n=1}^{\infty} u_n(h) \cdot \cos \frac{2\pi n\tau}{T_1(h)}, \tag{6.58}$$

where

$$u_n(h) = \left[\frac{2\pi n}{T_1(h)}\right]^2 x_n(h), \tag{6.59}$$

and $x_n(h)$ are the Fourier coefficients (6.55) for the charge.

Similarly for $h > 0$:

$$u(\tau,h) = \sum_{n=1}^{\infty} u_n(h) \cdot \cos \frac{2\pi(2n-1)\tau}{T_2(h)}, \tag{6.60}$$

where

$$u_n(h) = \left[\frac{2\pi(2n-1)}{T_2(h)}\right]^2 x_n(h), \tag{6.61}$$

and $x_n(h)$ are Fourier coefficients (6.57) for the charge.

The graphs of the spectral components of the dimensionless voltage (6.52) at the NC capacitor under negative and positive energy, constructed using formulae (6.59) and (6.61), for the same energy values as in Fig. 6.18 are shown in Fig. 6.19.

Since for any of the expansions (6.58) or (6.60) Parseval's equality is valid:

$$\frac{2}{T} \cdot \int_0^T u^2(\tau,h)\,d\tau = \sum_{n=1}^{\infty} u_n^2(h), \tag{6.62}$$

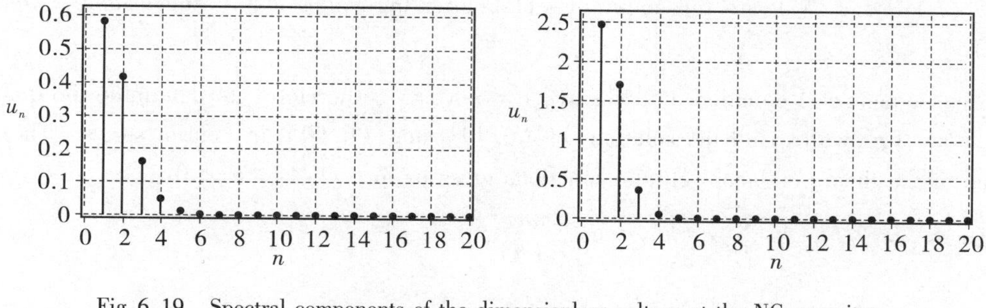

Fig. 6.19 Spectral components of the dimensionless voltage at the NC capacitor: on the left for $h = -0.05$, on the right for $h = 1.2$

then we can rewrite the expression (6.53) for CND as follows:

$$K_u(h) = \sqrt{2\frac{\overline{u^2}}{u_1^2} - 1}, \tag{6.63}$$

where the bar over the letter indicates the operation of averaging with respect to the oscillation period T:

$$\overline{u^2} \equiv \frac{1}{T} \cdot \int_0^T u^2 \mathrm{d}\tau. \tag{6.64}$$

The average value (6.64) can easily be calculated using the expression (6.52):

$$\overline{u^2} = \overline{x^2} - 2\overline{x^4} + \overline{x^6}. \tag{6.65}$$

Thus, the calculation of the CND was reduced to the determination of the average values $\overline{x^{2 \cdot l}}$ ($l = 1,2,3$) of the degrees of the dimensionless charge, and formulae (6.59) and (6.61) were simply necessary for determining the amplitude of the first harmonic $u_1(h)$.

We will calculate of average values over a period for even degrees of charge. Using formula (6.6), for the case that $-1/4 < h < 0$ we find:

$$\overline{x^{2l}} = \frac{1}{T_1} \cdot \int_0^{T_1} A^{2l} \mathrm{d}n^{2l}\left[\frac{2K(k_1)\tau}{T_1}, k_1\right] \mathrm{d}\tau, l = 1,2,3. \tag{6.66}$$

On inserting $w = sn\left[\frac{2K(k_1)\tau}{T_1}, k_1\right]$ the averages (6.66) reduce to the following integrals:

$$\overline{x^{2l}} = \frac{A^{2l}}{K(k_1)} \cdot \int_0^1 \frac{(1 - k_1^2 w^2)^l \mathrm{d}w}{\sqrt{(1 - w^2)(1 - k_1^2 w^2)}}, l = 1,2,3. \tag{6.67}$$

In their turn, the integrals in formula (6.67) are represented by linear combinations of elliptic integrals:

$$I_l(k) = \int_0^1 \frac{w^{2l}\mathrm{d}w}{\sqrt{(1 - w^2)(1 - k^2 w^2)}}. \tag{6.68}$$

In case that $l \geqslant 2$ these integrals follow the recurrence relations:

$$(2l - 1)k^2 \cdot I_l - (2l - 2)(k^2 + 1)I_{l-1} + (2l - 3)I_{l-2} = 0. \tag{6.69}$$

Since the initial conditions for the difference equation (6.69) are known: $I_0(k) = K(k)$, $I_1(k) = [K(k) - E(k)]/k^2$, where $E(k)$ denotes the complete elliptic integral of the second kind:

$$E(k) = \int_0^1 \sqrt{\frac{1 - k^2 w^2}{1 - w^2}} \mathrm{d}w, \tag{6.70}$$

Then

$$I_2(k) = \frac{(2 + k^2)K(k) - 2(1 + k^2)E(k)}{3k^4},$$

$$I_3(k) = \frac{(8 + 3k^2 + 4k^4)K(k) - (8 + 7k^2 + 8k^4)E(k)}{15k^6}, \tag{6.71}$$

and, using the formulae (6.67) – (6.71), we find the required average values $\overline{x^{2l}}$:

$$\overline{x^2} = A^2 \frac{E(k_1)}{K(k_1)}, \overline{x^4} = \frac{A^4}{3}\left[k_1^2 - 1 + 2(2 - k_1^2)\frac{E(k_1)}{K(k_1)}\right],$$

$$\overline{x^6} = \frac{A^6}{15}\left[-4 \cdot (2 - 3k_1^2 + k_1^4) + (23 - 23k_1^2 + 8k_1^4)\frac{E(k_1)}{K(k_1)}\right]. \tag{6.72}$$

Finally, combining formulae(6.59), (6.65)and(6.72), we calculate the CND (6.63) for the dimensionless energy lying in the interval $-1/4 < h < 0$.

The situation when the dimensionless energy $h > 0$ is considered in complete analogy. Using the formula (6.50), for average values of even powers of the dimensionless charge, we obtain:

$$\overline{x^{2l}} = \frac{1}{T_2}\int_0^{T_2} A^{2l} cn^{2l}\left[\frac{4 \cdot K(k_2)\tau}{T_2}, k_2\right] d\tau, l = 1, 2, 3. \tag{6.73}$$

On inserting $w = sn\left[\frac{4K(k_2)\tau}{T_2}, k_2\right]$, the averages (6.73) reduce to the following integrals:

$$\overline{x^{2l}} = \frac{A^{2l}}{K(k_2)}\int_0^1 \frac{(1 - w^2)^l dw}{\sqrt{(1 - w^2)(1 - k_2^2 w^2)}}, l = 1,2,3, \tag{6.74}$$

and are equal:

$$\overline{x^2} = \frac{A^2}{k_2^2}\left[k_2^2 - 1 + \frac{E(k_2)}{K(k_2)}\right], \overline{x^4} = \frac{A^4}{3k_2^4}\left[2 - 5k_2^2 + 3k_2^4 + 2(2k_2^2 - 1)\frac{E(k_2)}{K(k_2)}\right],$$

$$\overline{x^6} = \frac{A^6}{15k_2^6} \cdot \left[15k_2^6 - 34k_2^4 + 27k_2^2 - 8 + (23k_2^4 - 23k_2^2 + 8)\frac{E(k_2)}{K(k_2)}\right]. \tag{6.75}$$

CND (6.63) for dimensionless energy $h > 0$ is obtained by combining formulae (6.61), (6.65) and (6.75).

We will build graph of the nonlinear distortion coefficient and its asymptotics near the homoclinic eight. The final expression for CND is rather cumbersome; therefore, in Fig. 6.20 the CND graph is presented.

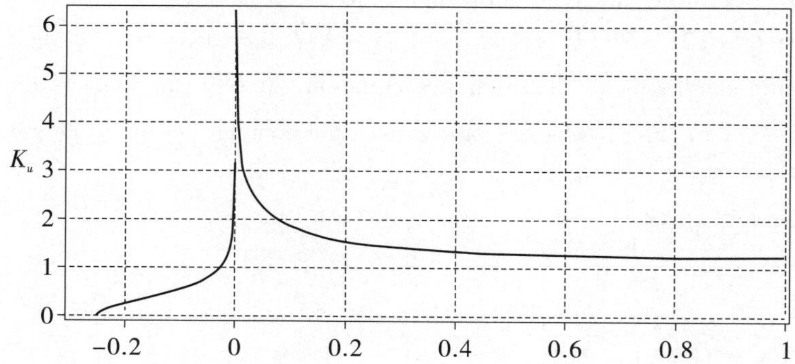

Fig. 6.20 The coefficient of nonlinear distortion depending on the energy h

Fig. 6.20 shows that $K_u(-1/4) = 0$. This is natural, since the Duffing Eq. (6.48), linearized near the equilibrium state $x = 1$, corresponding to the value $h = -1/4$, reduces to the harmonic oscillator equation, whose solution, as it is well known, does not incorporate higher harmonics.

For large values of energy $K_u(h) \approx 1$, that is, for $h \gg 1$, the squared amplitude of the first harmonic of the voltage (6.60) at the NC capacitor is approximately equal to the sum of the squared amplitudes of higher harmonics.

The nature of the CND inversion to infinity as the phase trajectory of Eq. (6.48) approaches the homoclinic eight corresponding to the value $h = 0$, can be easily determined from the following considerations: in case that $h \to 0$ the oscillation period $T(h) \to +\infty$ [6]. This means that when $h \to 0$ the squared voltage (6.52) at the NC capacitor can be calculated with respect to the charge $x(\tau) = \sqrt{2}/ch\tau$ at the homoclinic loop. In formula (6.64) the integration can be extended to infinity and the figure of one in expression (6.63) can be neglected.

Thus, we get that near the zero energy the CND is:

$$K_u(h) \approx \begin{cases} \dfrac{1}{4\pi^3}\sqrt{\dfrac{7}{30}}\left(\ln\dfrac{16}{|h|}\right)^{5/2}, & 0 < -h \ll 1, \\ \dfrac{1}{2\pi^3}\sqrt{\dfrac{7}{15}}\left(\ln\dfrac{16}{h}\right)^{5/2}, & 0 < h \ll 1. \end{cases} \tag{6.76}$$

Finally, since in accordance with the second of formulae (6.46) the dimensional voltage (6.27) at the NC capacitor is $U(t) = \dfrac{\alpha^{3/2}}{\beta^{1/2}}u(\tau)$, then the value (6.53) or (6.63) is the CND for the dimensional voltage (6.27) as well and it can be measured by existing network analyzers. The characteristic quantity of the dimensional voltage in the circuit $\alpha^{3/2}/\beta^{1/2} \sim 5$.

By the developed above method for determining the averages over a period, the CND with respect to the voltage and current in oscillatory circuits with other nonlinear elements, such as a single-layer ferroelectric capacitor with an operating temperature above its Curie temperature, the blocked $p-n$ junction of a semiconductor diode etc. can be calculated.

6.11 Generators of chaotic electrical oscillations on basis of ferroelectric capacitor with a negative capacitance

A number of modern schemes of data transmission employ dynamic chaos[42]. And at present there is considerable interest to these schemes because chaotic signals are naturally broadband signals. The broadband communication systems became dominant in the last-mile applications since they provide relatively high noise immunity, confidentiality, and data-transmission rates.

The nonlinear system that generates the chaotic signal is the most important component of data transmission system of this type. Development of material science leads to progressive broadening of variety of such systems. As an illustration of this trend one can say about the recent discovery of bilayer ferroelectric systems possessing by negative capacitance.

Voltage on NC-capacitors is the following function of its charge q [35,36]:

$$U_{NC} = -\alpha \cdot q + \beta \cdot q^3, \ \alpha > 0, \beta > 0. \tag{6.77}$$

The very significant peculiarity of NC-capacitors is alternating in sign of linear and cubic terms in formula (6.77). This peculiarity provides existence of the so-called the "homoclinic figure-eight" on phase plane of oscillatory circuit with NC capacitor[40] (see Fig. 6.1) But it is well investigated that presence of homoclinic loop on phase plane of the system under consideration is an evidence of possibility of existence of dynamic chaos in the system[43].

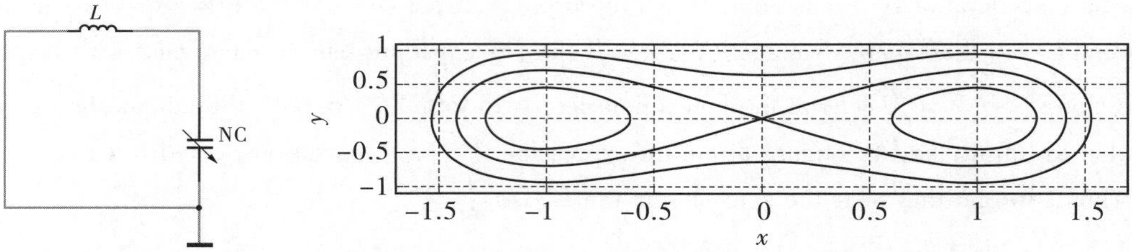

Fig. 6.21 Electrical scheme of oscillatory circuit with NC capacitor (on the left) and its phase portrait (on the right)

That is why let us now consider series combination of inductance L, resistance R and NC capacitor under the action of cosine input voltage with amplitude U_0 and frequency ω. The scheme of this electric circuit is shown on Fig. 6.22.

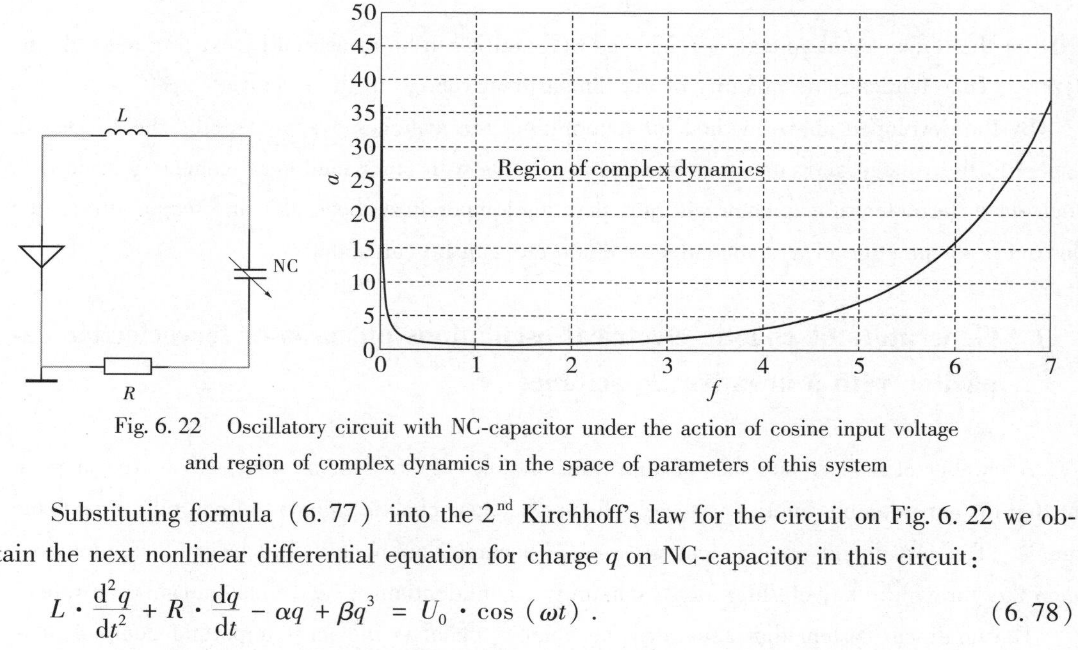

Fig. 6.22 Oscillatory circuit with NC-capacitor under the action of cosine input voltage and region of complex dynamics in the space of parameters of this system

Substituting formula (6.77) into the 2nd Kirchhoff's law for the circuit on Fig. 6.22 we obtain the next nonlinear differential equation for charge q on NC-capacitor in this circuit:

$$L \cdot \frac{d^2 q}{dt^2} + R \cdot \frac{dq}{dt} - \alpha q + \beta q^3 = U_0 \cdot \cos(\omega t). \tag{6.78}$$

Introducing the following dimensionless variables and parameters:

$$\tau = \sqrt{\frac{\alpha}{L}} \cdot t, \ x = \sqrt{\frac{\beta}{\alpha}} \cdot q, \ \delta = \frac{R}{2\sqrt{\alpha L}}, \ A_0 = \sqrt{\frac{\beta}{\alpha^3}} \cdot U_0, \ \Omega = \omega \cdot \sqrt{\frac{L}{\alpha}}, \tag{6.79}$$

we reduce input equation (6.78) to the form:

$$\ddot{x} - x + x^3 = -2\delta\dot{x} + A_0\cos(\Omega \cdot \tau), \tag{6.80}$$

where $\dot{x}$ is the derivative of dimensionless charge x with respect to dimensionless time τ.

Right side in equation (6.80) gives rise to destruction of the "homoclinic figure-eight". According to Ref. [43] one can calculate the Melnikov function:

$$\Delta(\tau_0) = \int_{-\infty}^{+\infty} \dot{X}(\tau) \cdot [-2\delta \cdot \dot{X}(\tau) + A_0\cos(\Omega\tau)]d\tau, \tag{6.81}$$

where $X(\tau) = \dfrac{\sqrt{2}}{\mathrm{ch}(\tau + \tau_0)}$ is analitic expression for "homoclinic figure-eight" and τ_0 is arbitrary temporal shift along the separatrix.

The value (6.81) has a meaning of the distance between stable and unstable manifolds of the saddle point (0, 0)[43]. It is equal to:

$$\Delta(\tau_0) = -\frac{8\delta}{3} - \frac{\sqrt{2}\pi A_0\Omega}{\mathrm{ch}(\pi\Omega/2)} \cdot \sin(\Omega\tau_0). \tag{6.82}$$

It is easy to see that if the inequality

$$a > \frac{\sqrt{2}}{3} \cdot \frac{\mathrm{ch}(f)}{f}, \tag{6.83}$$

where

$$a = \frac{A_0}{2 \cdot \delta}, f = \frac{\pi \cdot \Omega}{2}, \tag{6.84}$$

is true then the Melnikov function (6.82) is alternating in sign. Hence following to the Melnikov criterion[43] this fact gives an evidence of existence of transversal intersection of stable and unstable manifolds of fixed point in system (6.80) and serves an indication of presence of homoclinic structure and connected with it complex dynamics namely countable set of periodic orbits and continuum of nonperiodic trajectories. Boundary of the region of complex dynamics for the system (6.80) in its parametric space (f, a) plotted in accordance with inequality (6.7) is shown on Fig. 6.22.

And in our case dimensionless parameters (6.8) $a = \dfrac{U_0\sqrt{\beta L}}{\alpha R}$ and $f = \dfrac{\pi\omega}{2}\sqrt{\dfrac{L}{\alpha}}$ are expressed via physical parameters L, R, α, β, U_0 and ω of elements of this circuit.

In practice there is leakage of charge q on NC – capacitor[35,36]. Let us simulate this leakage by resistance R_n combined in parallel with NC-capacitor. Therefore oscillatory circuit under our investigation ought to be modified as it is shown on Fig. 6.23. It means that one ought to complicate the mathematical model of our circuit.

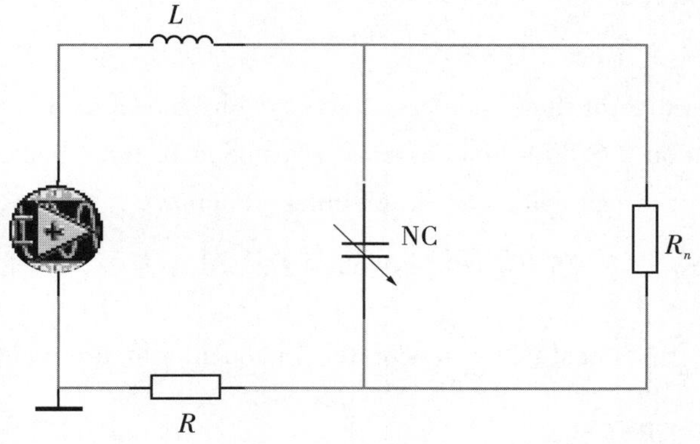

Fig. 6. 23 Oscillatory circuit with NC-capacitor with leakage under the action of cosine input voltage

The 1st Kirchhoff's law for the circuit on Fig. 6. 23 gives us that rate of changing of charge q on NC-capacitor is equal to:

$$\frac{dq}{dt} = I + I_n, \tag{6.85}$$

where I and I_n are currents via resistances R and R_n respectively. The 2nd Kirchhoff's law for both contours of circuit on Fig. 6. 3 reduces to the next pair of equations:

$$L \cdot \frac{dI}{dt} + RI + U_{NC}(q) = U_0 \cos(\omega t), \quad R_n I_n + U_{NC}(q) = 0. \tag{6.86}$$

Combining expressions (6. 77), (6. 85) and (6. 86) we derive the following generalization of equation (6. 78):

$$L \cdot \frac{d^2 q}{dt^2} + \left(R - \frac{\alpha L}{R_n}\right) \cdot \frac{dq}{dt} + \frac{3\beta L}{R_n} \cdot q^2 \cdot \frac{dq}{dt} + \left(1 + \frac{R}{R_n}\right) \cdot (-\alpha Q + \beta Q^3) = U_0 \cdot \cos(\omega t). \tag{6.87}$$

Suppose that $R \ll \alpha \cdot L/R \ll R_n$ and then introduce the next dimensionless values:

$$\bar{\tau} = \sqrt{\frac{\alpha}{L} \cdot \left(1 + \frac{R}{R_n}\right)} \cdot t, \quad x = \sqrt{\frac{\beta}{\alpha}} \cdot q, \quad \bar{\delta} = \frac{R}{2\sqrt{\alpha L}} \cdot \frac{1}{\sqrt{1 + R/R_n}} \cdot \left(1 - \frac{\alpha \cdot L}{R \cdot R_n}\right),$$

$$\bar{A}_0 = \sqrt{\frac{\beta}{\alpha^3}} \cdot \frac{U_0}{1 + R/R_n}, \quad \bar{\Omega} = \sqrt{\frac{L}{\alpha}} \cdot \frac{\omega}{\sqrt{1 + R/R_n}}, \quad \bar{\gamma} = \frac{\sqrt{\alpha L}}{R_n} \cdot \frac{1}{\sqrt{1 + R/R_n}}. \tag{6.88}$$

In variables (6. 88) equation (6. 87) can be rewritten as follows:

$$x'' - x + x^3 = -2\bar{\delta} x' - 3\bar{\gamma} x^2 x' + \bar{A}_0 \cos(\bar{\Omega}\bar{\tau}), \tag{6.89}$$

where x' is the derivative of dimensionless charge x with respect to dimensionless time $\bar{\tau}$.

In the same manner the Melnikov function for Eq. (6. 89) is equal to:

$$\bar{\Delta}(\bar{\tau}_0) = \int_{-\infty}^{+\infty} X'(\bar{\tau}) \cdot [-2 \cdot \bar{\delta} \cdot X'(\bar{\tau}) - 3 \cdot \bar{\gamma} \cdot X^2(\bar{\tau}) \cdot X'(\bar{\tau}) + \bar{A}_0 \cdot \cos(\bar{\Omega}\bar{\tau})] d\bar{\tau}. \tag{6.90}$$

Calculating of integral (6.90) we obtain that:

$$\bar{\Delta}(\bar{\tau}_0) = -\frac{8\bar{\delta}}{3} - \frac{16\bar{\gamma}}{5} - \frac{\sqrt{2}\pi \cdot \bar{A}_0\bar{\Omega}}{\mathrm{ch}(\pi \cdot \bar{\Omega}/2)} \cdot \sin(\bar{\Omega}\bar{\tau}_0). \tag{6.91}$$

It follows from formula (6.91) that the region of complex dynamics for the system (6.89) is represented by the next inequality:

$$\bar{a} > \frac{\sqrt{2}}{3} \cdot \frac{\mathrm{ch}(\bar{f})}{\bar{f}}, \tag{6.92}$$

where dimensionless parameters:

$$\bar{a} = \frac{a}{\sqrt{1 + R/R_n} \cdot \left(1 + \frac{7}{5} \cdot \frac{\alpha \cdot L}{R \cdot R_n}\right)}, \quad \bar{f} = \frac{f}{\sqrt{1 + R/R_n}} \tag{6.93}$$

are slightly expanded in comparison with parameters (6.84) of boundary (6.83) of the region of complex dynamics for the system (6.80). We also underline that shapes of curves (6.83) and (6.92) are exactly the same. It means that for electrical circuit on Fig. 6.23 possession by complex dynamics is the rough feature in the sense of the theory of dynamical systems[44].

6.12 On promising trends of research on fractals and textures

On the basis of the above mentioned author's works, let us try to bring into focus the most promising trends of fractal research in the field of progress in modern fundamental and applied sciences[5,7,17,45-55]:

1. Investigation of the capabilities of textural (spatial and spectral), fractal and entropy features for radar detection tasks.

2. Synthesis of new models of radar signals scattering by land covers based on the theory of deterministic chaos, strange attractors and fractal probability distributions – stable distributions.

3. Investigation of wave phenomena (propagation and scattering of waves, diffusion processes) in fractal inhomogeneous media based on fractional integro-differential operators. Further development of fractal electrodynamics.

4. Synthesis of channel models of radar and telecommunication system based on spatial fractal generalized correlators and fractal frequency coherence functions.

5. Investigation of the possibilities of identification of target shape or contours using fractal, textural and entropic features. Work on the singularities of the input function.

6. Investigation of the potential possibilities and limitations of fractal methods for processing radar and communication signals, including fractal modulation and demodulation, fractal coding and data compression, fractal image synthesis, fractal filters. Transition to fractal radio systems. Fractals in acoustic electronics.

7. Investigation of adaptive space-time signal processing based on fractional dimension and fractional operators.

8. Search and study of new combined methods for detecting and identification of low contrast target classes in high-intensity non-Gaussian noise.

9. Investigation of the possibilities of developing new media for transmitting information, multiple-band fractal absorbing materials, constructing fractal antennas and fractal frequency selective surfaces and volumes. Further development of the theory and technology of fractal impedances.

10. Synthesis of new classes of fractals and multifractals with a generalization of the concept of set measure.

11. Study of the type or sampling topology of a one-dimensional (multidimensional) signal, for example, for tasks of artificial intelligence in order to compile dictionaries of fractal features based on fractal primitives that are elements of the fractal language with fractal grammar, i. e., investigation of *the Problem of* "*dimensional sclerosis*" of physical signals and signatures. These concepts, introduced by the author, suggest the study of the topological features of each specific individual sampling, but not average implementations, which are often of different nature.

12. The forecast of the formation mechanisms and roughness characteristics in order to control the geometrical parameters of the micro relief to obtain the desired physicochemical and operational properties of products with modern non-equilibrium processing technologies of their surface layer. Fractals in nanotechnology. (In 2008, the author proposed a new concept, namely, "Scaling of a rough fractal layer and nanotechnology")

13. The development of fractal non-inertial relativistic radiolocation in curved space-time of connected structures, i. e., fractal geometry of space-time of deterministic structures. At present this fundamental scientific trend has acquired an imposing name of "Fractal Cosmology". [Our works with co-authors are listed in publications (*arXiv*) on this theoretical trend].

6.13 A new direction in the theory of statistical solutions and in statistical radio engineering

The rapid development of the theory of fractals in radiolocation and radiophysics has led to the establishment of a new theoretical direction in modern science and technology. It can also be described as "Statistical Theory of Fractal Radar"[5,7,17,45-55].

This direction includes (at least at the initial stage) the following fundamental questions:

(1) The theory of the integer and fractional measure.

(2) Caratheodory construction in the measure theory.

(3) Hausdorff measure and Hausdorff-Besicovitch dimension.

(4) The theory of topological spaces.

(5) The dimension theory.

(6) The line from the point of view of mathematician.

(7) Non-differentiable functions and sets.

(8) Fundamentals of the theory of probability.
(9) Stable probability distributions.
(10) The theory of fractional calculus.
(11) The classical Brownian motion.
(12) Generalized Brownian motion.
(13) Fractal sets.
(14) Anomalous diffusion.
(15) The main criteria for statistical decision theory in radar.
(16) Wave propagation in fractal random media.
(17) Wave scattering generalized Brownian surface.
(18) Wave scattering surface on the basis of non-differentiable functions.
(19) Difractals.
(20) Cluster analysis.
(21) Theory and circuitry of fractal detectors.
(22) Fractal-scaling or scale-invariant radar.
(23) The multi-radar.
(24) MIMO radar.
(25) Cognitive radar and quantum radar.

This list of studied questions, of course, is supposed to be expanded and refined in the future. The author has been dealing with it for 40 years of his scientific career[55].

In fact, this is a new program of university courses and lectures.

6.14 Conclusion

The professor A. Potapov created, developed and applied fractal-scaling methods for radiolocation problems and forming the foundations of fractal element base[5,7,17,45-55]. For the first time ever approaches to development of a fractal radar and a fractal MIMO-radar were considered. The author emphasizes that the synthesis of topological (fractal, textural, chaotic, etc.) detectors makes for a fresh look at the problem of detecting super weak actual signals. As a result of that, the professor A. Potapov discovery in the away-back 1980-ies takes the meaning of generalized detection. Thus, pure energy and pure topological detectors are not contrary to each other and they do not duplicate, but complement one another.

Due to topological detectors it is possible to see the process of energy detection in a new light and to find some essential faults in it. Consequently, topological detection becomes not less, if not more, valuable for theory and practice than energy detection. The theory of topological detection is formulated in Ref. [5, 7, 17, 45-55]. It is especially necessary for the purpose of reexamining the former theory and in that way producing new results that are not available to traditional concepts of radiolocation.

Thus, topological detection opens the door to a radically new field of statistical decision theory and provides an opportunity to correct ideas in this field, and even to create new ones, which is of great theoretical and practical importance. The sufficiently detailed reasoning reported here should contribute to a better understanding of proposed by the professor A. Potapov fundamentally new interpretation of the problem of radar (and other kinds of) detection. The proposed theory has much in common with cognitive radar.

Professor A. Potapov raised the foregoing problems as early as in 1980 and 40 years he has been successfully working on their solution and development. Thus, during more than 40 years, almost from scratch, fundamental bases of the theory that will be applied in the following decades were formed. Not results, not specific solutions are the most valuable, but namely the solution method, the approach to it. The created method is presented in Ref. [5, 7, 17, 45 – 55] and in this monograph.

References

Chapter 1

[1]POTAPOV A A. Fractals in radiophysics and radar[M]. Moscow:Logos,2002. (in Russian)

[2]POTAPOV A A. Fractals in radiophysics and radar:topology of a sample[M]. Moscow:Universitetskaya Kniga,2005. (in Russian)

[3]POTAPOV A A. Fractals and chaos as a base of new breakthrough technologies in modern radio systems[M] // CROWNOVER R M. Introduction to fractals and chaos. Moscow:Tekhnosfera,2006. (in Russian)

[4]BUNKIN B V,REUTOV A P,POTAPOV A A,et al. Aspects of perspective radiolocation[M]. Moscow:Radiotekhnika,2003. (in Russian).

[5]PODOSENOV S A,POTAPOV A A,SOKOLOV A A. Pulse electromagnetics of wideband radio systems and fields of coupled structures[M]. Moscow:Radiotekhnika,2003. (in Russian)

[6]Bystrov R P,POTAPOV A A,Sokolov A V. Millimeter radar combined with fractal processing[M]. Moscow: Radiotekhnika,2005. (in Russian)

[7]POTAPOV A A. Synthesis of images of earth covers in the optical and millimeter wave ranges[M]//Doctoral Dissertation in Mathematics and Physics. Moscow:IRE RAN,1994. (in Russian)

[8]SHIRMAN Y D,LOSEV Y I,MINERVIN N N,et al. Radioelectronic systems:principles of design and theory [M]. Moscow:MAKVIS,1998. (in Russian)

[9]FEDOROV I B. Informational technologies in radio engineering systems[M]. Moscow:MGTU,2003. (in Russian)

[10]BELOGLAZOV I N,DZHANDZHGAVA G I,CHIGIN G P. Fundamentals of navigation by geophysical fields [M]. Moscow:Nauka,1985. (in Russian).

[11]KREMER I Y. Spatial-temporal signal processing[M]. Moscow:Radio i Svyaz' ,1984. (in Russian)

[12]FAL'KOVICH S E,PONOMAREV V I,SHKVARKO Y V. Optimal reception of spatialtemporal signals in scattering radio channels[M]. Moscow:Radio i Svyaz' ,1989. (in Russian)

[13]POTAPOV A A. The use of modulated millimeter waves for image formation and identification[J]. Radiotekhnika,1989,12:61 -64.

[14]PAPOULIS A. Systems and transforms with applications in optics[M]. New York:McGraw-Hill,1968.

[15]GOODMAN J W. Introduction to fourier optics[M]. New York:McGraw-Hill,1968; Moscow:Mir,1970.

[16]ZVEREV V A. Radiooptics:signal transformation in radio and optics[M]. Moscow:Sov. Radio,1975. (in Russian)

[17]GOODMAN J W. Statistical optics[M]. New York:Wiley,1985.

[18]SMOKTII O I,FABRIKOV V A. Methods of theory of systems and transformations in optics[M]. Moscow: Nauka,1989. (in Russian)

[19]MANDEL L,WOLF E. Optical coherence and quantum optics[M]. Cambridge:Cambridge University Press, 1995.

[20] WEYL H. The propagation of electromagnetic waves over a plane conductor[J]. Ann Phys, 1919, 60: 481 – 500.

[21] KHURGIN Y I, YAKOVLEV V P. Methods of the theory of integer functions in radiophysics, the communications theory and optics[M]. Moscow: Fizmatgiz, 1962. (in Russian)

[22] AVDEEV V B, ASHIKHMIN A V. Modeling of miniature ultrawideband antennas[M]. Voronezh: Voronezh University Press, 2005. (in Russian)

[23] BASS F G, FUKS I M. Wave scattering by a statistically rough surface[M]. Moscow: Fizmatgiz, 1972. (in Russian)

[24] BREKHOVSKIKH L M, LYSANOV Y P. Fundamentals of ocean acoustics[M]. Leningrad: Gidrometeoizdat, 1982.

[25] POTAPOV A A. On the Indicatrixes of waves scattering from the random fractal anisotropic surface[M]// Brambila F. Fractal analysis: applications in physics, engineering and technology. Rijeka: InTech, 2017: 187 – 248.

[26] SHOROKHOVA E A, KASHIN A V. Some features of the scattering of electromagnetic waves on statistically uneven earth covers in the millimeter wavelength range[J]. Izv Vyssh Uchebn Zaved Radiofiz, 2005, 48 (6), 478 – 487.

[27] KENNEDY R S. Fading dispersive communication channels[M]. New York: Wiley-Interscience, 1969.

[28] FRENKS L. Signals theory[M]. New York: Prentice-Hall, 1969.

[29] VAN TREES H L. Detection, estimation and modulation theory[M]. New York: Wiley, 1968 – 1971. (in Russian)

[30] SHTAGER E A. Radio scattering by complex bodies[M]. Moscow: Radio i Svyaz' 1986. (in Russian)

[31] GORELIK G S. Oscillations and waves[M]. Moscow: Fizmatgiz, 1959. (in Russian)

[32] POTAPOV A A. Wave and diffraction 90: thes. x all-union symp. diffraction and propagation of waves[C]. Moscow: Fizich. o-vo SSSR, 1990: 183.

[33] POTAPOV A A. Statistical methods in the theory of transmission and transformation of information signals: Thes. Int. Sci. -Tech. Conf. [C]. Kiev: KIIGA, 1992: 85.

[34] ISHIMARU A. Wave propagation and scattering in random media: Vols. 1 and 2 [M]. New York: Academic, 1978.

[35] BERRY M V. The statistical properties of echoes diffracted from rough surfaces[J]. Philos Trans R Soc London Ser, 1973, 273 (1237), 611 – 654.

[36] BROWN G S. The average impulse responce of a rough surface and its applications[J]. IEEE Trans. Antennas Propag. 1977, 25 (1): 67 – 74.

[37] POTAPOV A A. Radiophysical effects in the interaction of electromagnetic radiation of the millimeter wave range with the environment[M]. Moscow: Zarubezh. Radioelektron. 1992, Pt. 1, No. 8: 36 – 76; 1992, Pt. 2, No. 9: 4 – 28; 1992, Pt. 3, No. 11: 23 – 48; 1993, Pt. 4, No. 3: 36 – 48; 1993, Pt. 5, No. 7 – 9: 32 – 49; 1994, Pt. 6, No. 7 – 8: 11 – 30; 1995, Pt. 7, No. 1: 27 – 36.

[38] STRATONOVICH R L. Selected problems of fluctuation theory in radio engineering[M]. Moscow: Sovetskoe Radio, 1961. (in Russian)

[39] KENDALL M G, STUART A. The advanced theory of statistics[M]. 4th ed. London: Griffin, 1977.

[40] CRAMER H. Mathematical methods of statistics[M]. Princeton: Princeton University Press, 1946.

[41] MALAKHOV A N. Cumulant analysis of random non-gaussian processes and their transformations[M]. Moscow: Sovetskoe Radio, 1978. (in Russian)

[42]STRATONOVICH R L. Nonlinear nonequilibrium thermodynamics[M]. Moscow:Nauka,1985. (in Russian)

[43]GARDINER C. Handbook of stochastic methods in physics,chemistry and natural sciences[M]. Heidelberg: Springer-Verlag,1983.

[44]MONIN A S,YAGLOM A M. Statistical fluid mechanics:Vol. 1[M]. Cambridge,Mass. :MIT,1971.

[45]GRIGORIU M. Applied non-gaussian process. NJ:Englewood Cliffs,1995.

[46]POTAPOV A A. Study of the effect of vegetation on the backscattered millimeter wave field[J]. Radiotekh Elektron,1991,36 (2):239 -246.

[47]POTAPOV A A,GERMAN V A. Effects of deterministic chaos and strange attractor in the radar of dynamic systems of the vegetative cover type[J]. Tech Phys Lett,2002,28 (7):586 -588.

[48]POTAPOV A A,GERMAN V A. Reconstruction of a strange attractor in reflected radar signals [M] // Aspects of Perspective Radar. Moscow:Radiotekhnika,2003. (in Russian)

[49]REUTOV A P,POTAPOV A A,GERMAN V A. Strange attractors and fractals as the basis of a new dynamic model of radar signals scattered by vegetation[J]. Nelineinyi Mir,2003,1 (1 -2):12 -27.

[50]POTAPOV A A. Generalized correlator of fields scattered by rough surfaces[J]. J Commun Technol Electron,1996,41 (9):759 -766.

[51]POTAPOV A A. Functional approach to correlation of radio waves scattered by an anisotropic rough surface [J]. Elektromagn. Volny I Elektron Sistemy,1996,1 (1):32 -37.

[52]ZUBKOVICH S G. Statistical characteristics of radio signals reflected from the earth's surface[M]. Moscow: Sovetskoe Radio,1968. (in Russian)

[53]RYTOV S M,KRAVTSOV YU A,TATARSKII V I. Principles of statistical radiophysics Vol. 1 -4[M]. Berlin:Springer-Verlag,1989.

[54]BAKUT P A,MANDROSOV V I,MATVEEV I N,et al. Theory of coherent images[M]. Moscow:Radio i Svyaz',1987. (in Russian)

[55]BALTES H P. Inverse scattering problems in optics[M]. Berlin:Springer-Verlag,1980.

[56]VASIL'EV V A,DOBROVIDOV A V,KOSHKIN G M. Nonparametric estimation of functionals of distributions of stationary sequences[M]. Moscow:Nauka,2004. (in Russian)

[57]ARMAND N A. On the correlation function of waves scattered by rough surfaces[J]. Radiotekh Elektron, 1985,30 (7):1307 -1311.

[58]OREKHOV B I,GARNAKER'YAN A A,BUKHARIN V D,et al. Spatial correlation of radio signals reflected from the sea surface[J]. Radiotekh Elektron,1988,33 (3):512 -517.

[59]GARNAKER'YAN A A,OREKHOV B I,BUKHARIN V D,et al. On the effect of slopes of a statistically rough surface on the magnitude of the reflected signal[J]. Radiotekh Elektron,1988,33 (12):2537 -2543.

[60]BUKHARIN V D,OREKHOV B I,GARNAKER'YAN A A,et al. Investigation of the energy characteristics of a signal reflected from a statistically uneven surface[J]. Radiotekhnika 1990,11:32 -35.

[61]ZHUKOVSKII A P,ONOPRIENKO E I,CHIZHOV V I. Theoretical fundamentals of radio altimetry[M]. Moscow:Sovetskoe Radio,1979. (in Russian)

[62]KIRILLOV N E. Noise-Immune message transmission over linear channels with randomly varying parameters [M]. Moscow:Svyaz',1971. (in Russian)

[63]LOBKOVA L M. Propagation of radio waves above sea surface[M]. Moscow:Radio i Svyaz',1991. (in Russian)

[64]SOKOLOVSKII V I,CHERKASHINA L N. Determination of the average intensity of the scattered field by averaging over the frequencies of the selected interval[J]. Radiotekh Elektron,1971,16 (8):1391 -1396.

[65] POTAPOV A A. On remote measurement of the height and characteristics of chaotic covers by the frequency coherence function, application of remote radiophysical methods to exploration of natural environment: Thes. All-Union Conf., Erevan[C]. Erevan: AN ArmSSR, 1990: 122 – 123.

[66] WEISSMAN D E. Two frequency radar interferometry applied to the measurement of ocean wave height[J]. IEEE Trans. Antennas Propag, 1973, 21: 649 – 656.

[67] WEISSMAN D E, JOHNSON J W. Dual frequency correlation radar measurements of the height statistics of ocean waves[J]. IEEE Trans Antennas Propag, 1977, 25: 74 – 83.

[68] MILLER L S. The application of near-nadir Δk radar techniques to geodetic altimetry and oceanografic remote sensing[J]. IEEE Trans Geosci Electron, 1983, 21 (1): 16 – 24.

[69] KUZNETSOVA T I. On the phase retrieval problem in optics[J]. Sov Phys. Usp, 1988, 31 (4): 364 – 371.

[70] PISKORZH V. Phase problem in channels with scattering[J]. Izv Vyssh Uchebn. Zaved Radiofiz, 1992, 35 (3 – 4): 252 – 263.

[71] POTAPOV A A. Scattering of millimeter radio waves by isotropic and anisotropic randomly irregular surfaces and spatial-frequency correlation functions[J]. Elektromagn Volny I Elektron Sistemy, 1997, 2 (5): 68 – 75.

[72] POTAPOV A A, SOKOLOV A V. Frequency function of coherence of wave field scattered by rough surface with large Rayleigh constant[J]. J Commun Technol Electron, 1996, 41 (11): 983 – 988.

[73] ISHIMARU A, AILES-SENGERS L, PHU P, et al. Pulse broadening and two-frequency mutual coherence function of the scattered wave from rough surfaces[J]. Waves Random Media, 1994, 4 (2): 139 – 148.

[74] ALEKSEEV G A. Frequency and space correlation of waves scattered by a rough surface[J]. Izv Vuzov Radiotekhnika, 1966, 9 (1): 137 – 140.

[75] ALEKSEEV G A. Scattering of regular and noise signals by an uneven surface[J]. Radiotekh Elektron, 1968, 13: 1683 – 1685.

[76] SHMELEV A B. On the frequency correlation of a sound field scattered by a statistically uneven surface when a directed spherical wave is incident[J]. Trudy Radiotekhn Instit AN SSSR, 1971, 5: 48 – 52.

[77] OPALENOV YU V, POTAPOV A A, FEDYUNIN S YU. Radiophysical measuring complex with a complex phase-manipulated signal in the millimeter wave range[J]. Radiotekhnika, 1991, 11: 67 – 70.

[78] POTAPOV A A, GALKINA T V, KOLESNIKOV A I, et al. Monte Carlo calculation of scattering of millimeter waves by earth covers with chaotic irregularities. 4th All-Union School on Propagation of Millimeter and Submillimeter Waves in Atmosphere, Sept. 3 – 10, 1991[C]. Nizhniy Novgorod: NIRFI, 1991: 106 – 107.

[79] POTAPOV A A, GERMAN V A. Methods of fractal processing of weak signals and low-contrast images[J]. Avtometriya (Novosibirsk) 2006, 42 (5): 3 – 25.

[80] POTAPOV A A. Synergetic principles of nonlinear dynamics and fractals in the development of new information technologies for modern radio systems[J]. Radiotekhnika, 2005, 8: 73 – 88.

[81] GULYAEV YU V, NIKITOV S A, POTAPOV A A, et al. Design of fractal radio systems: numerical analysis of electromagnetic properties of the sierpinski fractal antenna[J]. J Commun Technol Electron, 2005, 50 (9): 988 – 993.

[82] POTAPOV A A. Wave effects in inhomogeneous media. proc. VIII All-Rus. School-Sem., Krasnovidovo, May 26 – 31, 2002[C]. Moscow: Mosk Gos Univ, 2002: 69 – 70.

[83] POTAPOV A A, GERMAN V A. The frequency coherence function of a fractal radio channel. Electronics and Informatics 2002. Int. Sci. -Tech. Conf., Moscow-Zelenograd, Nov. 19 – 21, 2002[C]. Moscow: MIET, 2002: 221 – 222.

[84]POTAPOV A A,GERMAN V A. The possibility of applying theory deterministic chaos in models of channel radar and telecommunication systems. Electronics and Informatics 2002. Int. Sci. -Tech. Conf. , Moscow-Zelenograd,Nov. 19 – 21,2002[C]. Moscow:MIET,2002:219 – 220.

[85]CLIFFORD S F,BARRICK D E. Remote sensing of Sea State by Analysis of Backscattered Microwave Phase Fluctuations[J]. IEEE Trans Antennas Propag,1978,26 (5):699 – 705.

[86]WOODWARD P M. Probability and information theory with applications to radar [M]. New York:McGraw-Hill,1953.

[87]COOK C,BERNFELD M. Radar signal:an introduction to theory and application [M]. New York:Academic,1967.

[88]FUKS I M,SHTAGER E A. Investigation of the characteristics of scattering of frequency-modulated signals by a body having a complex shape[J]. Radiophysics and Quantum Electronics,1975,18 (1):77 – 80.

[89]FUKS I M,SHTAGER E A. The method of the generalized uncertainty function in problems of wave scattering on bodies of complex shape and rough surfaces[M]. Preprint. Kharkov,IRE AN USSR,1977:42. (in Russian)

[90]ABRAMOVICH Y I,KOSHEVOI V M,LAVRINENKO V P. Mutual uncertainty function of signals reflected from spatially distributed targets[J]. Radiotekh Elektron,1977,22 (10):2109 – 2114.

[91]RUBTSOV M I,CHETYRKIN N V. On the mutual function of uncertainty LFM-Signal when locating a long target[J]. Radiotekh Elektron,1979,24 (4):751 – 758.

[92]SOROKO L M. Holography and Coherent Optics[M]. Moscow:Nauka,1971.

[93]GUSAKOVSKII V E,SHTAGER E A. Generalized uncertainty function of a complex body[J]. Radiotekh Elektron,1983,28 (7):1317 – 1323.

[94]CHIZHOV V I. Detection of radio signals reflected from statistically long uneven surfaces[J]. Izv Vyssh Uchebn Zaved Radioelektronika,1982,25 (9):43 – 47.

[95]ZHUKOVSKII A P,KOSTROMINA N V,NUZHDIN V M. Signal uncertainty function by scattered in three dimensions[J]. Izv Vyssh Uchebn Zaved Radioelektronika,1985,28 (4):13 – 20.

[96]VINOKUROV V I,GENKIN V A,KALENICHENKO S P,et al. Sea radar[M]. Leningrad:Sudostroenie,1986. (in Russian)

[97]LOMANN A V,VIRNITTSER B. Triple correlations[J]. Proc IEEE,1984,72 (7):889 – 901.

[98]BARAKAT R,EBSTEIN S. Bispectral diffraction imagery. I. The bispectral optical transfer function[J]. J Opt Soc Am A,1987,4 (9):1756 – 1763.

[99]WIGNER E. On the quantum correction for thermodynamic equilibrium[J]. Phys Rev,1932,40:749 – 759.

[100]VOLTERRA V. Theory of functionals and of integral and integro-differential equations[M]. London:Blackie,1930.

[101]FADDEEVA V N,TERENT'EV N M. Tables of values of the probability integral of a complex argument [M]. Moscow:Gostekhteorizdat,1954. (in Russian)

[102]ABRAMOVITZ M,STEGUN I A. Handbook of mathematical functions,with formulas,graphs,and mathematical tables[M]. New York:Dover,1965.

[103]ARMAND N A. Signal recovery,ionospheric dispersion,distorted and lateral scattering on a rough surface [J]. Radiotekhnika,2005,8:32 – 39.

[104]BERRY M V,BLACKWELL Z V. Difractal echoes[J]. J Phys A,1981,14 (11):3101 – 3110 .

[105] BERRY M V. Difractals[J]. J Phys, 1979, 12 (6): 781 – 797.

[106] FUNG A K, MOORE R K. The correlation function in Kirchoff's method of solution of scattering of waves from statistically rough surfaces[J]. J Geophys Res, 1966, 71 (12): 2939 – 2943.

[107] KOLOSOV M A, POZHIDAEV V N. Statistical models of hydrometeor weakening. All-Union School on Propagation of Millimeter and Submillimeter Waves in Atmosphere, Moscow, Feb. 10 – 17, 1982[C]. Moscow: IRE AN SSSR, 1983: 101 – 121.

[108] POZHIDAEV V N, SOKOLOV A V. Methods for assessing the probability of attenuation of millimeter waves in the atmosphere in the presence of precipitation and water aerosols[J]. Preprint, 1987, 1(460): 67. (in Russian)

[109] TREBITS R N, HAYES R D, BOMAR L C. Mm-Wave reflectivity of land and sea[J]. Microwave J, 1978, 21(8): 49, 52, 53, 83.

[110] RICHARD V W, KAMMERER J E, WALLACE Y B. Rain backscatter measurements at millimeter wavelengths[J]. IEEE Trans Geosci Electron, 1988, 26 (3): 244 – 252.

[111] LIEBLE H J, HUFFORD G A, MANABE T A. A model for the complex permitivvity of water at frequencies below 1 THz[J]. Int J Infrared Millim Waves, 1991, 12 (7): 659 – 675.

[112] KADYGROV E N, POTAPOV A A. Some results of using radio waves millimeter range in the presence of precipitation[J]. Tr Central Aerological Observatory (Dolgoprudny), 1990, 168: 76 – 82.

[113] CHURILOV V P, SHARAPOV L I. Measurement of radar reflections from rain at 140 GHz. XIII All-Union Conf. on Propagation of Radio Waves, Gorki, Russia, June, 1981[C]. Moscow: Nauka, 1981: Pt. 2, 105 – 106.

[114] HULST H C. Light scattering by small particles[M]. New York: Wiley, 1957.

[115] DEIRMENDJIAN D. Electromagnetic scattering on spherical polydispersions[M]. New York: Elsevier, 1969.

[116] BOHREN C F, HUFFMAN D R. Absorption and scattering of light by small particles[M]. New York: Wiley, 1983.

[117] OLSEN R L, ROGERS D V, HODGE D B. The aR^b relation in the calculation for rain attenuation[J]. IEEE Trans Antennas Propag, 1978, 26 (2): 318 – 329.

[118] OKAMURA S, SONOHARA A. Atmospheric attenuation by rain at 140 GHz[J]. Electron Commun Japan, 1977, 60 (4): 94 – 100.

[119] GUSHCHINA I YA, ZRAZHEVSKII A YU, MALINKIN V G, et al. Attenuation of short millimeter radio waves in rain[J]. Radiotekh Elektron, 1980, 25: 1522 – 1524.

[120] POTAPOV A A. The scattering properties of biological objects in the microwave range. Biomedical and Environmental Instrument-Making. Int. Sci. -Tech. Conf., Ryazan, Russia, June 2 – 4, 1992[C]. Rjazan: RRTI, 1992: 34.

[121] SHULTZ F V, BURGENER R C, KING S. Measurement of the radar cross section of a man[J]. Proc IRE, 1958, 46 (46): 476 – 481.

[122] MITNIK L M, VIKTOROV S V. Radar probing of the earth's surface from space[M]. Leningrad: Gidrometeoizdat, 1990. (in Russian)

[123] POTAPOV A A. On meteorological interpretation of the mesostructures of microwave images. Statistical Methods in the Theory of Transmission and Transformation of Information Signals. Int. Sci. -Tech. Conf. Feb 25 – 27, 1992[C]. Kiev: KIIGA, 1992: 106 – 107.

[124] DOVIAK R. J, ZRNIC D S. Doppler Radar and Weather Observation[M]. Orlando: Academic, 1986.

[125] DAVIES P G, LANE J A. Review of propagation characteristics and prediction for satellite links at frequencies of 10 – 40 GHz[J]. IEE Proc, 1986, F133 (4): 420 – 428.

[126] KRASOVSKII A A. Handbook on the theory of automatic control[M]. Moscow: Nauka, 1987. (in Russian)

[127] ANDREEV G A, POTAPOV A A. Active orientation systems by geophysical fields[J]. Zarubezh Radioelektron, 1988, 9: 62 – 85.

[128] REED C G, HOGAN J J. Range correlation guidance for cruise missiles[J]. IEEE Trans Aerosp Electron Syst, 1979, 15 (4): 547 – 554.

[129] POTAPOV A A, OPALENOV Y V. Long-Term millimeter-wave measurements of the earth's mantle scattering characteristics using a helicopter[J]. Elektromagn Volny I Elektron Sistemy, 1997, 2 (3): 71 – 74.

[130] POTAPOV A A. Statistics of spatio-temporal characteristics of scattering of the earth's surface by millimeter waves[J]. Elektromagn Volny I Elektron Sistemy, 1997, 2 (1): 77 – 80.

[131] POTAPOV A A. A statistical approach to the description of images of textures of the earth's surface in the optical and radio range. Mathematical Methods of Pattern Recognition. Ⅳ All-Union Conf., Oct. 24 – 26, 1989[C]. Riga: MIPKRRiS, 1989: 150 – 151.

[132] POTAPOV A A. Radiophysical model for the formation of reference radar maps of heterogeneous terrain in the millimeter wave range[J]. Elektromagn Volny I Elektron Sistemy, 1997, 2 (4): 31 – 34.

[133] PAVEL'EV V A, POTAPOV A A. Influence of the ground surface on the structure of a pulse signal in millimeter wavelength band[J]. Radiotekh Elektron, 1994, 39 (4): 573 – 582.

[134] DMITRIEV S V, OPALENOV Y V, POTAPOV A A, et al. On remote sensing of the earth covers by combined radiophysical systems with a quasi-continuous noise-like phase-manipulated signal. Remote Sensing of Agropedological and Water Resources. All-Union Conf., Oct. 16 – 18, 1990[C]. Barnaul: Altaisk. Gos. Univ., 1990: 17 – 18.

[135] POTAPOV A A. New information technologies based on probabilistic textural and fractal signs in radar detection of low contrasting targets[J]. J Commun Technol Electron, 2003, 48 (9): 1012 – 1119.

[136] FUNG A K, EOM H. Coherent scattering of a spherical wave from an irregular surface[J]. IEEE Trans Antennas Propag, 1983, 31 (1): 68 – 72.

[137] POTAPOV A A, VYACHESLAVOVA O F. Qualitative and quantitative assessment of the surface of machine parts based on fractal dimensions and fractal signatures[J]. Obozrenie Prikladnoi I Promyshlennoi Matematiki, 2004, 11 (4): 901 – 903.

[138] POTAPOV A A. New information radiophysical technologies fractal in radiolocation: fractal and synergetic. Proc. 28th ESA Antenna Workshop on Space Antenna Systems and Technologies, the Netherlands, May 31 – June 3, 2005[C]. Noordwijk: ESTEC, 2005: 1047 – 1050.

[139] POTAPOV A A. Fractal methods of information transfer. Irreversible Processes in Nature and Technology. Ⅲ All-Rus. Conf., Jan. 24 – 26, 2005[C]. Moscow: N. Bauman MGTU, 2005: 252 – 253.

[140] POTAPOV A, POTAPOV V. Fractal radioelement's, devices and systems for radar and ruture telecommunications: antennas, capacitor, memristor, smart 2d frequencyselective surfaces, labyrinths and other fractal metamaterials[J]. Materials, Methods & Technologies, 2017, 11: 492 – 512.

[141] POTAPOV A A, POTAPOV A A, POTAPOV V A. Fractal radioelement's, devices and fractal systems for radar and telecommunications. 14th Sino-Russia Symp. Advanced Materials and Technologies. Sanya, Hainan Province, China: Nov. 28 – Dec. 1, 2017[C]. Beijing: Metallurgical Industry Press, 2017: 499 – 506.

[142] POTAPOV A A, LL'YIN E M, CHIGIN E P, et al. Development and structure of the first etalon dictionary of fractal properties of target classes[J]. Electromagn Phenom, 2005, 5 (2 (15)): 105 – 142.

[143] POTAPOV A A, BULAVKIN V V, GERMAN V A, et al. Fractal signature methods for profiling of processed surfaces[J]. Tech Phys, 2005, 50 (5): 560 – 575.

[144] BULAVKIN V V, POTAPOV A A, GERMAN V A, et al. Theory of fractals in the problem of formation and evaluation surface quality products[J]. Tyazheloe Mashinostroenie, 2005, 6: 19 – 25.

[145] POTAPOV A A, GERMAN V A, POTAPOV V A. Fractal antennas, fractal detectors of weak signals and fractal frequency-selective structures, as the basis of new electronic systems and devices, Basic and Applied Problems of Modern Physics. Thes. Mosc. Conf., Feb. 25 – 28, 2006[C]. Moscow: Lebedev Physical Institute RAN, 2006: 132 – 133.

[146] GULYAEV Y V, NIKITOV S A, POTAPOV A A, et al. Application of fractal theory, fractional measure and scaling effects in schemes of radio detector[J]. Nelineinyi Mir, 2006, 4 (4 – 5): 165 – 171.

[147] POTAPOV A A, GERMAN V A. Fractal nonparametric radio signal detector[J]. Radiotekhnika, 2006, 5: 30 – 36.

[148] POTAPOV A A. The current state of breakthrough fractal technologies in radio physics and radio electronics. Chaos and Structures in Nonlinear Systems. Theory and Experiment. Proc. 5th Int. Sci. Conf., June 15 – 17, 2006[C]. Astana, Kazakhstan: Gumilev ENU, 2006: 37 – 41.

[149] GULYAEV Y V, NIKITOV S A, POTAPOV A A, et al. Concepts of scaling and fractal dimension in the design of a fractal detector of radio signals[J]. J Commun Technol Electron, 2006, 51 (8): 909 – 916.

[150] POTAPOV A A. State of the art and tendencies in developing breakthrough fractal technologies in radiophysics and radio electronics[J]. Fiz Voln Protsessov Radiotekh Sist, 2006, 9 (3): 44 – 54. (in Russian)

[151] PAKHOMOV A A, POTAPOV A A. Digital processing of low contrast images distorted by a turbulent layer: theory and experiment[J]. Fiz Voln Protsessov Radiotekh Sist, 2008, 11 (3): 15 – 34. (in Russian)

[152] POTAPOV A A. Fractal antennas, fractal detectors and fractal frequency-selective structures: Development of a unified approach to a new class of radio systems, Global Information Systems. Problems and Tendencies of Development. Proc. 1 Int. Sci. Conf., Oct. 3 – 6, 2006[C]. Kharkov: KhNURE, 2006: 406 – 407.

[153] POTAPOV A A. Fractals, scaling, textures, fractional operators and deterministic chaos as the physical and mathematical components of the new conceptions and methods in radar and radio physics[J]. The Journal of Engineering, 2019 (20): 7204 – 7209.

[154] POTAPOV A A, GERMAN V A. Methods of measuring the fractal dimension and fractal signatures of a multidimensional stochastic signal[J]. J Commun Technol Electron, 2004, 49 (12): 1370 – 1391.

[155] POTAPOV A A. Development and structure of an ensemble of fractal features of target classes for linear and nonlinear radar problems[J]. Nelineinyi Mir, 2006, 4 (7 – 9): 361 – 386. (in Russian)

[156] POTAPOV A A, GERMAN V A. The main provisions of the theory and technology of fractal processing of radar information against the background of intense interference. Problems of the Efficiency and Safety of Complex Technical and Information Systems. Proc. XXV Interregional Sci.-Tech. Conf., Dedicated to 65th Birthday of Serpukhov's Military Inst. of Missile Forces, Oct. 19 – 20, 2006[C]. Serpukhov: Serpukhov MIRV, 2006: 195 – 199.

[157] POTAPOV A A. About fractal radiophysics and fractal radio electronics, innovation in radio engineering information-telecommunication technologies. Proc. Jubilee Sci.-Tech. Conf. Dedicated to the 60th Birthday of OAO Radiotekhnicheskii Inst. im. A. L. Mintsa and Faculty of Aircraft Electronics, Moscow Aviation Inst., Oct. 24 – 26, 2006[C]. Moscow: Ekstra Print, 2006: 66 – 84.

[158] POTAPOV A A. Development and structure of an ensemble of fractal features of target classes for linear and nonlinear radar problems[M]//Nonlinear radar. Moscow: Radiotekhnika, 2006: 19 – 44. (in Russian)

[159]POTAPOV A A. Fractals, scaling, and power distributions in new information technologies for information processing and management in emergency and extreme situations, Data Processing and Control in Extraordinary and Extremal Situations. Proc. 5th Int. Conf. , Oct. 24 – 26, 2006[C]. Belarus, Minsk: OIPI NAN, 2006: 7 – 12.

[160]POTAPOV A A. Modern classes of fractal antennas and fractal frequency selective surfaces and volumes, day on diffraction – 2006. Abstracts Int. Conf. , May 30 – June 2, 2006[C]. St. Petersburg: St. Petersburg Univ. , 2006: 84 – 85.

[161]POTAPOV A A. Synergetics and problems of radio electronics: fundamentals, methods, applied problems [J]. Sinergetika, 2006, 8: 163 – 179.

[162]POTAPOV A A, GALKINA T V, KOLESNIKOV A I, et al. Linearly modeled patterns in the correlation recognition of statistical texture fields. Statistical Methods in the Theory of Transmission and Transformation of Information Signals. Thes. X All-Union Sci. -Tech. Conf. , Oct. 4 – 6, 1988[C]. Kiev: KIIGA, 1988: 143 – 144.

[163]POTAPOV A A, KOLESNIKOV A I. Spectral characteristics of images of the Earth's surface[J]. Radiotekh. Elektron, 1993, 38 (10): 1851 – 1862.

[164]LIN N, LEE H P, LIM S P, et al. Wave scattering from fractal surfaces[J]. J Mod Opt, 1995, 42 (1): 225 – 241.

[165]CUMMING IAN G, WONG FRANK H. Digital signal processing of synthetic aperture radar data: algorithms and implementation[M]. Norwood, MA: Artech House, 2005.

[166]HERSHKOWITZ S J, AUGUST W R. Theory and practice of radar target identification[M]. Norwood, MA: Artech House, 2000.

Chapter 2

[1]LYASOFF A. Path integral methods for parabolic partial differential equations with examples from computational finance[J]. Mathematical Journal, 2004, 9(2): 399 – 422.

[2]RAJTER-CIRIC D. A note on fractional derivatives of colombeau generalized stochastic processes[J]. Novi Sad Journal Math, 2010, 40(1): 111 – 121.

[3]MARTIAS C. Stochastic integration on generalized function spaces and its applications[J]. Stochastics and Stochastic Reports, 1996, 57(3 – 4): 289 – 301.

[4]OBERGUGGENBERGER M, RAJTER-CIRIC D. Stochastic differential equations driven by generalized positive noise[J]. Publications de I'Institut Mathematique. Nouvelle Serie, 2005, 77 (91): 7 – 19.

[5]FOUKZON J. The solution classical and quantum feedback optimal control problem without the bellman equation[EB/OL]. (2009 – 04 – 27). http://arxiv.org/abs/0811.2170v4.

[6]FOUKZON J, POTAPOV A A. Homing Missile Guidance Law with Imperfect Measurements and Imperfect Information about the System[EB/OL]. (2012 – 10 – 08). http://arxiv.org/abs/1210.2933.

[7]FOUKZON J. Large deviations principles of non-Freidlin-Wentzell type[EB/OL]. (2008 – 03 – 14). http://arxiv.org/abs/0803.2072.

[8]COLOMBEAU J F. Elementary introduction to new generalized functions[M]. Amsterdam: North-Holland, 1985.

[9]COLOMBEAU J F. New generalized functions and multiplication of the distributions[M]. Amsterdam: North-Holland, 1983.

[10] BIAGIONI H A. A nonlinear theory of generalized functions[M]. Berlin: Springer-Verlag, 1990.

[11] BIAGIONI H A, COLOMBEAU J F. New generalized functionsand C^∞ functions with values in generalized complex numbers[J]. J London Math Soc, 1986, 33: 169 - 179.

[12] EGOROV Y V. A contribution to the theory of generalized functions[J]. Russian Mathematical Surveys, 1990, 45(5): 1.

[13] DELCROIX A, HASLER M F, PILIPOVIC S, et al. Algebras of generalized functions through sequence spacesalgebras. Functoriality and associations[J]. Int J Math Sci, 2002(1): 13 - 31.

[14] DELCROIX A, HASLER M F, PILIPOVIC S, et al. Generalized function algebras as sequence space algebras [J]. Proc. Amer. Math. Soc, 2004 (132): 2031 - 2038.

[15] IKEDA N, WATANABE S. Stochastic differential equations and diffusion processes[M]. Amsterdam: North Holland Publ. Co., 1981: 480.

[16] GIKHMAN I I, SKOROKHOD A V. Stochastic differential equations[M]. Berlin: Springer, 1972.

[17] KHASMINSKII R Z. Stochastic stability of differential equations[M]. Berlin: Springer-Verlag, 2012.

[18] DOOB J L. Martingales and one-dimensional diffusion. [J]. Trans. Amer. Math. Soc, 1955, 78: 168 - 208.

[19] DYNKIN E B. Markov processes[M]. Moscow: Fizmatgiz, 1963.

[20] GUTMAN S. On optimal guidance for homing missiles[J]. Journal of Guidance and Control, 1979, 2: 296 - 300.

[21] GLIZER V, TURETSKY V. Complete solution of a differential game with linear dynamics and bounded controls[J]. Applied Mathematics Research, 2008, 1.

[22] IDAN M, SHIMA T. Integrated sliding model autopilot-guidance for dual-control missiles[J]. Journal of Guidance, Control and Dynamics, 2007, 30(4).

[23] BLACK F, SCHOLES M. The pricing of options and corporate liabilities[J]. J Political Economy, 1973, 81: 637 - 659.

[24] HANSON F B, WESTMAN J J. optimal consumption and portfolio policies for important jump events: modeling and computational considerations. Proceedings of 2001 American Control Conference, June 25, 2001[C]. 2001: 4456 - 4661.

[25] HANSON F B, WESTMAN J J. Stochastic analysis of jump-diffusions for financial log-return processes[C]. Proceedings of Stochastic Theory and Control Workshop, New York: Springer-Verlag, 2002: 1 - 15.

[26] CONT R, TANKOV P. Financial modelling with jump processes[M]. New York: Chapman & Hall/CRC Press, 2003.

[27] LEHMANN J, REIMANN P, HANGGI P. Surmounting oscillating barriers: path-integral approach for weak noise[J]. Phys Rev, 2000, 84: 1639 - 1642.

[28] ARNOLD P. Symmetric path integrals for stochastic equations with multiplicative noise[J]. Phys Rev, 2000, 61: 6099 - 6102.

[29] NELSON E. Feynman integrals and the schrödinger equation[J]. J Math Phys, 1964, 5: 332 - 343.

[30] SORAGGI R L. Fourier analysis on Colombeau's algebra of generalized functions[J]. Mathématique, 1996, 69(1): 201 - 227.

[31] SUZUKI M. Relationship between d-dimensional quantal spin systems and $(d+1)$-dimensional Ising systems: equivalence, critical exponents and systematic approximants of the partition function and spin correlations[J]. Prog Theor Phys, 1976, 56: 1454 - 1469.

[32] SUZUKI M. Generalized trotter's formula and systematic approximants of exponential operators and inner derivations with applications to many-body problems[J]. Commun math Phys, 1976, 51: 183 - 190.

[33] WIO H S. On the solution of Kramers' equation by Trotter's formula[J]. J Chem Phys, 1988, 88: 52 – 51.

[34] GARETTO C. Pseudo-differential operators in algebras of generalized functions and global hypoellipticity [EB/OL]. (2005 – 02 – 14). http://arxiv.org/abs/math/0502283.

[35] GARETTO C. Pseudo-differential operators with generalized symbol sand regularity theory 2004[EB/OL]. http://www.mat.univie.ac.at/~diana/papers/thesis.pdf.

[36] Taylor M E. Pseudo-differential Operators[M]. New Jersey: Princeton Univ. Press, 1981.

[37] MASLOV V P. Operational methods[M]. Moscow: Mir, 1976.

[38] MASLOV V P. Complex markov chain and feynman continual integrals[M]. Moscow: Sci., 1976.

[39] NAZAIKINSKII V E, SHATALOV V E, STERNIN B Y. Methods of non-commutative analysis: theory and applications[M]. Berlin: Walter de Gruyter, 1995.

[40] DUBINSKII Y A. The algebra of pseudo-differential operators with analytic symbols and its applications to mathematicaphysics[J]. Russian Mathematical Surveys, 1982, 37(5): 109.

[41] CAVALHEIRO A C. Weighted sobolev spaces and degenerate elliptic equations[J]. Bol Soc Paran Mat, 2008, 26(1 – 2): 117 – 132.

[42] SHUN Z, MCCULLAGH P. Laplace approximation of high dimensional integrals[J]. Journal of the Royal Statistical Society: Ser. B (Methodological), 1995, 57: 749 – 760.

[43] FEDORYUK M V. Asymptotic methods in analysis[J]. Encyclopaedia of Mathematical Sciences, 1989, 13: 83 – 191.

[44] FEYNMAN R P, HIBBS A R. Quantum mechanics and path Integrals[M]. New York: McGraw-Hill, 1965.

[45] KLEINERT H. Path integrals in quantum mechanics, statistics, polymer physics, and financial markets[M]. Singapore: World Scientific, 2009.

[46] WEI C, JIAN Z W. Adaptive wavelet collocation methods for initial value boundary problems of nonlinear PDE's[R/OL]. http://www.science.gov/.

[47] PODOSENOV S A, POTAPOV A A, FOUKZON J, et al. Nonholonomic, fractal and bound structures in relativistic continuum, electrodynamics, quantum mechanics and cosmology. book 1. theory of pulse radiation and field interaction with holonomic and fractal objects[M]. Moscow: LENAND, 2016: 432. (in Russian)

[48] PODOSENOV S A, POTAPOV A A, FOUKZON J, et al. Nonholonomic, fractal and bound structures in relativistic continuum, electrodynamics, quantum mechanics and cosmology. book 2. Force fields in bound and nonholonomic structures [M]. Moscow: LENAND, 2016: 440. (in Russian)

[49] PODOSENOV S A, POTAPOV A A, FOUKZON J, et al. Nonholonomic, fractal and bound structures in relativistic continuum, electrodynamics, quantum mechanics and cosmology. book 3. asymptotic methods in problems of classical and fractal pulse electrodynamics and quantum cosmology in space-time of negative fractal dimension[M]. Moscow: LENAND, 2016: 256. (in Russian)

Chapter 3

[1] VIJAY R, SLICHTER D H, SIDDIQI I. Observation of quantum jumps in a superconducting artificial atom [J]. Phys Rev Lett, 2011, 106: 110502.

[2] PEIL S, GABRIELSE G. Observing the quantum limit of an electron cyclotron: QND measurements of quantum jumps between Fock states[J]. Phys Rev Lett, 1999, 83: 1287 – 1290 .

[3] NAGOURNEY W, SANDBERG J, DEHMELT H. Shelved optical electron amplifier: observation of quantum

jumps[J]. Phys Rev Lett, 1986, 56: 2797 – 2799.

[4] SAUTER T, NEUHAUSER W, BLATT R, et al. Observation of quantum jumps[J]. Phys Rev Lett, 1986, 57: 1696 – 1698.

[5] BERGQUIST J C, HULET R G, ITANO W M, et al. Observation of quantum jumps in a single atom[J]. Phys Rev Lett, 1986, 57: 1699 – 1702.

[6] BOHR N. XXXVII. On the constitution of atoms and molecules[J]. The London, Edinburgh, and Dublin Philosophical Magazine and Journal of Science, 1913, 26 (153): 476 – 502.

[7] DUM R, ZOLLER P, RITSCH H. Monte Carlo simulation of the atomic master equation for spontaneous emission[J]. Phys Rev A, 1992, 45 (7): 4879 – 4887.

[8] DALIBARD J, CASTIN Y, MØLMER K. Wave-function approach to dissipative processes in quantum optics [J]. Phys Rev Lett, 1992, 68 (5): 580 – 583.

[9] GATAREK D, GISIN N. Continuous quantum jumps and infinite-dimensional stochastic equations[J]. J Math Phys, 1991, 32 (8): 2152.

[10] MOLMER K, CASTIN Y, DALIBARD J. Monte Carlo wave-function method in quantum optics[J]. J Opt Soc Am, 1993, 10 (3): 524 – 538.

[11] GARDINER C W, PARKINS A S, ZOLLER P. Wave-function quantum stochastic differential equations and quantum-jump simulation methods[J]. Phys Rev, 1992, 46, 4363 – 4381.

[12] VOGT N, JESKE J, COLE J. Stochastic Bloch-Redfield theory: quantum jumps in a solid-state environment [J]. Phys Rev B, 2013, 88: 1 – 11.

[13] REINER J E, WISEMAN H M, MABUCHI H. Quantum jumps between dressed states: a proposed cavity-QED test using feedback[J]. Phys Rev A, 2003, 67: 042106.

[14] BERKELAND D J, RAYMONDSON D A, TASSIN V M. Tests for non-randomness in quantum jumps[J]. Phys Rev A, 2004, 69: 052103.

[15] WISEMAN H M, MILBURN G J. Interpretation of quantum jump and diffusion processes illustrated on the Bloch sphere[J]. Phys Rev, 1993, 47 (3): 1652 – 1666.

[16] Wiseman H M. Toombes G E. Quantum jumps in a two-level atom: simple theories versus quantum trajectories[J]. Phys. Rev, 1999, 60 (3): 2474 – 2490.

[17] STOJANOVIC M. Regularization for heat kernel in nonlinear parabolic equations[J]. Taiwanese Journal of Mathematics, 2008, 12(1): 63 – 87.

[18] GARETTO C. Fundamental solutions in the colombeau framework: applications to solvability and regularity theory[J]. Acta Appl Math, 2008, 102: 281 – 318.

[19] FOUKZON J, POTAPOV A A, PODOSENOV S A. Exact quasiclassical asymptotics beyond Maslov canonical operator[J]. International Journal of Recent advances in Physics, 2014, 3(4). http://arxiv.org/abs/1110.0098

[20] PODOSENOV S A, POTAPOV A A, FOUKZON J, et al. Nonholonomic, fractal and related structures in the relativistic continuous media, electrodynamics, quantum mechanics and cosmology: Asymptotic methods in problems of classical and fractal pulse electrodynamics[M]. In press. ISBN 978 – 5 – 9710 – 1525 – 3.

[21] NELSON E. Feynman integrals and the Schrödinger equation[J]. J Math Phys, 1964, 5: 332 – 343.

[22] MASLOV V P. Complex markov chains and continual Feynman integral[M]. Moscow: Nauka, 1976.

[23] APOSTOL T M. Mathematical analysis[M]. 2nd edition. Addison Wesley, 1974.

[24] CHEN W W L. Fundamentals of Analysis, Published by W. W. L. Chen via Internet, 2003.

[25] GUPTA S L, RANI N. Principles of real analysis[M]. New Delhi: Vikas Puplishing House, 1998.

[26] LEHMANN J, REIMANN P, HANGGI P. Surmounting oscillating barriers: path-integral approach for weak noise[J]. Phys Rev E, 2000, 62: 6282 – 6303.

[27] FEDORYUK M V. The method of steepest descent[M]. Moscow: Nauka, 1977 (in Russian)

[28] FEDORYUK M V. Asymptotic methods in analysis[M]//GAMKRELIDZE R V. Encyclopedia of mathematical sciences: Vol 13. Heidelberg: Springer, 1989: 83 – 191.

[29] SHUN Z, CULLAGH P M. Laplace approximation of high dimensional integrals[J]. Journal of the Royal Statistical Society. Series B (Methodological), 1995, 57(4): 749 – 760.

[30] FOUKZON J. Strong large deviations principles of non-Freidlin-Wentzell type-optimal control problem with imperfect information-jumps phenomena in financial markets[J]. Communications in Applied Sciences, 2014, 2(2): 230 – 363.

Chapter 4

[1] RITOV S M. Life, memories, interview, transactions, poetry, records[M]. Moscow: LENAND, 2012: 552. (in Russian)

[2] OLDHAM K B, SPANIER J. The fractional calculus[M]. New York: Academic Press, 1974: 234.

[3] POTAPOV A A. Fractals in radio physics and radar[M]. Moscow: Logos, 2002: 664. (in Russian)

[4] POTAPOV A A. Fractals in radio physics and radar: topology of a sample[M]. Moscow: Universitetskaya Kniga, 2005: 848. (in Russian)

[5] POTAPOV A A, CHERNYKH V A, LETNIKOV A. Fractional calculus in the physics of fractals[M]. Saarbrücken: LAMBERT Academic Publ., 2012: 688. (in Russian)

[6] ROGERS C A. Hausdorff measures[M]. London: Cambridge University Press, 1970: 179.

[7] BUNKIN B V, REUTOV A P, POTAPOV A A, et al. Aspects of Perspective Radiolocation[M]. Moscow: Radiotekhnika, 2003: 512. (in Russian)

[8] SKOLNIK M. Radar handbook[M]. Moscow: Tekhnosfera, 2014: Book 1: 672; Book 2: 680. (in Russian)

[9] POTAPOV A A. Synthesis of images of earth covers in the optical and millimeter wave ranges[M]//Doctoral dissertation in mathematics and physics. Moscow: IRE RAN, 1994. (in Russian)

[10] POTAPOV A A. Synthesis of the earth cover images in the optical and millimeter wave bands: doctoral dissertation in mathematics and physics[M]. Moscow: IREE of RAS, 1994: 436. (in Russian)

[11] HARALICK R M, SHANMUGAN K, DINSTEIN I. Textural features for image classification [J]. IEEE Trans, 1973, 6: 610 – 621.

[12] POTAPOV A A. The use of modulated millimeter waves for image generation and identification[M]. Moscow: Radiotekhnika, 1989: 61 – 64.

[13] POTAPOV A A. The textures, fractal, scaling effects and fractional operators as a basis of new methods of information processing and fractal radio systems designing [C]. Proceedings of SPIE-The International Society for Optical Engineering, 2009.

[14] POTAPOV A A. New information technology in radar detection of lowcontrast targets based on probabilistic texture and fractal features[J]. J Communications Technology and Electronics, 2003, 48: 1012 – 1029.

[15] POTAPOV A A, GALKINA T V, ORLOVA T I, et al. Dispersion detection method of determined object at texture optical and radar images of earth surface[J]. Radiotekhnika i electronica, 1990, 35(11): 2295 – 2301.

[16]POTAPOV A A, GALKINA T V, ORLOVA T I, et al. Edge enhancement method of extended deterministic object in stochastic fields[J]. Radiotekhnika i electronica, 1991, 36(11): 2240 – 2242.

[17]POTAPOV A A. Fractals and chaos as a base of new breakthrough technologies in modern radio systems [M]//CROWNOVER R M. Introduction to fractals and chaos. Moscow: Tekhnosfera, 2006: 374 – 479. (in Russian)

[18]POTAPOV A A. Radiophysical effects when coupling electromagnetic millimeter-wave radiation with environment[J]. Zarubezh. Radioelektron, 1992, 8: 36 – 76; 9: 4 – 28; 11: 23 – 48; 1993, 3: 36 – 48; 7 – 9: 32 – 49; 1994, 7/8: 11 – 30; 1995, 1: 27 – 36 (journal monograph version). (in Russian)

[19]POTAPOV A A. Investigation of plant cover influence on back scattered field of millimeter waves[J]. J Communications Technology and Electronics, 1991, 36(2): 239 – 246.

[20]PAVEL'EV V A, POTAPOV A A. Influence of the ground surface on the structure of a pulse signal in millimeter wavelength band[J]. Radiotekh. Elektron, 1994, 39(4): 573 – 582.

[21]POTAPOV A A. The theory of functionals of stochastic backscattered fields[J]. J Communications Technology and Electronics, 2007, 52: 245 – 292.

[22]POTAPOV A A, GERMAN V A. Detection of artificial objects with fractal signatures [J]. Pattern Recognition and Image Analysis, 1998, 8(2): 226 – 229.

[23]POTAPOV A A. Topology of sample[J]. Nelineiny Mir, 2004, 2(1): 4 – 13.

[24]POTAPOV A A, GERMAN V A. Fractals, fractal target selection and fractal antennas. Proc. 1st Int. Workshop on Mathematical Modeling of Physical Processes in Inhomogeneous Media, March 20 – 22, 2001 [C]. Guanajuato, 2001: 44 – 46.

[25]POTAPOV A A, GERMAN V A. Processihg of optic and radar images of the earth surface by fractal methods [J]. J Communications Technology and Electronics, 2000, 45(8): 853 – 860.

[26]POTAPOV A A. Fractals in remote sensing[J]. Zarubezh Radioelektron Usp Sovrem. Radioelectron, 2000 (6): 3 – 65.

[27]POTAPOV A A. Fractals in radiophysics and radar. elements of the theory of fractals: a review[J]. J Communications Technology and Electronics, 2000, 45(11): 1157 – 1164.

[28]POTAPOV A A. Fractals in radio physics and radiolocation: fractal analysis of signals [J]. J Communications Technology and Electronics, 2000, 46(3): 237 – 246.

[29]POTAPOV A A. Fractals in radiophysics and radar: fundamental theory of wave scattering by a fractal surface [J]. J Communications Technology and Electronics, 2002, 47(5): 461 – 487.

[30]POTAPOV A A. The fractal analysis in modern problems of radars and radio physics[J]. Radiotekhnika, 2003(8): 55 – 66.

[31]POTAPOV A A, GERMAN V A. Methods of measuring the fractal dimension and fractal signatures of a multidimensional stochastic signal[J]. J Communications Technology and Electronics, 2004, 49: 1370 – 1391.

[32]GULYAEV Y V, NIKITOV S A, POTAPOV A A, et al. Concepts of scaling and fractal dimension in the design of a fractal detector of radio signals[J]. J Communications Technology and Electronics, 2006, 51: 909 – 916.

[33]POTAPOV A A, GERMAN V A. Fractal processing of faint signals and low-contrast images[J]. Optoelectronics Instrumentation and Data Processing, 2006(5): 4 – 20.

[34]POTAPOV A A, GULYAEV Y V, NIKITOV S A, et al. Newest images processing methods[M]. Moscow: FIZMATLIT, 2008: 496. (in Russian)

[35]POTAPOV A A. Application of the fractal theory and scaling effects during processing of low-contrast images

and super weak signals in the presence of intensive noise. Abstracts Int. conf. " Zababakhin Scientific Talks", devoted to E. I. Zababakhin's 95-th anniversary, April 16 – 20, 2012 [C]. Snezhinsk: RFNC-VNIITF, 2012: 312.

[36] POTAPOV A A. Fractal method and fractal paradigm in modern natural science[M]. Nauchnaya Kniga: Voronezh, 2012: 108.

[37] POTAPOV A A. Fractal paradigm and fractal-scaling methods in fundamentally new dynamic fractal signal detectors. Proc. the Eighth Int. Kharkov Symposium on Physics and Engineering of Millimeter and SubMillimeter Waves-MSMW' 13, June 24 – 28, 2013[C]. Kharkov: IRE NASU, 2013: 644 – 647.

[38] POTAPOV A A. Fractals, scaling and fractional operators in modern physics and radio engineering. Book of Abstracts of the Int. Conf. XIV Khariton's Topical Scientific Reading "High-Power Pulsed Electrophysics", April 21 – 25, 2014[C]. Sarov: RFNC-VNIIEF, 2014: 81 – 82.

[39] POTAPOV A A. On the theory and justification of physical applications fractal labyrinths. Abstracts Int. conf. "Zababakhin Scientific Talks", June 2 – 6, 2014[C]. Snezhinsk: RFNC-VNIITF, 2014: 317 – 318.

[40] LEONOV K N, POTAPOV A A, USHAKOV P A. Application of invariant properties of chaotic signals in the synthesis of noiseimmune broadband systems for data transmission[J]. J Communications Technology and Electronics, 2014, 59(12): 1393 – 1411.

[41] POTAPOV A A, LAKTYUN'KIN A V. Frequency coherence function of a space-time radar channel forming images of an anisotropic fractal surface and fractal objects[J]. J Communications Technology and Electronics, 2015, 60: 962 – 969.

[42] PODOSENOV S A, POTAPOV A A, FOUKZON J, et al. Nonholonomic, fractal and bound structures in relativistic continuous medium, electrodynamics, quantum mechanics and cosmology: three volumes[M]. Moscow: LENAND, 2015: 1128. (in Russian) (Vol. 1. Pulse radiation theory and coupling of fields and holonomic and fractal objects. Vol. 2. Force fields in bound and nonholonomic structures. Vol. 3. Asymptotic methods in problems of classical and fractal pulse electrodynamics and quantum cosmology in space-time of negative fractal dimension.)

[43] POTAPOV A A. New trend in radio physics: fractal scaling or scale-invariant radiolocation[J]. Fiz Voln Prosessov Radiotekh Sist, 2015, 18(3): 44 – 54.

[44] POTAPOV A A. At the origins of the fractal-scaling or scale-invariant radiolocation (1980 – 2015)[J]. Radiotekhnika, 2015(8): 95 – 108.

[45] POTAPOV A A, RASSADIN A E. Feynman integrals, fractal paradigm and new point of view on hydroacoustics[J]. Eurasian Physical Technical Journal, 2015, 12(23): 3 – 13.

[46] POTAPOV A A. Fractal scaling or scale-invariant radiolocation principles and their application in radars with synthetic aperture. Collected papers III All-Russian Sci. Tech. Conf. " Development prospects of long-range detection radar, integrated systems, VKO infoware complexes and control and information processing complexes: "RTI VKO systems-2015", May 28, 2015[C]. Moscow: MGTU im. Baumana, 2015: 573 – 590.

[47] POTAPOV A A. Oscillator with fractional differential positive feedback as model of fractal dynamics[J]. J Computational Intelligence and Electronic Systems, 2014, 3(3): 236 – 237.

[48] POTAPOV A A. The global fractal method, fractal paradigm and the fractional derivatives method in fundamental radar problems and designing of revolutionary radio signals detectors. Zbornik radova Konferencije MIT-Matematicke i informacione tehnologije (Vrnjackoj Banji od 5. do 9. septembra i u Becicima od 10. do 14. septembra 2013. godine) [C]. Kosovska Mitrovica: Prirodno-matematicki fakultet Ulverziteta u Pristini (Serbia), 2014: 539 – 552.

[49]POTAPOV A A. New conception of fractal radio device with fractal antennas and fractal detectors in the MIMO-systems. Book of Abstracts Third Int. Scientific Symp. "The Modeling of Nonlinear Processes and Systems (MNPS-2015)",June 22 – 26,2015[C]. Moscow:Ianus-K,2015:33.

[50]POTAPOV A A. Fractal radar:towards 1980 – 2015. Proc. of CHAOS 2015 Int. Conf. ,May 26 – 29,2015 [C]. Paris:Henri Poincaré Institute,2015 :559 – 573.

[51]POTAPOV A A. Oscillations,waves,structures and systems illustrated by examples of the global fractal-scaling method (measure zero sets,singularity,scaling,sample topology,sprites,jets,elves,memristors,oscillators,fractal mazes,robust antenna arrays and fractal detectors)[J]. The Non-Linear World,2014,12(4):3 – 38.

[52]POTAPOV A A,POTAPOV ALEXEY A. Development and base ideas of fractals radio systems and fractal radio elements conception. Book of Abstracts 8nd Int. Conf. (CHAOS' 2015)on Chaotic Modeling,Simulation and Applications,May 26 – 29,2015[C]. Paris:Henri Poincaré Institute,2015:100 – 101.

[53]POTAPOV A A. Features of multi-fractal structure of the high-attitude lightning discharges in the ionosphere:elves,jets,sprites. Book of Abstracts 8nd Int. Conf. (CHAOS' 2015)on Chaotic Modeling,Simulation and Applications,May 26 – 29,2015[C]. Paris:Henri Poincaré Institute,2015:101 – 102.

[54]POTAPOV A A. Fractals and scaling in the radar:a look from 2015. Book of Abstracts 8nd Int. Conf. (CHAOS' 2015)on Chaotic Modeling,Simulation and Applications,May 26 – 29,2015[C]. Paris:Henri Poincaré Institute,2015:102.

[55]POTAPOV A A,RASSADIN A E. Fractals and path integrals in three-dimensional wave equation. Book abstracts of Int. Conf. -School "Infinite-dimensional dynamics,dissipative systems,and attractors",July 13 – 17,2015[C]. Nizhniy Novgorod:Lobachevsky State University of Nizhniy Novgorod,2015:26 – 27.

[56]POTAPOV A A,RASSADIN A E. Feynmanons in Korteveg-de Vriez equation. Book abstracts of Int. Conf. -School "Infinite-dimensional dynamics,dissipative systems,and attractors",July 13 – 17,2015[C]. Nizhniy Novgorod:Lobachevsky State University of Nizhniy Novgorod,2015:28 – 29.

[57]POTAPOV A A,RASSADIN A E. Feynmanons in gravitational waves in fluid // Book of abstracts of Int. Conf. -School "Dynamics,Bifurcations and Chaos 2015 (DBC Ⅱ)" Dedicated to 80th birthday of L. P. Shilnikov,July 20 – 24,2015[C]. Nizhniy Novgorod:Lobachevsky State University of Nizhniy Novgorod,2015:25 – 26.

[58]POTAPOV A A,RASSADIN A E. Plane waves,lie groups and Feynman integrals. Book of abstracts of Int. Conf. -School "Dynamics,Bifurcations and Chaos 2015 (DBC Ⅱ)" Dedicated to 80th birthday of L. P. Shilnikov,July 20 – 24,2015[C]. Nizhniy Novgorod:Lobachevsky State University of Nizhniy Novgorod,2015:26 – 27.

[59]POTAPOV A A. On the fractal-scaling or scale-invariant radio location and its methods. Book of Abstracts Int. Conf. -School "Shilnikov WorkShop 2015",December 17 – 19,2015[C]. Nizhniy Novgorod:Lobachevsky State University of Nizhniy Novgorod,2015:21.

[60]POTAPOV A A,RASSADIN A E. Feynman integrals and radiation of waves in two-and three-dimensional spaces. Book of abstracts of Int. Conf. -School "Dynamics,Bifurcations and Chaos 2016 (DBC Ⅲ)",July 18 – 22,2016[C]. Nizhniy Novgorod:Lobachevsky State University of Nizhniy Novgorod,2016:54.

[61]POTAPOV A A. New conception of fractal radio device with fractal antennas and fractal detectors in the MIMO Systems. Book of Abstracts 9th Int. Conf. (CHAOS' 2016)on Chaotic Modeling,Simulation and Applications,May 23 – 26,2016[C]. London:University of London,2016:85.

[62]POTAPOV A A,LAKTYUN'KIN A V. Frequency coherence function of a radar channel forming images of a

fractal surface and fractal objects. Book of Abstracts 9th Int. Conf. (CHAOS' 2016) on Chaotic Modeling, Simulation and Applications, May 23 - 26, 2016[C]. London: University of London, 2016: 85 - 86.

[63] POTAPOV A A, POTAPOV V A. Scaling of the fractals antennas. Book of Abstracts 9th Int. Conf. (CHAOS' 2016) on Chaotic Modeling, Simulation and Applications, May 23 - 26, 2016[C]. London: University of London, 2016: 86 - 87.

[64] Andrianov M N, POTAPOV A A. Increase efficiency of system of data transmission in fading channels using MIMO Systems. Proc. 2016 CIE Int. Conf. on Radar "Radar 2016", Oct. 10 - 12, 2016[C]. Beijing: Chinese Institute of Electronics (CIE), 2016: 119 - 123.

[65] POTAPOV A A, ZHANG W. Simulation of new ultra-wide band fractal antennas based on fractal labyrinths. Proc. 2016 CIE Int. Conf. on Radar "Radar 2016", Oct. 10 - 12, 2016[C]. Beijing: Chinese Institute of Electronics (CIE), 2016: 319 - 323.

[66] POTAPOV A A, GERMAN V A, PAHOMOV A A. Processing of images obtained from unmanned aerial vehicles in the regime of flight over inhomogeneous terrain with fractal-scaling and integral method. Proc. 2016 CIE Int. Conf. on Radar "Radar 2016", Oct. 10 - 12, 2016[C]. Beijing: Chinese Institute of Electronics (CIE), 2016: 585 - 587.

[67] POTAPOV A A, LAZKO FF. Gradients distribution matrices of and lacunarity in the capacity of texture measure of SAR and UAVs images. Proc. 2016 CIE Int. Conf. on Radar "Radar 2016", Oct. 10 - 12, 2016 [C]. Beijing: Chinese Institute of Electronics (CIE), 2016: 605 - 608.

[68] POTAPOV A A. Strategic directions in synthesis of new topological radar detectors of low-contrast targets against the background of high-intensity noise from the ground, sea and precipitations. Proc. 2016 CIE Int. Conf. on Radar "Radar 2016", Oct. 10 - 12, 2016[C]. Beijing: Chinese Institute of Electronics (CIE), 2016: 692 - 696.

[69] POTAPOV A A, GERMAN V A. A Local dispersion method of measuring a fractal dimension and fractal signatures. Proc. 2016 CIE Int. Conf. on Radar "Radar 2016", Oct. 10 - 12, 2016[C]. Beijing: Chinese Institute of Electronics (CIE), 2016: 799 - 803.

[70] KOVALEV A N, KOVALEV F N, KONDRAT'EV V V, et al. Measurements ambiguity resolution in phase direction finders of forward-scattering radar systems. Proc. 2016 CIE Int. Conf. on Radar "Radar 2016", Oct. 10 - 12, 2016[C]. Beijing: Chinese Institute of Electronics (CIE), 2016: 1036 - 1040.

[71] POTAPOV A A, POTAPOV A A (Jr.), POTAPOV V A. Fractal capacitor, fractional operators and fractal impedances[J]. Nelineinyi mir, 2006, 4(4 - 5): 172 - 187.

[72] POTAPOV A A. The fractal-scaling radiolocation: formation history 1980 - 2015 [J]. Chaotic Modeling and Simulation (CMSIM), 2016(3): 317 - 331.

[73] POTAPOV A A. On fractal dimension spectrum of new lightning discharge types in ionosphere: elves, jets and sprites[J]. Eurasian Physical Technical Journal, 2016, 13 (2): 26.

[74] POTAPOV A A. Fractality and scaling problems in radio location and radio physics with new methods of detection of low-contrast targets against a background of high intensity noise. Proc. of the XV Int. Academic Congress Fundamental and Applied Studies in the Modern World, September 6 - 8, 2016[C]. Oxford: Oxford University Press, 2016: 314 - 322.

[75] POTAPOV A A. Chaos theory, fractals and scaling in the radar: a look from 2015. [M]//SKIADAS C. The foundations of chaos revisited: from poincaré to recent advancements. Basel: Springer Int. Publ., 2016: 195 - 218.

[76] HAYKIN S. Cognitive radar: a way of the future[J]. IEEE Signal Processing Magazine, 2006, 23: 30 - 40.

Chapter 5

[1]FOLDY L L. The multiple scattering of waves. [J]. Phys Rev,1945,67:107 – 119.

[2]TVERSKY V. On propagation in random media of discrete scatterers[J]. Proc Sympos Appl Math,1964,16: 84 – 116.

[3]TVERSKY V. Theory and microwave measurements of higher statistical moments of randomly scattered fields. In: Electromagnetic Scattering. Proc. of the Interdisciplinary Conference held in June,1965, at the University of Massachusetts at Amherst, Amherst, MA USA[C]. New York: Gordon and Breach,1967:579 – 696.

[4]RYTOV S M, KRAVTSOV YU A, TATARSKII V I. Introduction to statistical radiophysics, part 2: random fields[M]. Moscow: Nauka,1978. (in Russian)

[5]ISIMARU A. Wave propagation and scattering in random media[M]. New York: Academic,1978.

[6]MISHCHENKO M I, TRAVIS L D, LACIS A A. Multiple scattering of light by particles: radiative transfer and coherent backscattering[M]. New York: Cambridge University Press,2006:507.

[7]MANDELBROT B. The fractal geometry of nature[M]. New York: Freeman and Company,1982:470.

[8]BERRY M V. Diffractals[J]. J Phys A: Math Gen,1979,12(6):781 – 797.

[9]JAGGARD D L. On fractal electrodynamics[M]//KRITIKOS H N, JAGGARD D L. Recent advances in electromagnetic theory. New York: Springer-Verlag,1990:183 – 224.

[10]JAGGARD D L. Fractal electrodynamics and modeling[M]//BERTONI H L, FELSEN L B. Directions in electromagnetic wave modeling. New York: Plenum,1991:435 – 446.

[11]JAGGARD D L. Fractal electrodynamics: wave interaction with discretely self-similar structures[M]// BAUM C, KRITIKOS H N. Electromagnetic symmetry. London: Taylor & Francis,1995:231 – 281.

[12]JAGGARD D L. Fractal electrodynamics: from super antennas to superlattices[M]//VEHE J L, LUTTON E, TRICOT C. Fractals in engineering. Berlin: Springer Verlag,1997:204 – 221.

[13]JAGGARD D L. Fractal electrodynamics: surfaces and superlattices [M]// WERNER D H, MITTRA R. Frontiers in electromagnetics. New York: IEEE Press,2000:2 – 47.

[14]POTAPOV A A. Fractals in radio physics and radiolocation[M]. Moscow: Logos,2002:664. (in Russian)

[15]POTAPOV A A. Fractals in radio physics and radiolocation: sampling topology[M]. Moscow: University Book,2005:848. (in Russian)

[16]POTAPOV A A. Fractals and chaos as basis of new challenging technologies in modern radio systems[M]// CROWNOVER R M. Introduction to fractals and chaos. Moscow: Tekhnosfera,2006:374 – 479.

[17]POTAPOV A A. Chaos theory, fractals and scaling in the radar: a look from 2015[M]//SKIADAS C. The foundations of chaos revisited: from poincaré to recent advancements. Switzerland, Basel: Springer Int. Publ., 2016:195 – 218.

[18]POTAPOV A A. On the indicatrixes of waves scattering from the random fractal anisotropic surface[M]// BRAMBILA F. Fractal analysis-applications in physics, engineering and technology. Rijeka: InTech,2017: 187 – 248.

[19]POTAPOV A A. Postulate "the topology maximum at the energy minimum" for textural and fractal-and-scaling processing of multidimensional super weak wignals against a background of noises. [M]//UVAROVA L A, NADYKTO A B, LATYSHEV A V. Nonlinearity: problems, solutions and applications: Vol. 2. New York: Nova Science Publ,2017:35 – 94.

[20]POTAPOV A A. Biography and publication index[M]. Moscow: Center of printing services "Rainbow",

2019:256. (Approved by the Academic Council of IREE V. A. Kotelnikov of the Russian Academy of Sciences 26. 12,2018).

[21] MISHCHENKO M I, HOVENIER J W, TRAVIS L D. Light scattering by nonspherical particles: theory, measurements, and applications[M]. London: Academic Press, 2000:690.

[22] KULIKOV D A, POTAPOV A A, RASSADIN A E, et al. Model for growth of fractal solid state surface and possibility of its verification by means of atomic force microscopy [C]. IOP Conf. Ser.: Mater. Sci. Eng, 2017, 256. NO. 012026:10.

[23] POTAPOV A A, RASSADIN A E, STEPANOV A V, et al. Nonlinear dynamics of fractals with cylindrical generatrix on surface of solid state[M]. Beijing: Metallurgical Industry Press, 2017:491 – 493.

[24] DOVIAK R, ZRNICH D. Doppler radars and meteorological observations[M]. Leningrad: Gidrometeoizdat, 1998:512. (in Russian)

[25] RASTOGI P K, SCHEUCHER K F. Range dependence of scattering from a fractal medium: Simulation results [J]. Radio Science, 1990, 25(5):1057 – 1063.

[26] POTAPOV A A. Waves in large disordered anisotropic fractal systems, in clusters of small-size space vehicles, in Synthesized Space Antenna Aggregations-Cluster Apertures, and in Radar. Int. Conf. – School "Shilnikov WorkShop 2017" (Nizhni Novgorod, Russia, December 15 – 16, 2017) [C]. Nizhni Novgorod: Lobachevsky State University of Nizhni Novgorod, 2017:55 – 56.

[27] POTAPOV A A. Waves in large disordered fractal systems: radar, nanosystems, and clusters of unmanned aerial vehicles and small-size spacecrafts[J]. Journal of Communications Technology and Electronics, 2018, 63(9):980 – 997.

[28] POTAPOV A A. Multiple scattering of waves in fractal discrete randomly-inhomogeneous media from the point of view of radiolocation of the self-similar multiple targets[J]. RENSIT, 2018, 10(1):3 – 22.

[29] Potapov Alexander A. Waves in large disordered anisotropic fractal systems, in clusters of drones or small-size space vehicles, in synthesized space antenna aggregations, and in radiolocation. the 12th Chaotic Modeling and Simulation Int. Conf. with Quantum Complexity and Nanotechnology Symp. (Chania, Crete, Greece: 18 – 22 June, 2019) [C]. Rome: International Society for the Advancement of Science and Technology, 2019:79 – 80.

[30] WANG Z S, LU B W. The scattering of electromagnetic waves in fractal media[J]. Waves in Random Media, 1994, 4(1):97 – 103.

[31] POTAPOV A A. Fractal and topological sustainable methods of overcoming expected uncertainty in the radiolocation of low-contrast targets and in the processing of weak multi-dimensional signals on the background of high-intensity noise: a new direction in the statistical decision theory[J]. Journal of Physics: Conf. Ser., 2017, 918(012015):19.

[32] POTAPOV A A. On the fractal theory application in adaptive population methods of formation of dynamical groups of unmanned aerial vehicles and in processing of incoming information in respect to its effective application theory[J]. Eurasian Physical Technical Journal, 2017, 1(27):6 – 17.

[33] CHENG H, PAGE J, OLSEN J. Dynamic mission control for UAV swarm via task stimulus approach[J]. American Journal of Intelligent Systems, 2012, 2(7):177 – 183.

[34] GONZALES D, HARTING S. Designing unmanned systems with greater autonomy: using a federated, partially open systems architecture approach. Santa Monica[M]. Calif: RAND, 2014:96.

[35] POTAPOV A A. Fractals, scaling, textures, fractional operators and deterministic chaos as the physical and mathematical components of the new conceptions and methods in radar and radio physics[J]. The Journal of Engineering, 2019:6.

Chapter 6

[1]GNEDENKO B V, KOLMOGOROV A N. Limit distributions for sums of independent random variables[M]. Moscow:Leningrad, 1949:264. (in Russian)

[2]OLDHAM K B, SPANIER J. The fractional calculus[M]. New York: Acad. Press, 1974: 225.

[3]MANDELBROT B. The fractal geometry of nature[M]. New York:Freeman and Company, 1982:470.

[4]SAMKO S G, KILBAS A A, MARICHEV O I. Integrals and derivatives of fractional order and some of their applications. Minsk: Science and Technology, 1987:688. (in Russian)

[5]POTAPOV A A. Fractals in radio physics and radiolocation. Moscow: Logos, 2002: 664.

[6]NAKHUSHEV A M. Fractional calculus and its application[M]. Moscow:Fizmatlit, 2003:272. (in Russian)

[7]POTAPOV A A. Fractals in radio physics and radiolocation: sampling topology[M]. Moscow: University Book, 2005:848.

[8]PSKHU A V. Partial fractional derivative equations[M]. Moscow:Science, 2005: 199. (in Russian)

[9]VOLTERRA V. The theory of functionals, integral and integro-differential equations[M]. Moscow:Nauka, 1982:304.

[10]UCHAYKIN V V. The method of fractional derivatives[M]. Ulyanovsk:Artichoke, 2008:512. (in Russian)

[11]POTAPOV A A, CHERNYKH V A. Fractional calculus of A. V. Letnikov in fractal physics[M]. Saarbrücken: LAMBERT Academic Publ., 2012:688.

[12]ROGERS C A. Hausdorff measures[M]. London: Cambridge Univ. Press, 1970:197.

[13]POTAPOV A A, GULYAEV Y V, NIKITOV S A, et al. The latest techniques of image processing[M]. Moscow: Fizmatlit, 2008:496.

[14]CRISTEA L L, STEINSKY B. Curves of infinite length in 4 × 4-labyrinth fractals[J]. Geom Dedicata, 2009,141:1 – 17.

[15]POTAPOV A A, Slezkin D V, POTAPOV V A. Fractal labyrinths as geometry base of new kinds of fractal antennas and fractal antenna arrays[J]. Radiotekhnika, 2013,8:31 – 36. (in Russian)

[16]POTAPOV A A, SLEZKIN D V. Fractal labyrinths as small antennas development base. Zbornik radova Konferencije MIT-Matematicke i informacione tehnologije (Vrnjackoj Banji od 5. do 9. septembra i u Becicima od 10. do 14. Sept 2013. godine)[C]. Kosovska Mitrovica: Prirodno-matematicki fakultet Ulverziteta u Pristini (Serbia), 2014:553 – 559.

[17]POTAPOV A A. Fractals and chaos as basis of new challenging technologies in modern radio systems[M]// CROWNOVER R M. Introduction to Fractals and Chaos. Moscow: Tekhnosfera, 2006:374 – 479.

[18]POTAPOV A A. On the fractal radio physics and fractal radio electronics. Proc. Jubilee Sci.-Tech. Conf. Dedicated to the 60th Birthday of OAO Radiotekhnicheskii Inst. im. A. L. Mintsa and Faculty of Aircraft Electronics, Moscow Aviation Inst. "Innovation in Radio Engineering Information-Telecommunication Technologies",October 24 – 26, 2006[C]. Moscow, 2006:66 – 84.

[19]POTAPOV A A. New information radiophysical technologies fractal in radiolocation: fractal and synergetic. Proc. of 28th ESA Antenna Workshop on pace Antenna Systems and Technologies (Noordwijk, The Netherlands,May 31 – June 03,2005)[C]. ESTEC: Noordwijk, 2005:1047 – 1050.

[20]JOANNOPULOS J D, JOHNSON S G, WIN J N, et al. Photonic crystals: molding the flow of light[M]. Princeton: Princeton Univ. Press, 2008.

[21] HOU B, LIAO X Q, POON J K S. Resonant infrared transmission and effective medium response of subwavelength H-fractal apertures[J]. Optics Express, 2010,18(4):3946 – 3951.

[22] POTAPOV A A. Fractal antennas, nano-technologies, resonances and plasmons[J]. Achievements of Modern Radioelectronics,2011,5:5 – 12.

[23] POTAPOV A A, BULAVKIN V V, GERMAN V A, et al. Fractal signature methods for profiling of processed surfaces[J]. Tech Phys,2005,50(5):560 – 575.

[24] POTAPOV A A, POTAPOV ALEXEY A, POTAPOV V A. Fractal radioelement's, devices and fractal systems for radar and telecommunications[M] Beijing: Metallurgical Industry Press, 2017:499 – 506.

[25] POTAPOV A A. On the issues of fractal radio electronics: Part 1. Processing of multidimensional signals, radiolocation, nanotechnology, radio engineering elements and sensors[J]. Eurasian Physical Technical Journal,2018, 2(30): 5 – 15.

[26] POTAPOV A A. On the issues of fractal radio electronics: part 2. distribution and scattering of radio waves, radio heat effects, new models, large fractal systems[J]. Eurasian Physical Technical Journal,2018, 2(30): 16 – 23.

[27] POTAPOV A, POTAPOV V. Fractal radioelement's, devices and systems for radar and future telecommunications: antennas, capacitor, memristor, smart 2d frequency-selective surfaces, labyrinths and other fractal metamaterials[J]. Materials, Methods & Technologies. 2017,11:492 – 512.

[28] CHUA L O. Memristor—the missing circuit element[J]. IEEE Transactions on Circuit Theory, 1971,18(5):507 – 519.

[29] JAMES M T, TAO H. The fourth element[J]. Nature, 2008,453(1):42 – 43.

[30] REKHVIASHVILI S SH, POTAPOV A A. Memristor and the integral quantum hall effect[J]. Journal of Communications Technology and Electronics, 2012,57(2): 189 – 191.

[31] POTAPOV A A. Oscillator with fractional differential positive feedback as model of fractal dynamics[J]. Journal of Computational Intelligence and Electronic Systems,2014, 3(3): 236 – 237.

[32] BOGOLYUBOV N N, METROPOL'SKIY Y A. Asymtotic methods in non-linear oscillations theory[M]. 4th ed. Moscow: Nauka, 1974. (in Russian)

[33] POTAPOV A A. Oscillations, waves, structures and systems illustrated by examples of the global fractal-scaling method (measure zero sets, singularity, scaling, sample topology, sprites, jets, elves, memristors, oscillators, fractal mazes, robust antenna arrays and fractal detectors)[J]. Non-Linear World. 2014,12(4):3 – 38. (in Russian)

[34] POTAPOV A A, POTAPOV A A (Jr.), POTAPOV V A. The fractal capacitor, fractal operators, and fractal impedances[J]. Nonlinear World, 2006,4(4 – 5):172 – 187. (in Russian)

[35] KHAN A I, BHOWMIK D YU P, KIM S J, et al. Experimental evidence of ferroelectric negative capacitance in nanoscale heterostructures[J]. Appl Phys Lett,2011,99(11):113501 – 3.

[36] APPLEBY D J R, PONON N K, KWA K S K, et al. Experimental observation of negative capacitance in ferroelectrics at room temperature[J]. Nano Lett,2014,14(7):3864 – 3868.

[37] SALAHUDDIN S, DATTA S. Use of negative capacitance to provide voltage amplification for low power nanoscale devices[J]. Nano Lett,2008,8:405 – 410.

[38] KOSTROMINA O, POTAPOV A, RAKUT I, et al. On nonstationary regimes in electric circuits with ferroelectric negative capacitance. Proc. of the Int. Siberian Conf. Control and Communications SIBCON – 2017 (Kazakhstan, Astana, June 29 – 30, 2017)[C]. Astana; Tomsk: S. Seifullin Kazakh Agrotechnical University, 2017:3.

[39] KOSTROMINA O S, POTAPOV A A, RAKUT I V, et al. The coefficient of nonlinear distortion on account

of voltage in an oscillating circuit with a ferroelectric capacitor with negative capacitance. Proc. 10th Int. Scientific Conf. Chaos and Structures in Nonlinear Systems. Theory and Experiment, devoted to the 75th anniversary of Prof. Z. Zhanabaev. (Kazakhstan, Almaty, 16 – 18 June, 2017) [C]. -Almaty, al-Farabi Kazakh National University, 2017:315 – 320.

[40] KOSTROMINA O S, POTAPOV A A, RAKUT I V, et al. Total harmonic distorsions in an oscillatory circuit with a ferro electric capacitor with negative capacitance[J]. Eurasian Physical Technical Journal, 2017, 2(28):14 – 21.

[41] MOROZOV A D. Resonances, cycles and chaos in quasi-conservative systems[M]. Moscow: Izhevsk, Regular and Chaotic Dynamics, 2005:420. (in Russian)

[42] LEONOV K N, POTAPOV A A, USHAKOV P A. Application of invariant properties of chaotic signals in the synthesis of noise-immune broadband systems for data transmission[J]. Journal of Communications Technology and Electronics, 2014, 59(12):1393 – 1411.

[43] MELNIKOV V K. On stability of center under periodic on time perturbations[J]. Proc of Moscow mathematical society, 1963, 12:3 – 52. (in Russian)

[44] ANDRONOV A A, PONTRYAGIN L S. Rough systems[J]. Reports of Academy of Sciences of the USSR, 1937, 14:247 – 251. (in Russian)

[45] POTAPOV A A. Chaos theory, fractals and scaling in the radar: a look from 2015[M]//SKIADAS C. The foundations of chaos revisited: from poincaré to recent advancements. Basel: Springer Int. Publ., 2016: 195 – 218.

[46] POTAPOV A A. On the indicatrixes of waves scattering from the random fractal anisotropic surface[M]// BRAMBILA F. Fractal analysis-applications in physics, engineering and technology. Rijeka: InTech, 2017:187 – 248.

[47] POTAPOV A A. Postulate "the topology maximum at the energy minimum" for textural and fractal-and-scaling processing of multidimensional super weak signals against a background of noises[M]//UVAROVA L A, NADYKTO A B, LATYSHEV A V. Nonlinearity: problems, solutions and applications, Vol. 2. New York: Nova Science Publ, 2017:35 – 94.

[48] POTAPOV A A. Fractal scaling or scale-invariant radar: a breakthrough into the future[J]. Universal Journal of Physics and Application (USA), 2017, 11(1):13 – 32.

[49] POTAPOV A A. Fractal and topological sustainable methods of overcoming expected uncertainty in the radiolocation of low-contrast targets and in the processing of weak multi-dimensional signals on the background of high-intensity noise: A new direction in the statistical decision theory [J]. Journal of Physics, 2017, 918: 19. No. 012015.

[50] POTAPOV A A. Thematic course: statistical theory of fractal radar. Book of Abstracts of the 11th Chaotic Modeling and Simulation Int. Conf. (Italy, Rome, June 5 – 8, 2018) [C]. Rome: International Society for the Advancement of Science and Technology, 2018:91 – 92.

[51] POTAPOV A A. Creation and development of the fundamental area "fractal radiophysics and fractal radio electronics: development of fractal radio systems". Part 1. Theory and main scientific prospects[J]. Eurasian Physical Technical Journal, 2019, 16(31):137 – 143.

[52] POTAPOV A A. Creation and development of the fundamental area "fractal radiophysics and fractal radio electronics: development of fractal radio systems". Part 2. Selected results and perspective trends[J]. Eurasian Physical Technical Journal, 2019, 16 (31):144 – 152.

[53] POTAPOV A A. Fractal electrodynamics: numerical modeling of small fractal antenna devices and fractal 3D microwave resonators for modern ultra-wideband or multiband radio systems [J]. Journal of Communica-

tions Technology and Electronics,2019,64(7):629 –663.

[54] GULYAEV Y V, POTAPOV A A. Application of fractal theory, fractional operators, textures, scaling effects, and nonlinear dynamics methods in the synthesis of new information technologies in radio electronics (specifically, radiolocation) [J]. Journal of Communications Technology and Electronics,2019,64(9):911 –925.

[55] GULYAEV Y V. POTAPOV A A. Biography and publication index[M]. Moscow: Center of printing services "Rainbow", 2019:256. (Approved by the Academic Council of IREE V. A. Kotelnikov of the Russian Academy of Sciences 26. 12,2018).

Appendix A

Proposition A1. [16]. Assume that (1) $\varphi(t), \alpha(t) \in L_1([0,T])$, $\sup_{t\in[0,T]} |\varphi(t)| < \infty$, $\sup_{t\in[0,T]} |\alpha(t)| < \infty$ and (2) the inequality

$$\varphi(t) \leqslant \alpha(t) + L\int_0^T \varphi(s)\,\mathrm{d}s \tag{A.1}$$

is satisfied. Then the inequality

$$\varphi(t) \leqslant \alpha(t) + L\int_0^T \mathrm{e}^{L(t-s)}\alpha(s)\,\mathrm{d}s \tag{A.2}$$

is satisfied.

Theorem A1. (I) Assume that: (1) let $\boldsymbol{x}_{t,n}(\omega), n = 1,2,\cdots$ be the solutions of the Ito's SDE's:

$$\mathrm{d}\boldsymbol{x}_{t,n} = \boldsymbol{b}_n(\boldsymbol{x}_{t,n},t)\,\mathrm{d}t + \boldsymbol{\sigma}_n(x_{t,n},t)\,\mathrm{d}\boldsymbol{W}(t,\omega), \tag{A.3}$$

$\boldsymbol{x}_{0,n} = \boldsymbol{x}(\omega), x \in \mathbf{R}^d$.

And let $\tilde{\boldsymbol{x}}_{t,n}(t), n = 1,2,\cdots$ be the solutions of the Ito's SDE's:

$$\mathrm{d}\tilde{\boldsymbol{x}}_{t,n} = \tilde{\boldsymbol{b}}_n(\tilde{\boldsymbol{x}}_{t,n},t)\,\mathrm{d}t + \tilde{\boldsymbol{\sigma}}_n(\tilde{\boldsymbol{x}}_{t,n},t)\,\mathrm{d}\boldsymbol{W}(t,\omega), \tag{A.4}$$

$\tilde{\boldsymbol{x}}_{0,n} = x(\omega), x \in \mathbf{R}^d$.

Here

$$\boldsymbol{\sigma}_n(\boldsymbol{x}_{t,n},t)\,\mathrm{d}\boldsymbol{W}(t,\omega) = \sum_{r=1}^{k} \sigma_{r,l,n}(\boldsymbol{x}_{t,n},t)\,\mathrm{d}\boldsymbol{W}_r(t,\omega),$$

$$\tilde{\boldsymbol{\sigma}}_n(\tilde{\boldsymbol{x}}_{t,n},t)\,\mathrm{d}\boldsymbol{W}(t,\omega) = \sum_{r=1}^{k} \tilde{\sigma}_{r,l,n}(\boldsymbol{x}_{t,n},t)\,\mathrm{d}\boldsymbol{W}_r(t,\omega),$$

$l = 1,2,\cdots,d.$

(2) The inequalities

$$\|\boldsymbol{b}_n(\boldsymbol{x},t)\|^2 + \|\sigma_n(\boldsymbol{x},t)\|^2 \leqslant K_n(1 + \|\boldsymbol{x}\|^2), \tag{A.5}$$

$$\|\boldsymbol{b}_n(\boldsymbol{x},t) - \boldsymbol{b}_n(\boldsymbol{y},t)\| + \|\boldsymbol{\sigma}_n(\boldsymbol{x},t) - \boldsymbol{\sigma}_n(\boldsymbol{x},t)\| \leqslant K_n\|\boldsymbol{x} - \boldsymbol{y}\|, \tag{A.6}$$

$$\|\tilde{\boldsymbol{b}}_n(\boldsymbol{x},t)\|^2 + \|\tilde{\boldsymbol{\sigma}}_n(\boldsymbol{x},t)\|^2 \leqslant K_n(1 + \|\boldsymbol{x}\|^2), \tag{A.7}$$

$$\|\tilde{\boldsymbol{b}}_n(\boldsymbol{x},t) - \tilde{\boldsymbol{b}}_n(\boldsymbol{y},t)\| + \|\tilde{\boldsymbol{\sigma}}_n(\boldsymbol{x},t) - \tilde{\boldsymbol{\sigma}}_n(\boldsymbol{x},t)\| \leqslant K_n\|\boldsymbol{x} - \boldsymbol{y}\|, \tag{A.8}$$

$$\|\boldsymbol{b}_n(\boldsymbol{x},t) - \tilde{\boldsymbol{b}}_n(\boldsymbol{x},t)\| \leqslant \delta_{1,n}\|\boldsymbol{x}\|, \tag{A.9}$$

$$\|\boldsymbol{\sigma}_n(\boldsymbol{x},t) - \tilde{\boldsymbol{\sigma}}_n(\boldsymbol{x},t)\| \leqslant \delta_{2,n}\|\boldsymbol{x}\|, \tag{A.10}$$

where $0 \leqslant t \leqslant T$, is satisfied.

Then the inequality

$$\sup_{0\leq t\leq T}\mathbf{E}\left(\|\boldsymbol{x}_{t,n}-\tilde{\boldsymbol{x}}_{t,n}\|^2\right)\leq e^{L_n}(T\delta_{1,n}^2+\delta_{2,n}^2)\mathbf{E}\left(\int_0^T\|\tilde{\boldsymbol{x}}_{t,n}\|^2 dt\right) \tag{A. 11}$$

with $L_n = 3(1+T)K_n$, is satisfied.

(Ⅱ) Let $\tau_{U_n}(\omega) = \tau_n(\omega)$ be the random variable equal to the time at which the sample function of the process $\tilde{\boldsymbol{x}}_{t,n}$ first leaves the bounded neighborhood $U_n \ni 0$, and let $\tau_n(\omega,t) = \min(\tau_n(\omega),t)$.

Assume that: (1) $\forall n: U_n \subset U_{n+1}$, $\cup U_n = \mathbf{R}^d$,

$$(2)\ \sup_{n\in\mathbf{N}}\left[\mathbf{E}\left(\int_0^T\|\tilde{\boldsymbol{x}}_{t,n}\|^2 dt\right)\right] < \infty. \tag{A. 12}$$

Then the inequality

$$\sup_{0\leq t\leq T}\left[\mathbf{E}\left(\|\boldsymbol{x}_{\tau_n(\omega,t),n}-\tilde{\boldsymbol{x}}_{\tau_n(\omega,t),n}\|^2\right)\right]\leq e^{L_n}(T\delta_{1,n}^2+\delta_{2,n}^2)\sup_{n\in\mathbf{N}}\left[\mathbf{E}\left(\int_0^T\|\tilde{\boldsymbol{x}}_{t,n}\|^2 dt\right)\right] \tag{A. 13}$$

with $L_n = 3(1+T)K_n$ is satisfied.

Proof. (I) From Eq. (A. 3) and Eq. (A. 4) one obtain

$$\boldsymbol{x}_{t,n}-\tilde{\boldsymbol{x}}_{t,n} = \boldsymbol{\xi}_n(t) + \int_0^t\left[\boldsymbol{b}_n(\boldsymbol{x}_{s,n},s)-\tilde{\boldsymbol{b}}_n(\tilde{\boldsymbol{x}}_{s,n},s)\right]ds +$$

$$\int_0^t[\boldsymbol{\sigma}_n(\boldsymbol{x}_{s,n},s)-\tilde{\boldsymbol{\sigma}}_n(\tilde{\boldsymbol{x}}_{s,n},s)]d\boldsymbol{W}(s). \tag{A. 14}$$

Here

$$\boldsymbol{\xi}_n(t) = \int_0^t[\boldsymbol{b}_n(\tilde{\boldsymbol{x}}_{s,n},s)-\tilde{\boldsymbol{b}}_n(\tilde{\boldsymbol{x}}_{s,n},s)]ds + \int_0^t[\boldsymbol{\sigma}_n(\tilde{\boldsymbol{x}}_{s,n},s)-\tilde{\boldsymbol{\sigma}}_n(\tilde{\boldsymbol{x}}_{s,n},s)]d\boldsymbol{W}(s). \tag{A. 15}$$

From Eq. (A. 15) and inequalities (A. 6) and (A. 8) one obtain the inequality

$$\mathbf{E}\left[\|\boldsymbol{x}_{t,n}-\tilde{\boldsymbol{x}}_{t,n}\|^2\right]\leq 3\mathbf{E}\left[\|\boldsymbol{\xi}_n(t)\|^2\right]+L_n\int_0^t\mathbf{E}\left[\|\boldsymbol{x}_{s,n}-\tilde{\boldsymbol{x}}_{s,n}\|^2\right]ds, \tag{A. 16}$$

with $L_n = 3(1+T)K_n$. Using Proposition 1, from inequality (A. 16) one obtain the inequality

$$\mathbf{E}\left[\|\boldsymbol{x}_{t,n}-\tilde{\boldsymbol{x}}_{t,n}\|^2\right]\leq 3\mathbf{E}\left[\|\boldsymbol{\xi}_n(t)\|^2\right]+L_n\int_0^t e^{L_n(t-s)}\mathbf{E}\left[\|\boldsymbol{x}_{s,n}-\tilde{\boldsymbol{x}}_{s,n}\|^2\right]ds. \tag{A. 17}$$

From inequality (A. 9) one obtain the inequality

$$\sup_{0\leq t\leq T}\left|\left|\mathbf{E}\left[\int_0^t[\boldsymbol{b}_n(\tilde{\boldsymbol{x}}_{s,n},s)-\tilde{\boldsymbol{b}}_n(\tilde{\boldsymbol{x}}_{s,n},s)]ds\right]\right|\right|^2\leq$$

$$T\int_0^T\mathbf{E}[\|\boldsymbol{b}_n(\tilde{\boldsymbol{x}}_{s,n},s)-\tilde{\boldsymbol{b}}_n(\tilde{\boldsymbol{x}}_{s,n},s)\|^2]ds\leq T\delta_{1,n}^2\int_0^T\mathbf{E}[\|\tilde{\boldsymbol{x}}_{s,n}\|^2]ds. \tag{A. 18}$$

From inequality (A. 10) one obtain the inequality

$$\mathbf{E}\left[\sup_{0\leq t\leq T}\left|\left|\int_0^t[\boldsymbol{\sigma}_n(\boldsymbol{x}_{s,n},s)-\tilde{\boldsymbol{\sigma}}_n(\tilde{\boldsymbol{x}}_{s,n},s)]d\boldsymbol{W}(s)\right|\right|^2\right]\leq$$

$$4\mathbf{E}\left[\int_0^T[\|\boldsymbol{\sigma}_n(\boldsymbol{x}_{s,n},s)-\tilde{\boldsymbol{\sigma}}_n(\tilde{\boldsymbol{x}}_{s,n},s)\|^2]ds\right]\leq$$

$$\delta_{2,n}^2\int_0^T\mathbf{E}[\|\tilde{\boldsymbol{x}}_{s,n}\|^2]ds. \tag{A. 19}$$

From Eq. (A. 15) and Inequalities (A. 18) – (A. 19) one obtain the inequality

$$\sup_{0\leqslant t\leqslant T}\mathbf{E}[\ \|\boldsymbol{\xi}_n(t)\|^2] \leqslant (T\delta_{1,n}^2+\delta_{2,n}^2)\int_0^T \mathbf{E}[\ \|\tilde{\boldsymbol{x}}_{s,n}\|^2]\mathrm{d}s. \tag{A. 20}$$

Substituting the inequality (A. 20) into inequality (A. 17) gives

$$\sup_{0\leqslant t\leqslant T}\mathbf{E}[\ \|\boldsymbol{x}_{t,n}-\tilde{\boldsymbol{x}}_{t,n}\|^2] \leqslant \mathrm{e}^{L_n}(T\delta_{1,n}^2+\delta_{2,n}^2)\mathbf{E}\left[\int_0^T \|\tilde{\boldsymbol{x}}_{t,n}\|^2\mathrm{d}t\right]. \tag{A. 21}$$

The inequality (A. 21) completed the proof.

Proof. (Ⅱ) Similarity to proof of the statement (Ⅰ).

Let $\mathfrak{C}_i=(\Omega_i,\Sigma_i,\mathbf{P}_i), i=1,2$ be a probability spaces such that: $\Omega_1\cap\Omega_2=\varnothing$. Let $\boldsymbol{W}(t,\omega)$ be a Wiener process on $\mathfrak{C}_1$ and let $\boldsymbol{W}(t,\bar{\omega})$ be a Wiener process on $\mathfrak{C}_2$.

Proposition A2. Assume that (1) $\varphi(t,\bar{\omega}),\alpha(t,\bar{\omega}) \in L_1((0,T])\mathbf{P}_2$ – o. s., $\sup_{t\in[0,T]}|\varphi(t,\bar{\omega})|<\infty\,\mathbf{P}_2$ – o. s., $\sup_{t\in[0,T]}|\alpha(t,\bar{\omega})|<\infty\,\mathbf{P}_2$ – *o. s.*, and (2) the inequality

$$\varphi(t,\bar{\omega}) \leqslant \alpha(t,\bar{\omega}) + L_{\bar{\omega}}\int_0^T \varphi(s,\bar{\omega})\mathrm{d}s \tag{A. 22}$$

$\mathbf{P}_2$ – o. s. is satisfied. Then the inequality

$$\varphi(t,\bar{\omega}) \leqslant \alpha(t,\bar{\omega}) + L_{\bar{\omega}}\int_0^T \mathrm{e}^{L_{\bar{\omega}}(t-s)}\alpha(s,\bar{\omega})\mathrm{d}s \tag{A. 23}$$

$\mathbf{P}_2$ – o. s. is satisfied.

Theorem A2. (Ⅰ) Assume that: (1) let $\boldsymbol{x}_{t,n}=\boldsymbol{x}_{t,n}(\omega,\bar{\omega}), n=1,2,\cdots$ be the solutions of the Ito's SDE's

$$\mathrm{d}\boldsymbol{x}_{t,n} = \boldsymbol{b}_n(\boldsymbol{x}_{t,n},t,\bar{\omega})\mathrm{d}t + \boldsymbol{\sigma}_n(\boldsymbol{x}_{t,n},t,\bar{\omega})\mathrm{d}\boldsymbol{W}(t,\omega), \tag{A. 24}$$

$$x_{0,n} = x(\omega,\bar{\omega}), x\in\mathbf{R}^d.$$

And let $\tilde{x}_{t,n}(t)=\tilde{x}_{t,n}(\omega,\bar{\omega}), n=1,2,\cdots$ be the solutions of the Ito's SDE's

$$\mathrm{d}\tilde{\boldsymbol{x}}_{t,n} = \tilde{\boldsymbol{b}}_n(\tilde{x}_{t,n},t,\bar{\omega})\mathrm{d}t + \tilde{\boldsymbol{\sigma}}_n(\tilde{x}_{t,n},t,\bar{\omega})\mathrm{d}\boldsymbol{W}(t,\omega), \tag{A. 25}$$

$$\tilde{\boldsymbol{x}}_{0,n} = \boldsymbol{x}(\omega,\bar{\omega}), \boldsymbol{x}\in\mathbf{R}^d.$$

Here

$$\boldsymbol{\sigma}_n(\boldsymbol{x}_{t,n},t,\bar{\omega})\mathrm{d}\boldsymbol{W}(t,\omega) = \sum_{r=1}^k \boldsymbol{\sigma}_{r,l,n}(\boldsymbol{x}_{t,n},t,\bar{\omega})\mathrm{d}\boldsymbol{W}_r(t,\omega),$$

$$\tilde{\boldsymbol{\sigma}}_n(\tilde{\boldsymbol{x}}_{t,n},t,\bar{\omega})\mathrm{d}\boldsymbol{W}(t,\omega) = \sum_{r=1}^k \tilde{\boldsymbol{\sigma}}_{r,l,n}(\boldsymbol{x}_{t,n},t,\bar{\omega})\mathrm{d}\boldsymbol{W}_r(t,\omega),$$

$l=1,2,\cdots,d.$

(2) The inequalities

$$\|\boldsymbol{b}_n(\boldsymbol{x},t,\bar{\omega})\|^2 + \|\boldsymbol{\sigma}_n(\boldsymbol{x},t,\bar{\omega})\|^2 \leqslant K_{n,\bar{\omega}}(1+\|x\|^2)\mathbf{P}_2 - \text{o. s.}, \tag{A. 26}$$

$$\|\boldsymbol{b}_n(\boldsymbol{x},t,\bar{\omega})-\boldsymbol{b}_n(\boldsymbol{y},t,\bar{\omega})\| + \|\boldsymbol{\sigma}_n(\boldsymbol{x},t,\bar{\omega})-\boldsymbol{\sigma}_n(\boldsymbol{x},t,\bar{\omega})\| \leqslant K_{n,\bar{\omega}}\|\boldsymbol{x}-\boldsymbol{y}\|\mathbf{P}_2 - \text{o. s.}, \tag{A. 27}$$

$$\|\tilde{\boldsymbol{b}}_n(\boldsymbol{x},t,\bar{\omega})\|^2 + \|\tilde{\boldsymbol{\sigma}}_n(\boldsymbol{x},t,\bar{\omega})\|^2 \leqslant K_{n,\bar{\omega}}(1+\|\boldsymbol{x}\|^2)\mathbf{P}_2 - \text{o. s.}, \tag{A. 28}$$

$$\|\tilde{\boldsymbol{b}}_n(\boldsymbol{x},t,\bar{\omega})-\tilde{\boldsymbol{b}}_n(\boldsymbol{y},t,\bar{\omega})\| + \|\tilde{\boldsymbol{\sigma}}_n(\boldsymbol{x},t,\bar{\omega})-\tilde{\boldsymbol{\sigma}}_n(\boldsymbol{x},t,\bar{\omega})\| \leqslant K_{n,\bar{\omega}}\|\boldsymbol{x}-\boldsymbol{y}\|\mathbf{P}_2 - \text{o. s.}, \tag{A. 29}$$

$$\| \boldsymbol{b}_n(\boldsymbol{x},t,\bar{\omega}) - \tilde{\boldsymbol{b}}_n(\boldsymbol{x},t,\bar{\omega}) \| \leqslant \delta_{1,n} \| \boldsymbol{x} \| \mathbf{P}_2 - \text{o. s.}, \tag{A. 30}$$

$$\| \boldsymbol{\sigma}_n(\boldsymbol{x},t,\bar{\omega}) - \tilde{\boldsymbol{\sigma}}_n(\boldsymbol{x},t,\bar{\omega}) \| \leqslant \delta_{2,n} \| \boldsymbol{x} \| \mathbf{P}_2 - \text{o. s.}, \tag{A. 31}$$

where $0 \leqslant t \leqslant T$, is satisfied. Then the inequality

$$\sup_{0 \leqslant t \leqslant T} \mathbf{E}_{\Omega_1} \left[\| \boldsymbol{x}_{t,n}(\omega,\bar{\omega}) - \tilde{\boldsymbol{x}}_{t,n}(\omega,\bar{\omega}) \|^2 \right] \leqslant$$

$$\mathrm{e}^{L_{n,\bar{\omega}}} (T\delta_{1,n}^2 + \delta_{2,n}^2) \mathbf{E}_{\Omega_1} \left[\int_0^T \| \tilde{\boldsymbol{x}}_{t,n}(\omega,\bar{\omega}) \|^2 \mathrm{d}t \right] \tag{A. 32}$$

with $L_{n,\bar{\omega}} = 3(1+T) K_{n,\bar{\omega}} \mathbf{P}_2 - \text{o. s.}$ is satisfied.

(Ⅱ) Let $\tau_{U_n}(\omega,\bar{\omega}) = \tau_n(\omega,\bar{\omega})$ be the random variable equal to the time at which the sample function of the process $\tilde{\boldsymbol{x}}_{t,n}$ first leaves the bounded neighborhood $U_n \ni 0$, and let $\tau_n(\omega,\bar{\omega}, t) = \min(\tau_n(\omega,\bar{\omega}), t)$.

Assume that: (1) $\forall n: U_n \subset U_{n+1}, \cup U_n = \mathbf{R}^d$, (2)

$$\sup_{n \in \mathbf{N}} \left(\mathbf{E}_{\Omega_1} \left[\int_0^T \| \tilde{\boldsymbol{x}}_{t,n}(\omega,\bar{\omega}) \|^2 \mathrm{d}t \right] \right) < \infty \, \mathbf{P}_2 - \text{o. s.} \tag{A. 33}$$

Then the inequality

$$\sup_{0 \leqslant t \leqslant T} \left(\mathbf{E}_{\Omega_1} \left[\| \boldsymbol{x}_{\tau_n(\omega,\bar{\omega},t),n} - \tilde{\boldsymbol{x}}_{\tau_n(\omega,\bar{\omega},t),n} \|^2 \right] \right) \leqslant \mathrm{e}^{L_{n,\bar{\omega}}} (T\delta_{1,n}^2 + \delta_{2,n}^2) \times$$

$$\sup_{n \in \mathbf{N}} \left(\mathbf{E}_{\Omega_1} \left[\int_0^T \| \tilde{\boldsymbol{x}}_{t,n}(\omega,\bar{\omega}) \|^2 \mathrm{d}t \right] \right) \mathbf{P}_2 - \text{o. s.} \tag{A. 34}$$

With $L_{n,\bar{\omega}} = 3(1+T) K_{n,\bar{\omega}}$ is satisfied.

Proof. (Ⅰ) From Eq. (A. 24) and Eq. (A. 25) one obtain

$$\boldsymbol{x}_{t,n}(\omega,\bar{\omega}) - \tilde{\boldsymbol{x}}_{t,n}(\omega,\bar{\omega}) = \boldsymbol{\xi}_n(t,\omega,\bar{\omega}) + \int_0^t [\boldsymbol{b}_n(\boldsymbol{x}_{s,n},s,\bar{\omega}) - \boldsymbol{b}_n(\tilde{\boldsymbol{x}}_{s,n},s,\bar{\omega})] \mathrm{d}s +$$

$$\int_0^t [\boldsymbol{\sigma}_n(\boldsymbol{x}_{s,n},s,\bar{\omega}) - \tilde{\boldsymbol{\sigma}}_n(\tilde{\boldsymbol{x}}_{s,n},s,\bar{\omega})] \mathrm{d}\boldsymbol{W}(s). \tag{A. 35}$$

Here

$$\boldsymbol{\xi}_n(t,\omega,\bar{\omega}) = \int_0^t [\boldsymbol{b}_n(\tilde{\boldsymbol{x}}_{s,n},s,\bar{\omega}) - \tilde{\boldsymbol{b}}_n(\tilde{\boldsymbol{x}}_{s,n},s,\bar{\omega})] \mathrm{d}s +$$

$$\int_0^t [\boldsymbol{\sigma}_n(\tilde{\boldsymbol{x}}_{s,n},s,\bar{\omega}) - \tilde{\boldsymbol{\sigma}}_n(\tilde{\boldsymbol{x}}_{s,n},s,\bar{\omega})] \mathrm{d}\boldsymbol{W}(s). \tag{A. 36}$$

From Eq. (A. 36) and inequalities (A. 27) and (A. 28) one obtain that the inequality

$$\mathbf{E}_{\Omega_1}(\| \boldsymbol{x}_{t,n} - \tilde{\boldsymbol{x}}_{t,n} \|^2) \leqslant 3\mathbf{E}(\| \boldsymbol{\xi}_n(t,\omega,\bar{\omega}) \|^2) +$$

$$L_n \int_0^t \mathbf{E}_{\Omega_1} [\| \boldsymbol{x}_{s,n}(t,\omega,\bar{\omega}) - \tilde{\boldsymbol{x}}_{s,n}(t,\omega,\bar{\omega}) \|^2] \mathrm{d}s, \tag{A. 37}$$

with $L_{n,\bar{\omega}} = 3(1+T) K_{n,\bar{\omega}} \cdot \mathbf{P}_2 - \text{o. s.}$ is satisfied. Using now Proposition 2, from inequality (A. 23) one obtain the inequality

$$\mathbf{E}_{\Omega_1} [\| \boldsymbol{x}_{t,n}(\omega,\bar{\omega}) - \tilde{\boldsymbol{x}}_{t,n}(\omega,\bar{\omega}) \|^2] \leqslant 3\mathbf{E}_{\Omega_1} [\| \boldsymbol{\xi}_n(t,\omega,\bar{\omega}) \|^2] +$$

$$L_{n,\bar{\omega}} \int_0^t \mathrm{e}^{L_{n,\bar{\omega}}(t-s)} \mathbf{E}_{\Omega_1} [\| \boldsymbol{x}_{s,n}(\omega,\bar{\omega}) - \tilde{\boldsymbol{x}}_{s,n}(\omega,\bar{\omega}) \|^2] \mathrm{d}s \mathbf{P}_2 - \text{o. s.} \tag{A. 38}$$

From inequality (A. 30) one obtain the inequality

$$\sup_{0\leqslant t\leqslant T}\left\|\mathbf{E}_{\Omega_1}\left[\int_0^t[\boldsymbol{b}_n(\tilde{\boldsymbol{x}}_{s,n},s,\bar{\omega})-\tilde{\boldsymbol{b}}_n(\tilde{\boldsymbol{x}}_{s,n},s,\bar{\omega})]\mathrm{d}s\right]\right\|^2\leqslant$$

$$T\int_0^T\mathbf{E}_{\Omega_1}[\ \|\boldsymbol{b}_n(\tilde{\boldsymbol{x}}_{s,n},s,\bar{\omega})-\tilde{\boldsymbol{b}}_n(\tilde{\boldsymbol{x}}_{s,n},s,\bar{\omega})\|^2]\mathrm{d}s\leqslant$$

$$T\delta_{1,n}^2\int_0^T\mathbf{E}_{\Omega_1}[\ \|\tilde{\boldsymbol{x}}_{s,n}\|^2]\mathrm{d}s\mathbf{P}_2-\text{o. s.}\qquad (\text{A. }39)$$

From inequality(A. 31) one obtain the inequality

$$\mathbf{E}_{\Omega_1}\left[\sup_{0\leqslant t\leqslant T}\left\|\int_0^t[\boldsymbol{\sigma}_n(\boldsymbol{x}_{s,n},s,\bar{\omega})-\tilde{\boldsymbol{\sigma}}_n(\tilde{\boldsymbol{x}}_{s,n},s,\bar{\omega})]\mathrm{d}\boldsymbol{W}(s)\right\|^2\right]\leqslant$$

$$4\mathbf{E}_{\Omega_1}\left[\int_0^T\left[\ \|\boldsymbol{\sigma}_n(\boldsymbol{x}_{s,n},s)-\tilde{\boldsymbol{\sigma}}_n(\tilde{\boldsymbol{x}}_{s,n},s)\|^2\right]\mathrm{d}s\right]\leqslant$$

$$\delta_{2,n}^2\int_0^T\mathbf{E}_{\Omega_1}[\ \|\tilde{\boldsymbol{x}}_{s,n}\|^2]\mathrm{d}s\mathbf{P}_2-\text{o. s.}\qquad (\text{A. }40)$$

From Eq. (A. 15) and inequalities(A. 18) – (A. 19) one obtain that the inequality

$$\sup_{0\leqslant t\leqslant T}\mathbf{E}_{\Omega_1}[\ \|\boldsymbol{\xi}_n(t)\|^2]\leqslant(T\delta_{1,n}^2+\delta_{2,n}^2)\int_0^T\mathbf{E}[\ \|\tilde{\boldsymbol{x}}_{s,n}\|^2]\mathrm{d}s.\qquad (\text{A. }41)$$

$\mathbf{P}_2$ – o. s. is satisfied. Substituting the inequality(A. 41) into inequality(A. 39) gives:

$$\sup_{0\leqslant t\leqslant T}\mathbf{E}_{\Omega_1}[\ \|\boldsymbol{x}_{t,n}-\tilde{\boldsymbol{x}}_{t,n}\|^2]\leqslant$$

$$\mathrm{e}^{L_n}(T\delta_{1,n}^2+\delta_{2,n}^2)\mathbf{E}_{\Omega_1}\left[\int_0^T\|\tilde{\boldsymbol{x}}_{t,n}\|^2\mathrm{d}t\right]\mathbf{P}_2-\text{o. s.}\qquad (\text{A. }42)$$

The inequality(A. 42) completed the proof.

Proof. (Ⅱ) Similarity to proof of the statement(Ⅰ).

Appendix B

Let us consider now regularized Feynman-Colombeau propagator $(K_\varepsilon(x,T\mid y,0))_\varepsilon$ given by Feynman path integral:

$$\widetilde{K}_\varepsilon(x,T\mid y,0;\sigma,l) =$$

$$\int_{q(0)=y}^{q(T)=x} D^+[q(t)]\exp\left(-\frac{1}{\hbar}\boldsymbol{S}_1(q,T;\sigma,l)\right)\exp\left(-\frac{1}{\hbar}\boldsymbol{S}_2(q(T),\boldsymbol{\lambda})\right)\exp\left(\frac{i}{\hbar}\boldsymbol{S}_\varepsilon(\dot{q},q,T)\right), \tag{B.1}$$

where $\hbar \in (0,1]$,

$$\boldsymbol{S}_1(q,T;\sigma,l) = \int_0^T \mathrm{d}t[\{[q(t)-\boldsymbol{\lambda}]^2;\sigma,l\}], \tag{B.2}$$

$$\boldsymbol{S}_2(q(T),\boldsymbol{\lambda}) = [q(T)-\boldsymbol{\lambda}]^2,\ \boldsymbol{\lambda} = (\lambda_1,\cdots,\lambda_d) \in \mathbf{R}^d, \tag{B.3}$$

$$\boldsymbol{S}_\varepsilon(\dot{q},q,T) = \int_0^T L(\dot{q}(t),q(t),t)\mathrm{d}t,\ L(\dot{q}(t),q(t),t) = \frac{m}{2}\dot{q}^2(t) -$$

$$V_\varepsilon(q(t),t),V_{\varepsilon=0}(x,t) = V(x,t), \tag{B.4}$$

$$V(x,t) = g_1(t)x + g_2(t)x^2 + g_3(t)x^3 + \cdots + g_\alpha(t)x^\alpha, \tag{B.5}$$

$$\alpha = (i_1,\cdots,i_d),x^\alpha = x_1^{i_1}\times\cdots\times x_d^{i_d},\ \|\alpha\| = \sum_{r=1}^{d} i_r,$$

$$V_\varepsilon(q(t),t) = V(q_\varepsilon(t),t),q_\varepsilon(t) = (q_{1,\varepsilon}(t),\cdots,q_{d,\varepsilon}(t)), \tag{B.6}$$

$$q_{i,\varepsilon}(t) = \frac{q_i(t)}{1+\varepsilon^{2k}|q(t)|^{2k}},\varepsilon \in (0,1],k \geqslant 1. \tag{B.7}$$

Here: (1) $\sigma \in (0,1]$, $\hbar \ll \sigma$ and (2) for each path $q(t)$ such that

$$q(t) = \sum_{n=1}^{\infty} a_n \sin\left(\frac{n\pi t}{T}\right) + u(t,T,y,x),u(0,T,y,x) = y,u(t,T,y,x) = x,$$

where $u(t,T,y,x)$ is a given function, operator $(p(t);\sigma,l)$ are

$$\{q(t);\sigma,l\} = \sum_{n=1}^{l}\sigma a_n\sin\left(\frac{n\pi t}{T}\right) + \sum_{n=l+1}^{\infty} a_n\sin\left(\frac{n\pi t}{T}\right). \tag{B.8}$$

(3) $D^+[q(t)]$ is a positive Feynman "measure".

Therefore regularized Colombeau solution of the Schrödinger equation corresponding to regularized propagator (B.1) are

$$(\Psi_\varepsilon(T,x;\sigma,l,\boldsymbol{\lambda}))_\varepsilon = \left(\int_{-\infty}^{\infty}\mathrm{d}y\Psi(y)\widetilde{K}_\varepsilon(x,T\mid y,0;\sigma,l)\right)_\varepsilon =$$

$$\left(\int_{q(T)=x} D^+[q(t)]\Psi[q(0)]\exp\left[-\frac{1}{\hbar}\boldsymbol{S}(q,T;\sigma,l,\boldsymbol{\lambda})\right]\exp\left[\frac{i}{\hbar}\boldsymbol{S}_\varepsilon(\dot{q},q,T)\right]\right)_\varepsilon =$$

$$\left(\int dy \int_{q(0)=y}^{q(T)=x} D^{+}[q(t)]\Psi[q(0)]\exp\left[-\frac{1}{\hbar}S(q,T;\sigma,l,\lambda)\right]\exp\left[\frac{i}{\hbar}S_{\varepsilon}(\dot{q},q,T)\right]\right)_{\varepsilon}. \tag{B.9}$$

Here

$$S(q,T;\sigma,l,\lambda) = S_1(q,T;\sigma,l) + S_2(q(T),\lambda). \tag{B.10}$$

Let us consider now regularized quantum average

$$(\langle \hat{x}_i,T;\sigma,l,\lambda,\varepsilon\rangle)_{\varepsilon} = \left(\int_{-\infty}^{\infty} dx x_i |\Psi_{\varepsilon}(T,x;\sigma,l,\lambda)|^2\right)_{\varepsilon}. \tag{B.11}$$

From Eqs. (B.5) and (B.11) one obtain

$$(|\hat{x}_i,T;\sigma,l,\lambda,\varepsilon|)_{\varepsilon} \leqslant$$

$$\left(\int dx\left\{\int_{q(T)=x} D^{+}[q(t)]\Psi[q(0)][|q_i(T)|]^{1/2}\exp\left[-\frac{1}{\hbar}S(q,T;\sigma,l,\lambda)\right]\cdot\right.\right.$$

$$\left.\left.\cos\left[\frac{i}{\hbar}S_{\varepsilon}(\dot{q},q,T)\right]\right\}^2\right)_{\varepsilon} + \left(\int dx\left\{\int_{q(T)=x} D^{+}[q(t)]\Psi[q(0)][|q_i(T)|]^{1/2}\cdot\right.\right.$$

$$\left.\left.\exp\left[-\frac{1}{\hbar}S(q,T;\sigma,l,\lambda)\right]\sin\left[\frac{i}{\hbar}S_{\varepsilon}(\dot{q},q,T)\right]\right\}^2\right)_{\varepsilon} \tag{B.13}$$

From Eqs. (B.5) – (B.13) one obtain

$$|\langle \hat{x}_i,T;\sigma,l,\lambda,\varepsilon\rangle - \lambda_i| = \left|\langle \hat{x}_i,T;\sigma,l,\lambda,\varepsilon\rangle - \lambda_i\int_{-\infty}^{\infty} dx|\Psi_{\varepsilon}(T,x;\sigma,l,\lambda)|^2\right| =$$

$$\left|\int_{-\infty}^{\infty} dx(x_i-\lambda_i)|\Psi_{\varepsilon}(T,x;\sigma,l,\lambda)|^2\right| \leqslant \int_{-\infty}^{\infty} dx|x_i-\lambda_i||\Psi_{\varepsilon}(T,x;\sigma,l,\lambda)|^2 =$$

$$\int dx\left\{\int_{q(T)=x} D^{+}[q(t)]\Psi[q(0)][|q_i(T)-\lambda_i|]^{1/2}\exp\left[-\frac{1}{\hbar}S(q,T;\sigma,l,\lambda)\right]\cdot\right.$$

$$\left.\cos\left[\frac{1}{\hbar}S_{\varepsilon}(\dot{q},q,T)\right]\right\}^2 + \int dx\left\{\int_{q(T)=x} D^{+}[q(t)]\Psi[q(0)][|q_i(T)-\lambda_i|]^{1/2}\cdot\right.$$

$$\left.\exp\left[-\frac{1}{\hbar}S(q,T;\sigma,l,\lambda)\right]\sin\left[\frac{1}{\hbar}S_{\varepsilon}(\dot{q},q,T)\right]\right\}^2 \tag{B.14}$$

Using replacement $q_i(t)-\lambda_i:=q_i(t), i=1,\cdots,d$ into RHS of the Eq. (B.9) one obtain

$$|\langle \hat{x}_i,T;\sigma,l,\lambda,\varepsilon\rangle - \lambda_i| \leqslant$$

$$\int dx\left\{\int_{q(T)=x} D^{+}[q(t)]\Psi[q(0)][|q_i(T)|]^{1/2}\exp\left[-\frac{1}{\hbar}S(q,T;\sigma,l,\lambda)\right]\cdot\right.$$

$$\left.\cos\left[\frac{1}{\hbar}S_{\varepsilon}(\dot{q},q+\lambda,T)\right]\right\}^2 + \int dx\left\{\int_{q(T)=x} D^{+}[q(t)]\Psi[q(0)][|q_i(T)|]^{1/2}\cdot\right.$$

$$\left.\exp\left[-\frac{1}{\hbar}S(q,T;\sigma,l,\lambda)\right]\sin\left[\frac{1}{\hbar}S_{\varepsilon}(\dot{q},q+\lambda,T)\right]\right\}^2 =$$

$$\int dx[I_1^2(x,T;\sigma,l,\lambda,\varepsilon)] + \int dx[I_2^2(x,T;\sigma,l,\lambda,\varepsilon)]. \tag{B.15}$$

Here

$$S(q,T;\sigma,l) = S_1(q,T;\sigma,l) + S_2(q(T)), S_1(q,T;\sigma,l) = \int_0^T dt(\{[q(t)]^2;\sigma,l\}),$$

$$S_2[q(T)] = [q(T)]^2, \lambda \in \mathbf{R}^d \tag{B.16}$$

and

$$I_1(x,T;\sigma,l,\lambda,\varepsilon) = \int_{q(T)=x} D^+[q(t)]\Psi[q(0)][|q_i(T)|]^{1/2}\exp\left[-\frac{1}{\hbar}S(q,T;\sigma,l)\right]\cdot$$

$$\cos\left[\frac{1}{\hbar}S_\varepsilon(\dot{q},q+\lambda,T)\right] \tag{B. 17}$$

$$I_2(x,T;\sigma,l,\lambda,\varepsilon) = \int_{q(T)=x} D^+[q(t)]\Psi[q(0)][|q_i(T)|]^{1/2}\exp\left[-\frac{1}{\hbar}S(q,T;\sigma,l)\right]\cdot$$

$$\sin\left[\frac{1}{\hbar}S_\varepsilon(\dot{q},q+\lambda,T)\right] \tag{B. 18}$$

Let us rewrite a function $V_\varepsilon(q(t)+\lambda,t)$ in the following equivalent form:

$$V_\varepsilon(q(t)+\lambda,t) = V_{\varepsilon,0}(q(t),t,\lambda) + V_{\varepsilon,1}(q(t),t,\lambda), \tag{B. 19}$$

$$V_{\varepsilon,0}(q(t),t,\lambda) = a_{\varepsilon,1}(q(t),t,\lambda)q(t) + a_{\varepsilon,2}(q(t),t,\lambda)q^2(t), \tag{B. 20}$$

$$V_{\varepsilon,1}(q(t),t,\lambda) = a_{\varepsilon,3}(q(t),t,\lambda)q^3(t) + \cdots + a_{\varepsilon,\alpha}(q(t),t,\lambda)q^\alpha(t), \tag{B. 21}$$

where

$a_{\varepsilon=0,1}(q(t),t,\lambda) = c_1(t,\lambda), a_{\varepsilon=0,2}(q(t),t,\lambda) = c_2(t,\lambda),\cdots,a_{\varepsilon=0,\alpha}(q(t),t,\lambda) = c_\alpha(t,\lambda)$.

Let us evaluate now path integral $I_1(T;\sigma,l,\lambda)$ given via Eq. (B. 17). Substituting Eq. (B. 19) into RHS of the Eq. (B. 17) gives

$$I_1(x,T;\sigma,l,\lambda,\varepsilon) = I_1^{(1)}(x,T;\sigma,l,\lambda,\varepsilon) + I_1^{(2)}(x,T;\sigma,l,\lambda,\varepsilon) =$$

$$\int_{q(T)=x} D^+[q(t)]\Psi[q(0)][|q_i(T)|]^{\frac{1}{2}}\exp\left[-\frac{1}{\hbar}S(q,T;\sigma,l)\right]\cos\left[\frac{1}{\hbar}S_{\varepsilon,1}(\dot{q},q+\lambda,T)\right]\cdot$$

$$\cos\left[\frac{1}{\hbar}S_{\varepsilon,2}(q+\lambda,T)\right] + \int_{q(T)=x} D^+[q(t)]\Psi[q(0)][|q_i(T)|]^{\frac{1}{2}}\cdot$$

$$\exp\left(-\frac{1}{\hbar}S(q,T;\sigma,l)\right)\sin\left(\frac{1}{\hbar}S_{\varepsilon,1}(\dot{q},q,\lambda,T)\right)\sin\left[-\frac{1}{\hbar}S_{\varepsilon,2}(q,\lambda,T)\right], \tag{B. 22. a}$$

$$I_2(x,T;\sigma,l,\lambda,\varepsilon) = I_2^{(1)}(x,T;\sigma,l,\lambda,\varepsilon) + I_2^{(2)}(x,T;\sigma,l,\lambda,\varepsilon) =$$

$$\int_{q(T)=x} D^+[q(t)]\Psi[q(0)][|q_i(T)|]^{\frac{1}{2}}\exp\left(-\frac{1}{\hbar}S(q,T;\sigma,l)\right)\cos\left(\frac{1}{\hbar}S_{\varepsilon,1}(\dot{q},q+\lambda,T)\right)\cdot$$

$$\sin\left(\frac{1}{\hbar}S_{\varepsilon,2}(q+\lambda,T)\right) + \int_{q(T)=x} D^+[q(t)]\Psi[q(0)][|q_i(T)|]^{\frac{1}{2}}\cdot$$

$$\exp\left(-\frac{1}{\hbar}S(q,T;\sigma,l)\right)\sin\left(\frac{1}{\hbar}S_{\varepsilon,1}(\dot{q},q,\lambda,T)\right)\cos\left(-\frac{1}{\hbar}S_{\varepsilon,2}(q,\lambda,T)\right), \tag{B. 22. b}$$

where

$$S_{\varepsilon,1}(\dot{q},q,\lambda,T) = \int_0^T L_\varepsilon(\dot{q}(t),q(t),t,\lambda)\,\mathrm{d}t,\ L_\varepsilon(\dot{q}(t),q(t),t,\lambda) =$$

$$\frac{m}{2}\dot{q}^2(t) - V_{\varepsilon,0}(q(t),t,\lambda), \tag{B. 23}$$

$$S_{\varepsilon,2}(q,\lambda,T) = \int_0^T V_{\varepsilon,1}(q(t),t,\lambda)\,\mathrm{d}t, \tag{B. 24}$$

$$I_1^{(1)}(x,T;\sigma,l,\lambda,\varepsilon) = \int_{q(T)=x} D^+[q(t)]\Psi[q(0)][|q_i(T)|]^{\frac{1}{2}}\exp\left[-\frac{1}{\hbar}S(q,T;\sigma,l)\right]\cdot$$

$$\cos\left[\frac{1}{\hbar}\boldsymbol{S}_{\varepsilon,1}(\dot{q},q,\lambda,T)\right]\cos\left[\frac{1}{\hbar}\boldsymbol{S}_{\varepsilon,2}(q,\lambda,T)\right], \tag{B.25.a}$$

$$I_2^{(1)}(x,T;\sigma,l,\lambda,\varepsilon) = \int_{q(T)=x} D^+[q(t)]\Psi[q(0)][|q_i(T)|]^{\frac{1}{2}}\exp\left[-\frac{1}{\hbar}\boldsymbol{S}(q,T;\sigma,l)\right]\cdot$$

$$\cos\left[\frac{1}{\hbar}\boldsymbol{S}_{\varepsilon,1}(\dot{q},q,\lambda,T)\right]\sin\left[\frac{1}{\hbar}\boldsymbol{S}_{\varepsilon,2}(q,\lambda,T)\right], \tag{B.25.b}$$

$$I_1^{(2)}(x,T;\sigma,l,\lambda,\varepsilon) = \int_{q(T)=x} D^+[q(t)]\Psi[q(0)][|q_i(T)|]^{\frac{1}{2}}\exp\left[-\frac{1}{\hbar}\boldsymbol{S}(q,T;\sigma,l)\right]\cdot$$

$$\sin\left[\frac{1}{\hbar}\boldsymbol{S}_{\varepsilon,1}(\dot{q},q,\lambda,T)\right]\sin\left[-\frac{1}{\hbar}S_{\varepsilon,2}(q,\lambda,T)\right]. \tag{B.26.a}$$

$$I_2^{(2)}(x,T;\sigma,l,\lambda,\varepsilon) = \int_{q(T)=x} D^+[q(t)]\Psi[q(0)][|q_i(T)|]^{\frac{1}{2}}\exp\left[-\frac{1}{\hbar}\boldsymbol{S}(q,T;\sigma,l)\right]\cdot$$

$$\sin\left[\frac{1}{\hbar}\boldsymbol{S}_{\varepsilon,1}(\dot{q},q,\lambda,T)\right]\cos\left(-\frac{1}{\hbar}\boldsymbol{S}_{\varepsilon,2}(q,\lambda,T)\right). \tag{B.26.b}$$

Let us evaluate now n-dimensional path integral $I_{1,n}^{(1)}(x,T;\sigma,l,\lambda,\varepsilon)$:

$$I_{1,n}^{(1)}(x,T;\sigma,l,\lambda,\varepsilon) = \int_{q(T)=x} D_n^+[q(t)]\Psi[q(0)][|q_i(T)|]^{\frac{1}{2}}\exp\left[-\frac{1}{\hbar}S(q,T;\sigma,l)\right]\cdot$$

$$\cos\left[\frac{1}{\hbar}\boldsymbol{S}_{\varepsilon,1}(\dot{q},q,\lambda,T)\right]\cos\left[\frac{1}{\hbar}\boldsymbol{S}_{\varepsilon,2}(q,\lambda,T)\right] = \int_{q(T)=x} D_n^+[q(t)]\Psi[q(0)][|q_i(T)|]^{\frac{1}{2}}\cdot$$

$$\exp\left[-\frac{1}{\hbar}\boldsymbol{S}(q,T;\sigma,l)\right]\left\{\cos\left[\frac{1}{\hbar}\boldsymbol{S}_{\varepsilon,1}(\dot{q},q,\lambda,T)\right]+1\right\}\cos\left[\frac{1}{\hbar}\boldsymbol{S}_{\varepsilon,2}(q,\lambda,T)\right]-$$

$$\int_{q(T)=x} D_n^+[q(t)]_n\Psi[q(0)][|q_i(T)|]^{\frac{1}{2}}\exp\left[-\frac{1}{\hbar}\boldsymbol{S}(q,T;\sigma,l)\right]\cos\left[\frac{1}{\hbar}\boldsymbol{S}_{\varepsilon,2}(q,\lambda,T)\right]. \tag{B.27}$$

From Eq. (B.27) one obtain the inequality

$$|I_{1,n}^{(1)}(x,T;\sigma,l,\lambda,\varepsilon)| \leqslant \left|\int D_n^+[q(t)]\Psi[q(0)][|q_i(T)|]^{\frac{1}{2}}\exp\left[-\frac{1}{\hbar}\boldsymbol{S}(q,T;\sigma,l)\right]\cdot\right.$$

$$\left.\left\{\cos\left[\frac{1}{\hbar}\boldsymbol{S}_{\varepsilon,1}(\dot{q},q,\lambda,T)\right]+1\right\}\right| - \int_{q(T)=x} D_n^+[q(t)]\Psi[q(0)][|q_i(T)|]^{\frac{1}{2}}\cdot$$

$$\exp\left[-\frac{1}{\hbar}\boldsymbol{S}(q,T;\sigma,l)\right]\cos\left[\frac{1}{\hbar}\boldsymbol{S}_{\varepsilon,2}(q,\lambda,T)\right] = \left|\int_{q(T)=x} D_n^+[q(t)]\Psi[q(0)]\cdot\right.$$

$$\left.[|q_i(T)|]^{\frac{1}{2}}\exp\left[-\frac{1}{\hbar}\boldsymbol{S}(q,T;\sigma,l)\right]\cos\left[\frac{1}{\hbar}\boldsymbol{S}_{\varepsilon,1}(\dot{q},q,\lambda,T)\right]\right| + \int_{q(T)=x} D_n^+[q(t)]\cdot$$

$$\Psi[q(0)][|q_i(T)|]^{\frac{1}{2}}\exp\left[-\frac{1}{\hbar}\boldsymbol{S}(q,T;\sigma,l)\right] - \int_{q(T)=x} D_n^+[q(t)]\Psi[q(0)]\cdot$$

$$[|q_i(T)|]^{\frac{1}{2}}\exp\left[-\frac{1}{\hbar}\boldsymbol{S}(q,T;\sigma,l)\right]\cos\left[\frac{1}{\hbar}\boldsymbol{S}_{\varepsilon,2}(q,\lambda,T)\right]. \tag{B.28}$$

From Eq. (B.28) one obtain the inequality

$$\left|I_{1,n}^{(1)}(x,T;\sigma,l,\lambda,\varepsilon)\right| \leqslant \left|\int_{q(T)=x} D_n^+[q(t)]\Psi[q(0)][|q_i(T)|]^{\frac{1}{2}}\exp\left[-\frac{1}{\hbar}\boldsymbol{S}(q,T;\sigma,l)\right]\cdot\right.$$

$$\left.\cos\left[\frac{1}{\hbar}\boldsymbol{S}_{\varepsilon,1}(\dot{q},q,\lambda,T)\right]\right| - \sum_{i=1}^{\infty}\frac{(-1)^i\hbar^{-2i}}{(2i)!}\int_{q(T)=x} D_n^+[q(t)]\Psi[q(0)][|q_i(T)|]^{\frac{1}{2}}\cdot$$

$$[S_{\varepsilon,2}(q,\lambda,T)]^{2i}\exp\left[-\frac{1}{\hbar}S(q,T;\sigma,l)\right] = |\mathbf{J}_{1,n}^{(1)}(x,T;\sigma,l,\lambda,\varepsilon,\hbar)| - \sum_{i=1}^{\infty}\frac{(-1)^i\hbar^{-2i}}{(2i)!}\cdot$$

$$\mathcal{R}_{\varepsilon}^{(i)}(x,T;\sigma,l,n), \tag{B.29}$$

where

$$\mathbf{J}_{1,n}^{(1)}(x,T;\sigma,l,\lambda,\varepsilon,\hbar) =$$

$$\int_{q(T)=x} D_n^+[q(t)]\Psi[q(0)][|q_i(T)|]^{\frac{1}{2}}\exp\left[-\frac{1}{\hbar}S(q,T;\sigma,l)\right]\cos\left[\frac{1}{\hbar}S_{\varepsilon,1}(\dot{q},q,\lambda,T)\right], \tag{B.30}$$

$$\mathcal{R}_{\varepsilon}^{(i)}(x,T;\sigma,l,n) = \int_{q(T)=x} D_n^+[q(t)]\Psi[q(0)][|q_i(T)|]^{\frac{1}{2}}[S_{\varepsilon,2}(q,\lambda,T)]^{2i}\cdot$$

$$\exp\left[-\frac{1}{\hbar}S(q,T;\sigma,l)\right]. \tag{B.31}$$

Using replacement $q_i(t) := \hbar^{\frac{1}{2}}q_i(t), t\in[0,T], i=1,\cdots,d$ into RHS of the Eq. (B. 31) one obtain

$$\mathcal{R}_{\varepsilon}^{(i)}(x,T;\sigma,l,n) = \hbar^{1/4}\int_{q(T)=\frac{x}{\sqrt{\hbar}}}\breve{D}_n^+[q(t)]\Psi[\hbar^{\frac{1}{2}}q(0)][|q_i(T)|]^{\frac{1}{2}}[S_{\varepsilon,2}\cdot$$

$$(\hbar^{1/2}q,\lambda,T)]^{2i}\exp\left[-\frac{1}{\hbar}S(\hbar^{1/2}q,T;\sigma,l)\right] = \hbar^{1/4}\hbar^{i/2}\int dy\int_{q(0)=\frac{y}{\sqrt{\hbar}}}^{q(T)=\frac{x}{\sqrt{\hbar}}}\breve{D}_n^+[q(t)]\Psi[q(0)]\cdot$$

$$[|q_i(T)|]^{\frac{1}{2}}[\hat{S}_{\varepsilon,2}(q,\lambda,T,\hbar)]^{2i}\exp[-S(q,T;\sigma,l)] = \hbar^{1/4}\hbar^{i/2}\hat{R}_{\varepsilon}^{(i)}(x,T;\sigma,l,n), \tag{B.32}$$

where

$$\breve{D}_n^+[(q(t)] = D_n^+[\hbar^{\frac{1}{2}}q(t)], t\in[0,T], \breve{\Psi}[q(0)] = \frac{\eta^{d/4}}{(2\pi)^{d/4}\hbar^{d/4}}\exp\left[\frac{\eta q^2(0)}{2}\right],$$

see Eq. (3. 1) and

$$\hat{S}_{\varepsilon,2}(q,\lambda,T,\hbar) = \int_0^T\hat{V}_{\varepsilon,1}(q(t),t,\lambda,\hbar)\mathrm{d}t, \tag{B.33}$$

$$\hat{V}_{\varepsilon,1}(q(t),t,\lambda,\hbar) = a_{\varepsilon,3}(q(t),t,\lambda)q^3(t) + \cdots + \hbar^{\frac{\alpha-3}{2}}a_{\varepsilon,\alpha}(q(t),t,\lambda)q^{\alpha}(t). \tag{B.34}$$

$$\hat{\mathcal{R}}_{\varepsilon}^{(i)}(x,T;\sigma,l,n) = \int dy\int_{q(0)=\frac{y}{\sqrt{\hbar}}}^{q(T)=\frac{x}{\sqrt{\hbar}}}\breve{D}_n^+[q(t)]\breve{\Psi}[q(0)][|q_i(T)|]^{\frac{1}{2}}[\hat{S}_{\varepsilon,2}(q,\lambda,T,\hbar)]^{2i}\cdot$$

$$\exp[-S(q,T;\sigma,l)]. \tag{B.35}$$

From (B. 29) – (B. 35) one obtain

$$|I_{1,n}^{(1)}(x,T;\sigma,l,\lambda,\varepsilon)| \leqslant |\mathbf{J}_{1,n}^{(1)}(x,T;\sigma,l,\lambda,\varepsilon,\hbar)| - \hbar^{\frac{1}{4}}\sum_{i=1}^{\infty}\frac{(-1)^i\hbar^i}{(2i)!}\hat{\mathcal{R}}_{\varepsilon}^{(i)}(x,T;\sigma,l,n)$$

$$\leqslant |\mathbf{J}_{1,n}^{(1)}(x,T;\sigma,l,\lambda,\varepsilon,\hbar)| - \hbar^{\frac{1}{4}}\Xi_{\varepsilon}(x,T;\sigma,l,n), \tag{B.36}$$

where

$$\Xi_{\varepsilon}(x,T;\sigma,l,\hbar,n) = \sum_{i=1}^{\infty}\frac{(-1)^i\hbar^i}{(2i)!}\hat{\mathcal{R}}_{\varepsilon}^{(i)}(x,T;\sigma,l,n). \tag{B.37}$$

Proposition B. 1. [23]-[25]. Let $\{s_{n,m}\}_{n,m=1}^{n,m=\infty}$ be a double sequence $s:\mathbf{N}\times\mathbf{N}\to\mathbf{C}$. Let

$\lim_{n,m\to\infty} s_{n,m} = a$. Then the iterated limit: $\lim_{n\to\infty}(\lim_{m\to\infty} s_{n,m})$ exist and equal to a if and only if $\lim_{m\to\infty} s_{n,m}$ exists for each $n \in \mathbf{N}$.

Proposition B. 2. Let $I_1^{(1)}(x,T;\sigma,l,\lambda,\varepsilon,\hbar) = I_1^{(1)}(x,T;\sigma,l,\lambda,\varepsilon)$, where $I_1^{(1)}(x,T;\sigma,l,\lambda,\varepsilon)$ is given via Eq. (B. 25) and let $I_1^{(2)}(x,T;\sigma,l,\lambda,\varepsilon,\hbar) = I_1^{(2)}(x,T;\sigma,l,\lambda,\varepsilon)$, where $I_1^{(2)}(x,T;\sigma,l,\lambda,\varepsilon)$ is given via Eq. (B. 26). Then $I_2^{(2)}(x,T;\sigma,l,\lambda,\varepsilon) =$

(1) $\lim_{\substack{\varepsilon\to 0\\ \sigma\to 0}}\lim_{\hbar\to 0}\int dx\,[I_1^{(1)}(x,T;\sigma,l,\lambda,\varepsilon)]^2 \leqslant$

$$\lim_{\hbar\to 0}\int dx\left\{\int_{q(T)=x} D^+[q(t)]\Psi[q(0)][|q_i(T)|]^{\frac{1}{2}}\cos\left[\frac{1}{\hbar}\mathbf{S}_1(\dot{q},q,\lambda,T)\right]\right\}^2,$$

(2) $\lim_{\substack{\varepsilon\to 0\\ \sigma\to 0}}\lim_{\hbar\to 0}\int dx\,[I_1^{(2)}(x,T;\sigma,l,\lambda,\varepsilon)]^2 = 0,$

(3) $\lim_{\substack{\varepsilon\to 0\\ \sigma\to 0}}\lim_{\hbar\to 0}\int dx[I_1^{(1)}(x,T;\sigma,l,\lambda,\varepsilon)I_1^{(2)}(x,T;\sigma,l,\lambda,\varepsilon)] = 0,$

(4) $\lim_{\substack{\varepsilon\to 0\\ \sigma\to 0}}\lim_{\hbar\to 0}\int dx\,[I_2^{(2)}(x,T;\sigma,l,\lambda,\varepsilon)]^2 \leqslant$

$$\lim_{\hbar\to 0}\int dx\left\{\int_{q(T)=x} D^+[q(t)]\Psi[q(0)][|q_i(T)|]^{\frac{1}{2}}\sin\left[\frac{1}{\hbar}\mathbf{S}_1(\dot{q},q,\lambda,T)\right]\right\}^2,$$

(5) $\lim_{\substack{\varepsilon\to 0\\ \sigma\to 0}}\lim_{\hbar\to 0}\int dx\,[I_2^{(1)}(x,T;\sigma,l,\lambda,\varepsilon)]^2 = 0,$

(6) $\lim_{\substack{\varepsilon\to 0\\ \sigma\to 0}}\lim_{\hbar\to 0}\int dx[I_2^{(2)}(x,T;\sigma,l,\lambda,\varepsilon)I_2^{(1)}(x,T;\sigma,l,\lambda,\varepsilon)] = 0.$

Here

$$\mathbf{S}_1(\dot{q},q,\lambda,T) = \mathbf{S}_{\varepsilon=0,1}(\dot{q},q,\lambda,T) = \int_0^T L_{\varepsilon=0}(\dot{q}(t),q(t),t,\lambda)\,dt,$$

$$L_{\varepsilon=0}(\dot{q}(t),q(t),t,\lambda) = \frac{m}{2}\dot{q}^2(t) - V_{\varepsilon=0,0}(q(t),t,\lambda).$$

Proof (I) Let us to choose an sequence $\{\hbar_m\}_{m=1}^{\infty}$ such that

(1) $\lim_{m\to\infty}\hbar_m = 0$ and

(2) $\lim_{m,n\to\infty}\int dx\,\{\Xi_\varepsilon^{(m)}(x,T;\sigma,l,\hbar_m,n)\}^2 =$

$$\lim_{m,n\to\infty}\int dx\left\{\sum_{i=1}^{m}\frac{(-1)^i\hbar_m^i}{(2i)!}\hat{\mathcal{R}}_\varepsilon^{(i)}(x,T;\sigma,l,n)\right\}^2 = 0.$$

We note that from (2) follows that: perturbative expansion

$$\int dx\,\{\Xi_\varepsilon(x,T;\sigma,l,\hbar_m,n)\}^2 = \hbar_m^{1/4}\int dx\left\{\sum_{i=1}^{\infty}\frac{(-1)^i\hbar_m^i}{(2i)!}\hat{\mathcal{R}}_\varepsilon^{(i)}(x,T;\sigma,l,n)\right\}^2.$$

vanishes in the limit $m,n\to\infty$. From (B. 36) and Schwarz's inequality using Proposition B. 1, one obtain

$$\lim_{m,n\to\infty}\int dx\,[I_{1,n}^{(1)}(x,T;\sigma,l,\lambda,\varepsilon,\hbar_m)]^2 \leqslant$$

$$\lim_{m,n\to\infty}\int dx\left\{\left|\mathbf{J}_{1,n}^{(1)}(x,T;\sigma,l,\lambda,\varepsilon,\hbar_m)\right|^2-\hbar_m^{1/4}\Xi_\varepsilon(x,T;\sigma,l,\hbar,n)\right\}^2\leqslant$$

$$\lim_{m,n\to\infty}\int dx\left\{\mathbf{J}_{1,n}^{(1)}(x,T;\sigma,l,\lambda,\varepsilon,\hbar_m)\right\}^2+$$

$$\lim_{m,n\to\infty}\left\{2\hbar_m^{\frac{1}{4}}\sqrt{\left[\int dx\,\{\mathbf{J}_{1,n}^{(1)}(x,T;\sigma,l,\lambda,\varepsilon,\hbar_m)\}^2\int dx\,\{\Xi_\varepsilon(x,T;\sigma,l,\hbar_m,n)\}^2\right]}+\right.$$

$$\left.\int dx\,\{\Xi_\varepsilon(x,T;\sigma,l,\hbar_m,n)\}^2\right\}=$$

$$\lim_{m,n\to\infty}\int dx\,\{\mathbf{J}_{1,n}^{(1)}(x,T;\sigma,l,\lambda,\varepsilon,\hbar_m)\}^2=\lim_{\hbar\to0}\lim_{n\to\infty}\int dx\,\{\mathbf{J}_{1,n}^{(1)}(x,T;\sigma,l,\lambda,\varepsilon,\hbar)\}^2=$$

$$\lim_{\hbar\to0}\int dx\left\{\int_{q(T)=x}D^+[q(t)]\Psi[q(0)][|q_i(T)|]^{\frac{1}{2}}\cos\left[\frac{1}{\hbar}S_1(\dot q,q,\lambda,T)\right]\right\}^2.\qquad(B.38)$$

Let us to choose now an subsequence $\{\hbar_{m_k}\}_{m_k=1}^{\infty}$ such that the limit:

$$\lim_{k,n\to\infty}\int dx\,[I_{1,n}^{(1)}(x,T;\sigma,l,\lambda,\varepsilon,\hbar_{m_k})]^2$$

exist and

$$\lim_{k,n\to\infty}\int dx\,[I_{1,n}^{(1)}(x,T;\sigma,l,\lambda,\varepsilon,\hbar_{m_k})]^2=\lim_{m,n\to\infty}\int dx\,[I_{1,n}^{(1)}(x,T;\sigma,l,\lambda,\varepsilon,\hbar_m)]^2\qquad(B.39)$$

From (B. 39) and Proposition B. 1 one obtain

$$\lim_{k,n\to\infty}\int dx\,[I_{1,n}^{(1)}(x,T;\sigma,l,\lambda,\varepsilon,\hbar_{m_k})]^2=\lim_{k\to\infty}\left\{\lim_{n\to\infty}\int dx\,[I_{1,n}^{(1)}(x,T;\sigma,l,\lambda,\varepsilon,\hbar_{m_k})]^2\right\}.$$

(B. 40)

From (B. 39), (B. 40) and (B. 38) one obtain

$$\lim_{\substack{\varepsilon\to0\\ \sigma\to0}}\lim_{\hbar\to0}\int dx\,[I_1^{(1)}(x,T;\sigma,l,\lambda,\varepsilon)]^2\leqslant\lim_{\substack{\varepsilon\to0\\ \sigma\to0}}\lim_{k\to\infty}\left\{\lim_{n\to\infty}\int dx\,[I_{1,n}^{(1)}(x,T;\sigma,l,\lambda,\varepsilon,\hbar_{m_k})]^2\right\}=$$

$$\lim_{\substack{\varepsilon\to0\\ \sigma\to0}}\lim_{k\to\infty}\int dx\,[I_1^{(1)}(x,T;\sigma,l,\lambda,\varepsilon,\hbar_{m_k})]^2\leqslant$$

$$\lim_{\hbar\to0}\int dx\left\{\int_{q(T)=x}D^+[q(t)]\Psi[q(0)][|q_i(T)|]^{\frac{1}{2}}\cos\left[\frac{1}{\hbar}S_1(\dot q,q,\lambda,T)\right]\right\}^2.\qquad(B.41)$$

The inequality (B. 41) completed the Proof of the statement (1).

(Ⅱ) Let us estimate now n-dimensional path integral

$$I_{1,n}^{(2)}(x,T;\sigma,l,\lambda,\varepsilon)=$$

$$\int_{q(T)=x}D_n^+[q(t)]\Psi[q(0)][|q_i(T)|]^{\frac{1}{2}}\exp\left[-\frac{1}{\hbar}S(q,T;\sigma,l)\right]\sin\left[\frac{1}{\hbar}S_{\varepsilon,1}(\dot q,q,\lambda,T)\right]\cdot$$

$$\sin\left[-\frac{1}{\hbar}S_{\varepsilon,2}(q,\lambda,T)\right].\qquad(B.42)$$

From Eq. (B. 42) one obtain the inequality

$$|I_{1,n}^{(2)}(x,T;\sigma,l,\lambda,\varepsilon)|\leqslant$$

$$\int_{q(T)=x}D_n^+[q(t)]\Psi[q(0)][|q_i(T)|]^{\frac{1}{2}}\exp\left[-\frac{1}{\hbar}S(q,T;\sigma,l)\right]\left|\sin\left[\frac{1}{\hbar}S_{\varepsilon,2}(q,\lambda,T)\right]\right|\leqslant$$

$$\sum_{i=0}^{\infty}\frac{\hbar^{-(2i+1)}}{(2i+1)!}\int_{q(T)=x}D_n^+[q(t)]\Psi[q(0)][|q_i(T)|]^{\frac{1}{2}}[|S_{\varepsilon,2}(q,\lambda,T)|]^{(2i+1)}$$

$$\exp\left[-\frac{1}{\hbar}S(q,T;\sigma,l)\right]=\sum_{i=0}^{\infty}\frac{\hbar^{-(2i+1)}}{(2i+1)!}£_{\varepsilon}^{(i)}(x,T;\sigma,l,n) \tag{B.43}$$

where

$$£_{\varepsilon}^{(i)}(x,T;\sigma,l,n)=$$

$$\int_{q(T)=x}D_n^+[q(t)]\Psi[q(0)][\,|q_i(T)|\,]^{\frac{1}{2}}[\,|S_{\varepsilon,2}(q,\lambda,T)|\,]^{(2i+1)}\exp\left[-\frac{1}{\hbar}S(q,T;\sigma,l)\right]. \tag{B.44}$$

Using replacement $q_i(t):=\hbar^{\frac{1}{2}}q_i(t),t\in[0,T],i=1,\cdots,d$ into RHS of the Eq. (B.44) one obtain

$$£_{\varepsilon}^{(i)}(x,T;\sigma,l,n)=$$

$$\hbar^{1/4}\int_{q(T)=\frac{x}{\sqrt{\hbar}}}\breve{D}^+[q(t)]_n\Psi[\hbar^{\frac{1}{2}}q(0)][\,|q_i(T)|\,]^{\frac{1}{2}}[\,|S_{\varepsilon,2}(q,\lambda,T)|\,]^{(2i+1)}\cdot$$

$$\exp\left[-\frac{1}{\hbar}S(\hbar^{1/2}q,T;\sigma,l)\right]=\hbar^{1/4}\hbar^{i/2}\int dy\int_{q(T)=\frac{x}{\sqrt{\hbar}}q(T)=\frac{y}{\sqrt{\hbar}}}\breve{D}_n^+[q(t)]\breve{\Psi}[q(0)][\,|q_i(T)|\,]^{\frac{1}{2}}\cdot$$

$$[\,|\hat{S}_{\varepsilon,2}(q,\lambda,T)|\,]^{(2i+1)}\exp[-S(q,T;\sigma,l)]=\hbar^{1/4}\hbar^{i/2}\hat{£}_{\varepsilon}^{(i)}(x,T;\sigma,l,n), \tag{B.45}$$

where

$$\breve{D}_n^+(q(t))=D_n^+[\hbar^{\frac{1}{2}}q(t)],t\in[0,T],\breve{\Psi}[q(0)]=\frac{\eta^{d/4}}{(2\pi)^{d/4}\hbar^{d/4}}\exp\left[\frac{\eta q^2(0)}{2}\right],$$

see Eq. (3.1) and

$$\hat{S}_{\varepsilon,2}(q,\lambda,T,\hbar)=\int_0^T\hat{V}_{\varepsilon,1}(q(t),t,\lambda,\hbar)\,dt, \tag{B.46}$$

$$\hat{V}_{\varepsilon,1}(q(t),t,\lambda,\hbar)=a_{\varepsilon,3}(q(t),t,\lambda)q^3(t)+\cdots+\hbar^{\frac{\alpha-3}{2}}a_{\varepsilon,\alpha}(q(t),t,\lambda)q^{\alpha}(t). \tag{B.47}$$

$$£_{\varepsilon}^{(i)}(x,T;\sigma,l,n)=\int dy\int_{q(0)=\frac{y}{\sqrt{\hbar}}}^{q(T)=\frac{x}{\sqrt{\hbar}}}\breve{D}_n^+[q(t)]\breve{\Psi}[q(0)][\,|q_i(T)|\,]^{\frac{1}{2}}[\,|\hat{S}_{\varepsilon,2}(q,\lambda,T)|\,]^{(2i+1)}$$

$$\exp[-S(q,T;\sigma,l)]. \tag{B.48}$$

From (B.43) – (B.48) one obtain

$$|I_{1,n}^{(2)}(x,T;\sigma,l,\lambda,\varepsilon)|\leqslant\hbar^{\frac{1}{4}}\sum_{i=0}^{\infty}\frac{\hbar^{2(i+1)}}{(2i+1)!}\hat{£}_{\varepsilon}^{i}(x,T;\sigma,l,n)=\Theta_{\varepsilon}(x,T;\sigma,l,\hbar,n). \tag{B.49}$$

Let us to choose an sequence $\{\hbar_m\}_{m=1}^{\infty}$ such that

(1) $\lim\limits_{m\to\infty}\hbar_m=0$ and

(2) $\lim\limits_{m,n\to\infty}\int dx\,\{\Theta_{\varepsilon}^{(m)}(x,T;\sigma,l,\hbar_m,n)\}^2=$

$$\lim_{m,n\to\infty}\int dx\left[\sum_{i=0}^{m}\frac{\hbar_m^{2(i+1)}}{(2i+1)!}£_{\varepsilon}^{(i)}(x,T;\sigma,l,n)\right]^2=0.$$

We note that from (2) follows that: perturbative expansion

$$\int dx\,[\Theta_{\varepsilon}(x,T;\sigma,l,\hbar_m,n)]^2=\hbar_m^{1/4}\int dx\left[\sum_{i=0}^{\infty}\frac{\hbar^{2(i+1)}}{(2i+1)!}\hat{£}_{\varepsilon}^{(i)}(x,T;\sigma,l,n)\right]^2,$$

vanishes in the limit $m,n\to\infty$. From (B.49) one obtain

$$\lim_{m,n\to\infty}\int dx\,[I_{1,n}^{(2)}(x,T;\sigma,l,\lambda,\varepsilon,\hbar_m)]^2 \leqslant \lim_{m,n\to\infty}\int dx\,[\Theta_\varepsilon(x,T;\sigma,l,\hbar_m,n)]^2. \quad (B.50)$$

Let us to choose now an subsequence $\{\hbar_{m_k}\}_{m_k=1}^{\infty}$ such that the limit:

$$\lim_{k,n\to\infty}\int dx\,[I_{1,n}^{(2)}(x,T;\sigma,l,\lambda,\varepsilon,\hbar_{m_k})]^2$$

exists and

$$\lim_{k,n\to\infty}\int dx\,[I_{1,n}^{(2)}(x,T;\sigma,l,\lambda,\varepsilon,\hbar_{m_k})]^2 = \lim_{m,n\to\infty}\int dx\,[I_{1,n}^{(2)}(x,T;\sigma,l,\lambda,\varepsilon,\hbar_m)]^2 \quad (B.51)$$

From (B.51) and Proposition B.1 one obtain

$$\lim_{k,n\to\infty}\int dx\,[I_{1,n}^{(2)}(x,T;\sigma,l,\lambda,\varepsilon,\hbar_{m_k})]^2 = \lim_{k\to\infty}\left\{\lim_{n\to\infty}\int dx\,[I_{1,n}^{(2)}(x,T;\sigma,l,\lambda,\varepsilon,\hbar_{m_k})]^2\right\}. \quad (B.52)$$

From (B.50), (B.51) and (B.52) one obtain

$$\lim_{\substack{\varepsilon\to0\\ \sigma\to0}}\lim_{\hbar\to0}\int dx\,[I_1^{(2)}(x,T;\sigma,l,\lambda,\varepsilon)]^2 \leqslant$$

$$\lim_{\substack{\varepsilon\to0\\ \sigma\to0}}\lim_{k\to\infty}\left\{\lim_{n\to\infty}\int dx\,[I_{1,n}^{(2)}(x,T;\sigma,l,\lambda,\varepsilon,\hbar_{m_k})]^2\right\} =$$

$$\lim_{\substack{\varepsilon\to0\\ \sigma\to0}}\lim_{k\to\infty}\int dx\,[I_1^{(2)}(x,T;\sigma,l,\lambda,\varepsilon,\hbar_{m_k})]^2 = 0.$$

Proof of the statements (3) – (6) is similarly to the proof of the statements (1) – (2).

Theorem B.1. Let

$I_1(x,T;\sigma,l,\lambda,\varepsilon,\hbar) = I_1(x,T;\sigma,l,\lambda,\varepsilon)$, $I_2(x,T;\sigma,l,\lambda,\varepsilon,\hbar) = I_2(x,T;\sigma,l,\lambda,\varepsilon)$,

where

$I_1(x,T;\sigma,l,\lambda,\varepsilon)$ is given via Eq. (B.22a) – Eq. (B.22b). Then

$$\lim_{\substack{\varepsilon\to0\\ \sigma\to0}}\lim_{\hbar\to0}\int dx[I_1^2(x,T;\sigma,l,\lambda,\varepsilon)] \leqslant$$

$$\lim_{\hbar\to0}\int dx\left\{\int_{q(T)=x} D^+[q(t)]\Psi[q(0)][|q_i(T)|]^{\frac{1}{2}}\cos\left[\frac{1}{\hbar}\mathbf{S}_1(\dot q,q,\lambda,T)\right]\right\}^2, \quad (B.53.a)$$

$$\lim_{\substack{\varepsilon\to0\\ \sigma\to0}}\lim_{\hbar\to0}\int dx[I_2^2(x,T;\sigma,l,\lambda,\varepsilon)] \leqslant$$

$$\lim_{\hbar\to0}\int dx\left\{\int_{q(T)=x} D^+[q(t)]\Psi[q(0)][|q_i(T)|]^{\frac{1}{2}}\sin\left[\frac{1}{\hbar}\mathbf{S}_1(\dot q,q,\lambda,T)\right]\right\}^2, \quad (B.53.b)$$

Here

$$\mathbf{S}_1(\dot q,q,\lambda,T) = \mathbf{S}_{\varepsilon=0,1}(\dot q,q,\lambda,T) = \int_0^T L_{\varepsilon=0}(\dot q(t),q(t),t,\lambda)\,dt, \quad (B.54)$$

$$L_{\varepsilon=0}(\dot q(t),q(t),t,\lambda) = \frac{m}{2}\dot q^2(t) - V_{\varepsilon=0,0}(q(t),t,\lambda). \quad (B.55)$$

Proof. We remain that

$$I_1(x,T;\sigma,l,\lambda,\varepsilon,\hbar) = I_1^{(1)}(x,T;\sigma,l,\lambda,\varepsilon,\hbar) + I_1^{(2)}(x,T;\sigma,l,\lambda,\varepsilon,\hbar). \quad (B.56)$$

From Eq. (B.56) we obtain

$$\int \mathrm{d}x[I_1^2(x,T;\sigma,l,\lambda,\varepsilon,\hbar)] \leqslant \int \mathrm{d}x\,[I_1^{(1)}(x,T;\sigma,l,\lambda,\varepsilon,\hbar)]^2 +$$

$$\int \mathrm{d}x\,[I_1^{(2)}(x,T;\sigma,l,\lambda,\varepsilon,\hbar)]^2 + 2\int \mathrm{d}x[\,|I_1^{(1)}(x,T;\sigma,l,\lambda,\varepsilon)I_1^{(2)}(x,T;\sigma,l,\lambda,\varepsilon,\hbar)|\,] =$$

$$\int \mathrm{d}x\,[I_1^{(1)}(x,T;\sigma,l,\lambda,\varepsilon,\hbar)]^2 + \int \mathrm{d}x\,[I_1^{(2)}(x,T;\sigma,l,\lambda,\varepsilon,\hbar)]^2 +$$

$$2\sqrt{\int \mathrm{d}x\,[I_1^{(1)}(x,T;\sigma,l,\lambda,\varepsilon,\hbar)]^2 \int \mathrm{d}x\,[I_1^{(2)}(x,T;\sigma,l,\lambda,\varepsilon,\hbar)]^2}. \qquad \text{(B. 57)}$$

Let us to choose now an sequences $\{\hbar_m\}_{m=1}^{\infty}$, $\{\varepsilon_k\}_{k=1}^{\infty}$, $\{\sigma_l\}_{l=1}^{\infty}$ such that:

(1) $\lim\limits_{m\to\infty}\hbar_m = 0, \lim\limits_{k\to\infty}\varepsilon_k = 0, \lim\limits_{l\to\infty}\sigma_l = 0$

(2) $$\lim_{\substack{k\to\infty\\ l\to\infty}}\lim_{m\to\infty}\int \mathrm{d}x\,[I_1^{(2)}(x,T;\sigma_l,l,\lambda,\varepsilon_k,\hbar_m)]^2 =$$

$$\lim_{\substack{k\to\infty\\ l\to\infty}}\lim_{m\to\infty}\int \mathrm{d}x\,[I_1^{(2)}(x,T;\sigma_l,l,\lambda,\varepsilon_k,\hbar_m)]^2 = 0, \qquad \text{(B. 58)}$$

(3) $$\lim_{\substack{k\to\infty\\ l\to\infty}}\lim_{m\to\infty}\int \mathrm{d}x\,[I_1^{(1)}(x,T;\sigma_l,l,\lambda,\varepsilon_k,\hbar_m)]^2 \leqslant$$

$$\lim_{\hbar\to 0}\int \mathrm{d}x\left\{\int_{q(T)=x} D^+[q(t)]\Psi[q(0)][\,|q_i(T)|\,]^{\frac{1}{2}}\cos\left[\frac{1}{\hbar}S_1(\dot{q},q,\lambda,T)\right]\right\}^2. \qquad \text{(B. 59)}$$

Therefore from inequality (B. 57), Eq. (B. 58) and inequality (B. 59) we obtain

$$\lim_{\substack{\varepsilon\to 0\\ \sigma\to 0}}\lim_{\hbar\to 0}\int \mathrm{d}x[I_1^2(x,T;\sigma,l,\lambda,\varepsilon,\hbar)] \leqslant \lim_{\substack{k\to\infty\\ l\to\infty}}\lim_{m\to\infty}\int \mathrm{d}x[I_1^2(x,T;\sigma_l,l,\lambda,\varepsilon_k,\hbar_m)] \leqslant$$

$$\lim_{\substack{k\to\infty\\ l\to\infty}}\lim_{m\to\infty}\int \mathrm{d}x\,[I_1^{(1)}(x,T;\sigma_l,l,\lambda,\varepsilon_k,\hbar_m)]^2 + \lim_{\substack{k\to\infty\\ l\to\infty}}\lim_{m\to\infty}\int \mathrm{d}x\,[I_1^{(2)}(x,T;\sigma_l,l,\lambda,\varepsilon_k,\hbar_m)]^2 +$$

$$2\lim_{\substack{k\to\infty\\ l\to\infty}}\lim_{m\to\infty}\sqrt{\int \mathrm{d}x\,[I_1^{(1)}(x,T;\sigma_l,l,\lambda,\varepsilon_k,\hbar_m)]^2\int \mathrm{d}x\,[I_1^{(2)}(x,T;\sigma_l,l,\lambda,\varepsilon_k,\hbar_m)]^2} =$$

$$\lim_{\substack{k\to\infty\\ l\to\infty}}\lim_{m\to\infty}\int \mathrm{d}x\,[I_1^{(1)}(x,T;\sigma_l,l,\lambda,\varepsilon_k,\hbar_m)]^2 =$$

$$\lim_{\hbar\to 0}\int \mathrm{d}x\left\{\int_{q(T)=x} D^+[q(t)]\Psi[q(0)][\,|q_i(T)|\,]^{\frac{1}{2}}\cos\left[\frac{1}{\hbar}S_1(\dot{q},q,\lambda,T)\right]\right\}^2. \qquad \text{(B. 60)}$$